MÖSSBAUER SPECTROSCOPY

Proceedings of the Fifth Seeheim Workshop,
held in Seeheim, Germany,
21–25 May 2002

Edited by

P. GÜTLICH

University of Mainz, Germany

B. W. FITZSIMMONS

University of Essex, Colchester, UK

R. RÜFFER

ESRF, Grenoble, France

and

H. SPIERING

University of Mainz, Germany

Reprinted from *Hyperfine Interactions*
Volume 144, Nos. 1–4 (2002)
Volume 145, Nos. 1–4 (2002)

SPRINGER SCIENCE+BUSINESS MEDIA, LLC

A C.I.P. Catalogue record for this book is available from the Library of Congress.

ISBN 978-94-010-3993-2 ISBN 978-94-010-0045-1 (eBook)

DOI 10.1007/978-94-010-0045-1

Printed on acid-free paper

Table of Contents

Scientific Committee

F. Berry, Milton Keynes
P. Gütlich, Mainz (Chair)
W. Keune, Duisburg
C. McCammon, Bayreuth
D. L. Nagy, Budapest
R. Rüffer, Grenoble
H. Spiering, Mainz
A. X. Trautwein, Lübeck
G. Wortmann, Paderborn

Local Organizers

J. Ensling
M. Ghafari
P. Gütlich
H. Spiering
U. Stumm

Hyperfine Interactions **144/145:** 1–2, 2002.

Preface

The series of the "Seeheim Workshops on Mössbauer Spectroscopy" was continued with the 5th Meeting which was held on May 21–25, 2002, once again in the very attractive Training Center of Deutsche Lufthansa in Seeheim, Germany. More than 120 scientists from some 22 countries attended the meeting.

The participants were treated to five invited lectures of one hour's duration on topics ranging from a very timely presentation on advances in nuclear resonant inelastic X-ray scattering to molecular magnetism through geoscientific studies aided by Mössbauer spectroscopy.

There were 26 invited talks of which a representative set of authors were invited to submit a printed version and these are reproduced here together with some 8 papers from those authors who had originally submitted them for oral presentation.

From the outset, the poster sessions at Seeheim have been lively sessions during which the subject matter gets mulled over and discussed warmly.

We wish to express our sincere thanks for financial support from:

Deutsche Forschungsgemeinschaft
European Science Foundation (Program "Molecular Magnets")
Johannes-Gutenberg-Universität Mainz
Freunde der Universität Mainz e.V
Karl-Gückinger-Stiftung der Universität Mainz
Materialwissenschaftliches Forschungszentrum der Universität Mainz
Ministerium für Wissenschaft, Weiterbildung, Forschung und Kultur des Landes Rheinland-Pfalz

and

BHF-Bank Frankfurt
Max-Planck-Institut für Chemische Physik fester Stoffe, Dresden
Schott Glas, Mainz
Stadt-und Kreissparkasse Darmstadt
SusTech GmbH & Co. KG, Darmstadt

We are also grateful to Deutsche Lufthansa for accommodating us again in their very pleasant Training Center in Seeheim. We particularly appreciate the friendly cooperation of the staff there.

I sign, with some nostalgia, on behalf of the scientific committee, the local organizers, and my fellow co-editors Brian Fitzsimmons, Rudolf Rüffer and Hartmut Spiering.

PHILIPP GÜTLICH
Guest Editor

Hyperfine Interactions **144/145:** 3–20, 2002.

Vibrational Dynamics Studies by Nuclear Resonant Inelastic X-Ray Scattering

E. E. ALP[1], W. STURHAHN[1], T. S. TOELLNER[1], J. ZHAO[1], M. HU[2] and D. E. BROWN[3]
[1]*Advanced Photon Source, Argonne National Laboratory, Argonne, Illinois 60439, USA*
[2]*HP-CAT and Carnegie Institute of Washington, Advanced Photon Source, Argonne, Illinois 60439, USA*
[3]*Department of Physics, Northern Illinois University, De Kalb, Illinois 60115, USA*

Abstract. Nuclear resonant inelastic X-ray scattering of synchrotron radiation is being applied to ever widening areas ranging from geophysics to biophysics and materials science. Since its first demonstration in 1995 using the ^{57}Fe resonance, the technique has now been applied to materials containing ^{83}Kr, ^{151}Eu, ^{119}Sn, and ^{161}Dy isotopes. The energy resolution has been reduced to under a millielectronvolt. This, in turn, has enabled new types of measurements like Debye velocity of sound, as well as the study of origins of non-Debye behavior in presence of other low-energy excitations. The effect of atomic disorder on phonon density of states has been studied in detail. The flux increase due to the improved X-ray sources, crystal monochromators, and time-resolved detectors has been exploited for reducing sample sizes to nano-gram levels, or using samples with dilute resonant nuclei like myoglobin, or even monolayers. Incorporation of micro-focusing optics to the existing experimental setup enables experiments under high pressure using diamond-anvil cells. In this article, we will review these developments.

Key words: synchrotron radiation, inelastic X-ray scattering, nuclear resonant scattering, speed-of-sound, Debye sound velocity.

1. Introduction

The application areas of nuclear resonant inelastic X-ray scattering (NRIXS) of synchrotron radiation are widening. Since the first experiments in 1995 [1–3] using ^{57}Fe resonance, the inelastic scattering technique is now applied to ^{83}Kr [4], ^{151}Eu [5], ^{119}Sn [6, 7], and ^{161}Dy [8, 9] isotopes. It can be further applied to other isotopes like ^{129}I, ^{40}K, ^{149}Sm, ^{121}Sb, and ^{61}Ni, provided that issues regarding the background coming from natural radioactivity are addressed for the first two and regarding the high resolution optics working above 30 keV for the last two isotopes. Special and unique applications like samples under high pressure exceeding 100 GPa [10] and the potential to reach high temperatures simultaneously are of interest from geological point of view. Isotopically decorated thin multilayers, in which the resonant isotope is selectively deposited to a well-defined location helps differentiate interfaces in thin films and multilayers. Vibrational studies performed on oriented single crystals or textured samples allow unambiguous mode specific vibrational analysis [11–14]. The study of samples with dilute nuclear resonant nuclei as an impurity [15] provide a new avenue about learning the dynamical behavior of hosts that are not necessarily made up of a suitable Mössbauer isotope. Among applications that are more recent, one can point to the study of vibrational characteristics of myoglobin [16, 17] and model compounds such as porphyrins. These studies lead the way to explore the role of ligands attached to the active sites of metalloproteins.

While these developments may indicate a level of maturity, there still exists confusion regarding the name of the technique, and whether it is an X-ray scattering or Mössbauer spectroscopy method. The NRIXS, as the name implies, is an X-ray scattering technique measuring energy loss or gain during scattering processes when X-rays are scattered inelastically, i.e., when an exchange of energy and momentum takes place. However, there are difficulties associated with measuring small energy changes that require a resolution at the level of 10^8, and hence, special nuclei with a few nano-seconds or longer lifetime are used as built-in analyzers. These nuclei with stable ground state, and low lying excited energy states are more prominently known as Mössbauer nuclei. Thus, NRIXS may be considered as an X-ray scattering technique, developed by Mössbauer spectroscopists. Mössbauer effect is commonly defined as "recoilless absorption and emission of gamma rays". While the zero-phonon exchange nature of the Mössbauer effect received worldwide recognition, there were early proposals [18–20], as well as experiments with partial success using Te to measure complete vibrational excitation spectrum [21]. These ideas of using phonon-exchange to excite or detect a Mössbauer transition have stimulated the research in this area, culminating with the practical demonstration of the technique using synchrotron radiation [1–3].

2. The method of NRIXS

A typical experimental setup, as shown in Figure 1, is more or less universally used at various synchrotron radiation sources around the world. An intense X-ray source, e.g., a magnetic insertion device (undulator) placed on a low electron emittance storage ring produces X-rays with low angular divergence, typically 10 (vertical) and 20 (horizontal) μrad, covering a large energy spectrum [22]. The energy spectrum is wide enough to cover many Mössbauer nuclei as listed in Table I. The isotopes shown here are chosen such that there is a realistic chance of observing the nuclear resonance given the energy range of current synchrotron radiation sources, and the time resolution of the existing detectors. Some measured transition energies are different than the tabulated values found in the literature [23]. A deliberate effort to measure the nuclear transition energies with highest accuracy to date indicates the potential to use the Mössbauer transition energies as a standard length scale at interatomic distances [24, 25]. After reducing the bandpass to a few electron volts with a double-crystal (cryogenically cooled silicon or water-cooled diamond) monochromator, the energy width is further reduced to the meV level by

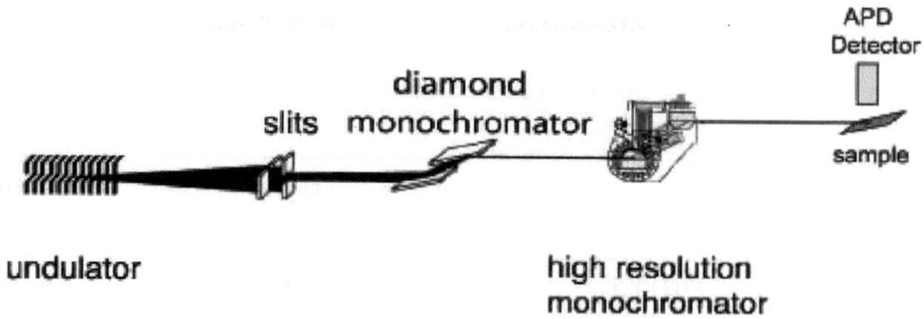

Figure 1. The schematic layout of the NRIXS experiment at 3-ID beamline of the Advanced Photon Source (APS) is shown. It consists of an undulator source (2.7 cm period, 4.5 m long), a primary monochromator (diamond) to reduce the energy bandwidth from a few hundred keV to a few eV, and a secondary tunable monochromator with a bandpass of meV (different for each isotope, see Table II). A nanosecond time-resolved detector is used to differentiate nuclear decay and/or nuclear decay induced electronic fluorescence from other elastic or inelastic scattering processes. There can be additional focusing components to facilitate small samples of a few microns in size, as well as different sample environments. The total length of the setup can vary between 30–70 m.

high-resolution monochromators [26]. A short list of present day monochromators used at the 3-ID beamline of the Advanced Photon Source (APS) is given in Table II. In addition to efficient reduction in energy bandpass, reducing the size of the X-ray beam in both directions to micrometers level is possible by using focusing optical elements like a pair of bent-grazing incidence total reflection mirrors, commonly known as Kirkpatrick–Baez [27] geometry, and with lesser efficiency by a Fresnel zone-plate [28], or a compound refractive lens [29]. The radiation is then incident on a sample, and scattered incoherent radiation is collected by a set of fast detectors, over as large a solid angle as possible. Clearly, this setup has very little to do with a conventional Mössbauer spectrometer, which consists of a radioactive source mounted on an electromagnetic Doppler shifter, a sample, and a detector. Of course, numerous versions of both experimental setups are used that employ different types of detection schemes, like conversion electron and time-resolved X-ray detectors. However, there is an essential difference in that, the scale and the types of instruments have a lot more in common with X-ray scattering than Mössbauer spectroscopy. While this difference, fortunately, did not stop the early pioneers [30, 31] from trying and eventually succeeding [32] to observe the nuclear resonance at a synchrotron, many found the technique too elaborate and difficult, particularly in terms of getting access to a facility to carry out systematic research. However, this situation has changed for the better now, after the availability of dedicated beamlines, such as 3-ID at the APS [33] and fully or partly dedicated beamlines listed in Table III.

The NRIXS technique involves excitation of a nuclear resonance simultaneous with exchange of phonons. The nuclear resonance here is more than a convenience to detect small energy changes compared to the incident beam energy (meV out of 10 keV), as it also provides element and isotope selectivity. Hence, NRIXS is unique among inelastic scattering methods. The technique has been described in detail [34, 35]. Here we will briefly discuss the cross-section and explain the

Table I. The Mössbauer isotopes within the spectral reach of modern synchrotron radiation sources. The current tunable high-resolution monochromators are built for transitions less than 30 keV. However, there are other optics possibilities for higher transitions like exact back-scattering [55], and nuclear light-house effect [56]

Isotope	Tabulated E_0 (eV)	Measured E_0 (eV)	Energy width Γ (neV)	Half-lifetime, $\tau_{1/2}$ (ns)	Observation technique
^{181}Ta	6238	6215.5	0.075	6050	Incoherent scattering
^{169}Tm	8401	8401.3	114	4	NBS
^{83}Kr	9401	9403.5	3.1	147	NFS, NRIXS
^{73}Ge	13263	–	0.15	2953	
^{57}Fe	14413	14412.5	4.67	97.8	NBS, NFS, NLE, NRIXS
^{151}Eu	21532	21541.4	47	9.7	NFS, NRIXS
^{149}Sm	22490	22496	64.1	7.1	NLE
^{119}Sn	23870	23879.5	25.7	17.75	NFS, NRIXS
^{161}Dy	25655	25651.4	16.2	28.2	NFS, NRIXS
^{129}I	27770	–	27.2	16.8	
^{40}K	29560	29834	107.3	4.25	Incoherent scattering
^{145}Nd	67100	–	6.8	67.1	
^{61}Ni	67400	67419	89	5.1	NFS, Inelastic scattering
^{193}Ir	73000	–	72.3	6.3	
^{133}Cs	81000	–	71.5	6.4	
^{67}Zn	93300	–	0.049	9200	

The acronyms in the last column are: NBS: nuclear Bragg scattering; NFS: nuclear forward scattering; NRIXS: nuclear resonant inelastic X-ray scattering with high enough resolution to study phonon excitations; NLE: nuclear light-house effect; Incoherent scattering refers to observation of nuclear excitation with synchrotron radiation in the incoherent channel without sufficient energy resolution, but with time discrimination.

measurable quantities from a NRIXS experiment and give examples from a number of general research topics.

3. The cross section of NRIXS and its information content

The incoherent inelastic interaction cross section of X-rays with a medium containing resonant nuclei when the incident beam energy is away from the resonance can be represented by

$$\sigma(\mathbf{k}, E) = \frac{\pi}{2}\sigma_0 \Gamma S(\mathbf{k}, E). \tag{1}$$

Table II. Characteristics of some monochromators developed for several Mössbauer isotopes in the accessible energy range of 9–27 keV at the 3-ID beamline of the APS. An important technological advance is the ability to link two crystals into a "virtual" channel-cut, which allows great flexibility in reaching meV level monochromaticity from 8 to 30 keV range [56]

Isotope	Energy (eV)	ΔE $E/\Delta E(\times 10^7)$	Flux (GHz)	Type	Notes
^{83}Kr	9.404	1 meV 0.94	6		Two dispersive channel-cut crystals, linked artificially, inline geometry suitable for microfocusing.
^{57}Fe	14.413	0.8 meV 1.8	1.0		Two separate crystal, high-throughput, but not suitable for microfocusing.
^{57}Fe	14.413	1.0 meV 1.4	2		Two dispersive channel-cut crystals, linked artificially, suitable for microfocusing.
^{151}Eu	21.541	0.8 meV 2.6	0.4		Nested geometry, outer crystals are linked artificially, suitable for microfocusing.
^{119}Sn	23.880	0.85 meV 2.8	0.2		Nested geometry, outer crystals are linked artificially, suitable for microfocusing.
^{119}Sn	23.880	0.14 meV 17	0.004		Two separate crystal, high-throughput, but not suitable for microfocusing.
^{161}Dy	25.651	0.5 meV 5.1	0.1		Nested geometry, outer crystals are linked artificially, suitable for microfocusing.

Here σ_0 is the nuclear resonance cross section, and Γ is energy width of the first nuclear excited state. The absorption probability of a photon with energy E and momentum $\mathbf{k}$ due to exchange of energy with a collective excitation is given by $S(\mathbf{k}, E)$. This quantity should not be confused with $S(\mathbf{q}, \omega)$, the cross section which can be obtained in momentum transfer, or $\mathbf{q}$-resolved experiment, like coherent inelastic neutron scattering. Equation (1) indicates that the interaction cross-section of incident photons with the sample during the phonon exchange is related to the nuclear level width and the phonon excitation probability. The overall

Table III. A list of the dedicated nuclear resonant scattering beamlines at various synchrotron radiation sources around the world. Only those facilities with timing or bunch structures suitable for nuclear resonant studies for at least significant part of the time are listed. Other modern facilities do not support suitable timing pattern, and therefore, they are not listed here. Further details regarding energy and size of the electron beam, and the X-ray photon brightness are given elsewhere [22]

Location$^{(*)}$	Beamline	Isotopes studied	Other capabilities
APS	3-ID	Kr, Fe, Eu, Sn, Dy	Undulator source, half-dedicated microfocusing, quarter-wave phase plate for circular polarization, DAC, IR-laser
ESRF	ID-18	Ta, Fe, Eu, Sn, Dy	Undulator source, microfocusing, high magnetic fields, high pressure DAC cells, low temp. fully dedicated
	ID-22N	Ta, Fe, Eu, Sn, Dy	Undulator source, same as ID-18, Half-dedicated
SPring-8	BL09XU	Fe, Eu, Sn, Dy, K	Undulator source, half dedicated, DAC, low temperature
	BL11XU		
	BL 35XU	Fe, Dy	Undulator source, dedicated to inelastic X-ray scattering, capable of nuclear resonant scattering, DAC
KEK-AR	NE#4	Fe, Sn, Dy, Eu	Undulator source, single-bunch mode exceeding 40 mA per bunch, half-dedicated station
HASYLAB	DORIS F4	Fe, Eu	Bending magnet source, dedicated station, low temperature
	DORIS BW4	Fe, Eu, Sn	Wiggler source (2.7 m long), half-dedicated station, low temperature, high magnetic field
	PETRA	Fe, Dy, Eu, Sn, Ni	Undulator source (4 m), half-dedicated station, low temperature, high magnetic field

$^{(*)}$APS: Advanced Photon Source, Argonne National Lab, Argonne, Illinois, USA. ESRF: European Synchrotron Radiation Facility, Grenoble, France. SPring-8, Super Photon Ring, Hyogo, Japan. KEK-AR: National Laboratory for High Energy Physics, Accumulator Ring, Tsukuba, Japan. HASYLAB: Hamburg Synchrotron Radiation Laboratory, Hamburg, Germany.

resolution of the spectrum will be given by the monochromator resolution rather than internal line width of the nuclear transition, which is negligibly smaller. This formula also implies that there can be anisotropic dependence of the cross section for single crystal or oriented samples, a reference axis provided by the incident photon direction. The real intensity measured will depend also on other intrinsic electronic and nuclear factors like the fluorescence yield, internal conversion coefficient, and experimental factors like detector efficiency, and isotopic concentration in the sample. These details are discussed extensively [34, 35], and they will not be repeated here. The absorption probability due to phonons in Equation (1) is $S(\mathbf{k}, E)$, and it is related to self correlation function, $G_s(\mathbf{r}, t)$ [36] by a double Fourier transform in space and time:

$$S(\mathbf{k}, E) = \frac{1}{2\pi} \int \mathrm{d}t\, \mathrm{d}^3\mathbf{r} \mathrm{e}^{\mathrm{i}(\mathbf{kr}-\omega t)} G_s(\mathbf{r}, t). \tag{2}$$

The information content in $S(\mathbf{k}, E)$ is related to inelastic absorption due to creation or annihilation of one or more phonons. In the harmonic approximation, the one-phonon density of states can be obtained from the inelastic nuclear absorption probability through an iterative process described earlier [1]. The one-phonon term, $S_1(E)$ is related to projected partial phonon density of states:

$$S_1(E) = \frac{E_R \cdot g(E)}{E(1 - \mathrm{e}^{-E/kT})}. \tag{3}$$

Here, E_R is the recoil energy of a free nucleus, k is the Boltzmann constant, and E is the transferred energy, and T is the temperature. The normalized density of states, $g(E)$, is integrated over phonon momentum in the first Brillouin zone and normalized per unit atomic volume. In a harmonic solid, thermodynamic properties are additive and they can be obtained from the knowledge of the density of states. In a compound containing resonant and non-resonant nuclei, only partial density of states is measured. The "partial" character is due to origin of signal coming from a Mössbauer isotope. This may be considered as a restriction because the number of Mössbauer nuclei suitable for synchrotron radiation work is limited. However, in systems like proteins, where there are many atoms per molecule, it is a great advantage to be sensitive to the vibrational modes of the Mössbauer atom only. Similarly, isotopically decorated interfaces provide an unambiguous signal originating from a particular section of a sample and makes NRIXS unique among other vibrational spectroscopic methods.

Physical properties that can be obtained directly or derived from NRIXS data are listed in Table IV. Most of these thermodynamic or elastic properties can be theoretically calculated or simulated by a variety of modern computational methods like density functional theory, molecular dynamics, or obtained from empirical force constants derived from other experimental techniques. This fact makes NRIXS even more interesting, as the core information gained is related to the interatomic potential.

The important distinguishing characteristics of NRIXS are the element and isotope selectivity, relative independence from sample size, form and geometry, the rich information content in terms of vibrational, thermodynamic and elastic properties, and the potentially universal applications when the Mössbauer isotope is included as a probe in a matrix. We will discuss some these characteristics next.

Table IV. Properties that can be extracted directly or indirectly from NRIXS data

Property	Information content
Lamb–Mössbauer factor, or recoil-free fraction	f_{LM}, recoil free fraction obtained from density of states, $g(E)$: $f_{LM} = \exp\left(-E_R \int \frac{g(E)}{E} \cdot \coth \frac{\beta E}{2}\, \mathrm{d}E\right)$
Second-order Doppler shift	$\delta_{SOD} = -E_0 \frac{\langle v^2 \rangle}{2c^2}$
Average kinetic energy	Extracted from second moment of energy spectrum: $T = \frac{1}{4E_R} \langle (E - E_R)^2 \rangle$
Average force constant	Extracted from third moment of energy spectrum: $\frac{\partial^2 U}{\partial z^2} = \frac{m}{2\hbar^2} \langle E^3 \rangle$
Phonon density of states	Extracted one-phonon absorption probability, $S_1(E)$: $g(E) = \frac{E}{E_R} \tanh \frac{\beta E}{2} (S_1(E) + S_1(-E))$
Specific heat (vibrational part only)	$C_V = 3k_B \int_0^\infty \left(\frac{\beta E}{2}\right)^2 \operatorname{csc} h(\beta E) g(E)\, \mathrm{d}E$
Vibrational entropy	$S_V = 3k_B \int_0^\infty \left\{\frac{\beta E}{2} \coth(\beta E) - \ln[2 \sinh(\beta E)]\right\} g(E)\, \mathrm{d}E$
Debye sound velocity (aggregate sound velocity)	From low-energy portion of the density of states: $g(E) = \frac{3V}{2\pi \hbar^3 v_D^3} E^2$
Mode specific vibrational amplitude	Contribution of mode α of atom j to zero-point fluctuation [11, 12]: $\langle r_{j\alpha}^2 \rangle_0 = \frac{\hbar^2}{2m_j \omega_\alpha^2} e_{j\alpha}^2$
Mode specific Gruneisen constant	From pressure dependence of phonon frequencies ω_α of acoustic or optical modes: $\gamma_\alpha = -\frac{\partial \ln \omega_\alpha}{\partial \ln V}$
Temperature of the sample	From detailed balance between phonon occupation probability

3.1. ELEMENT AND ISOTOPE SELECTIVITY

The NRIXS technique should not be considered limited because it is only applicable to some of the selected isotopes listed in Table I. To the contrary, this feature has already been exploited in such a way that made the technique unique and powerful. For example, in an atomically layered structure, one might be interested in differentiating between behavior of atoms at the interface or inside. In such a case, it is possible to selectively deposit the Mössbauer active element's isotope at the desired location by sputtering, and performing the experiments on each sample separately. This feature has been exploited in Fe/Cr thin films [37]. ^{57}Fe has been selectively deposited at the interface or inside a layer in a Fe/Cr multilayers system. The results show a reduction in the dispersion of $[0\,0\,\xi]$ phonons near the H point of the Brillouin zone.

Perhaps, one of the best examples of the use of element selectivity is the set of experiments conducted on protein mimic compounds of Fe-porphyrins [12–14] and Fe-containing proteins like myoglobin [16, 17]. The vibrational modes of iron containing proteins are of interest to those trying to connect the effect of ligands attached to the active sites of metalloproteins to the bending of heme group, and its allosteric effects. Whether one uses a model compound like iron-tetraphenylporphyrin, or a real heme-protein like myoglobin, the difficulty of observing and assigning vibrational modes to the experimentally measured peaks is due to the shear number of modes, and lack of optical activity in some modes due to symmetry considerations, which may put limits to established techniques like Raman or infrared spectroscopy. However, element selectivity of NRIXS combined with lack of symmetry requirements provides a new avenue to identify displacement and frequency of modes associated with all the atoms in the molecule, through their impact on the iron atom. This can be seen in Figure 2, which shows the vibrational spectrum of single crystal [Fe(TPP) (2-MeIm)], (2methylimidazole) (tetraphenyl-porphyrinato) iron(II). This compound serves as a model for deoxy-heme proteins. The vibrational spectrum of a single crystal compound taken in two directions with respect to the porphyrin plane (in-plane and out-of-plane) is compared to the polycrystalline sample. The detailed breakdown of each mode assignment has been accomplished [14]. From the analysis, it is possible to learn the frequency, amplitude, and relative contribution of each eigenmode to the total potential energy [12]. When such studies are conducted systematically as the types of ligands attached to the iron varies (like CO, NO, OH, O_2), then a more complete picture of iron dynamics in metalloproteins may emerge. Another important application of NRIXS is in the study of spin-crossover complexes. These vibrational frequency changes in complexes that undergo a high-spin to low-spin transition upon cooling [38] can be monitored with high selectivity due to the ability of NRIXS to focus on a few modes [11, 39]. Thus, NRIXS may be considered a complementary spectroscopic tool to infrared or Raman spectroscopy. Particularly, when these studies are combined with density functional calculations similar to the

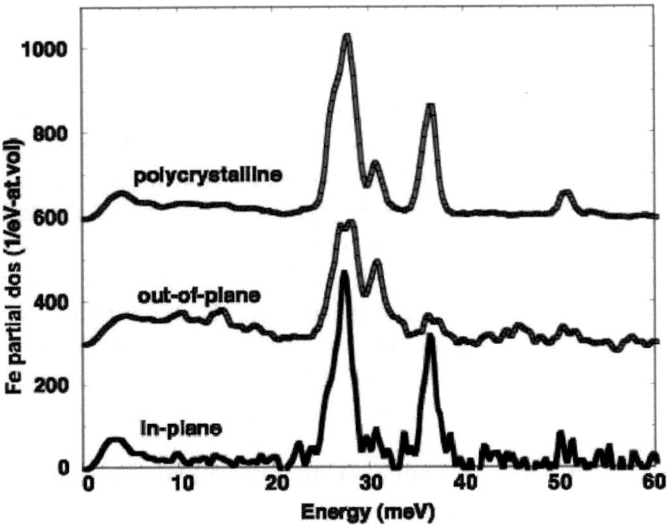

Figure 2. The measured Fe partial phonon density of states of [Fe(TPP)(2-MeIm)], ((2-methylimidazole) (tetraphenylporphyrinato) iron(II)). The single crystal whose porphyrin plane was oriented either parallel or perpendicular to X-ray propagation axis helped unambiguous identification of in-plane and out-of-plane vibrations. The data is taken with an overall resolution of 0.8 meV, two-crystal monochromator as described in [26]. The detailed breakdown of peak analysis has been accomplished [14].

ones performed on CO-binding heme proteins and its model compounds, functional properties of proteins can be better addressed [40].

3.2. SAMPLE SIZE

Focusing X-rays to a few micrometers in size with great efficiency is possible using horizontally and vertically focusing total external reflection mirrors [27] This, in turn, enables different kinds of experiments. Small samples with a total weight of nanogram and samples under extreme conditions like high pressures only achievable inside a diamond-anvil cell (DAC) can be studied. For this type of work, typically samples of a few micrometers are placed inside a gasket material and squeezed between diamond culets of a few hundred-micrometer in diameter to reach pressures exceeding a Mbar [10, 41]. In our set-up, X-rays are focused to a diameter less than 5 μm using a pair of elliptically bent mirrors. The sample thickness is also similar, giving a total weight of 1 nanogram. An example is provided to demonstrate the change in the density of states of iron at three different pressures in Figure 3. The high-pressure measurements are an important part of determination of equation of state of materials under extreme conditions, providing a wealth of information through density of states. Particularly, extraction of the aggregate speed-of-sound, mode-specific Grüneisen constant, which measures change in the vibrational frequency as a function of volume can be determined.

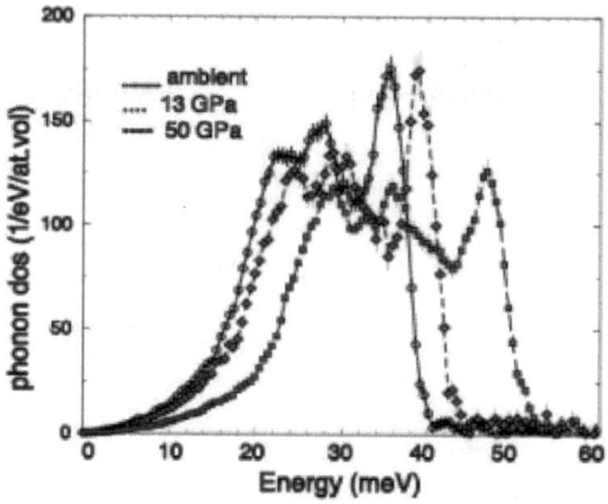

Figure 3. The phonon density of states of iron metal, recorded with 2 meV overall resolution, as a function of pressure in a diamond-anvil cell at room temperature. The data for ambient were taken from a thick iron foil that is outside the pressure cell. A more systematic study on a larger pressure range, albeit less statistics, allowed determination of aggregate speed of sound propagation in iron up to 150 GPa [10]. Note that there is a structural phase transition between 13 and 20 GPa, during which iron transforms from a body centered cubic phase to hexagonal close packed structure.

These thermodynamic and elastic parameters can be simulated using density functional theory. This aspect of the NRIXS should then provide important clues to find the active vibrational modes involved in a phase transition, whether due to a change in pressure or temperature.

3.3. SAMPLE FORMS: SINGLE CRYSTAL, POLYCRYSTALLINE OR AMORPHOUS

The NRIXS can be applied to different forms of aggregation. Single crystals, polycrystals in the form of powders or thin films, and amorphous phases can be studied. The information content can be quite different and revealing, as well as challenging to calculate in each case. We present two examples to illustrate this point. The vibrational spectra of ^{57}Fe in amorphous and crystalline Tb_2Fe are compared in Figures 4 (raw data) and 5 (partial density of states) [42]. As the near-neighbor environment for ^{57}Fe changes from well defined to a distribution, the force constants follow the atomic distribution change, resulting in a phonon spectrum that is energy dispersive, with higher frequency modes as well as heavily damped modes. The recoil free fraction obtained from the NRIXS data yields $f = 0.68$ for crystalline phase, and $f = 0.57$ for the amorphous phase, respectively. Without the benefit of having phonon density of states, this may be interpreted as softening of the lattice. However, it is clear that in addition to soft modes, some high-energy modes result from crystalline-to-amorphous transition. Thus, concepts like Debye temperature, or average displacement may not be that informative in the case of amorphous phases. This is a relevant finding for the correct interpretation of EXAFS data. In the standard formulation, the damping of the EXAFS signal at higher energies is accounted for by a single (e.g., Einstein) temperature, which may be misleading,

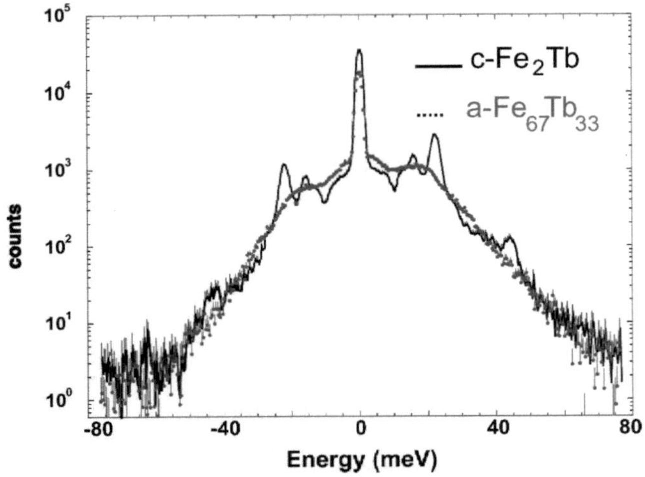

Figure 4. Raw NRIXS data as collected by energy scanning of the time-delayed nuclear resonance-induced Fe K_{α} fluorescence signal of 6.4 keV, integrated over a time window of 20–150 nsec. The solid line belongs to the crystalline form of Fe_2Tb, while the dotted data belong to the amorphous alloy of the same composition. Both samples were prepared in thin-film form. The logarithmic plot is intended to show that signal-to-noise ratio that exceeds 4 orders of magnitude. This feature of NRIXS can be exploited in many different ways in terms of small samples and very low isotope concentration. It should also be mentioned that the signal here represents a convolution of one, two and three phonon processes, as it is particularly visible at ±44 meV, which is a two phonon image of the peak at ±22 meV.

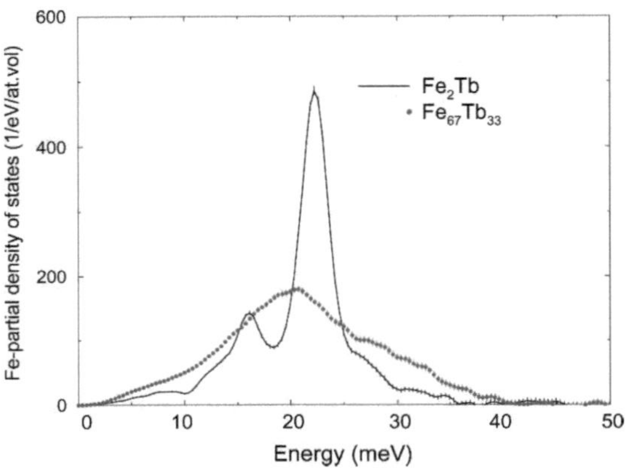

Figure 5. The vibrational density-of-states of iron are given for crystalline (solid) and amorphous (dotted) forms of Fe_2Tb. The sharp feature at 22 meV in the crystalline form can be attributed to a dispersion-free optical mode. The emergence of low-energy modes, as well as high-energy modes upon amorphization seems to be common characteristics of crystalline-to-amorphous [37] or order–disorder transformation [43, 44]. The change in phonon density of states shown above can be attributed to imbalance between Tb and Fe near neighbor coordination numbers, which creates a range of force constants between Fe and its neighbors.

since there is a distribution of vibrational frequencies. Similar features are also observed in earlier experiments in crystalline intermetallic compounds of Fe_3Al and Pt_3Fe undergoing order–disorder phase transition [43, 44]. Thus, the NRIXS data provide a way to check calculations that will include distribution of force constants and the dynamics of disordered and amorphous systems.

The data for single crystals or oriented samples can be as interesting and useful and this was shown earlier in the discussion of metalloproteins (Figure 2). The NRIXS cross-section given in Equation (1) is dependent on the momentum transfer, **k**, which is defined by the direction of the incident beam, and its magnitude by the energy of the nuclear transition. Thus, oriented samples or single crystals which form a well defined angle with **k** will have projected partial density of vibrational states along this direction. The case for anisotropic crystals was formulated in [45] and applied to $FeBO_3$. In the general case, the density of states is given by

$$g(\mathbf{s}, \omega) = \frac{1}{N} \sum_{m\kappa} \left|\mathbf{s} \cdot \mathbf{e}_{pm}^{(\kappa)}\right|^2 \delta\left(\omega - \omega_m^{(\kappa)}\right). \tag{4}$$

Here, $\omega_m^{(\kappa)}$ is the energy phonon modes m, with (quasi) momentum κ and $\mathbf{e}_{pm}^{(\kappa)}$ is the polarization vector at atom p in the unit cell. Vector **s** defines the direction of X-rays. The experimental procedure to derive local density of states and anisotropy functions for different Bravais lattices are also described [34]. Hence, one needs to be aware of presence of any anisotropy induced during phase transitions, and sample manipulation. Even for powder samples, the resulting density of states may prove to be sensitive to orientation preference occurred during packing of the sample, as it was the case for SnO powder samples packed on a tape [46]. This issue becomes sensitive for high-pressure studies where induced anisotropy may affect the speed of sound measurements in hcp-iron above 20 GPa [47].

3.4. SAMPLE GEOMETRY: THIN FILMS, BURIED INTERFACES, MONOLAYERS

Another unique characteristic of NRIXS is its ability to study thin films and buried interfaces. The low vertical divergence of the undulator radiation (10 μrad) and further focusing to micrometer levels enables studies of very small thin-film samples. Early studies on iron thin films provided some insight into phonon damping [48]. The current interest in thermal conductivity of thin films is due to nano-scale electronics components that are expected to transport heat. NRIXS studies provide a tool to measure phonon lifetime broadening. As the energy resolution improves from meV to sub-meV levels, this approach may become more effective. The best resolution obtained, 0.14 meV at 23.88 keV of ^{119}Sn nuclear resonance [49], and 0.12 meV at ^{57}Fe nuclear resonance [50], implies ability to measure frequency of less than a THz. When thin films are prepared in a special form that the layer of interest is squeezed between two layers with larger electron density, a standing wave can form inside the layer containing the nuclear resonant isotope. An experimental demonstration of this has been accomplished by synthesizing a sample of

$Pd/FeBO_3/Pd$, in which a 10-fold enhancement of Fe Kα fluorescence yield was obtained [51] This wave-guide effect enables measuring phonons in films as thin as monolayer.

The ability to study thin films has some other interesting consequences. The composition range of amorphous phase formation in transition metal/rare-earth (TM–RE) metallic glasses can be extended on TM or RE-rich side of the composition spectrum substantially when the film is prepared in thin-film form. For example, in Fe_xTb_{1-x} system, $0.3 < x < 0.8$ can be attained [52]. Systematic NRIXS studies carried out in this and similar systems may shed light onto the low frequency dynamics of disordered systems, such as the origin of the Boson peak.

3.5. DETERMINATION OF COMPLETE DENSITY OF STATES FROM PARTIAL VALUES

The NRIXS technique reveals position correlation information that is weighted by the nuclear isotope. The self-correlation function, $G_s(\mathbf{r}, t)$ given in Equation (2) can be explicitly written as follows:

$$G_s(\mathbf{r}, t) = \left\langle \mathrm{e}^{-\mathrm{i}k\mathbf{r}(t)} \mathrm{e}^{\mathrm{i}k\mathbf{r}(0)} \right\rangle, \tag{5}$$

where $\mathbf{r}$ is the displacement operator of each atom. When this operator is written explicitly for each atom j, the inelastic absorption probability $S(\mathbf{k}, E)$ becomes:

$$S(\mathbf{k}, E) = \frac{1}{N} \sum_{j=1}^{N} S^j(\mathbf{k}, E). \tag{6}$$

Therefore, experiments on alloys may require more than one isotope to be present in the alloy, for complete extraction of density of states. We have conducted experiments in $^{57}Fe_2^{161}Dy$, where both elements have suitable isotopes for NRIXS studies. The partial and reconstructed density-of-states are given in Figure 6. It is also instructive to see that the iron density of states in Laves phases of Fe_2Tb and Fe_2Dy are strikingly similar. In both cases, Fe atoms are surrounded by heavier rare-earth atoms with a crystallographically unique site. The dispersion free optical mode of Fe disappears with the introduction of chemical or topological disorder, as it was the case given in Figure 7.

3.6. NUCLEAR RESONANT ISOTOPE AS AN IMPURITY

The advances made by working with small samples can also be exploited to study dilute samples, in which Mössbauer nuclei are placed as a probe of the host matrix. This potential was demonstrated earlier in ^{57}Fe-doped Cu and Al [15]. The ability to work with dilute samples brings the possibility of measuring speed of sound in

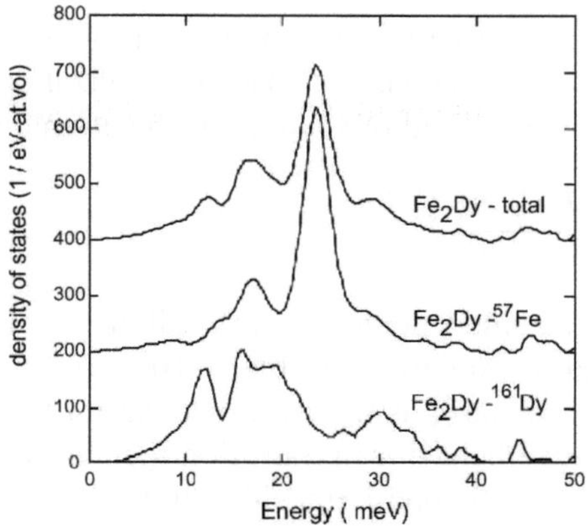

Figure 6. The partial density of states of Fe in Dy in Fe_2Dy sample obtained through ^{161}Dy (*bottom*) and ^{57}Fe (*middle*) by NRIXS techniques. The reconstructed total density of states is also shown (*top*). The energy resolution for ^{57}Fe was 2 meV, while energy-resolution for ^{161}Dy was 0.5 meV.

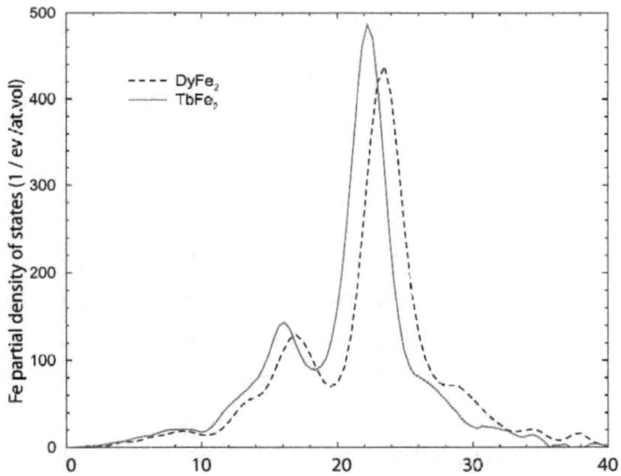

Figure 7. There is a remarkable similarity between phonon density of states of Laves phases of Fe_2Dy and Fe_2Tb. Fe_2Dy was a single crystal, while Fe_2Tb was in thin film form. C15 structure has one crystallographically equivalent site for Fe, with 6 iron and 6 Tb or Dy near neighbors. The width of the sharp peak is 3.5 meV, which is larger than the resolution function of 2 meV. This could be due to lack of perfect chemical order in the crystalline phase.

any matrix [53]. The low-energy part of phonon density of states has an energy dependence given by the Debye model:

$$g(E) = \frac{V}{2\pi\hbar^3 c_s^3} E^2. \tag{7}$$

For a cubic crystal

$$c_s = \left(\frac{E^2}{2\pi^2 g(E)}\right)^{1/3} \cdot \frac{a}{\hbar} \tag{8}$$

and for a nuclear resonant impurity with mass m', in a host matrix with mass m, a modified form to take into account of mass normalization [53]. Here, a is the lattice constant, and c_s is the speed of sound, n is the atomic density

$$c_s = \frac{1}{\hbar}\left[\frac{m'}{m}\frac{E^2}{2\pi^2 \cdot n \cdot g(E)}\right]^{1/3}. \tag{9}$$

Preliminary results obtained points toward validity of this hypothesis, in absence of other low-energy excitations in the 0–10 meV energy range. However, this may not always be the case, as it was shown earlier that there are significant deviations from Debye-like behavior in transition metal/rare-earth amorphous materials [37] and near a magnetic phase transition in FeO under pressure [54]. Same limitation applies when significant anisotropy is present [46, 47].

The quality of signal for NRIXS is another topic worthy of discussion. A raw data set obtained on Fe–Tb alloy is given in Figure 4. Here, we would like to emphasize the importance of time-discrimination in improving the signal-to-noise ratio. The raw data shows that the background is 4 orders of magnitude below the signal at 80 meV away from the elastic peak, in this iron-rich compound. Thus, even with small amount of impurity in a matrix, NRIXS studies can be conducted.

4. Summary

The relatively straightforward experiments of NRIXS provide a plethora of information about thermodynamic and elastic properties. Some of these properties are given in Table IV, and they are discussed previously in detail [34, 35]. However, as time goes on, we find out that NRIXS data are more versatile and richer than it appears. The ability to measure temperature and the use of resonant nuclei as a probe atom in a host matrix, which are not discussed here, should further provide newer applications. Huge potential in molecular vibrational spectroscopy as a complementary method to Raman or infrared spectroscopy are emerging with the ongoing experiments. As the theoretical machinery of modern computational physics is turned on, we expect that the NRIXS will become a method of choice for a certain class of experiments in such diverse fields as biology, geology, chemistry, and physics.

Acknowledgements

Many of the data presented here are born out of the experiments conducted at the 3-ID beamline of APS. Although proper credit are given for published results, we would like to acknowledge R. Röhlsberger, W. Keune, T. Sage, S. Durbin, B. Fultz, H.-K. Mao, G. Shen, and V. Struzhkin for continued scientific interactions. This work at Argonne National Laboratory is supported by US DOE BES under contract number W-31-109-ENG-38, and at Northern Illinois University is supported by State of Illinois under HECA.

References

1. Sturhahn, W., Toellner, T. S., Alp, E. E., Zhang, X. W., Ando, M., Yoda, Y., Kikuta, S., Seto, M., Kimball, C. W. and Dabrowski, B., *Phys. Rev. Lett.* **74** (1995), 3832.
2. Seto, M., Yoda, Y., Kikuta, S., Zhang, X. W. and Ando, M., *Phys. Rev. Lett.* **74** (1995), 3828.
3. Chumakov, A. I., Rüffer, R., Grünsteudel, H., Grünsteudel, H. F., Grübel, G., Metge, J., Leupold, O. and Goodwin, H. A., *Europhys. Lett.* **30** (1995), 3832.
4. Zhao, J. Y., Toellner, T. S., Hu, M. Y., Sturhahn, W., Alp, E. E., Shen, G. and Mao, H. K., *Rev. Sci. Instrum.* **73** (2002), 1608.
5. Koyama, I., Yoda, Y., Zhang, X. W., Ando, M. and Kikuta, S., *Japan. J. Appl. Phys.* **35** (1996), 6297.
6. Chumakov, A. I., Barla, A., Rüffer, R., Metge, J., Grünsteudel, H., Grünsteudel, H. F., Plessel, J., Winkelmann, H. and Abd-Elmeguid, M. M., *Phys. Rev. B* **58** (1998), 254.
7. Hu, M. Y., Toellner, T. S., Sturhahn, W., Hession, P. M., Sutter, J. P. and Alp, E. E., *Nucl. Instr. Meth. A* **430** (1999), 271.
8. Chumakov, A. I., Rüffer, R., Leupold, O., Barla, A., Thiess, H., Gil, J. M., Alberto, H. V., Vilão, R. C., Ayres de Campos, N., Kohn, G., Gerken, M. and Lucht, M., *Phys. Rev. B* **63** (2001), 172301.
9. Baron, A. Q. R., Tanaka, Y., Ishikawa, D., Miwa, D., Yabashi, M. and Ishikawa, T., *J. Synchrotron Rad.* **8** (2001), 1127.
10. Mao, H. K., Xu, J., Struzhkin, V. V., Shu, J., Hemley, R. J., Sturhahn, W., Hu, M. Y., Alp, E. E., Vocadlo, L., Alfe, D., Price, G. D., Gillan, M. J., Schwoerer-Böhning, M., Häusermann, D., Eng, P., Shen, G., Giefers, H., Lübbers, R. and Wortmann, G., *Science* **292** (2001), 914.
11. Paulsen, H., Winkler, H., Trautwein, A. X., Grünsteudel, H., Rusanov, V. and Toftlund, H., *Phys. Rev. B* **59** (1999), 975.
12. Sage, J. T., Paxson, C., Wyllie, G. R. A., Sturhahn, W., Durbin, S. M., Champion, P. M., Alp, E. E. and Scheidt, W. R., *J. Phys. Cond. Matter* **13** (2001), 1.
13. Rai, B., Durbin, S. M., Prohofsky, E. W., Sage, J. T., Wyllie, G. R. A., Scheidt, W. R., Sturhahn, W. and Alp, E. E., *Biophysics J.* **82** (2002), 2951.
14. Rai, B., Durbin, S. M., Prohofsky, E. W., Sage, J. T., Wyllie, G. R. A., Ellison, M. K., Scheidt, W. R., Sturhahn, W. and Alp, E. E., *Phys. Rev. E* **66** (2002), 0519014.
15. Seto, M., Kobayashi, Y., Kitao, S., Haruki, T., Mitsui, T., Yoda, Y., Nasu, S. and Kikuta, S., *Phys. Rev. B* **61** (2000), 11420.
16. Sage, J. T., Durbin, S. M., Sturhahn, W., Wharton, D. C., Champion, P. M., Hession, P., Sutter, J. and Alp, E. E., *Phys. Rev. Lett.* **86** (2001), 4966.
17. Achterhold, K., Keppler, K. C., Ostermann, A., van Bürck, U., Sturhahn, W., Alp, E. E. and Parak, F., *Phys. Rev. E* **65** (2002), 134305.
18. Singwi, K. S. and Sjölander, A., *Phys. Rev. B* **120** (1960), 1093.
19. Ingliss, D., In: H. Frauenfelder and W. A. Benjamin (eds), *The Mössbauer Effect*, New York, 1963, p. 169.
20. Vischer, V. M., *Ann. Phys.* **9** (1960), 194.
21. Weiss H. and Langhoff, H., *Phys. Lett. A* **69** (1979), 448.
22. Mülhaupt, G., Rüffer, R., *Hyp. Interact.* **123/124** (1999), 13.
23. Stevens, J. (ed), *Mössbauer Effect Reference and Data Journal*, http://www.unca.edu/medc.
24. Shvydko, Y., Lerche, M., Jäschke, J., Lucht, M., Gerdau, E., Gerken, M., Rüter, H. D., Wille, H. C., Becker, P., Alp, E. E., Sturhahn, W., Sutter, J. and Toellner, T. S., *Phys. Rev. Lett.* **85** (2000), 495.
25. Shvydko, Y. V., Lucht, M., Gerdau, E., Lerche, M., Alp, E. E., Sturhahn, W., Sutter, J. and Toellner, T. S., *J. Synchrotron Rad.* **9** (2002), 17.
26. Toellner, T. S., *Hyp. Interact.* **125** (1999), 3.
27. Eng, P. J., Newille, M., Rivers, M. L. and Sutton, S. R., *SPIE Proc.* **3449** (1998), 145.

28. Yun, W. B., Lai, B., Cai, Z., Maser, J., Legnini, D., Gluskin, E., Chen, Z., Krasnoperova, A. A., Vladimirsky, Y., Cerrina, F., DiFabrizio, E. and Gentili, M., *Rev. Sci. Instr.* **70** (1999), 2238.
29. Lengeler, B., Schroer, C. G., Benner, B., Günzler, T. F., Kuhlmann, M., Tümmler, J., Simionovici, A. S., Drakapoulos, M., Snigirev, A. and Sinigreva, I., *NIM A* **467–468** (2001), 944.
30. Ruby, S. L., *J. Physique* **35** (1974), C6-209.
31. Chechin, A. I., Andronova, N. V., Zelepukhin, M. V., Artem'ev, A. N., Stepanov, E. P., *JETP Lett.* **37** (1983), 633.
32. Gerdau, E., Rüffer, R., Winkler, H., Tolksdorf, W., Klages, C. P. and Hannon, J. P., *Phys. Rev. Lett.*, 54 (1985), 835.
33. Alp, E. E., Mooney, T. M., Toellner, T. S. and Sturhahn, W., *Hyp. Interact.* **90** (1994), 323.
34. Sturhahn, W. and Kohn, V., *Hyp. Interact.* **123/124** (1999), 367.
35. Chumakov, A. I. and Sturhahn, W., *Hyp. Interact.* **123/124** (1999), 781.
36. van Hove, L., *Phys. Rev.* **95** (1954), 249.
37. Ruckert, T., Keune, W., Sturhahn, W., Hu, M. Y., Sutter, J. P., Toellner, T. S. and Alp, E. E., *Hyp. Interact.* **126** (2000), 363.
38. Gütlich, P., Hauser, A. and Spiering, H., *Angew. Chem.* **106** (1994), 2109.
39. Grünsteudel, H., Paulsen, H., Winkler, H., Trautwein, A. X. and Toftlund, H., *Hyp. Interact.* **123/124** (1999), 841.
40. Kozlowski, P. M., Vogel, K. M., Zgiersky, M. Z. and Spiro, T. G., *J. Porphyrins and Phathalocyanines* **5** (2001), 312.
41. Lübbers, R., Wortmann, G. and Grünsteudel, H., *Hyp. Interact.* **123/124** (1999), 529.
42. Sturhahn, W., Röhlsberger, R., Alp, E. E., Ruckert, T., Schör, H. and Keune, W., *J. Magn. Mag. Mat.* **198** (1999), 590.
43. Fultz, B., Stephens, T. A., Sturhahn, W., Toellner, T. S. and Alp, E. E., *Phys. Rev. Lett.* **80** (1998), 3304.
44. Fultz, B., Stephens, T. A., Alp, E. E., Hu, M. Y., Sutter, J. P., Toellner, T. S. and Sturhahn, W., *Phys. Rev. B* **61** (2000), 1007.
45. Kohn, W., Chumakov, A. I. and Rüffer, R., *Phys. Rev. B* **58** (1998), 8437.
46. Giefers, H. and Wortmann, G., 2002, private communication.
47. Giefers, H., Lübbers, R., Rupprecht, K., Wortmann, G., Alfe, D. and Chumakov, A. I., *High Pres. Res.* **22** (2002), 501.
48. Röhlsberger, R., Sturhahn, W., Toellner, T. S., Quast, K. W., Hession, P., Hu, M. Y., Sutter, J. and Alp, E. E., *J. Appl. Phys.* **86** (1999), 584.
49. Toellner, T. S., Hu, M., Sturhahn, W., Bortel, G., Alp, E. E. and Zhao, J., *J. Synchrotron Rad.* **8** (2001), 1082.
50. Yabashi, M., Tamasaku, K. and Ishikawa, T., *Phys. Rev. Lett.* **87** (2001), 140801.
51. Röhlsberger, R., Sturhahn, W., Toellner, T. S., Quast, K. W., Alp, E. E., Bernhard, A., Metge, J., Rüffer, R. and Burkel, E., *Physica B* **263** (1999), 581.
52. Keune, W. and Sturhahn, W., *Hyp. Interact.* **123/124** (1999), 847.
53. Hu M. Y., Sturhahn, W., Toellner, T. S., Mannheim, P., Brown, D. E., Zhao, J. and Alp, E. E., *Phys. Rev. B* (2003), 094304.
54. Struzhkin, V., Mao, H. K., Hu, J., Schwoerer-Bohning, M., Shu, J., Hemley, R. J., Sturhahn, W., Hu, M. Y., Alp, E. E., Eng, P. and Shen, G., *Phys. Rev. Lett.* **87** (2001), 255501.
55. Shvydko, Y. and Gerdau, E., *Hyp. Interact.* **123/124** (1999), 741.
56. Röhlsberger, R., Quast, K. W., Toellner, T. S., Lee, P. L., Sturhahn, W., Alp, E. E. and Burkel, E., *Phys. Rev. Lett.* **87** (2001), 047601.
57. Shu, D., Han, Y., Toellner, T. S. and Alp, E. E., *SPIE Proceedings* **4771** (2002), 78.

Hyperfine Interactions **144/145:** 21–32, 2002.

Hyperfine Spectroscopy with Nuclear Resonant Scattering of Synchrotron Radiation

O. LEUPOLD
European Synchrotron Radiation Facility (ESRF), BP 220, F-38043 Grenoble, France
On leave from: Fachbereich Physik, Universität Rostock, D-18055 Rostock, Germany

Abstract. Since its observation in 1985 nuclear resonant scattering of synchrotron radiation has become an excellent tool to study hyperfine interactions in solids. It combines the advantages of both local probe experiments and scattering techniques and gives valuable information on magnetic and electronic structures in case of NFS experiments. Experiments benefit from the outstanding beam quality of 3rd generation synchrotron radiation sources, as the small beam size and divergence.

Key words: synchrotron radiation, nuclear resonant scattering, ^{57}Fe, ^{151}Eu, magnetic hyperfine interaction.

1. Introduction

Since the pioneering work of Gerdau *et al.* [1], which showed for the first time nuclear resonant scattering of synchrotron radiation (short NRS) this field has dynamically developed from an exotic "playground" – with the need of highly isotopic enriched perfect single crystals – to an established method. Since then the technical development, like dedicated synchrotron radiation sources, high-resolution X-ray optics and fast detectors, has made a rapid development possible. Nowadays, third-generation synchrotron radiation sources such as the European Synchrotron Radiation Facility (ESRF) in Grenoble/France, the Advanced Photon Source (APS) in Chicago/USA and SPring-8 in Harima/Japan provide high brilliant beams to perform efficiently NRS under extreme sample conditions, like small sample size, grazing incidence geometry or high pressure. A number of recent reviews on the development and on various aspects of the NRS technique can be found in [2].

There exist mainly two branches: nuclear forward scattering (short NFS) and nuclear inelastic scattering. NFS can be considered to be a complementary technique to Mössbauer spectroscopy, since it means in most cases hyperfine spectroscopy, combining the energy resolution of Mössbauer resonances with the outstanding properties of third generation sources. Due to the linear polarization of synchrotron radiation, NFS is especially sensitive to the direction of magnetic hyperfine fields.

The following chapters will give a general introduction to nuclear forward scattering and a short beamline layout for NRS experiments. We will treat NFS utilizing the two nuclei ^{57}Fe and ^{151}Eu.

2. Nuclear forward scattering of synchrotron radiation

The first NFS experiments were performed by Hastings *et al.* [3] and van Bürck *et al.* [4] on magnetized α-Fe foils. An overview on NFS can be found in recent review articles, as, for example, in [5, 6], which, however, cover only the ^{57}Fe resonance. In the following we will give a short introduction to the NFS method with special focus on the isotopes ^{57}Fe and ^{151}Eu and their magnetic hyperfine interaction.

2.1. BASICS OF NFS

The basics of NFS are visualized in Figure 1.

- Figure 1(a) gives a very schematic sketch of a NFS experiment. The black dots represent the electron bunches in the storage ring. A typical bunch length is ~100 ps. The highly collimated, pulsed synchrotron radiation impinges on the sample, the transmitted and forward scattered radiation is recorded by a fast detector.
- Figure 1(b) shows how the time evolution of this intensity can be used to discriminate the resonant from the nonresonant intensity. The black spikes correspond to the exciting pulses, which define the "time zero" of the nuclear excitation and will be suppressed by electronic means in the course of the experiment. At the ESRF in 16-bunch mode the distance between two successive pulses is 176 ns corresponding to the electron bunch distance. The shaded areas in between these spikes visualize the *nuclear forward scattered* intensity, which is delayed with respect to the exciting pulse due to the life time of the nuclear excited states. Typical life times are 141 ns for ^{57}Fe and 14.1 ns for ^{151}Eu.
- Figures 1(c,d) displays on two examples the time behaviour of the reemitted photons in forward direction, which is, in general, not just an exponential decay with the corresponding life time of the excited nuclear state. The time decay shows characteristic modulations, which are on the one hand caused by the hyperfine interactions (*quantum beats*) and on the other hand due to multiple scattering in thick samples (*dynamical beats*).
- In Figure 1(c) we show the simplest case, without hyperfine splitting the energetically broad synchrotron radiation pulse excites the nucleus (wide up-arrow), the reemitted radiation in forward direction (down-arrow) contains only γ-quanta of one energy. In the case of excitation of a single nucleus this results in a simple exponential decay (dotted line). In a real experiment one hardly deals with single nuclei, but uses samples with finite thickness. Coherent multiple scattering occurs in a thick sample, the forward scattered intensity exhibits a characteristic time modulation, the so-called *dynamical beats* (full line).
- Figure 1(d) shows the case of a hyperfine split excited nuclear state, where two transitions are coherently excited (wide up-arrow). Now the forward reemitted

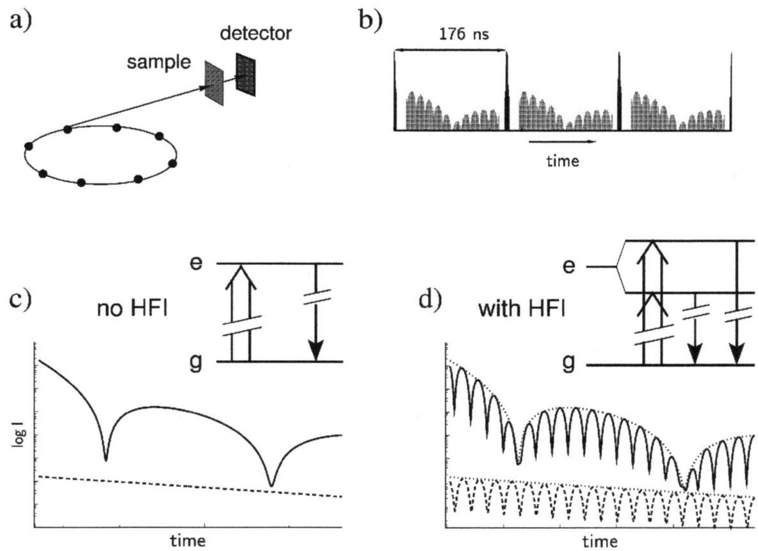

Figure 1. (a) Schematic of a NFS experiment. (b) Time structure of the detected radiation. (c) NFS time patterns for unsplit nuclear levels. (d) NFS time patterns for hyperfine split nuclear levels; the forward scattering process from two close lying nuclear levels gives rise to the chracteristic modulation of the time pattern. The dashed lines in (c), (d) correspond to a single nucleus response, the full lines result from a thick sample. The dotted lines in (d) correspond to the intensity pattern of the unsplit nuclear levels that determine the envelope of the quantum beat pattern. For more details see text.

radiation contains γ-quanta with two different energies (two down-arrows) which can interfere and lead to the so-called *quantum beats*. The dashed line shows again the forward scattering from a single nucleus, the beat frequency is determined by the energy splitting of the excited nuclear state, the envelope (dotted line) is an exponential which corresponds to the nuclear life time. In a thick sample (full line) the quantum beat pattern is superimposed by dynamical beats, the corresponding envelope (dotted line) has the same shape as the NFS pattern of a single line transition (Figure 1(c)) from a sample of half the thickness.

Like in conventional energy domain Mössbauer spectroscopy ^{57}Fe is the most widely used isotope in nuclear resonant scattering of synchrotron radiation. NFS experiments with other isotopes, i.e., ^{169}Tm [7] ,^{119}Sn [8–10] ^{83}Kr [11, 12], ^{181}Ta [13], and ^{151}Eu [14] have been treated in a recent review [16]. Others, which are especially important for magnetism, are, e.g., ^{161}Dy [17] or ^{149}Sm [18].

3. Beamline layout for NFS

Synchrotron radiation sources and "standard" beam line optics deliver a very intense photon beam with a bandwidth in the order of some eV, which is very large as

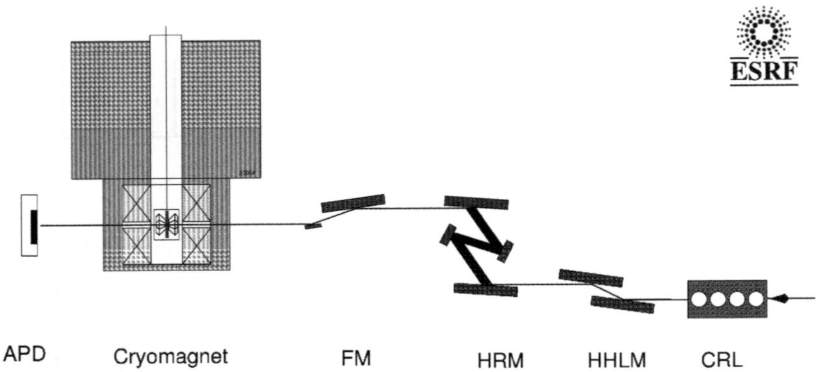

Figure 2. Experimental setup for nuclear forward scattering of synchrotron radiation. CRL: compound refractive lens; HHLM: high heat load monochromator; HRM: high-resolution monochromator; FM: focusing monochromator; APD: avalanche photo diode (for details see text).

compared to the extremely narrow widths of nuclear resonances, which are usually in the order of 1 neV to 100 neV in the case of Mössbauer isotopes. More sophisticated optics were designed, in order to reduce the energy bandwidth and by that the load on the detectors and the timing electronics. A typical experimental setup for nuclear forward scattering, e.g., at the ESRF beamline ID18 [19], is displayed in Figure 2.

The X-ray beam is monochromatized in two steps. A high heat load monochromator (HHLM) typically consisting of two cryogenically cooled Si(1 1 1) single crystals yields an energy resolution of some eV. The high-resolution monochromator (HRM) utilizes two Si channel cut crystals in nested configuration as proposed in [20] and delivers an energy band of some meV. For each isotope a special monochromator solution has to be found. The beam can be focused or collimated by a compound refractive lens CRL [21] and/or a focusing monochromator (FM) [22] to $\approx 100 \times 100\ \mu m^2$ FWHM, in order to improve the throughput and the energy resolution of the monochromators, as well as to reduce the beam size at the sample. The cryomagnetic system with a superconducting split coil magnet can reach magnetic fields up to 6T in a temperature range from ≈ 2 K to ≈ 400 K. This system has a large sample space (diameter 48 mm) in order to mount high-pressure cells or samples for grazing incidence experiments. The cryostat is mounted on a two circle goniometer for adjusting tilt and glancing angle, respectively. As fast detector serves an avalanche photo diode (APD) [23, 24] with a time resolution below 1 ns. For a review on monochromatization cf. [25], on fast detectors cf. [26].

Recently, a pair of Kirkpatrick–Baez mirrors has been successfully applied to focus the monochromatized beam down to $\sim 10 \times 15\ \mu m^2$ at the NRS stations ID18 and ID22N at the ESRF. Nuclear inelastic scattering studies in the Mbar range were performed at the APS with spot sizes of $\sim 5 \times 5\ \mu m^2$ using Kirkpatrick–Baez mirrors [27].

4. NFS and hyperfine interactions

4.1. MAGNETIC HYPERFINE INTERACTION IN α-IRON

As first example we show in Figure 3 NFS spectra of a magnetized α-^{57}Fe foil. The $I_e = 3/2 \rightarrow I_g = 1/2$ Mössbauer transition in ^{57}Fe (I_e, I_g are the spin quantum numbers of the excited and ground state, respectively) is a magnetic dipole transition. The magnetic hyperfine interaction completely lifts the degeneracy of the nuclear levels, the selection rules for dipole radiation allow $\Delta m = 0, \pm 1$ transitions, i.e., 6 transitions. Spectra were measured at a temperature $T \approx 100$ K in a helium bath cryostat in external fields of 1 T in plane and 3 T perpendicular to the plane of the foil. Figure 3 shows the pronounced dependence of the NFS time patterns on the external magnetic field direction. This is caused both by the angular characteristics of the dipole transition and the linear polarization of the synchrotron radiation, which is depicted in the coordinate system in Figure 3. We call it σ-polarization, when the electric field vector $\vec{E}$ is in the storage ring plane. The relevant Mössbauer transitions, the corresponding polarization eigenstates of

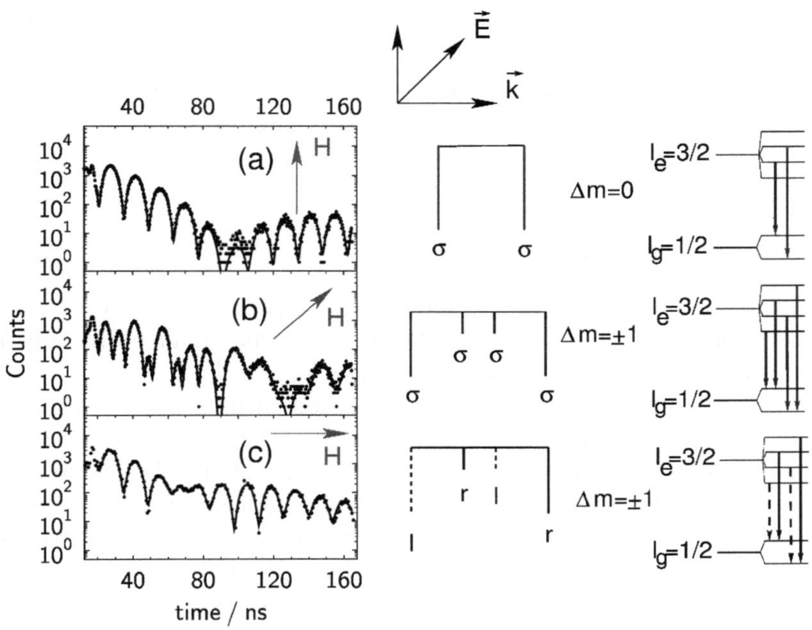

Figure 3. NFS spectra of a 3 μm thin α-^{57}Fe foil in external magnetic fields of different directions as indicated. Measurements were done at the nuclear resonance station BW4 [28] at HASYLAB/DESY. $\vec{k}$ is the direction of the photon, $\vec{E}$ the direction of the polarization vector of the beam (σ-polarization). The stick diagrams represent the corresponding Mössbauer spectra, the transition schemes on the right assume a linear polarized source. l and r denote left and right circular polarization, respectively.

the transitions and the change of the magnetic quantum number Δm are given in the figure, as well.

In case (a) the external magnetic field is perpendicular to the storage ring plane, only the $\Delta m = 0$ transitions are σ-polarized and can be excited. The two possible hyperfine transitions visualized by the grey line in the stick diagram and in the transition scheme give rise to the fast quantum beat modulation superimposed by the slower dynamical beat with its intensity minimum around 90 ns.

In case (b) the external magnetic field is parallel to the polarization of the synchrotron radiation, now the $\Delta m = \pm 1$ transitions have the proper polarization and can be excited. The interference of now four possible transitions with different amplitudes gives a more complicated beat pattern.

In case (c), where $\vec{H}$ is parallel to $\vec{k}$, the eigenpolarization states of the nuclear transitions are left and right circular. All 4 transitions can be excited, because the polarization of the incoming beam can be described as a linear combination of left and right circular polarization. A rather simple quantum beat pattern appears, due to the fact that $\Delta m = +1$ can not interfere with $\Delta m = -1$ transitions, and there is only one energy difference between transitions belonging to the same polarization state.

The two cases (a) and (b) in Figure 3, where the hyperfine field is aligned perpendicular to the beam by the external field, would not be distinguishable in Mössbauer spectroscopy with a conventional unpolarized source. Polarized single line Mössbauer sources are the exception rather than the rule. We want to mention the concept [29] and recent applications utilizing circular polarized [30–32] and linear polarized [33] Mössbauer sources.

4.2. EXAMPLES FOR NFS ON ^{57}Fe

Experiments on samples in confined geometry, like under high pressure, surfaces/thin films or microcrystals, especially benefit from the small beam size. In the following we will give some selected examples.

High pressures in the Mbar range are reached in diamond anvil cells, with sample diameters of about 50–100 μm. The first successful high-pressure NFS experiment was performed on the $\alpha \rightarrow \epsilon$ transition of iron [34] demonstrating the power of the NFS technique in high pressure studies. Another example of a high-pressure-induced phase transition is the transition from the antiferromagnetic to a nonmagnetic state in $FeBO_3$, discovered by NFS at about 45 GPa [35]. The application of external magnetic fields, which is very difficult in high-pressure Mössbauer spectroscopy, bears no problems in NFS with synchrotron radiation and was very useful for the analysis of magnetic NFS spectra in various RFe_2 Laves phases (R = Y, Gd, Sc, Lu, ...), which were studied up to pressures of 105 GPa [36, 37].

NRS experiments on surfaces and multilayers are usually performed in grazing incidence geometry close to the critical angle of total reflection, where the best signal-to-background ratio is achieved [38]. Then the effective sample height is $\sim$50 μm. A well-focused beam allowed one to investigate sub-monolayer coverage of ^{57}Fe on a tungsten surface, as demonstrated in [39], where the surprising result of a magnetisation perpendicular to the layer was found. The magnetic field history dependent change of magnetic domain sizes was possible to resolve by off-specular NRS in grazing incidence geometry on a ^{57}Fe/Cr multilayer on MgO [40], where it was shown that a magnetic field induced bulk-spin-flop was accompanied by a coarsening of the domain sizes.

Very recently, the above-mentioned Kirkpatrick–Baez focusing mirror optics enabled an experiment on single tiny microcrystals of only 5×5 μm^2 area [41].

4.3. MAGNETIC AND ELECTRIC HYPERFINE INTERACTIONS IN Eu COMPOUNDS

One very interesting nucleus in the field of magnetism is ^{151}Eu with a resonance energy of 21.5417 keV. The $I_e = 7/2 \rightarrow I_g = 5/2$ Mössbauer transition in ^{151}Eu is a magnetic dipole transition, as well. If a magnetic hyperfine interaction acts on the nucleus this results in 18 transitions. The numerous spectroscopic applications of this resonance benefit from the two valence states of europium and their large difference in isomer shift and different magnetic properties. The Eu^{2+} ion with the magnetic $4f^7(^8S_{7/2})$ state is an excellent example of pure spin magnetism, therefore, magnetic properties can be easily monitored with the ^{151}Eu resonance. Together with the Eu^{3+} ion with its non-magnetic $4f^6(^7F_0)$ configuration of the 4f shell, europium compounds offer the possibility to study, e.g., temperature and pressure dependent valence transitions and intermediate valence systems. ^{151}Eu Mössbauer spectroscopy was very successful to study these phenomena [42], general reviews on ^{151}Eu Mössbauer spectroscopy [43]. The resonance was first observed in nuclear forward scattering at the ESRF [14] and in nuclear inelastic scattering geometry at TRISTAN (KEK, Tsukuba) [15]. A recent review on NRS with this isotope and on instrumentation is given in [16].

As an introductory example the effect of magnetic hyperfine interaction on the ^{151}Eu nuclei is depicted in Figure 4, where we show experimental results on a polycrystalline sample of EuS, which is ferromagnetically ordered below $T_C =$ 16.5 K. The experimental setup at ID18 at the ESRF corresponds to that in Figure 2 given above. By proper choice of the direction of an external magnetic field one can select either $\Delta m = \pm 1$ or $\Delta m = 0$ transitions, when studying ferromagnetic compounds, and simplify considerably the time spectra as in the case of ^{57}Fe. The time spectra depend in a very pronounced manner on the external magnetic field direction. Note, that without an external field all 18 transitions contribute to the time spectrum (Figure 4, case(a)), while with external field in both cases (b) and (c) only the $\Delta m = \pm 1$ transitions can be excited. Again in the case that $\vec{H}$ is parallel to $\vec{k}$ the eigenpolarization states are left and right circular and a rather simple quantum beat pattern appears, due to the fact that $\Delta m = +1$ can not interfere with $\Delta m =$

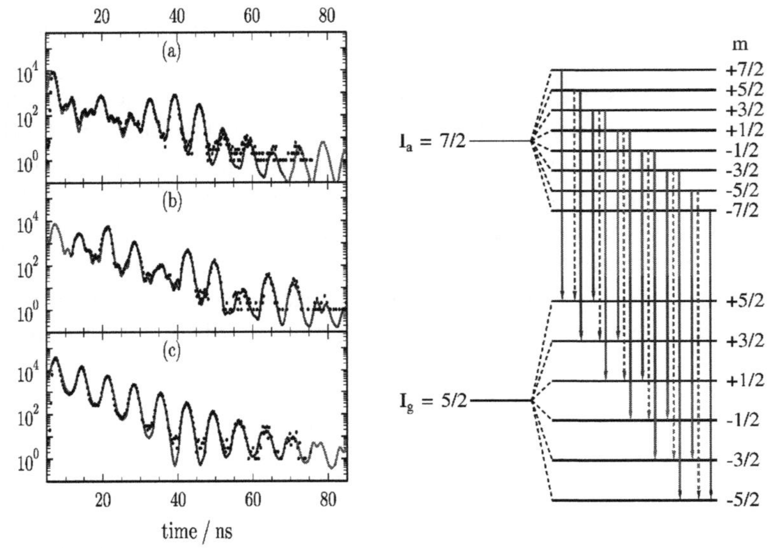

Figure 4. *Left*: NFS spectra of a EuS powder sample measured at ID18/ESRF; (a) $T = 4$ K, $H_{ext} = 0$; (b) $T = 7$ K, $H_{ext} = 3$ T parallel to $\vec{E}$; (c) $T = 4$ K $H_{ext} = 3.66$ T parallel to $\vec{k}$ (as defined in Figure 3). *Right*: nuclear level scheme and possible transitions assuming pure magnetic hyperfine interaction; dashed lines: $\Delta m = 0$ transitions; full lines: $\Delta m = \pm 1$ transitions.

-1 transitions (see Figure 4(c)). When $\vec{H}$ is parallel to the polarization vector $\vec{E}$ of the synchrotron radiation beam, all $\Delta m = \pm 1$ transitions have the same polarization and can interfere (Figure 4(b)).

After this first basic experiment several applications at the ^{151}Eu resonance were performed at the ID18 beam line. As mentioned above, the large difference in isomer shift between the Eu^{2+} and the Eu^{3+} valence states makes NFS on europium compounds an almost ideal tool to determine the valence of the Eu atom/ion, by using reference samples of well-known isomer shift upstream or downstream the sample under study. This technique was successfully applied at the ^{57}Fe resonance [44, 45]. It is worthwile to note that – in contrast to conventional Mössbauer spectroscopy – a direct determination of the isomer shift is not possible in a single NFS measurement without reference sample, due to the broadband excitation by the synchrotron radiation pulse. In the case of ^{151}Eu there exist two standard compounds, the divalent EuS and the trivalent EuF_3. For demonstration of the effect of the isomer shift relative to a reference sample, powder samples of EuS and EuF_3 were measured both alone and sandwiched together [46]. The results are shown in Figure 5.

Figure 5(a): The time spectrum of a EuS powder sample can be fitted with a single resonance line, assuming a slight thickness inhomogeneity, which smears out the expected Bessel minimum of the dynamical beat structure.

Figure 5(c): Time spectrum of a EuF_3 powder sample. This spectrum cannot be fitted with a single line only. The Bessel minimum is pronounced and

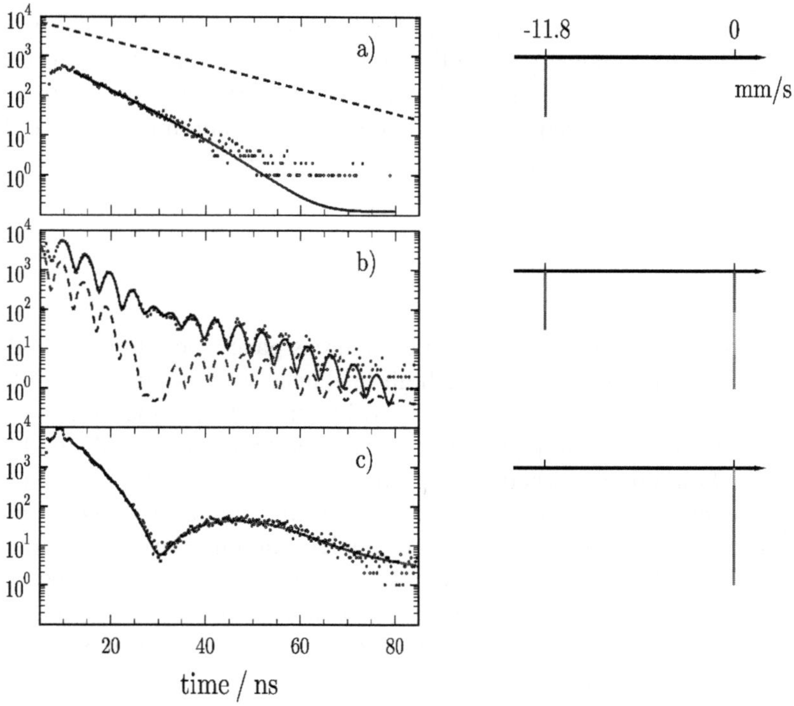

Figure 5. (a) Time spectrum of an EuS powder sample. The dashed line corresponds to a decay with the natural life time. (b) Time spectrum of an EuS/EuF$_3$ powder sample "sandwich". The dashed line resembles the shape of the spectrum of an "ideal sandwich" with two samples of the same thickness. (c) Time spectrum of an EuF$_3$ powder sample. The stick diagrams on the right visualize the position of the resonance lines in energy scale. The (small) line broadening in EuF$_3$ due to quadrupole splitting is neglected in the stick diagram.

excludes the possibility of a large thickness distribution. The best fit result was obtained assuming a quadrupole interaction parameter of the nuclear ground state, $eQ_gV_{zz} = (3.4 \pm 0.5)$ mm/s (where Q_g is the quadrupole moment of the ground state of the ^{151}Eu nucleus) together with an asymmetry parameter of the electric field gradient (EFG) tensor, $\eta = 0.8$. The sign of V_{zz}, the main component of the EFG-tensor, can not be determined, again due to the broad bandwidth of the exciting beam compared with the nuclear level width. The unavoidable line broadening of a standard SmF$_3$ Mössbauer source obscured this splitting of the EuF$_3$ resonance in conventional Mössbauer spectra [47]. Figure 5(b): The time spectrum of the EuS/EuF$_3$ sandwich clearly reveals the large isomer shift between the Eu^{2+} and the Eu^{3+} states, exhibiting fast quantum beats with a period of $\approx$5 ns. The fitted difference in isomer shift $\Delta(\text{IS}) = (11.8 \pm 0.05)$ mm/s corresponds to an energy of $8.5 \cdot 10^{-7}$ eV $\Leftrightarrow$ 205 MHz. With the use of the EuS isomer shift reference the determination of the sign of the quadrupole interaction parameter for EuF$_3$ is possible, in principle. But, at large η the spectrum of EuF$_3$ is almost symmetric, which makes an unambiguous determination difficult. However, the χ^2 of the fits favours the negative sign. The intensity at the Bessel minimum around 30 ns does not reach the zero level, due to the different thickness of the two samples. It only reduces the quantum beat contrast. The dashed line shows the simulation of the time delayed intensity of an "ideal sandwich" with two samples of the same thickness, keeping all other parameters fixed.

5. Summary

The third-generation synchrotron radiation sources gave a new impact on the field of hyperfine spectroscopy, due to the high brilliance and the wide tunability. Especially, the small beam size and the high collimation of the synchrotron radiation beam give applications in high pressure, surfaces and interfaces, and small crystals new impetus. Polarizing magnetic fields simplify the NFS spectra and can be used to unravel the magnetic ordering type. Synchrotron radiation is tunable in energy over the entire energy range of Mössbauer nuclei. Although most experiments are still performed at the resonance of the ^{57}Fe nucleus, which seems to be the ideal probe in solids, other Mössbauer isotopes, especially above 20 keV, like ^{151}Eu, are nowadays routinely available for nuclear resonance experiments at 3rd generation synchrotron radiation sources.

References

1. Gerdau, E., Rüffer, R., Winkler, H., Tolksdorf, W., Klages, C. P. and Hannon, J. P., *Phys. Rev. Lett.* **54** (1985), 835.
2. Gerdau, E. and de Waard, H. (eds), *Nuclear Resonant Scattering of Synchrotron Radiation (Part A), Hyp. Interact.* **123/124** (1999); *Nuclear Resonant Scattering of Synchrotron Radiation (Part B), Hyp. Interact.* **125** (2000).
3. Hastings, J. B., Siddons, D. P., van Bürck, U., Hollatz, R. and Bergmann, U., *Phys. Rev. Lett.* **66** (1991), 770.
4. van Bürck, U., Siddons, D. P., Hastings, J. B., Bergmann, U. and Hollatz, R., *Phys. Rev. B* **46** (1992), 6207.
5. Smirnov, G. V., *Hyp. Interact.* **97/98** (1996), 551.
6. Smirnov, G. V., *Hyp. Interact.* **123/124** (1999), 31.
7. Sturhahn, W., Gerdau, E., Hollatz, R., Rüffer, R., Rüter, H. D. and Tolksdorf, W., *Europhys. Lett.* **14** (1991), 821.
8. Alp, E. E., Mooney, T. M., Toellner, T., Sturhahn, W., Witthoff, E., Röhlsberger, R. and Gerdau, E., *Phys. Rev. Lett.* **70** (1993), 3351.
9. Kikuta, S., *Hyp. Interact.* **90** (1994), 335.
10. Barla, A., Sanchez, J. P., Ni, B., Doyle, B. P., Vulliet, P., Leupold, O., Rüffer, R., Kaczorowski, D., Plessel, J. and Abd-Elmeguid, M. M., *Phys. Rev. B* **66** (2002), 094425.
11. Johnson, D. E., Siddons, D. P., Larese, J. Z. and Hastings, J. B., *Phys. Rev. B* **51** (1995), 7909.
12. Baron, A. Q. R., Chumakov, A. I., Ruby, S. L., Arthur, J., Brown, G. S., van Bürck, U. and Smirnov, G. V., *Phys. Rev. B* **51** (1995), 16384.
13. Chumakov, A. I., Baron, A. Q. R., Arthur, J., Ruby, S. L., Brown, G. S., Smirnov, G. V., van Bürck, U. and Wortmann, G., *Phys. Rev. Lett.* **75** (1995), 549.

14. Leupold, O., Pollmann, J., Gerdau, E., Rüter, H. D., Faigel, G., Tegze, M., Bortel, G., Rüffer, R., Chumakov, A. I. and Baron, A. Q. R., *Europhys. Lett.* **35** (1996), 671.
15. Koyama, I., Yoda, Y., Zhang, X. W., Ando, M. and Kikuta, S., *Jpn. J. Appl. Phys.* **35** (1996), 6297.
16. Leupold, O., Chumakov, A. I., Alp, E. E., Sturhahn, W. and Baron, A. Q. R., *Hyp. Interact.* **123/124** (1999), 611.
17. Shvyd'ko, Yu. V., Gerken, M., Franz, H., Lucht, M. and Gerdau, E., *Europhys. Lett.* **56** (2001), 309.
18. Röhlsberger, R., Quast, K. W., Toellner, T. S., Lee, P. L., Sturhahn, W., Alp, E. E. and Burkel, E., *Phys. Rev. Lett.* **87** (2001), 047601.
19. Rüffer, R. and Chumakov, A. I., *Hyp. Interact.* **97/98** (1996), 589.
20. Ishikawa, T., Yoda, Y., Izumi, K., Suzuki, C. K., Zhang, X. W., Ando, M. and Kikuta, S., *Rev. Sci. Instrum.* **43** (1992), 824.
21. Snigirev, A. A., Kohn, V. G., Snigireva, I. I. and Lengeler, B., *Nature* **384** (1996), 49.
22. Freund, A. K., Comin, F., Hazemann, J., Hustache, P., Jenninger, B., Lieb, K. and Pierre, M., *Proc. SPIE* **3448** (1998), 1.
23. Kishimoto, S., *Rev. Sci. Instruments* **63** (1992), 824.
24. Baron, A. Q. R., *Nucl. Inst. Meth. A* **352** (1995), 665.
25. Toellner, T. S., *Hyp. Interact.* **125** (2000), 3.
26. Baron, A. Q. R., *Hyp. Interact.* **125** (2000), 29.
27. Mao, H. K., Xu, J., Struzhkin, V. V., Shu, J., Hemley, R. J., Sturhahn, W., Hu, M. Y., Alp, E. E., Vocadlo, L., Alfè, D., Price, G. D., Gillan, M. J., Schwoerer-Böhning, M., Häusermann, D., Eng, P., Shen, G., Giefers, H., Lübbers, R. and Wortmann, G., *Science* **292** (2001), 914.
28. Leupold, O., Gerdau, E., Rüter, H. D., Meyer-Klaucke, W., Trautwein, A. X. and Winkler, H., HASYLAB Annual Report, Part I, 1996, p. 73.
29. Gonser, U. and Fischer, H., In: *Topics in Current Physics, Mössbauer Spectroscopy II*, Vol. 25, Springer, Berlin, 1981, p. 99.
30. Szymanski, K., Dobrzynski, L., Prus, B. and Cooper, M. J., *Nucl. Instr. Meth. B* **119** (1996), 438.
31. Szymanski, K., Dobrzynski, L., Prus, B., Mal'tsev, Yu., Rogozev, B. and Silin, M., *Hyp. Interact. C* **3** (1998), 265.
32. Szymanski, K., Satula, D. and Dobrzynski, L., *J. Phys.: Condensed Matter* **11** (1999), 881.
33. Jäschke, J., Rüter, H. D., Gerdau, E., Smirnov, G. V. and Sturhahn, W., *Nucl. Instr. Meth. B* **155** (1999), 189.
34. Grünsteudel, H. F., Hesse, H.-J., Rüffer, R., Wortmann, G., Chumakov, A. I., Baron, A. Q. R., Grünsteudel, H., Leupold, O. and Metge, J., *Austral. J. Phys.* **51** (1998), 45.
35. Troyan, I. A., Gavrilyuk, A. G., Sarkisyan, V. A., Lyubutin, I. S., Rüffer, R., Leupold, O., Barla, A., Doyle, B. and Chumakov, A. I., *JETP Lett.* **74** (2001), 24.
36. Lübbers, R., Wortmann, G. and Grünsteudel, H. F., *Hyp. Interact.* **123/124** (1999), 529.
37. Lübbers, R., Rupprecht, K. and Wortmann, G., *Hyp. Interact.* **128** (2000), 115.
38. Röhlsberger, R., *Hyp. Interact.* **123/124** (1999), 301.
39. Röhlsberger, R., Bansmann, J., Senz, V., Jonas, K. L., Bettac, A., Leupold, O., Rüffer, R., Burkel, E. and Meiwes-Broer, K. H., *Phys. Rev. Lett.* **86** (2001), 5597.
40. Nagy, D. L., Bottyán, L., Croonenborghs, B., Deák, L., Degroote, B., Dekoster, J., Lauter, H. J., Lauter-Pasyuk, V., Leupold, O., Major, M., Meersschaut, J., Nikonov, O., Petrenko, A., Rüffer, R., Spiering, H. and Szilágyi, E., *Phys. Rev. Lett.* **88** (2002), 157202.
41. Wills, A. *et al.*, to be published.
42. Nowik, I., *Hyp. Interact.* **13** (1983), 89.
43. Barton, C. M. P. and Greenwood, N. N., In: J. G. Stevens and V. E. Stevens (eds), *Mössbauer Effect Data Index*, Plenum Press, New York, 1973, p. 395; Grandjean, F. and Long, G. J., In: G. J. Long and F. Grandjean (eds), *Mössbauer Spectroscopy Applied to Inorganic Chemistry*,

Vol. 3, Plenum Press, New York, 1989, p. 513. Harmatz, B., *Nuclear Data Sheets* **19** (1976), 33.

44. Alp, E. E., Sturhahn, W. and Toellner, T., *Nucl. Instr. Meth. B* **97** (1995), 526.
45. Leupold, O., Grünsteudel, H., Meyer, W., Grünsteudel, H. F., Winkler, H., Mandon, D., Rüter, H. D., Metge, J., Realo, E., Gerdau, E., Trautwein, A. X. and Weiss, R., In: I. Ortalli (ed), *Conference Proceedings "ICAME-95"*, Vol. 50, SIF, Bologna, 1996, p. 857.
46. Pleines, M., Lübbers, R., Strecker, M., Wortmann, G., Leupold, O., Shvyd'ko, Yu. V., Gerdau, E. and Metge, J., *Hyp. Interact.* **120/121** (1999), 181.
47. Pleines, M., Diplomarbeit, Universität Paderborn, 1998.

Hyperfine Interactions **144/145:** 33–44, 2002.

Imaging the Magnetic Structure of Thin Films and Nanoparticles

R. RÖHLSBERGER*
Universität Rostock, Fachbereich Physik, August-Bebel-Str. 55, 18055 Rostock, Germany

Abstract. Nuclear resonant scattering of synchrotron radiation is applied to investigate the magnetic spin structure of nanoscale systems. The outstanding brilliance of modern synchrotron radiation sources allows for sensitivities in the sub-monolayer regime. Due to the isotopic sensitivity of the scattering process, ultrathin probe layers of Mössbauer isotopes can be used to map out the magnetic structure of thin films with sub-nm spatial resolution. This technique is applied to determine the magnetic spin structure of an exchange-coupled bilayer system. Moreover, a procedure is introduced to determine the magnetic structure of nanoscale islands on a single-crystalline surface. An algorithm is developed that allows one to extract the magnetic structure function from a series of measured time spectra. These techniques can be extended to a number of Mössbauer isotopes that are constituents of modern magnetic materials.

Key words: synchrotron radiation, spin structure, exchange spring, thin films, grazing incidence, coherent scattering.

1. Introduction

The magnetic structure of nanoscale devices plays an increasingly important role in many branches of applied physics and materials science. This is particularly true for the rising field of magnetoelectronics where electronic transport properties are controlled by the magnetization of thin films. The most prominent example is the effect of giant magnetoresistance (GMR) that forms the basis for data readout in magnetic storage media [1]. The introduction of GMR read heads led to a significant increase in the storage density of magnetic hard disk drives. Another example is the magnetic random access memory (MRAM) that combines high-speed data processing with the property of non-volatile storage. The functionality of all these devices depends critically on the magnetic spin structure within thin films as it develops in external magnetic fields. With increasing complexity of novel magnetic nanostructures there is a growing need for efficient methods to characterize their magnetic properties on the atomic scale. In recent years one could observe an impressive progress in the development of such methods. Scanning probe methods like magnetic force microscopy (MFM) [2] and spin-polarized scanning tunneling

* Present address: Technische Universität München, Physik-Department E13, James-Franck-Str. 1, D-85747 Garching; e-mail: rroehlsb@ph.tum.de.

microscopy (SP-STM) [3], for example, allow one to directly image the magnetic structure of surfaces. While these methods offer a spatial resolution down to the atomic level, they cannot access the magnetic structure of buried layers and interfaces. This is the domain of scattering methods with polarization sensitive probes like photons [4] and neutrons [5] that penetrate into the sample. The availability of high-brilliance synchrotron radiation has recently opened new avenues for the application of X-ray scattering methods in this field [4]. If the X-ray energy is tuned to an electronic resonance of one of the constituting atomic species, these methods become element specific and the magnetic properties of individual layers can be probed, see, e.g., [6]. However, the scattering signal is derived from the whole layer and thus integrates over the magnetic structure within the layer. To some extent, the use of evanescent waves in grazing incidence geometry allows to restrict the probing depth to near-surface regions. Nevertheless, a direct measurement of the depth resolved magnetic spin structure is not possible using this approach. Only special reconstruction techniques may allow for an indirect determination of depth resolved magnetic properties as demonstrated recently via polarized neutron scattering [7]. This problem can be overcome if the interaction of the probing particle with the sample exhibits isotopic sensitivity: Magnetic properties are studied with very high spatial resolution using ultrathin probe layers of a Mössbauer isotope. This was successfully applied in a number of studies involving conversion electron Mössbauer spectroscopy [8–11]. While measurements of this kind are time consuming even with the use of strong radioactive sources, this technique has gained new momentum in recent years due to the availability of high-brilliance synchrotron radiation sources [12, 13].

In this article I will give examples of recent applications that demonstrate the potential of this method for the study of nanoscale magnetism. After an introduction into the scattering basics I will show how ultrathin probe layers can be used to image the depth-dependent magnetic structure of exchange-coupled thin films. In a second example I will introduce an algorithm to extract the magnetic structure function of nanoscale particles on surfaces from the measured data.

2. Scattering basics

Compared to classical Mössbauer spectroscopy, nuclear resonant scattering of synchrotron radiation exhibits qualitatively new features. First of all, synchrotron radiation is complementary to a radioactive source because of its broad energy spectrum and its pulsed time structure. This suggests to perform hyperfine spectroscopy on a time scale rather than on an energy scale: The simultaneous excitation of the hyperfine-split nuclear energy levels by a pulse of synchrotron radiation leads to temporal beats in the evolution of the subsequent nuclear decay. The analysis of this beat pattern allows a precise determination of the magnitude and the orientation of magnetic fields in the sample, as will be shown below. Second, the measured signal results from *coherent* scattering rather than from *incoherent* absorption.

Therefore, relative phase factors between different nuclear sites may lead to diffraction and other interference effects. Moreover, the brilliance of undulator radiation from modern synchrotron radiation source exceeds that of a radioactive source by several orders of magnitude. This makes this technique particularly suited for the investigation of low-dimensional systems like ultrathin films and nanostructures. It is a fortunate circumstance that the specular reflection from such systems at a grazing angle φ can be treated in the kinematical limit, i.e., the first-order Born approximation. In this case, the resonant reflectivity of a single resonant layer of thickness d on a substrate is given by [14]:

$$\mathbf{R}(\omega) \approx r_{02} e^{i\mathbf{f}_1(\omega)|t_{02}|^2 d} \quad \text{with } \mathbf{f}_1(\omega) = \frac{2\pi}{k_{0z}} \sum_j \rho_j \mathbf{M}_j(\omega), \tag{2.1}$$

where the sum runs over all atomic species j in the sample with number densities ρ_j. The subscripts 0, 1, and 2 label the vacuum, film and substrate, respectively. $\mathbf{f}_1(\omega)$ is the 2×2 matrix of the nuclear forward scattering amplitude, as introduced below. f_i is the corresponding scalar quantity for the nonresonant layer i. r_{ij} and t_{ij} are the Fresnel reflection and transmission coefficients for the boundary between the two media i and j, respectively. This means, the reflection from an ultrathin film can be treated like forward scattering with an "effective" thickness of $d' = d|t_{02}|^2$. The expression $|t_{02}|^2$ is proportional to the intensity of the radiation field at the surface of the substrate. It assumes its maximum value at the critical angle φ_c for electronic charge scattering.

It is instructive to expand Equation (2.1) in powers of the unit vector $\hat{m}$ that defines the magnetic quantization axis of the atom in the sample. The resonant scattering length for an electric dipole transition can then be written as:

$$\begin{aligned} [\mathbf{M}(\omega)]_{\mu\nu} = -\frac{3}{4\pi}\Big\{ & (\hat{\epsilon}_\mu \cdot \hat{\epsilon}_\nu)[F_{+1} + F_{-1}] \\ & - \mathrm{i}(\hat{\epsilon}_\mu \times \hat{\epsilon}_\nu) \cdot \hat{m}[F_{+1} - F_{-1}] \\ & + (\hat{\epsilon}_\mu \cdot \hat{m})(\hat{\epsilon}_\nu \cdot \hat{m})[2F_0 - F_{+1} - F_{-1}]\Big\}, \end{aligned} \tag{2.2}$$

where $\hat{\epsilon}_\mu$ and $\hat{\epsilon}_\nu$ are the unit vectors of the orthogonal polarization basis chosen to describe the scattering process. The functions $F_M = F_M(\omega)$ are the energy-dependent resonant strengths for dipole transitions with a change of $M = m_e - m_g$ in the magnetic quantum number:

$$\begin{aligned} F_M(\omega) &= K \sum_{m_i} \frac{C^2(I_g 1 I_e; m_i M)}{\hbar(\omega - \omega_{m_i M}) + \mathrm{i}\Gamma_0/2} \quad \text{with} \\ K &= \frac{2\pi f_{LM} \Gamma_0}{k_0(1+\alpha)(2I_g + 1)} \end{aligned} \tag{2.3}$$

where the sum runs over all ground state levels with magnetic quantum numbers m_i. The $C(I_g 1 I_e; m_i M)$ are the Clebsch–Gordan coefficients in the notation of

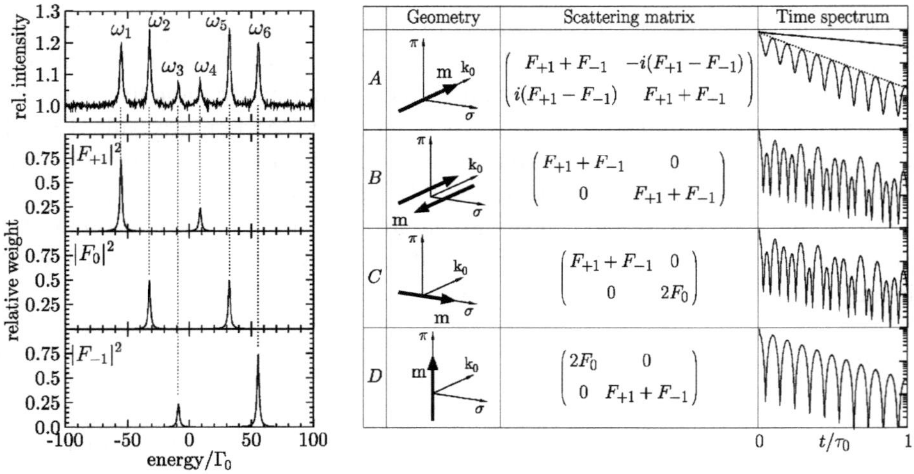

Figure 1. Left: The six dipole allowed transitions of the 14.4 keV resonance of ferromagnetic α-^{57}Fe decompose into three different polarization characteristics. Their individual energy dependencies are given by the functions F_{+1}, F_0, and F_{-1}, corresponding to left-circular, linear, and right-circular polarization, respectively. *Right*: For selected orientations of the magnetic hyperfine field the scattering matrices **f** and the time spectra of nuclear resonant scattering from a 2 nm thick film of ^{57}Fe on a total reflecting surface are shown. The identical shape of the time spectra in B and C demonstrates that the spin structure of the sample cannot be determined from just a single measurement.

Rose [15] that describe the relative strength of the transitions. The functions $F_M(\omega)$ for $M = -1, 0, +1$ are shown in Figure 1. The energetic positions $\omega_{m_i M}$ of the resonance lines are determined by the magnetic hyperfine interaction, i.e.,

$$\hbar\omega_{m_i M} = \hbar\omega_0 - (m_e\mu_e - m_g\mu_g)B, \tag{2.4}$$

where B is the magnetic hyperfine field. m_e and m_g are the magnetic quantum numbers of the ground and excited state, respectively, and μ_g and μ_e are the corresponding magnetic moments. In case of a magnetic dipole transition the role of the electric and magnetic fields of the electromagnetic wave are interchanged, i.e., one has to transform the polarization vectors in Equation (2.2) according to $\hat{\epsilon} \rightarrow \hat{\epsilon} \times \hat{k}_0$ with $\hat{k}_0 = \vec{k}_0/k_0$. The temporal evolution of the resonantly scattered amplitude is obtained via Fourier transformation of Equation (2.1):

$$\tilde{\mathbf{R}}(t) = \delta(t) - \tilde{\mathbf{f}}(t)\chi e^{-\chi t/2\tau_0} \quad \text{with } \chi = \frac{1}{4}\rho\sigma_0 f_{LM}|t_{02}|^2\frac{d}{\varphi}, \tag{2.5}$$

where $\tilde{\mathbf{f}}(t)$ is the Fourier transform of $\mathbf{f}(\omega)$, and σ_0 is nuclear absorption cross section at resonance. The quantity χ describes the speedup of the nuclear decay that results from multiple scattering processes in the sample [16].

In the most common scattering geometry the radiation is reflected vertically, i.e., the sample plane coincides with the plane of the storage ring which is the $(\hat{\sigma}, \hat{k}_0)$ plane. Assuming no polarization analysis in the detection process, we obtain for the resonantly reflected intensity:

$$I(t) = \left|\tilde{\mathbf{R}}_{11}\right|^2 + \left|\tilde{\mathbf{R}}_{12}\right|^2. \tag{2.6}$$

For some selected scattering geometries the matrices $\mathbf{M}(\omega)$ according to Equation (2.2) and the scattered intensities $I(t)$ are shown in Figure 1. Very often, the shape anisotropy of thin films forces the magnetization, i.e., the local quantization axis $\hat{m}$, into the film plane. Then the third term in Equation (2.2) vanishes, so that one obtains [14]:

$$I(t) = \left\{\left|\tilde{F}_{+1} + \tilde{F}_{-1}\right|^2 + \left(\hat{k}_0 \cdot \vec{\mu}\right)^2 \left|\tilde{F}_{+1} - \tilde{F}_{-1}\right|^2\right\} \chi^2 e^{-\chi t/\tau_0}. \tag{2.7}$$

This means that only transitions with a change in the magnetic quantum number of $M = \pm 1$ can be excited. Although other kinds of anisotropies may overcome the shape anisotropy and force the magnetization out of plane [13], the examples to be discussed here will deal with in-plane magnetization. The quantity $\vec{\mu}$ is the magnetic structure function of the sample that is given by:

$$\vec{\mu} = \sum_j p_j \hat{m}_j \mathrm{e}^{\mathrm{i}\vec{q}\cdot\vec{R}_j} \quad \text{with} \sum_j p_j = 1, \ |\hat{m}_j| = 1, \ \text{and } 0 \leqslant p_j \leqslant 1, \tag{2.8}$$

where the sum runs over all magnetic sublattices $\hat{m}_j$ and the phase factor accounts for the position $\vec{R}_j$ of the atoms. $\vec{q} = \vec{k} - \vec{k}'$ is the momentum transfer in the scattering process. For specular reflection in grazing incidence geometry, $\vec{q} \approx 0$ is a good approximation.

The experiments described here were performed at the nuclear-resonance beamline ID 18 of the ESRF [17]. Time spectra were recorded in the single-bunch filling mode of the storage ring. An energy bandwidth reduction to 6.5 meV by a high-resolution monochromator prevented the detector from being overloaded by the huge intensity of nonresonant photons. The samples were mounted on diffractometer stages to enable high-precision control of the angular adjustment.

3. The magnetic structure of exchange-spring bilayers

The exchange coupling between different magnetic materials plays an important role in many micromagnetic systems. For example, nanophase materials consisting of hard- and soft-magnetic phases are promising candidates for new magnetic materials with very-high-energy products beyond 1 MJ/m^3 [18]. Bilayers consisting of hard- and soft-magnetic thin films are ideal model systems to investigate the magnetic coupling between such phases. Such systems are characterized by a pinning of the soft-magnetic moments at the interface of the hard-magnetic film as a result of the exchange interaction. With increasing distance from the interface, the exchange coupling becomes weaker and the magnetization may rotate under the action of an external field. If, for example, the external field is applied orthogonal to the magnetization direction of the hard layer, the magnetic moments in the soft layer arrange in a spiral structure along the normal, as shown schematically in

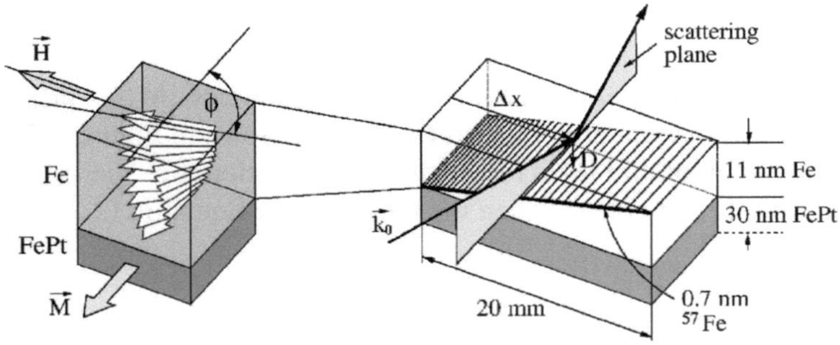

Figure 2. Left: Scheme of the spin structure in an exchange-spring bilayer consisting of a soft-magnetic layer (Fe) on a hard-magnetic layer (FePt) with uniaxial anisotropy. An external field $\vec{H}$ is applied perpendicular to the remanent magnetization $\vec{M}$ of the hard layer. *Right*: To image the resulting magnetic spiral, a tilted ultrathin probe layer of ^{57}Fe is deposited within the soft layer. The magnetic properties in depth D of the film are probed by adjusting the transverse coordinate Δx of the sample relative to the incident beam [23].

Figure 2. Due to the reversible nature of this rotation, this is called the exchange-spring effect [19]. While a number of micromagnetic models have been developed to described this behaviour [20–22], direct measurements of the magnetic structure of such systems have not been reported as of yet. In the experiment described here the depth dependent spin structure of exchange-coupled thin films is directly measured via the use of ultrathin probe layers.

The sample was a bilayer consisting of 11 nm soft-magnetic Fe on 30 nm hard-magnetic FePt. To minimize oxidation of the Fe layer, it was coated with a 3 nm thick Ag layer. All layers were prepared by rf-magnetron sputtering in a high-vacuum system with a base pressure of 2×10^{-7} mbar. The preparation of the layer system and its characterization are described in [23]. To measure the depth dependence of the magnetization rotation, an ultrathin tilted probe layer of ^{57}Fe was deposited within the Fe layer, as shown in Figure 2. To map out the depth dependence of the rotation angle, the sample was displaced transversely to the incident beam. The FePt layer was magnetized remanently along the direction of the incident wave vector $\vec{k}_0$. Time spectra taken at various positions Δx are shown in Figure 3 for an external field of $H = 160$ mT. The characteristic changes in the beat pattern reflect the rotation of the magnetization from an almost parallel alignment at $\Delta x = 0$ mm towards increasing perpendicular orientation with increasing Δx. From each of these spectra the rotation angle ϕ of the magnetization, as defined in Figure 2, was deduced by a fit of Equation (2.7) to the data (solid lines). The resulting depth dependence of the rotation angle is shown in Figure 3 for external fields of 160, 240 and 500 mT. Solid lines are simulations according to a one-dimensional micromagnetic model as outlined in [21], assuming the magnetic properties to be homogeneous throughout the Fe layer. However, close to the Fe/Ag interface the simulation clearly deviates from the measured data. This can be explained by a reduction of the exchange constant in the first 3 nm of the Fe layer, most probably induced by diffusion of oxygen through the Ag capping layer. The corresponding

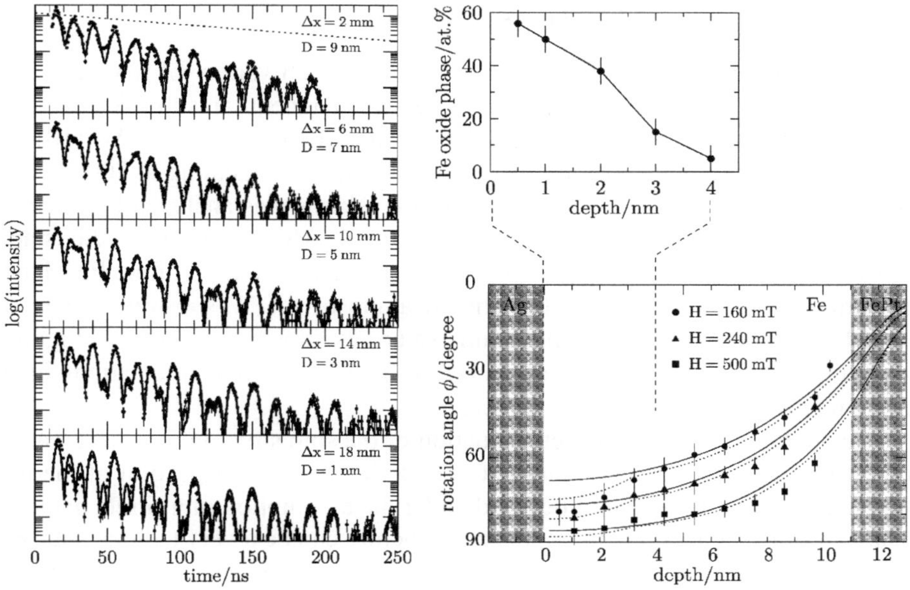

Figure 3. Left: Measured time spectra of the sample shown in Figure 2 in a perpendicular magnetic field of $B = 160$ mT at various lateral positions Δx. From the theoretical simulation of the beat pattern the rotation angle ϕ of the magnetization at the probe layer position was derived. This dependence is shown on the right side for various external fields. Solid lines are micromagnetic simulations assuming homogeneous magnetic properties in the Fe film. The deviations from the simulations near the Ag/Fe boundary are due to the formation of a native Fe oxide. Its relative fraction decreases with increasing depth, as shown in the upper graph [23]. Corresponding simulations are shown as dotted lines.

micromagnetic simulations are shown as dotted lines in Figure 3. Details of the simulation can be found in [23].

The experiment greatly benefitted from a significant increase of the signal from the ^{57}Fe probe layer due to an X-ray interference effect: at certain angles of incidence between the critical angles of Fe and FePt, the Ag/Fe/FePt layer system acts as an X-ray waveguide [24] where the radiation is trapped inside the Fe layer. The superposition of waves that are multiply reflected at the Ag/Fe/FePt boundaries leads to a significant X-ray flux enhancement inside the Fe layer. This effect yielded resonant (delayed) countrates of up to 100 s^{-1}, so that time spectra with very good statistical quality could be obtained within a few minutes. Thus, the magnetic properties of such layer systems can be investigated systematically within comparatively short data acquisition times.

The evaluation of the time spectra shown above was rather straightforward because the magnetization was unidirectional with a well-defined hyperfine field. Very often, however, the nuclei in the sample are subject to distributions of hyperfine interaction parameters. This applies for the magnitude as well as for the direction of hyperfine fields. In the following, I want to discuss an example where the spin structure of the sample consists of more than one magnetic sublattice.

4. The spin structure of Fe islands on W(110)

Ultrathin Fe films on the W(110) surface display a rich variety of magnetic phenomena due to the interplay of different types of magnetic anisotropies [25, 26]. In this experiment, single-crystalline, island-shaped Fe particles on a W(110) surface were produced by deposition of Fe and subsequent annealing at 700 K under ultra-high vacuum conditions. The shape of the islands was analyzed by scanning tunneling microscopy (STM), as shown in Figure 4. After preparation, the islands were capped with an Ag film of about 2 nm to prevent oxidation of the Fe islands. The sample was illuminated at the critical angle of the W substrate where the delayed radiation assumed its maximum value, as shown in Figure 4. Figure 5 shows a series of time spectra recorded at different azimuthal orientations of the sample relative to the incident wavevector.

To avoid ambiguities in the determination of the spin structure it is necessary to record time spectra at various sample orientations. A close inspection of the time spectra reveals that they cannot be fitted under the assumption of a single magnetic sublattice. Since "guessing" the right spin structure can be a very tedious approach to solve this inversion problem, it is more desirable to do this in an algorithmic manner. In the following I will introduce an procedure to extract the magnetic structure function $\vec{\mu}$ from the measured time spectra. This algorithm is based on a spectral analysis of the measured data. Previous attempts in this field [27–29] did not yield quantitative results, because the time spectra were strongly perturbed by effects of dynamical diffraction. On the other hand, if the scattering process can be treated in the kinematical limit, the beat pattern $I(t)$ can be described by an

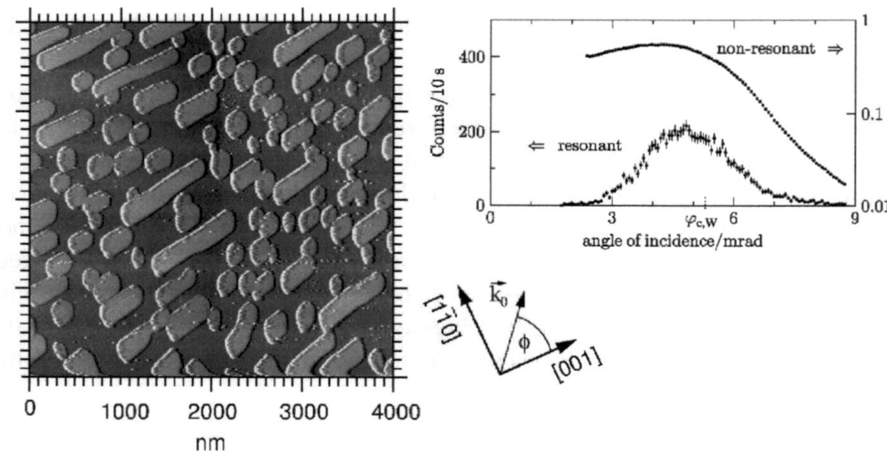

Figure 4. Left: Scanning tunneling microscopy image of Fe islands on W(110). The islands were produced after deposition of about 5 monolayers of Fe and subsequent annealing for 10 min at 700 K. The elongated islands are aligned along the in-plane (001) direction. *Right*: Rocking curves of the resonant and nonresonant reflectivity of the sample. The resonant reflectivity peaks near the critical angle $\varphi_{c,\mathrm{W}}$ of the W substrate [14].

analytical expression, derived from Equation (2.7):

$$\begin{aligned} I(t) &= e^{-\chi t/\tau_0}\big[G(0,0,\Omega_1)\big(1+S^2(\phi)\big) \\ &\qquad + G(\Omega_1+\Omega_2,\Omega_1-\Omega_2,\Omega_2)\big(1-S^2(\phi)\big)\big] \end{aligned} \tag{4.1}$$

with $\Omega_1 = \omega_4 - \omega_1$ and $\Omega_2 = \omega_3 - \omega_1$, where the ω_i are the frequencies of the resonance lines as defined in Figure 1. The function G is defined as:

$$G(c_1, c_2, c_3) = \cos c_1 t + \frac{1}{9}\cos c_2 t + \frac{2}{3}\cos c_3 t, \tag{4.2}$$

and the quantity $S(\phi)$ is given by:

$$S(\phi) = \vec{k}_0 \cdot \big[\mathbf{D}(\phi)\vec{\mu}\big] \tag{4.3}$$

with $\mathbf{D}(\phi)$ being the 2×2 rotation matrix that describes the azimuthal sample rotation. With the definitions $\vec{k}_0 = (1, 0)$ and $\vec{\mu} =: (m_x, m_y)$, the function S is given according to Equation (4.3) by:

$$S(\phi) = m_x \cos\phi + m_y \sin\phi. \tag{4.4}$$

Since Equation (4.1) is a superposition of damped harmonic functions, its Fourier transform is a superposition of Lorentzians $L(\Omega_i)$ centered around frequencies Ω_i with weight factors $g(\Omega_i)$:

$$\begin{aligned} I(\omega) &= \sum_{i=0}^{4} g(\Omega_i)L(\Omega_i) \\ &= \big(1+S^2\big)\big(1+a^2\big)L(\Omega_0) + 2a\big(1+S^2\big)L(\Omega_1) + 2a\big(1-S^2\big)L(\Omega_2) \\ &\qquad + \big(1-S^2\big)L(\Omega_3) + a^2\big(1-S^2\big)L(\Omega_4) \end{aligned} \tag{4.5}$$

with $\Omega_0 = 0$, $\Omega_3 = \Omega_1 + \Omega_2$, and $\Omega_4 = \Omega_1 - \Omega_2$. This means, $S(\phi)$ can be determined from the weights of the Lorentzians $L(\Omega_1)$ and $L(\Omega_2)$ as follows:

$$S = \sqrt{\frac{g(\Omega_1) - g(\Omega_2)}{g(\Omega_1) + g(\Omega_2)}}. \tag{4.6}$$

The right side of Figure 5 shows the Fourier transform of the measured time spectra. Four spectral components show up as clear peaks in the data at frequencies Ω_1, Ω_2, $\Omega_1 + \Omega_2$, and $\Omega_1 - \Omega_2$. These frequencies already indicate that the magnetic moments of the Fe islands must be oriented in the plane of the sample. If out-of-plane components were present, additional frequency components would appear in the spectra. The solid lines represent a sum of Lorentzians that were fitted to the data. From the area of these Lorentzians the weights $g(\Omega_i)$ were determined from which the quantity S was derived. The dependence of S on the azimuthal angle is shown in Figure 6. A fit of Equation (4.4) to the measured data yields

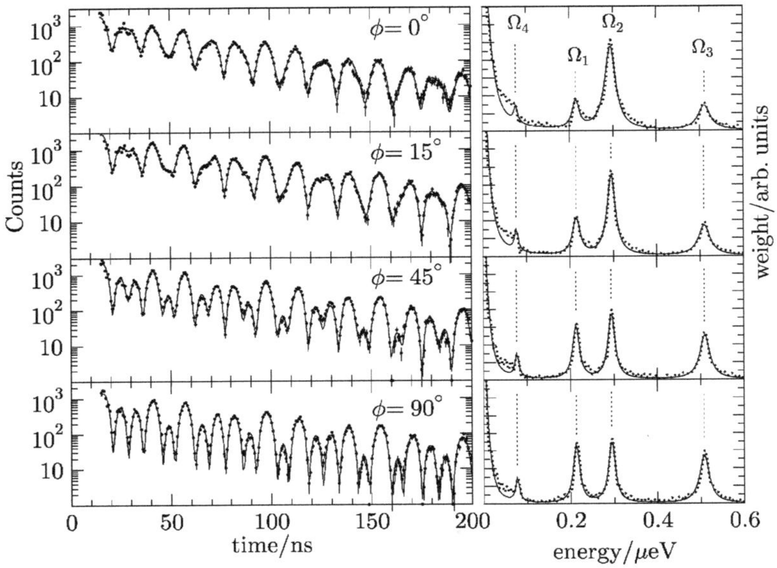

Figure 5. Time spectra of nuclear resonant scattering from Ag-capped Fe islands on W(110) for various azimuthal orientations. The solid lines are fits according to the theory outlined in the text. The right panel shows the Fourier transform of the measured time spectra to reveal the spectral components that contribute to the beat pattern. The solid lines are Lorentzian fits to the peaks.

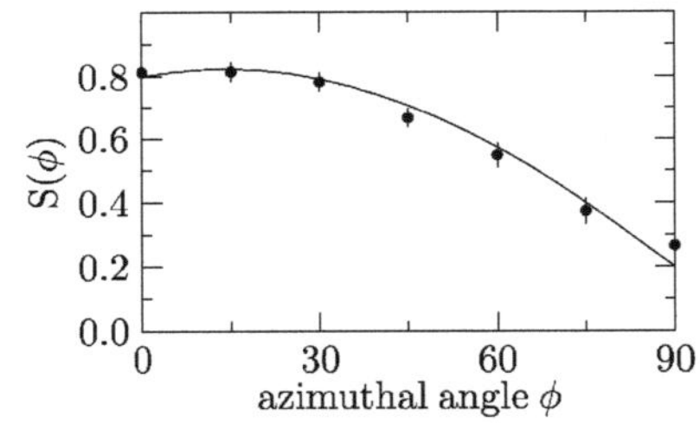

Figure 6. Measured values of the function $S(\phi)$, determined from evaluation of the time spectra shown in Figure 3. The solid line is a fit according to Equation (4.4), yielding $m_x = 0.77(3)$ and $m_y = 0.19(3)$.

$m_x = 0.77(3)$ and $m_y = 0.19(3)$. A reconstruction of the magnetic structure function $\vec{\mu}$ has to be guided by the conditions that are mentioned in Equation (2.8). This already excludes the magnetization of the Fe islands to be simply unidirectional (i.e., a single magnetic sublattice), because $m_x^2 + m_y^2 < 1$. The next-higher degree of complexity is the assumption of two magnetic sublattices. In fact, this leads to the following reconstruction:

$$\vec{\mu} = 0.8\hat{u}_{[0\,0\,1]} + 0.2\hat{u}_{[1\,\bar{1}\,0]}, \tag{4.7}$$

where $\hat{u}_{[0\,0\,1]}$ and $\hat{u}_{[1\,\bar{1}\,0]}$ are unit vectors along the in-plane $[0\,0\,1]$ and $[1\,\bar{1}\,0]$ directions, respectively. Although this is not a unique reconstruction, it is the most meaningful one, because the elongated shape of the large islands certainly forces the largest part of the magnetization into the [0 0 1] direction [14]. This particular magnetic structure is most probably the result of an interplay between shape- and surface anisotropies in the system. While the shape anisotropy forces the magnetization into the in-plane [0 0 1] direction, the surface anisotropy favours the in-plane [1 $\bar{1}$ 0] direction. These contributions, however, are dependent on the size of the islands, so that the dominant contribution for the small islands is the surface anisotropy. It can thus be concluded that the Fe islands are laterally separated, single magnetic domains, oriented along two orthogonal in-plane directions.

5. Conclusion and outlook

The magnetic structure of exchange-coupled thin films and separated islands was determined by nuclear resonant scattering of synchrotron radiation. Due to the very high brilliance of modern synchrotron radiation sources, the method is sensitive to coverages in the sub-monolayer regime. This opens unique applications in the field of nanoscale magnetism: In combination with other X-ray scattering methods, the relation between structural properties and the magnetic behaviour of deposited clusters can be explored. Moreover, due to the isotopic selectivity of the scattering process, magnetic properties can be determined with atomic resolution by selectively doping the sample with the Mössbauer nuclei [9]. Another promising application is the determination of internal magnetic fields in oriented biomolecules on surfaces and other magnetic nanostructures that are considered as future magnetic storage devices. Moreover, this scattering technique can be extended to other Mössbauer isotopes that are constituents of modern magnetic materials.

Acknowledgements

The experiments described here benefitted from the contributions of many enthusiastic coworkers whom I want to express my sincere thanks. While the work presented in Section 3 was performed together with H. Thomas, K. Schlage and K. Quast, the results discussed in Section 4 were obtained in collaboration with J. Bansmann, V. Senz and K. H. Meiwes-Broer. Special thanks go to A. Bettac and K. L. Jonas for their contributions in preparation of the Fe islands and the characterization via STM. The experiments were performed at the nuclear resonance beamline ID18 of the ESRF with help of O. Leupold and R. Rüffer who always provided an excellent environment for the experiments. Finally, I want to acknowledge the support of E. Burkel during the course of these experiments. This work was funded by the German BMBF under contract 05 KS1HRA 8.

References

1. Fert, A., Grünberg, P., Barthelemy, A., Petroff, F. and Zinn, W., *J. Magn. Magn. Mater.* **140–144** (1995), 1.

2. Martin Y. and Wickramasinghe, H. K., *Appl. Phys. Lett.* **50** (1987), 1455; Sáenz, J. J. *et al.*, *J. Appl. Phys.* **62** (1987), 4293.
3. Wiesendanger, R., Güntherodt, H. J., Güntherodt, G., Gambino, R. J. and Ruf, R., *Phys. Rev. Lett.* **65** (1990), 247.
4. Lovesey, S. W. and Collins, S. P., *X-ray Scattering and Absorption by Magnetic Materials*, Clarendon Press, Oxford, 1996.
5. Ankner, J. F. and Felcher, G. P., *J. Magn. Magn. Mater.* **200** (1999), 741.
6. Dürr, H. A., Guo, G. Y., van der Laan, G., Lee, J., Lauhoff, G. and Bland, J. A. C., *Science* **277** (1997), 213.
7. O'Donovan, K. V., Borchers, J. A., Majkrzak, C. F., Hellwig, O. and Fullerton, E. E., *Phys. Rev. Lett.* **88** (2002), 067201.
8. Tyson, J., Owens, A. H., Walker, J. C. and Bayreuther, G., *J. Appl. Phys.* **52** (1981), 2487.
9. Korecki, J. and Gradmann, U., *Phys. Rev. Lett.* **55** (1985), 2491.
10. Lugert, G. and Bayreuther, G., *Phys. Rev. B* **38** (1988), 11068.
11. Przybylski, M., Kaufmann, I. and Gradmann, U., *Phys. Rev. B* **40** (1989), 8631.
12. Niesen, L., Mugarza, A., Rosu, M. F., Coehoorn, R., Jungblut, R. M., Roozeboom, F., Baron, A. Q. R., Chumakov, A. I. and Rüffer, R., *Phys. Rev. B* **58** (1998), 8590.
13. Röhlsberger, R., Bansmann, J., Senz, V., Jonas, K. L., Bettac, A., Leupold, O., Rüffer, R., Burkel, E. and Meiwes-Broer, K. H., *Phys. Rev. Lett.* **86** (2001), 5597.
14. Röhlsberger, R., Bansmann, J., Senz, V., Bettac, A., Jonas, K. L., Meiwes-Broer, K. H., Leupold, O. and Rüffer, R., *Phys. Rev. B*, in print (2003).
15. Rose, M. E., *Elementary Theory of Angular Momentum*, Wiley, New York, 1957.
16. van Bürck, U., Siddons, D. P., Hastings, J. B., Bergmann, U. and Hollatz, R., *Phys. Rev. B* **46** (1992), 6207.
17. Rüffer, R. and Chumakov, A. I., *Hyp. Interact.* **97/98** (1996), 589; see also `http://www.esrf.fr/UsersAndScience/Experiments/HRRS/NRG/`.
18. Skomski, R. and Coey, J. M. D., *Phys. Rev. B* **48** (1993), 15812.
19. Kneller, E. F. and Hawig, R., *IEEE Trans. Magn.* **27** (1991), 3588.
20. Camley, R. E., *Phys. Rev. B* **35** (1987), 3608.
21. Fullerton, E. E., Jiang, J. S., Grimsditch, M., Sowers, C. H. and Bader, S. D., *Phys. Rev. B* **58** (1998), 12193.
22. Jiang, J. S., Kaper, H. G. and Leaf, G. K., *Discrete and Continuous Dynamical Systems – Series B* **1** (2001), 219.
23. Röhlsberger, R., Thomas, H., Schlage, K., Burkel, E., Leupold, O. and Rüffer, R., *Phys. Rev. Lett.* **89** (2002), 237201.
24. Feng, Y. P., Sinha, S. K., Deckman, H. W., Hastings, J. B. and Siddons, D. P., *Phys. Rev. Lett.* **71** (1993), 537.
25. Gradmann, U., Korecki, J. and Waller, G., *Appl. Phys. A* **39** (1986), 101.
26. Gradmann, U., In: K. H. J. Buschow (ed), *Handbook of Magnetic Materials*, Vol. 7, Elsevier, Amsterdam, 1993.
27. Grover, J. R., Siddons, D. P., Hastings, J. B., Faigel, G., Berman, L. E. and Haustein, P. E., *Hyp. Interact.* **62** (1990), 35.
28. Kikuta, S., In: G. Materlik, C. J. Sparks and K. Fischer (eds), *Resonant Anomalous X-Ray Scattering*, Proceedings of the International Conference on Anomalous Scattering (Malente, Germany, 1992), Elsevier, Amsterdam, 1994.
29. Andreeva, M. A., Semenov, V. G., Häggström, L., Lindgren, B., Kalska, B., Chumakov, A. I., Leupold, O., Rüffer, R., Prokhorov, K. A., Salashchenko, N. N., *The Physics of Metals and Metallography* **91** (2001), 22.

Hyperfine Interactions **144/145:** 45–52, 2002.

Recent Developments in Synchrotron Mössbauer Reflectometry

L. DEÁK[1], L. BOTTYÁN[1], M. MAJOR[1], D. L. NAGY[1], H. SPIERING[2], E. SZILÁGYI[1] and F. TANCZIKÓ[1]
[1]*KFKI Research Institute for Particle and Nuclear Physics, Budapest, Hungary*
[2]*Institute für Anorganische und Analytische Chemie, Johannes Gutenberg Universität, Mainz, Germany*

Abstract. Synchrotron Mössbauer Reflectometry (SMR), the grazing incidence nuclear resonant scattering of synchrotron radiation, can be applied to perform depth-selective phase analysis and to determine the isotopic and magnetic structure of thin films and multilayers. Principles and methodological aspects of SMR are briefly reviewed. Off-specular SMR provides information from the lateral structure of multilayers. In anti-ferromagneticly coupled systems the size of magnetic domains can be measured.

Key words: Mössbauer reflectometry, magnetic multilayers, off-specular scattering, antiferromagnetic domains, anisotropic optics.

1. Introduction

Although features of nuclear resonant scattering (NRS) of thin films were realized and dealt with from the early 60s [1], it has only been recently that the high brilliance and high degree of polarization of synchrotron radiation (SR), monochromator and detector techniques at 3rd generation sources allowed NRS to develop into a spectroscopic method in surface and thin film magnetism. For brevity, we call NRS of SR on thin films Synchrotron Mössbauer Reflectometry (SMR). SMR combines the sensitivity of Mössbauer spectroscopy to hyperfine interactions with the depth information yielded by reflectometry. SMR is established in the time and angular regime. Time differential (TD) SMR gives local (beating due to hyperfine interaction) information at a given incidence angle, time integral (TI) SMR gives an integral hyperfine interaction depth profile and superstructure information. Off-specular scattering and incoherent scattering offer novel applications.

The general description of specular reflection of grazing incidence Mössbauer radiation was given by Hannon *et al.* [2–5]. Starting from the quantum theory of γ-radiation, they formulated the dynamic theory of Mössbauer optics. Unfortunately, the dynamic theory provides rather slow algorithms for calculating reflectivity spectra; therefore, it is less efficient in spectrum fitting. In the grazing incidence limit, an optical model was derived from the dynamical theory [3, 5], which has been implemented in numerical calculations [6]. As it was shown later [7], this optical method is equivalent to that of Ref. [8].

The rigorous derivation of the general formulae for the transmissivity and the reflectivity of γ-radiation in the forward scattering and the grazing incidence case, respectively, were given by Deák *et al.* [8]. Like in Refs. [9–12], the Afanas'ev–Kagan nucleon current density expression of the dielectric tensor [13] and the covariant anisotropic optical formalism [14] were used. Instead of calculating the susceptibility tensor χ from the current densities of the nucleons, however, the problem was reduced to the calculation of the coherent forward scattering amplitude f. In the case of forward scattering, this general approach led to the theory of Blume and Kistner [15]. Using no intuitive pre-assumptions, Ref. [8] represents a firm basis of the Blume–Kistner theory [15] and of the Andreeva approximation [9–12]. The obtained reflectivity formulae in [8] are also suitable for fast numerical calculations in order to actually *fit* the experimental data [7].

The first successful grazing incidence NRS experiment with SR was performed by Grote *et al.* [16] in 1991. Chumakov *et al.* observed a pure nuclear reflection of SR from an isotopically periodic ^{57}Fe/Sc/^{56}Fe/Sc multilayer [17]. Alp *et al.* reported on the observation of nuclear resonant specular reflection with ^{119}Sn resonance [18]. An important step towards the realization of SMR was the observation of the *total reflection peak* [19, 20], i.e. the high number of *delayed* photons appearing close to the critical angle of the *electronic* reflectivity. The first SMR experiment aiming at the study of the magnetic structure of an anti-ferromagnetic (AF) ^{57}Fe/Cr multilayer was done by Toellner *et al.* [21]. The last years saw an increasing number of SMR experiments as a standard method for studying multilayers and thin films [22, 23].

2. Principles of specular SMR

In NRS of SR the low-lying levels of an ensemble of identical nuclei are coherently excited by the synchrotron radiation pulse. Since the levels are, as a rule, split by hyperfine interactions, the spatial and temporal coherence of the scattering results in characteristic patterns both of the angular distribution and the time evolution of the scattered radiation, which bear simultaneous and correlated information about topology and internal fields in the sample under study. SR is scattered both by nuclei and by electrons and these two processes interfere with each other, as well. Conventional Mössbauer spectroscopy and NRS of SR, although delivering similar information on hyperfine interaction and lattice dynamics, are complementary rather than equivalent to each other. The principal difference is that when the energy spectrum is scanned by the Doppler-shifted radiation of a γ-source, the recorded signal presents the incoherent sum of the spectral components of the transmitted radiation. In case of time domain NRS of SR, however, the response is formed by the coherent sum of the spectral components of the scattered radiation.

The specular SMR measurement is performed in θ–2θ geometry, in either time integral (TISMR) or time differential regime (TDSMR). TISMR means recording the total number of delayed photons from t_1 to t_2 as a function of θ. The time t_1 is a few nanoseconds and is determined by the bunch quality of the radiation source and by the dead time of the detector and the electronics, t_2 is a value somewhat below the bunch repetition time of the storage ring. As a rule, a θ–2θ scan of the *prompt* photons (conventionally called X-ray reflectometry) is recorded along with a delayed time integral SMR scan. TDSMR is a time response measurement in a fixed θ–2θ geometry performed at different values of θ. Like in the forward scattering case, hyperfine interaction results in quantum beats of the time response. The first step of an SMR measurement is usually to take a TISMR scan to select θ values of high enough delayed count rates at which TDSMR measurements can be performed. These regions are found near the total reflection peak [19, 20] and, in case of electronic or nuclear periodicity, near the electronic or nuclear Bragg reflections. A full SMR measurement consists of a prompt, a delayed time integral specular reflectivity scan, and a set of time response reflectivity measurements of the delayed photons. To extract the depth profile of hyperfine interactions with confidence, all these data should be evaluated *simultaneously*. If a full SMR measurement is not feasible for intensity reasons, a TISMR scan may still contain valuable information on the structure of the thin film.

Giving qualitative picture from the method we show a simple example on Figure 1, the simulation of a hypothetical two-dimensional energy-domain reflectivity spectrum of $[^{57}Fe(3.0\ nm)/^{56}Fe(1.5\ nm)]_{10}$ isotope periodical multilayer on glass as substrate.

Mössbauer reflectometry (MR) is a unification of reflectometry and Mössbauer spectroscopy. Indeed, taking the cross-sections at a fixed energies E on Figure 1, we get reflectometry spectra, as shown on Figure 2. For the given energies E having the index of reflection n from the expression of Lax [24] and using the op-

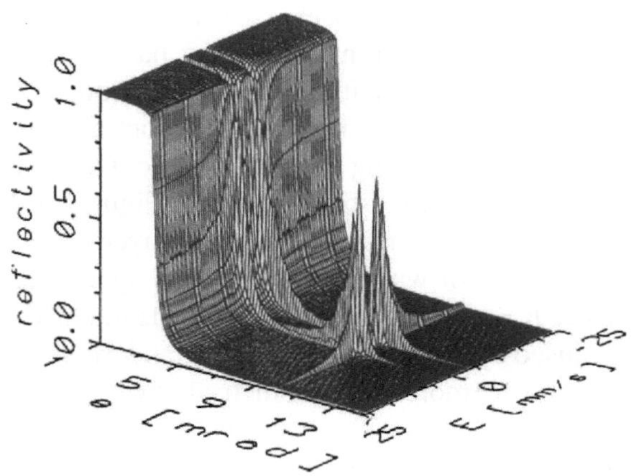

Figure 1. Energy-domain reflectivity of $^{57}Fe/^{56}Fe$ multilayer.

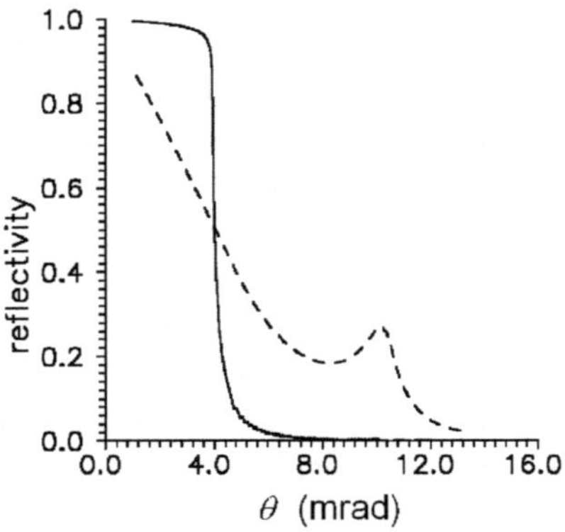

Figure 2. Reflectivity curves at resonance ($E = 3.072$ mm/s, dashed line) and off resonance at $E =$ 25 mm/s (solid line).

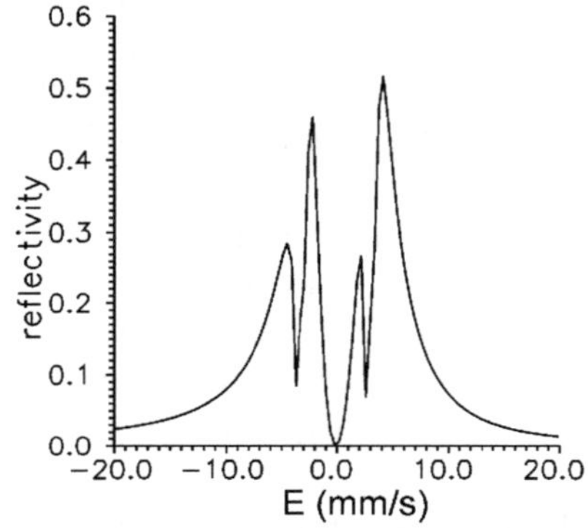

Figure 3. Mössbauer spectrum at the Bragg position ($\theta = 10.123$ mrad).

tical picture [6, 8], reflectometry spectra can be calculated as generalized Fresnel-formulae [6, 8]. Because of the isotopic periodicity we have a Bragg-peak at $\theta =$ 10.123 mrad on Figures 1 and 2 for energies near the resonance. However, far from the resonances, only the electronic scattering has considerable probability, so the spectrum is the X-ray reflectivity curve. From point of view of electronic scattering, the isotopic structure does not give any contrast, and the Bragg-peak disappears. On the other hand, taking the cross-sections at fixed grazing angles θ we get the energy dependent Mössbauer spectra (Figure 3). In the model, the hyperfine field was parallel to the surface of the multilayer and transversal electric, *viz.* σ-polarized, incident beam was assumed. In this special case there are only two Mössbauer lines at $E = \pm 3.072$ mm/s. The broadening and asymmetry of the individual lines are caused by the multiple scattering and the strong dispersion near the nuclear resonance; furthermore the asymmetry between the lines are caused by the electronic scattering.

Using SR as source, the simultaneous broadband coherent excitation results in time-domain SMR spectra, those are the Fourier-transformed reflectivity curves

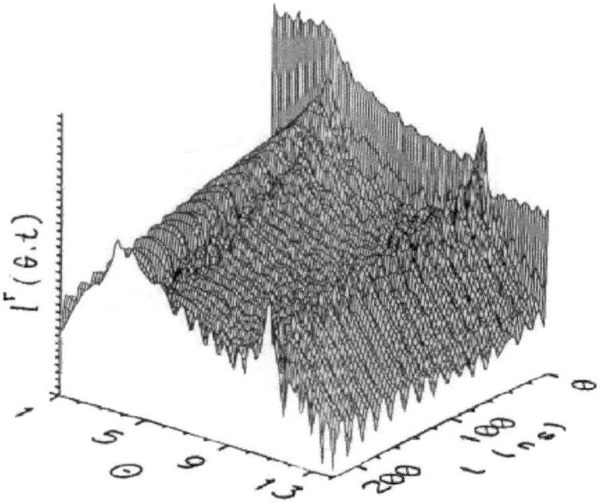

Figure 4. Time-domain reflectivity spectrum of ^{57}Fe/^{56}Fe multilayer.

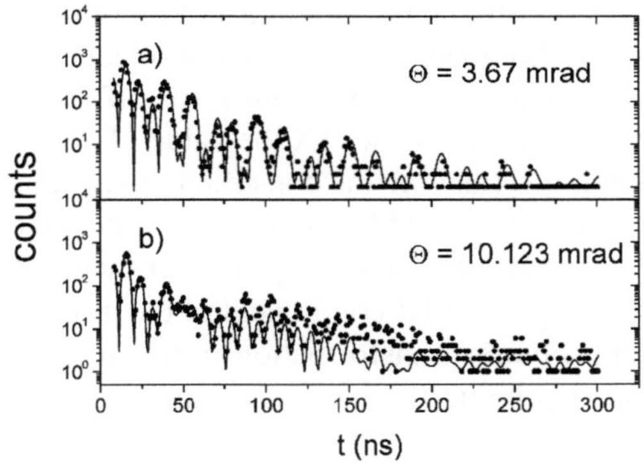

Figure 5. Experimental and theoretical TDSMR spectra of an ^{57}Fe/^{56}Fe multilayer at $\theta = 3.67$ mrad (near the critical angle) (a) and at $\theta = 10.123$ mrad (Bragg position) (b).

of the complex energy-domain, as shown on Figure 4. The TDSMR spectra, we measure (Figure 5), are the cross-sections at a fixed grazing angle θ of the two-dimensional surface in Figure 4. TISMR means integration of the TDSMR spectra from t_1 to t_2 as a function of θ, where t_1 and t_2 define fix time interval (Figure 6), as it was explained in the previous paragraph.

On the TISMR spectrum (Figure 6) we also see the Bragg peak ($\theta = 9$ mrad) [25] and an additional peak at the critical angle, $\theta = 3.67$ mrad, this latter being the interference effect of the electronic and nuclear scattering of photons [19, 20]. There is no Bragg peak on the X-ray reflectivity curve (Figure 6), as it was explained earlier.

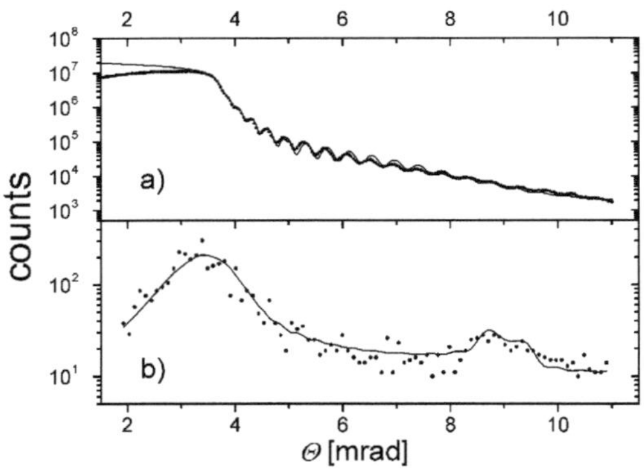

Figure 6. Non-resonant X-ray reflectivity (a) and TISMR (b) spectra of an $^{57}Fe/^{56}Fe$ multilayer.

3. Off-specular SMR

The specularly reflected radiation from layered systems does not depend on the lateral structure; it depends only on the lateral averages of material parameters [24, 26]. For studying lateral inhomogeneities, such as magnetic domains, etc., one can apply off-specular reflectometry. One possibility for off-specular reflectometry is the geometry called "ω-scan", where we fix the value of 2Θ (i.e. the detector) and vary the sample orientation ω (leaving the constraint of specular reflection $\omega = \Theta$).

Starting from the general theory of Lax [24] the off-specular intensity I_{off} can be expressed by the Fourier transformed depth profile of the coherent field inside the layers $T(\mathbf{k}'_\perp)$ ($\mathbf{k}'_\perp$ is the perpendicular component of the momentum of the scattered wave) and by the lateral Fourier transformation of the autocorrelation function of susceptibilities $C(\mathbf{K}_{\text{II}})$ ($\mathbf{K}_{\text{II}}$ being the later component of the momentum transfer vector)

$$I_{\text{off}} \propto \text{Tr}\left\lfloor T^{+}\left(\mathbf{k}'_\perp\right) C(\mathbf{K}_{\text{II}}) T\left(\mathbf{k}'_\perp\right)\rho\right\rfloor,$$

where ρ is the polarization density matrix of the incident radiation [27]. As 0th approximation we may assume that the autocorrelation function is exponential and so $I_{\text{off}} \propto C(\mathbf{K}_{\text{II}})$ is Lorentzian [27]. It follows that the correlation length is inversely proportional to the widths of the Lorentzian.

Figure 7 shows the off-specular X-ray reflectometry ("prompt") and off-specular SMR ("delayed") measurements on MgO(001)[^{57}Fe(26 Å)/Cr(13 Å)]$_{20}$ AF multilayer at the AF-Bragg position. The X-ray ω-scan resulted in a narrow line, indicating that there were no structural inhomogeneities in the lateral direction, additionally the wide Lorentzian on the delayed ω-scan shows the lateral inhomogeneity in the hyperfine magnetic field as a result of the magnetic domains. From the Lorentzian fit (solid line in Figure 7) we get 0.8 μm average domain size.

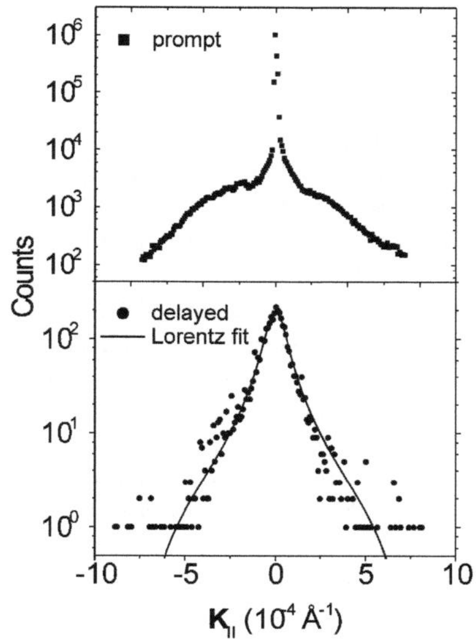

Figure 7. Prompt and delayed off-specular scans on MgO(001)[^{57}Fe(26 Å)/Cr(13 Å)]$_{20}$ AF multilayer at the AF-Bragg position. The solid line shows the Lorentzian fit.

4. Conclusion

A short overview of specular and off-specular SMR was given. Off-specular SMR is sensitive to the lateral structure of the hyperfine fields, so it can be used for studying magnetic domains in multilayers. Using the method we found 0.8 μm average domain size in a MgO(001)[^{57}Fe(26 Å)/Cr(13 Å)]$_{20}$ AF multilayer. The off-specular data evaluation will be soon available in EFFI (Environment For FItting).

Because of the underlying common optical approach the same theory can be applied for SMR, X-ray reflectometry, spin polarized neutron reflectometry [26] and X-ray resonance exchange scattering [28, 29] for both specular and off-specular reflexion. The computer program EFFI based on this calculus is freely available [7, 30].

Acknowledgements

This work was partly supported by the Hungarian Scientific Research Fund (OTKA) under Contract No. T029409 and by the Hungarian Academy of Sciences (Contact No. AKP 97-104 2,2/17). The financial support by the Deutsche Forschungsgemeinschaft (DFG) is gratefully acknowledged. The access to SR was generously made possible by ESRF Grenoble and HASYLAB Hamburg.

References

1. Bernstein, S. and Campbell, E. C., *Phys. Rev.* **132** (1963), 1625.
2. Hannon, J. P. and Trammell, G. T., *Phys. Rev.* **169** (1968), 315.
3. Hannon, J. P. and Trammell, G. T., *Phys. Rev.* **186** (1969), 306.
4. Hannon, J. P., Hung, N. V., Trammell, G. T., Gerdau, E., Mueller, M., Rüffer, R. and Winkler, H., *Phys. Rev. B* **32** (1984), 5068.
5. Hannon, J. P., Trammell, G. T., Mueller, M., Gerdau, E., Rüffer, R. and Winkler, H., *Phys. Rev. B* **32** (1985), 6363.
6. Röhlsberger, R., Ph.D. thesis, University Hamburg, 1994.
7. Spiering, H., Deák, L. and Bottyán, L., *Hyp. Interact.* **125** (2000), 197.
8. Deák, L., Bottyán, L., Nagy, D. L. and Spiering, H., *Phys. Rev. B.* **53** (1996), 6158.
9. Irkaev, S. M., Andreeva, M. A., Semenov, V. G., Beloserskii, G. N. and Grishin, O. V., *Nucl. Instrum. Methods B* **74** (1993), 554.
10. Andreeva, M. A., Irkaev, S. M. and Semenov, V. G., *Sov. Phys. JETP* **78** (1994), 965.
11. Andreeva, M. A. and Kuz'min, R. N., *Messbauerovskaya Gamma-Optika*, Moscow University, 1982.
12. Andreeva, M. A. and Rosete, K., *Poverkhnost'* **9** (1986), 145; *Vestnik Mosk. Univ., Ser. 3. Fiz. Astron.* **27** (1986), 57.
13. Afanas'ev, A. M. and Kagan, Yu., *Sov. Phys. JETP* **21** (1965), 215.
14. Borzdov, G. M., Barskovskii, L. M. and Lavrukovich, V. I., *Zh. Prikl. Spektrosk.* **25** (1976), 526.
15. Blume, M. and Kistner, O. C., *Phys. Rev.* **171** (1968), 417.
16. Grote, M., Röhlsberger, R., Dimer, M., Gerdau, E., Hellmich, R., Hollatz, R., Jäschke, J., Lüken, E., Metge, J., Rüffer, R., Rüter, H. D., Sturhahn, W., Witthoff, E., Harsdorff, M., Pfützner, W., Chambers, M. and Hannon, J.-P., *Europhys. Lett.* **14** (1991), 707.
17. Chumakov, A. I., Smirnov, G. V., Baron, A. Q. R., Arthur, J., Brown, D. E., Ruby, S. L., Brown, G. S. and Salashchenko, N. N., *Phys. Rev. Lett.* **71** (1993), 2489.
18. Alp, E. E., Mooney, T. M., Toellner, T., Sturhahn, W., Witthoff, E., Röhlsberger, R., Gerdau, E., Homma, H. and Kentjana, M., *Phys. Rev. Lett.* **70** (1993), 3351.
19. Baron, A. Q. R., Arthur, J., Ruby, S. L., Chumakov, A. I., Smirnov, G. V. and Brown, G. S., *Phys. Rev. B* **50** (1994), 10354.
20. Deák, L., Bottyán, L. and Nagy, D. L., *Hyp. Interact.* **92** (1994), 1083.
21. Toellner, T. L., Sturhahn, W., Röhlsberger, R., Alp, E. E., Sowers, C. H. and Fullerton, E. E., *Phys. Rev. Lett.* **74** (1995), 3475.
22. Chumakov, A. I., Nagy, D. L., Niesen, L., Alp, E. E., *Hyp. Interact.* **123/124** (1999), 427.
23. Röhlsberger, R., *Hyp. Interact.*, **123/124** (1999), 455.
24. Lax, M., *Rev. Mod. Phys.* **23** (1951), 287.
25. The Bragg position is different, because the model system differs by this measurement and simulation. The sample was: float glass/^{57}Fe(4.6 nm)/[^{56}Fe(2.9 nm)/^{57}Fe(2.3 nm)]□15/ Al(5.7 nm).
26. Deák, L., Bottyán, L., Nagy, D. L. and Spiering, H., *Physica B* **297** (2001), 113.
27. Deák, L., to be published.
28. Hannon, J. P., Trammell, G. T., Blume, M. and Doon Gibs, *Phys. Rev. Lett.* **61** (1988), 1245.
29. Deák, L. *et al.*, to be published (not yet implemented into our data evaluation program, EFFI).
30. The computer program EFFI is available from ftp://iacgu7.chemie.uni-mainz.de/pub/effi.

Hyperfine Interactions **144/145:** 53–64, 2002.

Magnetic Properties of Metallic Multilayered Systems

TERUYA SHINJO* and KO MIBU**
Institute for Chemical Research, Kyoto University, Uji, Kyoto-fu 611-0011, Japan

Abstract. The giant magnetoresistance (GMR) effect was an epoch-making discovery in the field of magnetic materials research. In this article, studies on magnetic multilayered systems relating to GMR are briefly surveyed and the role of interlayer coupling in GMR phenomena is argued. The usefulness of Mössbauer spectroscopy as a tool to investigate GMR systems is introduced. Although the GMR effect was observed for the first time in Fe/Cr multilayers, magnetic properties of ultrathin Cr layers were not well known until quite recently. Magnetic behaviors of multilayers including ultrathin Cr layers have been investigated from ^{119}Sn Mössbauer spectroscopy and neutron diffraction. In Cr/Sn multilayers, it was found that the spin-density-wave structure of Cr layer is systematically modified by the insertion of monatomic Sn layers.

Key words: GMR, interlayer coupling, Mössbauer spectroscopy, hyperfine field, ultrathin Cr layers.

1. Introduction for the GMR effect

Grünberg *et al.* have investigated magnetic behaviors of Fe/Cr/Fe trilayer system and found that an interlayer exchange coupling exists between Fe layers across an intervening Cr layer [1]. The coupling strength becomes rather strong when the Cr thickness is around 10 Å. In order to study the features of interlayer coupling in more details, Baibich *et al.* prepared Fe/Cr multilayers with various artificial periodicities. Each Fe layer is an independent ferromagnet but, because of the antiferromagnetic interlayer coupling, magnetizations in adjacent layers are oriented antiparallel. By applying strong external fields, Fe layer magnetizations are oriented parallel and, associating with the magnetic reorientation, the electric resistance was found to be reduced drastically. That is the giant magnetoresistance (GMR) effect discovered in 1988 [2] (Figures 1(a) and 2(a)). The decrease of resistance at low temperature by applying an external field, 2 T, has reached to be almost 50% in [Fe(30 Å)/Cr(9 Å)]$\times$60. Even at room temperature, the decrease of magnetoresistance (MR) was about 17%, which is much larger than the con-

* Present Address: International Institute for Advanced Studies, Kizu, Soraku-gun, Kyoto-fu 619-0225, Japan.

** Present Address: Research Center for Low Temperature and Materials Sciences (Uji), Kyoto University, Uji, Kyoto-fu 611-0011, Japan.

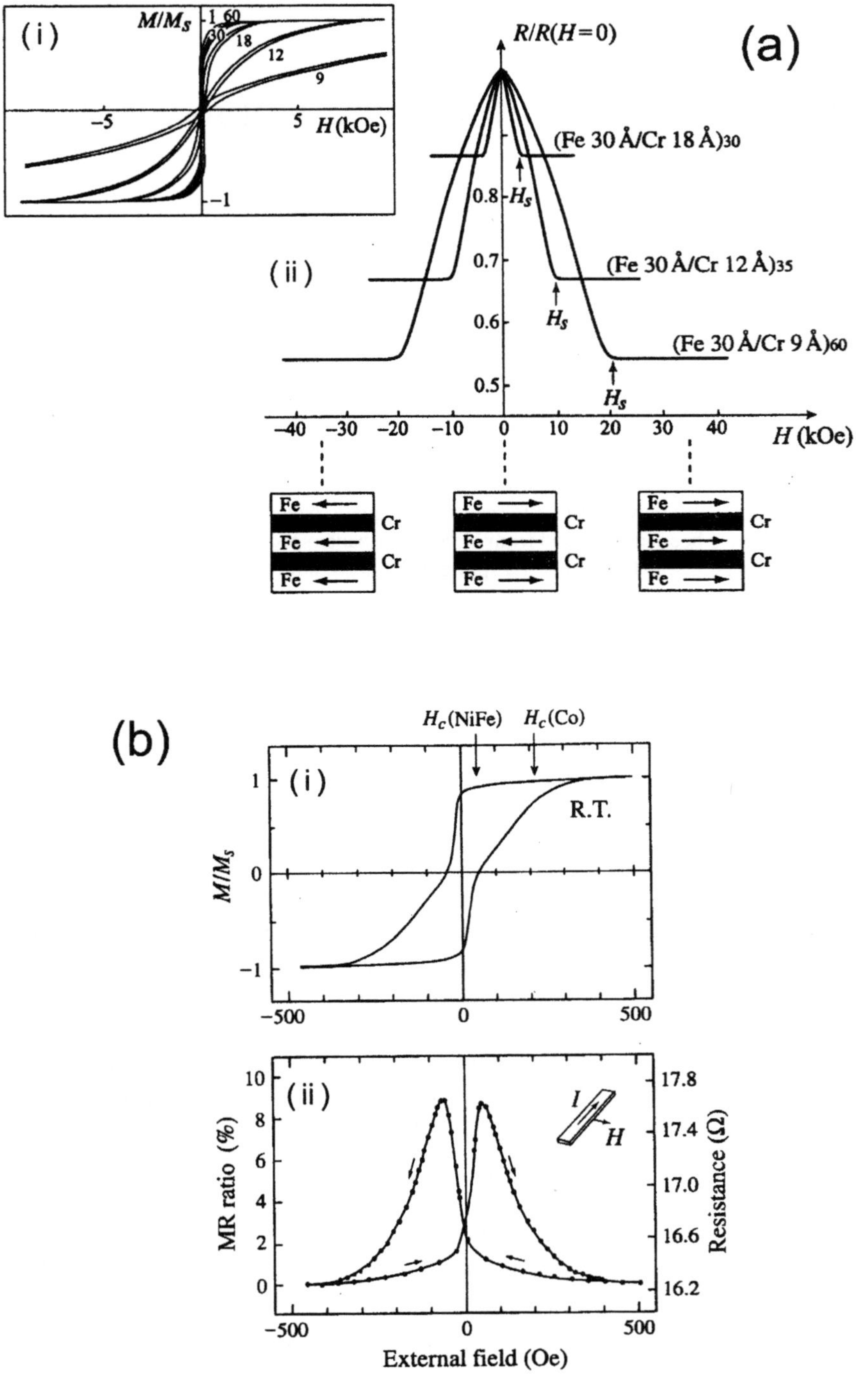

Figure 1. Examples of the GMR effect. Magnetization and magnetoresistance curves are shown for (a) Fe/Cr multilayers at 4.2 K [2], and (b) non-coupled Co/Cu/NiFe/Cu multilayers at 300 K [3].

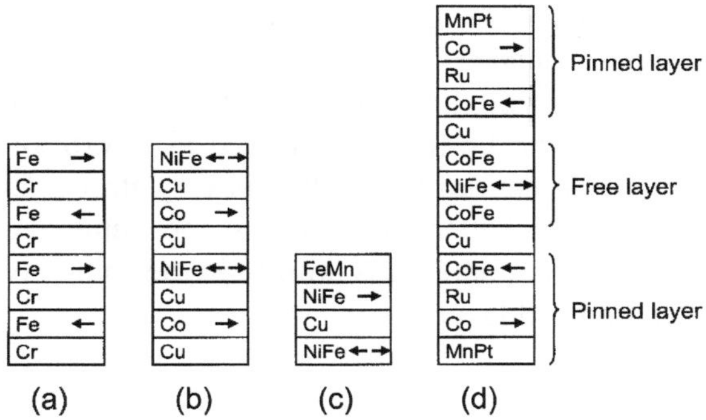

Figure 2. Schematic illustration of multilayered structures for the GMR effect; (a) coupled multilayers, (b) non-coupled multilayers, and (c), (d) spin valves.

ventional anisotropic magnetoresistance effect (AMR). The novel MR effect has attracted great attention from both fundamental and technical aspects.

By the discovery of GMR, two major issues have been raised up, spin-dependent scattering and interlayer coupling. In the argument of transport phenomena in general, it was not necessary to pay attention to the spin polarization of conduction electrons. Magnetic structures of normal magnetic materials are so rigid and cannot be converted for instance from antiferromagnetic to ferromagnetic. On the other hand, the magnetic structure of multilayers with moderate interlayer couplings can be varied from antiparallel to parallel alignments rather easily. Because of the spin dependent scattering, the resistance in an antiferromagnetic configuration becomes much higher than that in a ferromagnetic one. The large MR effect caused by the spin-dependent scattering is very attractive for technical applications, such as magnetic recording heads. However, if the external field to induce the magnetic structural change is too high, the field sensitivity of MR becomes very low and then the possibilities for technical applications would be very limited. Sooner or later, non-coupled type GMR systems were found to depend sensitively on external fields and the potential of GMR for technical applications is revealed [3] (Figures 1(b) and 2(b)).

Although the GMR was discovered because of the existence of interlayer coupling, technical applications require to suppress the interlayer coupling to be negligible. The first commercial product of magnetic recording head using GMR principle has been achieved in the three-layer form, ferromagnet/Cu/ferromagnet, called the spin-valve system [4] (Figure 2(c)). For the spacer material, Cu has been selected and it has been verified that a Cu layer with only 20 Å in thickness is enough to insulate the interlayer coupling. Two kinds of magnetic layers, "free" and "pinned" layers, are included in the structure. The pinned layer must be magnetically hard, having a high coercive force. In order to enhance the coercive force, the ferromagnetic layer is in contact with an antiferromagnetic layer. Due to the coupling with an antiferromagnet, the coercive force of the pinned layer is greatly enhanced. Although the spin-valve system has started with the simple three-layer structure, in order to improve the function, lately proposed are spin-valve structures including more than 10 layers (Figure 2(d)). It is interesting to point out that the interlayer coupling is playing an important role in these spin-valve structures. An artificial antiferromagnetic layer is synthesized as a form of Co/Ru/Co layer. Since the antiferromagnetic interlayer coupling is very strong, Co/Ru/Co behaves as an antiferromagnet, having better qualities than conventional antiferromagnetic materials. Thus, the interlayer coupling is utilized in the spin-valve systems, not as a spacer layer but for a pinned layer.

2. Role of interlayer coupling in GMR phenomena

After the discovery of the GMR effect in Fe/Cr system, MR properties have been investigated extensively for many multilayered systems consisting of magnetic and non-magnetic layers and similar GMR phenomena as those of Fe/Cr multilayers were observed in many multilayered systems. A typical example, exhibiting the largest MR ratios, is Co/Cu multilayers. Although Cu is a typical non-magnetic noble metal, an interlayer coupling is working between ferromagnetic layers separated by a Cu layer. The coupling through a Cu layer is in general smaller than that through a Cr layer, but if the Cu layer thickness is appropriate, the interlayer coupling strength is large enough to realize antiparallel magnetization arrangements and eventually a large MR effect occurs.

A fascinating discovery was the oscillatory dependence of the interlayer coupling on the spacer layer thickness and therefore the large MR effect appears at specific layer thicknesses of the non-magnetic spacer layer [5]. Concentrated studies on interlayer couplings have been carried out and it was found to be rather general that the interlayer coupling oscillatorily depends on the spacer layer thickness. This phenomenon suggests quantum well formations of conduction electrons in non-magnetic spacer layers and the oscillation period was interpreted as expected from the Fermi surface topology of the spacer metal. At least in noble metals, if the thickness is not less than 10 Å, the band structure of a thin layer is not greatly modified but is rather close to that of the bulk state. The understanding of electronic structures in metallic ultrathin layers has been considerably progressed by studying the role of the spacer layers in the GMR systems.

The existence of interlayer coupling indicates that there are magnetically polarized electrons in the spacer layer, although the material of the spacer is intrinsically non-magnetic. There have been several studies on the spin polarization by scanning electron microscopy and other techniques in vacuum atmosphere and oscillatory features were observed as a function of layer thickness, for example, for Cr [6]. However, a direct observation of polarized spins in buried spacer layers is very difficult and therefore the spatial distribution of spins or the energy dependence of spin polarization, for instance, cannot be studied. A magnetization caused by the spin polarizations in a non-magnetic spacer layer is too small to measure by conventional techniques. Particularly, because a non-magnetic spacer layer is sandwiched in between two ferromagnetic layers, it is practically impossible to estimate the static magnetization caused by the conduction electron spin polarization in a non-magnetic spacer layer. The feasibility of Mössbauer spectroscopy as a tool to study the spin polarization in a non-magnetic layer is argued in the next chapter.

3. Mössbauer spectroscopy for magnetic multilayer studies

3.1. ^{57}Fe MÖSSBAUER SPECTROSCOPIC STUDIES

Mössbauer spectra are easily obtained as long as a Mössbauer isotope is included in a sample, and furnish various kinds of information. If a sample does not include a Mössbauer isotope, a doping with an appropriate isotope is required for the measurements. If the doping can be done depth-selectively, unique information will be obtained from the Mössbauer spectra. For example, the results on Co/Ru measured by Hamada [7] are reproduced in Figure 3. Co/Ru multilayer systems are known to exhibit a large interlayer coupling [8]. In order to apply ^{57}Fe Mössbauer spectroscopy, where ^{57}Fe is the most convenient Mössbauer isotope, ^{57}Fe probing layers were introduced in Co/Ru multilayers. The thickness of ^{57}Fe layers was nominally 1 Å, roughly corresponding to one monolayer. For the study of chemical and magnetic interface profile of the ferromagnetic Co layer with the thickness of 20 Å, ^{57}Fe layer (1 Å) was deposited on the bottom interface of Co layer contacting with Ru layer, and at the distance of 2 Å from the bottom interface. Namely, in the second sample, Co(2 Å), ^{57}Fe(1 Å), Co(18 Å), and then Ru(15 Å) were successively deposited (Figure 3(b)). The Mössbauer absorption spectra at room temperature for the two samples are shown in Figure 3(a). Both

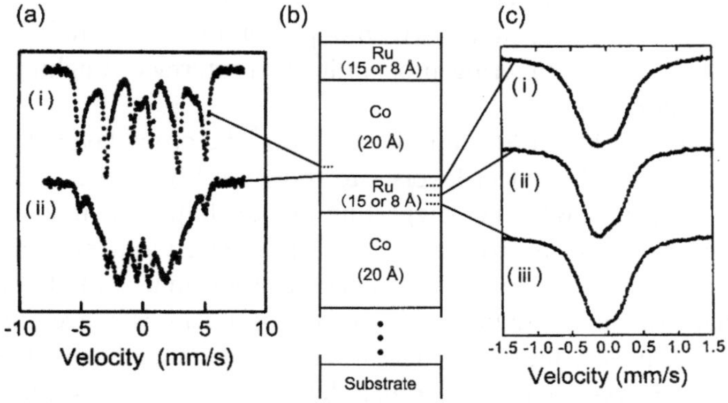

Figure 3. (a) ^{57}Fe Mössbauer spectra at 300 K for (i) [Co(2 Å)/^{57}Fe(1 Å)/Co(18 Å)/Ru(15 Å)]$\times n$ and (ii) [Co(20 Å)/^{57}Fe(1 Å)/Ru(15 Å)]$\times n$. (b) Positions of ^{57}Fe probes (1 Å) in a Co/Ru multilayered structure. (c) ^{57}Fe spectra at 300 K at various positions in a Ru(8 Å) layer sandwiched in between Co(20 Å) layers. The nominal structures of the samples are (i) [Co(20 Å)/Ru(6 Å)/^{57}Fe(1 Å)/Ru(2 Å)]$\times n$, (ii) [Co(20 Å)/Ru(4 Å)/^{57}Fe(1 Å)/Ru(4 Å)]$\times n$, and (ii) [Co(20 Å)/Ru(2 Å)/^{57}Fe(1 Å)/Ru(6 Å)]$\times n$.

are consisting basically of 6-line spectrum indicating the existence of magnetic order but the profiles are considerably different. In both spectra, the intensities of the No. 2 and 5 lines among 6 are stronger, showing that the preferred orientation of the magnetization is in the film plane. The difference of the profile in these spectra means that the Mössbauer spectroscopy has a good sensitivity to study the interface. The spectrum for the bottom interface (Figure 3(a)(ii)) is very broad because of the hyperfine field distribution. In the bottom interface layer, there should be large variations in the local atomic arrangement. The hyperfine field at ^{57}Fe greatly depends on the local arrangement, i.e., the numbers of Ru, Co and Fe atoms occupying the nearest neighbor sites. It is noted that a minor portion shows a sharp 6-lines with the same hyperfine field as that of Figure 3(a)(i). The nominal depth for ^{57}Fe probes in Figure 3(a)(i) is 2–3 Å from the interface contacting with Ru, namely corresponding to the second or third atom layer. The result indicates that the hyperfine field distribution is very broad at the bottom interface layer, but almost no distribution exists in the second layer, so that it is concluded that almost all the nearest neighbors surrounding a ^{57}Fe atom are Co at the nominal depth of 2 Å.

In the spectrum of Figure 3(a)(i), non-magnetic fraction is negligible. As shown in Figure 3(c), ^{57}Fe atoms diffused into Ru should show single line spectra with no hyperfine splitting. The absence of non-magnetic fraction certifies that the diffusion of Fe atoms into Ru layer does not occur. At the ferromagnet/Ru interfaces in the present sample, the effect of Ru on the ^{57}Fe hyperfine field is visible only at the bottom interface layer (Figure 3(a)(ii)) and actually no influence of Ru is observed at the second atom layer (Figure 3(a)(i)). This result means that the compositional profile sharply changes at the interface and the mutual mixing between Ru and Co (or Fe) is very limited. If such a model sample as shown here, with ^{57}Fe doping layers, is available, an interface chemical profile is able to measure and information on an atomic scale is furnished. However, it is difficult to measure intrinsic transport properties using such model samples with doped atoms and therefore the relation between spin-dependent scattering probability and interface roughness cannot be systematically studied.

Mössbauer spectra for Co/Ru multilayers with ^{57}Fe probe layers (1 Å) inserted at various positions in Ru spacer layers, are shown in Figure 3(c). Each spectrum is a broad asymmetric single line, suggesting that the hyperfine field is very small. The small hyperfine field means that the local magnetic moment of Fe is actually zero. This result seems to be reasonable since Fe atoms in usual bulk Fe–Ru alloys with hexagonal structures lose the local magnetic moment. However, in the present case, the Ru layer is very thin (8 Å in total) and the distance from a ferromagnetic Co(Fe) layer to ^{57}Fe probes is smaller than the Ru layer thickness. Furthermore, in a Ru layer, there exists a definite conduction electron spin polarization, which is indicated from the strong interlayer coupling across a Ru layer. Therefore, it could not be straightforwardly expected that Fe atoms in an ultrathin Ru layer sandwiched in between ferromagnetic layers lose their local magnetic moments.

Usually Fe atoms are strongly ferromagnetic and therefore ^{57}Fe is not a suitable probe to investigate magnetic properties of non-magnetic layers. Intrinsic magnetic properties will be greatly disturbed by the doping with magnetic Fe probes. It is rather unique that Fe atoms in Ru layers lose local magnetic moments and can be used as a non-magnetic probe to study the properties of non-magnetic layers (Ru). For the spectra in Figure 3(c), the observed hyperfine field estimated from the overall line width is around 1 T. This result confirms that no local magnetic moment exists at an Fe atom in a Ru layer with 8 Å thickness. At the same time it is suggested that the hyperfine field caused by the conduction electron spin polarization is fairly small (less than 1 T). The three spectra in Figure 3(c) have similar line profiles and similar distributions of hyperfine field. A slight difference is the asymmetry of the spectrum. The hyperfine fields are too small to derive a reasonable result in the deconvolution of the spectra. In addition, there might exist some quadrupole interaction, which can also be an origin of asymmetric broad line profile, and therefore further analyses on the line profile cannot make sense.

3.2. ^{119}Sn MÖSSBAUER SPECTROSCOPIC STUDIES

Usually ^{57}Fe atoms have their own magnetic moments and are not appropriate probes to study the magnetic behaviors of non-magnetic layers. For the study of non-magnetic spacer layers, a non-magnetic Mössbauer isotope is appropriate as a probe. The authors' group has carried out some magnetic multilayer studies using ^{197}Au and ^{119}Sn probes. ^{197}Au Mössbauer measurements have been performed by Kobayashi *et al.* for the multilayered samples consisting of Au and ferromagnetic elements with various layer thicknesses [9]. In summary, a sizable hyperfine splitting, whose hyperfine fields depend on the ferromagnetic species, Fe, Co or Ni, has been observed only in the vicinity of interface, within a few atom layers from the ferromagnetic layer. When the distance is larger, the hyperfine field is almost negligible. Appreciable hyperfine fields exist only on the sites directly contacting with ferromagnetic layers. Thus, the hyperfine fields on the Au nuclear sites caused by conduction electron polarizations were concluded to be very small. It is impossible to develop the argument on the spatial distribution of spin polarization from the analysis of hyperfine field.

Similar studies were performed also using ^{119}Sn probes. Several studies have been carried out by preparing multilayered samples including ^{119}Sn monolayers in the non-magnetic layers. In contrast to the case of ^{197}Au, the ^{119}Sn probes have a merit in selecting the positions (depth-selectively). If the impurity effect caused by the insertion of Sn layer is not serious, the spin polarization is able to be measured as a function of the distance from the interface. According to the spectra so far measured [10], the results are more or less similar to the results from ^{197}Au probes. Large hyperfine fields are observed when the probes are located close to the ferromagnetic layer, and decrease monotonously as the position goes farther from the interface. Figure 4 shows examples of ^{119}Sn spectra for Fe/Au, Fe/Ag,

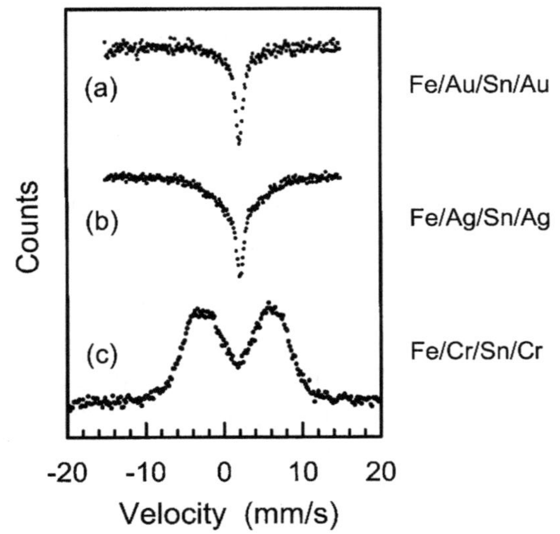

Figure 4. ^{119}Sn Mössbauer spectra at 300 K by absorption geometry for [Fe(20 Å)/Au(20 Å)/Sn(1.5 Å)/Au(20 Å)]$\times n$ and [Fe(20 Å)/Ag(20 Å)/Sn(1.5 Å)/Ag(20 Å)]$\times n$, and spectrum by emission geometry for [Fe(10 Å)/Cr(20 Å)/Sn(2 Å)/Cr(20 Å)]$\times n$. The spectrum by emission geometry has a mirror image structure of absorption one. The meaning of the hyperfine structure of the spectrum is the same in both geometries.

and Fe/Cr multilayers. Figures 4(a) and 4(b) correspond to the spectra for ^{119}Sn at the center of Au(40 Å) or Ag(40 Å) layers. The spectra indicate that the hyperfine field for the most part is actually zero. The hyperfine field caused by conduction spin polarizations is very small at 20 Å from the interface in these samples. From the line profile, it is suggested that a minor fraction has some hyperfine field. However, the hyperfine field is too small and the distribution is too broad to analyze quantitatively. Although the measurements for other samples with different depths also have been carried out, obtained spectra had similar profiles. Thus it is not yet successful to speculate the oscillatory spatial distribution of spin polarization in a non-magnetic layer from the analysis of Mössbauer spectra.

On the other hand, in Figure 4(c), a spectrum for ^{119}Sn located at the center of Cr(40 Å) layers in Fe/Cr multilayers is shown, which has a definite magnetic hyperfine splitting in contrast to the results in Figures 4(a) and 4(b). This is an undoubted evidence of the magnetic ordering in Cr layers, and at the same time, the usefulness of ^{119}Sn Mössbauer spectroscopy for the study of ultrathin Cr layers is revealed.

4. Magnetism of Cr/Sn multilayers

Grünberg *et al.* initiated the Fe/Cr/Fe trilayer studies, having an interest on the magnetic behavior of ultrathin Cr layers. As was mentioned already, the GMR effect was observed in Fe/Cr multilayers where the interlayer coupling exists between

Fe layers separated by a Cr layer. Similar interlayer couplings were also found in many non-magnetic spacer metal layers and the mechanism of coupling through a non-magnetic layer was more or less understood. However, the role of Cr's own magnetic order has not been clear for a rather long time. The Néel temperature of Cr metal was a source of confusion. Since the Néel temperature of a bulk Cr metal (311 K) is near room temperature, Cr layers were regarded as non-magnetic in the arguments at the initial stage. However, the existence of antiferromagnetic order in thin Cr layers was observed in gamma-ray perturbed angular correlation studies using Cd probes for Cr layers thicker than 60 Å [11] and in the neutron diffraction studies even for thinner Cr layers [12, 13]. These results suggest that the Néel temperature of Cr films can be higher than the bulk value and pieces of information on the magnetism of ultrathin Cr layers, especially in the situation sandwiched in between ferromagnetic Fe layers, are of great importance. In order to use ^{119}Sn Mössbauer spectroscopy as a tool to investigate magnetic properties of Cr layers, ^{119}Sn probe layers with the nominal thickness of 2 Å were inserted in ultrathin Cr layers.

Antiferromagnetic properties of bulk Cr metal were studied using ^{119}Sn Mössbauer spectroscopy by Dubiel [14]. He observed that the hyperfine fields at Sn probes dissolved in a Cr matrix were greatly distributed, typically from 0 to 8 T. This result is reasonable if the Cr magnetic structure is a spin-density wave (SDW) and the spin density at each atom site is different. Sn atoms are regarded as "passive" probes which detect the local magnetization without perturbing the magnetism of the matrix. On the other hand, when monatomic Sn layers are inserted periodically in a Cr(001) film, the Mössbauer spectra indicate that the hyperfine field is not much distributed and the value, 10–12 T, is considerably larger than that for Sn impurities in a bulk Cr metal [15]. This is an undoubted evidence that the ultrathin Cr layers have a magnetic order and the Néel temperature is much higher than room temperature.

Figure 5 shows typical ^{119}Sn conversion electron Mössbauer spectra for epitaxial Cr(001)/Sn multilayers at room temperature with varying the thickness of Cr layers. A large hyperfine field is observed for all these samples. Sn layers are thought to be substituting a monatomic Cr(001) layer and forming a strained bcc structure together with Cr layers. The large hyperfine field indicates that a large local magnetic moment is induced at a Cr atom in the neighbor sites of Sn layer. In contrast to Sn impurities in Cr, having the smaller and distributed hyperfine fields, Sn monolayers inserted in between (001) Cr layers serve as "active probes", and the Cr magnetic structure is modified by Sn layers.

As shown in Figure 5, the magnitude of hyperfine field at Sn probe layer does not greatly depend on the Cr layer thickness. Regardless of the Cr layer thickness, the local magnetic moment at Cr atom in the neighbor of Sn layer seems to be almost the same in the Cr layer. When the Cr layer thickness in a Cr/Sn multilayer is larger than about 40 Å, it was found from neutron diffraction that a SDW structure is established in the Cr layer at low temperatures [16]. Although the magnetization

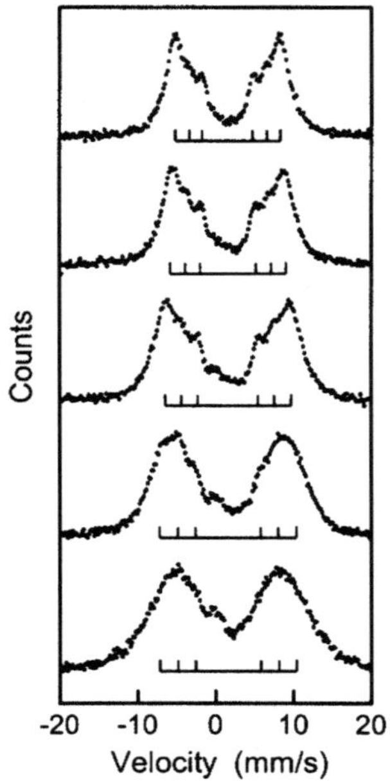

Figure 5. ^{119}Sn Mössbauer spectra at 300 K for [Cr(t_{Cr})/Sn(2 Å)]×n (t_{Cr} = 80, 40, 20, 10, and 5 Å).

becomes spatially inhomogeneous, the hyperfine field at the Sn sites was found to be unique and large. Therefore it is suggested that the SDW is structured so that the interface site with Sn layer becomes the antinode. The wavelength is determined by the distance between Sn layers, that is the thickness of the Cr layer. By varying the thickness of Cr layer in Cr/Sn multilayers, the SDW in Cr is modified. In other words, the spin density wavelength is artificially tuned in some extent.

For the study of magnetism in Fe/Cr multilayers, ^{119}Sn monolayers are inserted in the Cr layers [17]. The nominal structures consist of repetition of Fe/Cr/Sn/Cr. The Mössbauer spectra are shown in Figure 6(a). When the Cr layer thickness is moderate, the spectra are similar to those for Cr/Sn multilayers. When the Cr layers become extremely thin, the hyperfine field becomes smaller, suggesting that the spin density at the neighbor Cr atoms decreases as a function of Cr layer thickness. The Fe layers actively influence the magnetism of Cr layers. When the Cr layer thickness is 5 Å, the hyperfine field becomes very small but is apparently not zero. The ultrathin Cr has still an antiferromagnetic order. The temperature dependence indicates that the Néel temperature is much higher than room temperature, around 400 K or more (Figure 6(b)). Therefore the small hyperfine field is not simply a result of the decreasing of the Néel temperature. The size of magnetic moment of Cr atoms would be determined by the influences from Fe and Sn layers. Although the Sn layers are active probes, they are effective to see some common features of the magnetism for both Fe/Cr and Fe/Cr/Sn/Cr multilayers. It has been thus

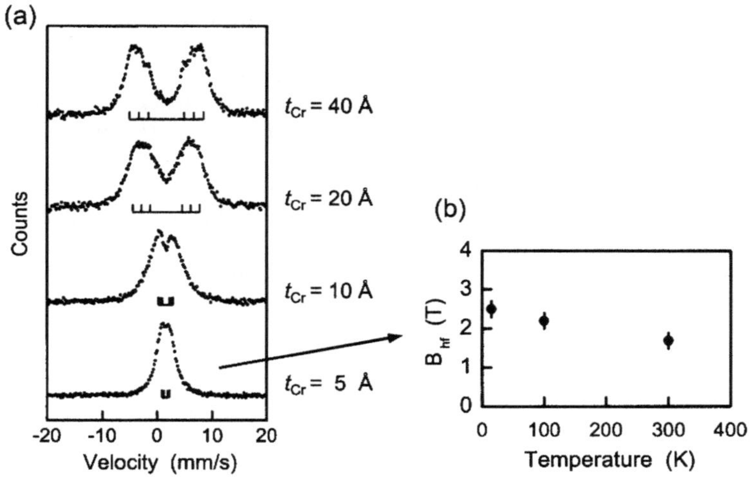

Figure 6. (a) ^{119}Sn Mössbauer spectra at 300 K for [Fe(10 Å)/Cr(t_{Cr})/Sn(2 Å)/Cr(t_{Cr})]×n (t_{Cr} = 40, 20, 10, and 5 Å). (b) The temperature dependence of the hyperfine field for the multilayer with t_{Cr} = 5 Å.

concluded that ultrathin Cr layers in Fe/Cr GMR systems have antiferromagnetic orders with small magnetic moments. For the understanding of the GMR behaviors, the role of Cr magnetism has to be taken into account. However, since the Néel temperature is fairly high, there has been no experimental work to elucidate the difference of transport features below and above the Néel temperature.

Acknowledgements

The authors would like thank Dr. N. Hosoito, Dr. T. Emoto and Dr. S. Hamada for their experimental collaboration. This work was partly supported by COE Elements Science and NEDO project.

References

1. Grünberg, P., Schreiber, R., Pang, Y., Brodsky, M. B. and Sowers, H., *Phys. Rev. Lett.* **57** (1986), 2442.
2. Baibich, M. N., Broto, J. M., Fert, A., Nguyen van Dau, F., Petroff, F. Etienne, P., Creuzet, G., Friederich, A. and Chazelas, J., *Phys. Rev. Lett.* **61** (1988), 2472.
3. Shinjo, T. and Yamamoto, H., *J. Phys. Soc. Jpn.* **59** (1990), 3061.
4. Dieny, B., Speriosu, V. S., Parkin, S. S. P., Gurney, B. A., Wilhoit, D. R. and Mauri, D., *Phys. Rev. B* **43** (1991), 1297.
5. Parkin, S. S. P., More, N. and Roche, K. P., *Phys. Rev. Lett.* **64** (1990), 2304.
6. Ungris, J., Celotta, R. J. and Pierce, D. T., *Phys. Rev. Lett.* **69** (1992), 1125.
7. Hamada, S., unpublished.
8. Parkin, S. S. P., *Phys. Rev. Lett.* **67** (1991), 3597.
9. Kobayashi, Y., Nasu, S., Emoto, T. and Shinjo, T., *Hyp. Interact.* **111** (1998), 129.
10. Emoto, T., Hosoito, N. and Shinjo, T., *J. Magn. Magn. Mater.* **189** (1998), 136.

11. Meersschaut, J., Dekoster, J., Schad, R., Beliën, P. and Rots, R. *Phys. Rev. Lett.* **75** (1995), 1638.
12. Fullerton, E. E., Bader, S. D. and Robertson, J. L., *Phys. Rev. Lett.* **77** (1996), 1382.
13. Schreyer, A., Majkrzak, C. F., Zeidler, Th., Schmitte, T., Bödeker, P., Theis-Bröhl, K., Abromeit, A., Dura, J. A. and Watanabe, T., *Phys. Rev. Lett.* **79** (1997), 4914.
14. Dubiel, S. M., *J. Magn. Magn. Mater.* **124** (1993), 31.
15. Mibu, K., Tanaka, S. and Shinjo, T., *J. Phys. Soc. Jpn.* **67** (1998), 2633.
16. Mibu, K., Takeda, M., Suzuki, J., Nakanishi, A., Kobayashi, T., Endoh, Y. and Shinjo, T., *Phys. Rev. Lett.* **89** (2002), 287202.
17. Mibu, K., Almokhtar, M., Tanaka, S., Nakanishi, A., Kobayashi, T. and Shinjo, T., *Phys. Rev. Lett.* **84** (2000), 2243.

Hyperfine Interactions **144/145:** 65–76, 2002.

Atomic Vibrational Dynamics of Thin Films Studied by Nuclear Resonant Inelastic X-Ray Scattering: Amorphous $Tb_{1-x}Fe_x$ Alloys

T. RUCKERT[1], W. KEUNE[1,*], B. SAHOO[1], W. STURHAHN[2],
T. S. TOELLNER[2], E. E. ALP[2] and R. RÖHLSBERGER[3]
[1] *Institut für Physik, Gerhard-Mercator-Universität Duisburg, D-47048 Duisburg, Germany; e-mail: keune@uni-duisburg.de*
[2] *Advanced Photon Source, Argonne National Laboratory, Argonne, IL 60439, USA*
[3] *Fachbereich Physik, Universität Rostock, D-18055 Rostock, Germany*

Abstract. Nuclear resonant inelastic X-ray scattering (NRIXS) of 14.4125 keV synchrotron radiation was used to measure directly the partial vibrational density of states (VDOS), $g(E)$, of vapor-quenched amorphous $Tb_{1-x}{}^{57}Fe_x$ thin film alloys ($0.20 \leqslant x \leqslant 0.82$). The structure of the samples was characterized by X-ray diffraction and Mössbauer spectroscopy. The behavior of the Lamb–Mössbauer factor obtained from measured vibrational excitation probability densities indicates vibrational softening with increasing Tb content. A plot of $g(E)/E^2$ versus E proves the existence of non-Debye-like vibrational excitations below about 10 meV, becoming more pronounced with rising Tb content.

Key words: nuclear resonance absorption, inelastic X-ray scattering, Mössbauer effect, amorphous Tb–Fe thin films, phonon density of states.

1. Introduction

Since the discovery of nuclear resonant inelastic X-ray scattering (NRIXS) of 14.4125 keV synchrotron radiation by ^{57}Fe nuclei [1–3], this technique has evolved as a powerful method to explore atomic vibrational properties of condensed matter not available otherwise [4, 5]. NRIXS provides a direct measurement of the frequency distribution of atomic vibrations, i.e., the vibrational density of states (VDOS), $g(E)$, for the vibrating resonant isotope in matter. In other words, it is the resonant-isotope projected "partial" VDOS that is directly measured by NRIXS. $g(E)$ is a key quantity from which important thermodynamic properties may be deduced [5]. Recently, NRIXS has opened a new domain, namely lattice dynamics, in thin-film, interface and multilayer research [6–10].

In the present work NRIXS was employed to measure directly the Fe-projected VDOS of vapor-quenched amorphous $Tb_{1-x}{}^{57}Fe_x$ (a-$Tb_{1-x}{}^{57}Fe_x$) thin film alloys. Amorphous systems are of considerable interest because they show an anomalous

* Author for correspondence.

behavior in the low-energy part of the vibrational spectrum [11]. In particular, there is a debate about the so-called "boson peak" [12–15], which appears in $g(E)$ as an excess contribution with respect to the usual Debye behavior [$g(E) \propto E^2$]. An enhancement of the VDOS at low exitation energies as compared to the Debye law is evident in earlier results obtained by coherent inelastic neutron scattering on bulk samples of binary metallic glasses, such as a-$Mg_{70}Zn_{30}$ [15]. In the present work, $g(E)$ was obtained over a wide concentration range, because vapor-quenched a-$Tb_{1-x}Fe_x$ alloy thin films with thicknesses between 175–800 Å were found to be metastable at room temperature (RT) in the regime $0.20 \leqslant x \leqslant 0.82$.

2. Experimental details

The experiments were performed at the undulator beam line 3-ID of SRI-CAT at the Advanced Photon Source. The bandwidth of the highly collimated beam of X-rays from the undulator is brought down to a value of the order of eV by a high-heat-load monochromator that uses two symmetric diamond (111)-Bragg reflections in a nondispersive arrangement. Further monochromatization down to 2.3 meV bandwidth (FWHM) is achieved by a high-resolution, nested Si crystal monochromator [16, 17]. The monochromatized synchrotron beam illuminated the ^{57}Fe-containing thin films under a grazing angle of $\approx$4 mrad. The energy was tuned in steps of 0.4 meV around the 14.4125 keV nuclear resonance of ^{57}Fe. An avalanche photodiode was mounted above the thin film and detected the time-delayed characteristic 6.4 keV fluorescence X-rays of iron emitted after the nuclear de-excitation. The measurements were performed with the samples usually at room temperature, and in few cases at 10 K. The collection times were 2–10 h per spectrum, depending on film thickness and Fe content. The NRIXS data evaluation and extraction of the VDOS was performed by using the computer program PHOENIX described elsewhere [18].

All a-$Tb_{1-x}{}^{57}Fe_x$ alloy thin films were prepared in ultrahigh vacuum (base pressure $\sim 5 \cdot 10^{-10}$ mbar) by thermal co-evaporation of Tb metal (99.99 at.% purity) and metallic ^{57}Fe (99.95 at.% purity, 95.5% enriched in ^{57}Fe) from two independently computer-controlled Knudsen cells [19]. The pressure during evaporation was $<1 \cdot 10^{-9}$ mbar. The deposition rates of Tb and ^{57}Fe were measured independently by two calibrated quartz-crystal oscillators and controlled by a personal computer. The desired alloy composition was achieved by selecting a constant ratio of the Tb and ^{57}Fe deposition rates, while maintaining the individual deposition rates constant. Depending on the alloy composition the deposition rates ranged between 0.01–0.70 Å/s. All alloy compositions, x, given here are the nominal compositions as determined from the ratio of the two deposition rates. The error in x is estimated to be ± 0.01. The alloy film thickness, t, was usually 175 Å for the Fe-rich films, and was increased up to 800 Å for the Fe-poor samples in order to optimize the NRIXS measurement time. The amorphous films were grown at 300 K on oxidized Si(001) wafers that usually were initially coated by a 200 Å thick

Table I. Survey of the different a-$Tb_{1-x}{}^{57}Fe_x$ alloy thin films and their thickness

Composition x	Film thickness (Å)
0.20	720
0.25	547
0.30	800
0.30	175
0.33	389
0.35	361
0.37	336
0.44	266
0.50	222
0.60	175
0.67	175
0.77	175
0.82	800
0.82	175

polycrystalline Pt buffer layer. Reflection of the synchrotron beam at the Pt layer in grazing incidence enhances the 6.4 keV fluorescence signal of inelastic nuclear resonance absorption in 175 Å thick a-$Tb_{1-x}{}^{57}Fe_x$ overlayers due to a waveguide effect (GIAR effect) [20]. All samples were coated with a 50–100 Å thick Si layer for protection. A survey of all a-$Tb_{1-x}{}^{57}Fe_x$ samples and their thickness is given in Table I.

The structure of the samples was characterized by (θ–2θ) X-ray diffraction (XRD) and ^{57}Fe conversion electron Mössbauer spectroscopy (CEMS). Cu-K_α-radiation and a graphite monochromator were used for XRD. CEM spectra were taken at room temperature by using a conventional He–CH_4 proportional counter with the film surface perpendicular to the incident 14.4 keV γ-ray of the ^{57}Co (Rh matrix) source. The CEM spectra were least-squares fitted by using the NORMOS computer program by Brand [21]. Isomer shift (δ) values are given relative to a bcc (α-) Fe absorber at room temperature.

3. Results and discussion

3.1. STRUCTURAL INVESTIGATIONS

The amorphous structure of the quench-condensed $Tb_{1-x}{}^{57}Fe_x$ thin films was confirmed by XRD. Some typical XRD diagrams are shown in Figure 1, where only Bragg peaks of the 200-Å thick polycrystalline Pt buffer layer and a weak indication of the (suppressed) Si(004) reflection of the Si(001) substrate at $2\theta = 69.2°$

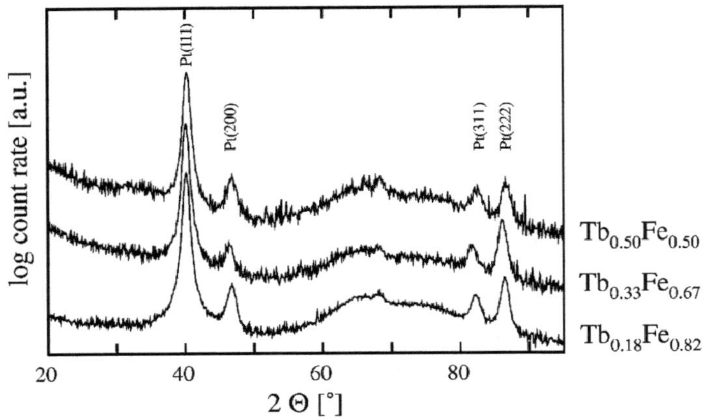

Figure 1. X-ray diffraction (θ–2θ) diagrams (Cu-K_α-radiation) of a-$Tb_{1-x}{}^{57}Fe_x$ films on Pt/Si(001) substrate for $x = 0.50$ (222 Å thick), 0.67 (175 Å thick) and 0.82 (175 Å thick), respectively. (In order to suppress the Si(004) reflection of the Si(001) substrate at $2\theta = 69.2°$ the film surface was tilted by 5° relative to the X-ray beam.)

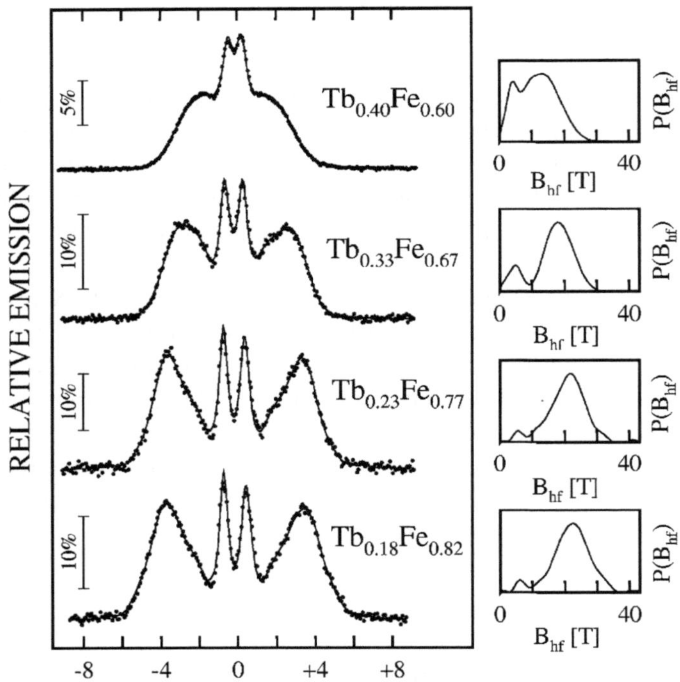

Figure 2. Room-temperature CEM spectra of 175-Å thick a-$Tb_{1-x}{}^{57}Fe_x$ films for $x = 0.60$, 0.67, 0.77 and 0.82 (from top to bottom, respectively). Right-hand side: corresponding hyperfine-field distributions $P(B_{\rm hf})$.

are observed. The absence of Bragg reflections from hcp-Tb or intermetallic Tb–Fe compounds (e.g., such as cubic $TbFe_2$) near $2\theta = 30$–35°, and from bcc Fe near $2\theta = 44.8°$, is evident. Similar XRD diagrams as in Figure 1 were observed at lower Fe concentrations down to $x = 0.20$ (not shown). The absence of sharp Bragg reflections prove the amorphous structure of the $Tb_{1-x}{}^{57}Fe_x$ thin films.

Typical CEM spectra of a-$Tb_{1-x}{}^{57}Fe_x$ films are shown in Figures 2 and 3. The spectra for $0.60 \leqslant x \leqslant 0.82$ in Figure 2 are similar to those reported earlier [19, 22–24] and confirm the amorphous structure. They have been analyzed by least-squares fitting in terms of a broad distribution of magnetic hyperfine fields $P(B_{hf})$, assuming a weak linear correlation between hyperfine field B_{hf} and isomer shift δ. Films with $x \geqslant 0.60$ show perpendicular magnetic anisotropy. From the line intensity ratio $3 : R_{23} : 1 : 1 : R_{23} : 3$ of the basic Zeeman sextets in the $P(B_{hf})$ distribution one can calculate the average tilting angle $\langle\theta\rangle$ between the film normal and the Fe spin direction (or B_{hf} direction) [19]. A least-squares fit of the spectra and the relative line intensities results in nearly zero average quadrupole line shift and in $\langle\theta\rangle$ values ranging between 25–29°. Accordingly, the average Fe spin direction is preferentially perpendicular to the film plane, in agreement with earlier reports [19, 22–24]. The measured average isomer shift values are given in Figure 4.

Contrary to the magnetic films with $x \geqslant 0.60$, a-$Tb_{1-x}{}^{57}Fe_x$ films with $x \leqslant 0.50$ are paramagnetic at RT, as is demonstrated by the quadrupole-split spectra in Figure 3. The apparent linewidth of these doublets is very large due to a distribution of electric quadrupole splittings, $P(\Delta E_Q)$. This is a typical feature of nonmagnetic amorphous binary Fe alloys [25–27]. Good fits to the slightly asymmetric doublet spectra were obtained via a distribution $P(\Delta E_Q)$, and assuming a linear correlation between quadrupole splitting ΔE_Q and isomer shift. The concentration dependence of the average isomer shift $\langle\delta\rangle$ and average quadrupole splitting $\langle\Delta E_Q\rangle$ is shown in Figure 4. The observed $\langle\Delta E_Q\rangle$ values are not too different from those found in amorphous Zr-rich Zr–Fe alloys [25, 26] and in a-Sc_3Fe [27].

3.2. NUCLEAR RESONANCE INELASTIC X-RAY SCATTERING (NRIXS) STUDY

Figure 5 exhibits typical NRIXS spectra (excitation probability versus energy transfer E) measured on a-$Tb_{1-x}{}^{57}Fe_x$ films at RT for different alloy compositions x. The spectra show a dominant central elastic peak at the nuclear transition energy E_o (energy transfer $E = 0$ meV) and side bands at lower and higher energy. X-rays with less energy than E_o excite the nuclear resonance by net annihilation of vibrational quanta (low-energy side band). The high-energy side band is produced by creation of vibrational quanta. The observed asymmetry in the spectra reflects the "detailed balance" [28] due to the Boltzmann factor. The vibrational side bands in Figure 5 have no sharp peaks and little structure, which is typical for the amorphous state [7]. The instrumental resolution function with a full width at half maximum

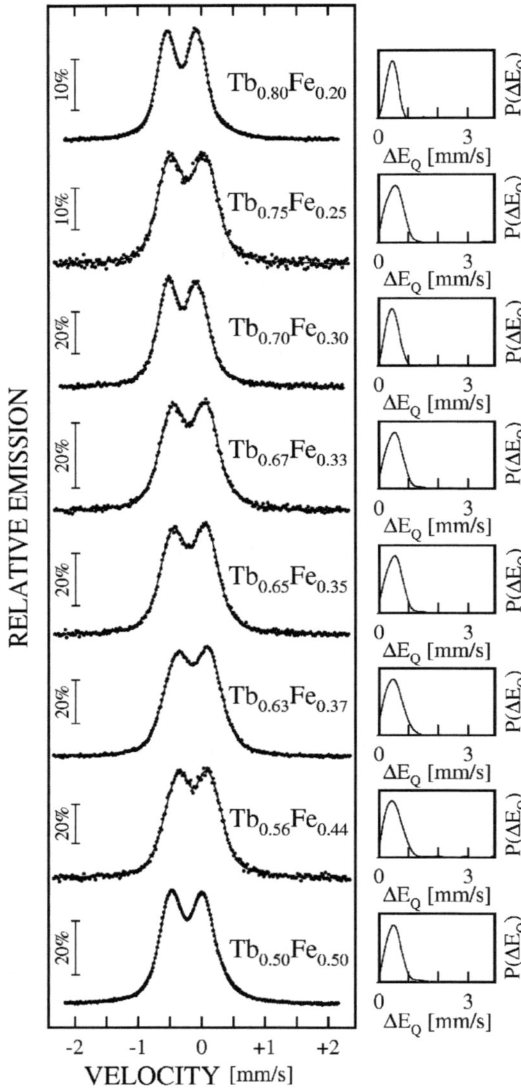

Figure 3. Room-temperature CEM spectra of a-$Tb_{1-x}{}^{57}Fe_x$ films for $x = 0.20, 0.25, 0.30$ (800 Å thick), 0.33, 0.35, 037, 0.44 and 0.50 (from top to bottom, respectively). Right-hand side: corresponding distributions $P(\Delta E_Q)$ of quadrupole splittings.

(FWHM) of 2.3 meV is also depicted in Figure 5. The resolution function is nearly symmetrical and falls off rapidly from the center without having extended wings.

After subtraction of the central elastic peak from the measured NRIXS spectra the data were normalized according to standard procedures [2, 5, 18], yielding the vibrational excitation probability per unit energy, $W(E)$. By integration of $W(E)$ over all energies the quantity $1 - f$ is obtained in a model independent way [18],

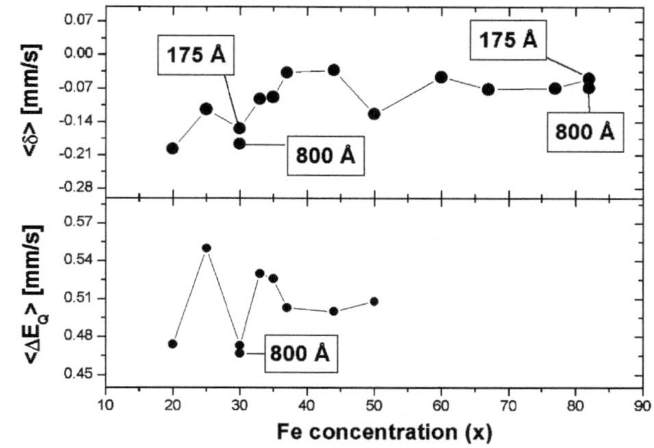

Figure 4. Average isomer shift $\langle\delta\rangle$ (relativ to α-Fe at 300 K) (*top*) and average quadrupole splitting $\langle\Delta E_Q\rangle$ (*bottom*) of a-$Tb_{1-x}{}^{57}Fe_x$ alloy thin films at room temperature versus Fe content x.

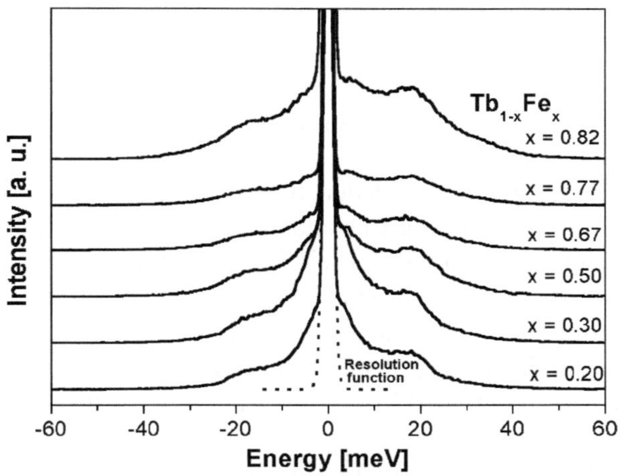

Figure 5. Intensity versus energy transfer obtained at 300 K from a-$Tb_{1-x}{}^{57}Fe_x$ films with $x = 0.20$ (720 Å thick), 0.30 (800 Å thick), 0.50 (222 Å thick), 0.67 (175 Å thick), 0.77 (175 Å thick) and 0.82 (800 Å thick) (from bottom to top, respectively). Also shown is the instrumental resolution function with FWHM = 2.3 meV (dashed line). (For clarity the spectra are vertically displaced.)

where f is the Lamb–Mössbauer factor (or f-factor). Figure 6 shows the concentration dependence of the f-factor of the a-$Tb_{1-x}{}^{57}Fe_x$ alloy films. A tendency for a decrease of f with decreasing Fe content is evident. At the highest Fe concentration ($x = 0.82$) f is found to be independent of the film thickness. In the regime of the lowest Fe concentrations fluctuations of f values appear, possibly due to different film thicknesses and/or possibly related to the fact that the thermodynamic phase diagram for crystalline Tb–Fe alloys shows a eutectic point in that regime [29], and small deviations of x might strongly affect also the properties of the amorphous state there. Compared with the amorphous $Tb_{0.33}{}^{57}Fe_{0.67}$ alloy, the f-factor of

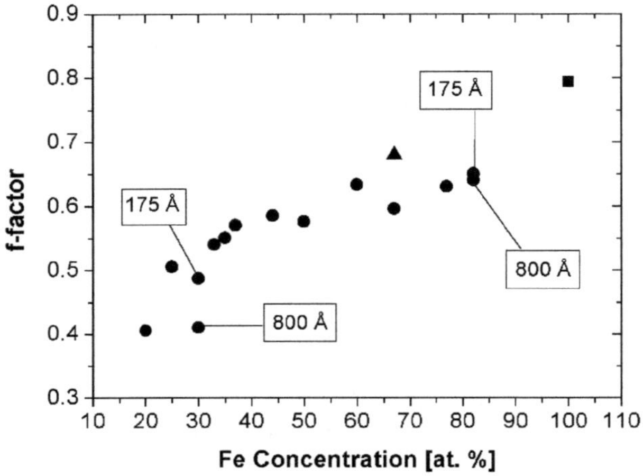

Figure 6. Lamb–Mössbauer factor (f-factor) at 300 K as a function of Fe concentration for a-$Tb_{1-x}{}^{57}Fe_x$ alloy films (full circles). For comparison: f-factor for bulk bcc Fe at 300 K (full square) and for cubic $Tb^{57}Fe_2(110)$ film (800 Å) at 300 K (full triangle).

a crystalline $Tb^{57}Fe_2(110)$ epitaxial film of the same composition is found to be somewhat larger (Figure 6).

The partial VDOS ($g(E)$) was deduced from the measured vibrational excitation probability density according to standard procedures [2, 5, 18]. Figure 7 exhibits $g(E)$ of various amorphous $Tb_{1-x}{}^{57}Fe_x$ films of different composition. The VDOS represents a structureless broad feature, as anticipated for such structurally and atomically disordered materials, where van-Hove singularities are smeared out. Qualitatively similar energy distributions (more exactly: generalized energy distributions) have been obtained by coherent inelastic neutron scattering on bulk samples of binary metallic glasses, e.g., a-$Mg_{70}Zn_{30}$ [15]. There is a tendency that the maximum energy of the VDOS in Figure 7 shifts to lower energies with decreasing Fe content x of the amorphous films, i.e. the Fe vibrations soften with decreasing x.

The low-energy modes in $g(E)$ below ~10 meV in Figure 7 are of particular interest because of the boson-peak phenomenon [12–14]. The usual way to observe a deviation from Debye-like behavior is to plot the reduced VDOS, $g(E)/E^2$, versus E. In this representation strict Debye-like behavior is reflected by a horizontal line which intersects the $g(E)/E^2$ axis at a value that is proportional to c_S^{-3} (c_S = sound velocity) [5]. In Figure 8 we have plotted $g(E)/E^2$ versus E for some representative examples: for two a-$Tb_{1-x}{}^{57}Fe_x$ films and, for comparison, for two crystalline systems, i.e. a 800-Å thick cubic $Tb^{57}Fe_2(110)$ epitaxial film and bulk bcc Fe. In Figure 8, a striking and unambiguous rise in $g(E)/E^2$ up to about 2.5 $(\text{at.vol.})^{-1}(\text{eV})^{-1}(\text{meV})^{-2}$ below ~10 meV is observed for the Tb-rich a-$Tb_{70}{}^{57}Fe_{30}$ film. Further, a peak near 3 meV seems to exist for this sample; however, because of the limited energy resolution of 2.3 meV (FWHM) it is not

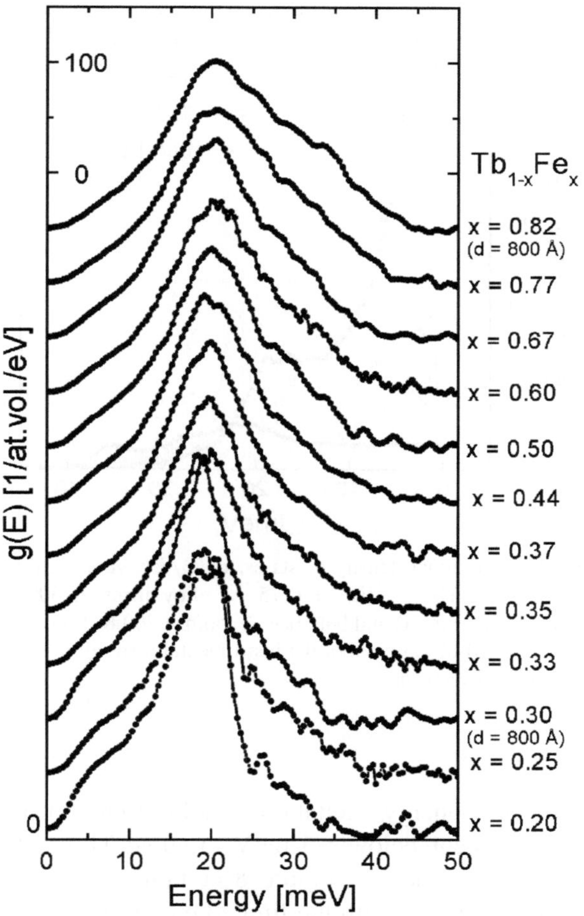

Figure 7. Partial vibrational density of states (VDOS), $g(E)$, of a-$Tb_{1-x}{}^{57}Fe_x$ thin film alloys at 300 K for different composition x. Energy resolution: 2.3 meV. (For clarity neighboring curves are vertically displaced by 50 $(\text{at.vol.})^{-1}(\text{eV})^{-1}$.)

clear at present whether this peak is real or an artifact of the NRIXS raw data evaluation, which includes subtraction of the elastic peak. (At 3.5 meV there is an about 10% contribution of the wing of the elastic peak to the measured NRIXS intensity; at 4 meV this contribution is practically zero.) For the less Tb containing a-$Tb_{0.33}{}^{57}Fe_{0.67}$ film the rise in $g(E)/E^2$ below ~10 meV is clearly much weaker [only up to ~0.9 $(\text{at.vol.})^{-1}(\text{eV})^{-1})(\text{meV})^{-2}$] than for the Tb-rich sample. Again, one can argue whether the peak observed near 3 meV is real or an artifact.

By contrast, the crystalline $Tb^{57}Fe_2$ film shows a much smaller value of ~0.5 $(\text{at.vol.})^{-1}(\text{eV})^{-1}(\text{meV})^{-2}$ near ~3 meV than the amorphous films, Figure 8. (The peaks at ~22.5 and ~16 meV, and the dip near 10 meV, result from extrema in the VDOS due to optical and acoustic phonons of the cubic Laves phase [30],

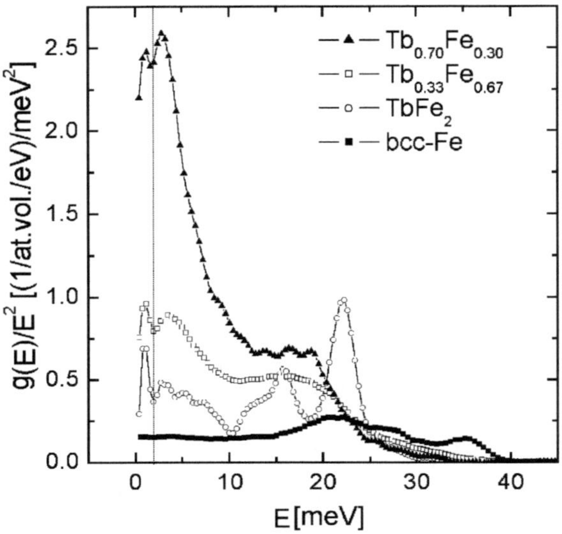

Figure 8. Reduced partial vibrational density of states, $g(E)/E^2$, versus energy E, of amorphous $Tb_{0.70}{}^{57}Fe_{0.30}$ (800 Å) and a-$Tb_{0.33}{}^{57}Fe_{0.67}$ (175 Å) alloy films at 300 K, and of epitaxial $Tb^{57}Fe_2(110)$ Laves phase film (800 Å) and bulk bcc Fe, both at 300 K, for comparison. The vertical dotted line indicates the FWHM of 2.3 meV of the instrumental resolution function (i.e. data points below 2.3 meV are physically irrelevant).

and are *not* related to low-energy excitations leading to a "boson peak".) Debye-like behavior seems to occur below ~5 meV in cubic $TbFe_2$. (The peaks in all $g(E)/E^2$ curves below 2.5 meV in Figure 8 certainly are artifacts of the peak subtraction technique combined with the limited energy resolution.) For bulk bcc Fe, Debye-like behavior is observed in $g(E)/E^2$ below ~15 meV (Figure 8), with $g(E)/E^2 \approx 0.13$ (at.vol.)$^{-1}$(eV)$^{-1}$(meV)$^{-2}$.

Although we cannot claim that we have observed the full "boson peak", Figure 8 clearly demonstrates the existence of non-Debye-like vibrational excitations below about 10 meV in quench-condensed a-$Tb_{1-x}{}^{57}Fe_x$ alloy films. The probability of excitation of these anomalous atomic vibrations increases with decreasing excitation energy and rising Tb content. They are not observed in crystalline systems such as cubic $TbFe_2$ and bcc Fe.

The microscopic origin of the observed anomalous low-energy excitations is unclear at present. For metallic glasses, either movements of atoms on the surface of very small voids or movements in lower density "defect" regions of the metallic glass have been suggested [15] as an explanation for low-energy excitations in these systems. We speculate that the larger atomic diameter of Tb (3.52 Å) as compared to that of Fe (2.48 Å) might be an important factor in the a-$Tb_{1-x}Fe_x$ system. In the structurally disordered network of large Tb atoms in Tb-rich amorphous alloys voids on an atomic scale might exist, which, when occupied by Fe atoms, provide enough space for the smaller Fe atoms to vibrate in shallow potentials with low frequencies. Such voids do not exist in crystalline systems.

4. Summary

The Fe-projected (partial) vibrational density of states (VDOS), $g(E)$, in vapor-quenched amorphous $Tb_{1-x}{}^{57}Fe_x$ thin film alloys ($0.20 \leqslant x \leqslant 0.82$) was measured by nuclear resonant inelastic X-ray scattering (NRIXS) of 14.4125 keV synchrotron radiation with 2.3 meV energy resolution. The amorphous structure of the films was confirmed by X-ray diffraction and ^{57}Fe conversion electron Mössbauer spectroscopy (CEMS) at 300 K. For $x \leqslant 0.60$ the magnetically-split spectra are characterized by a distribution of magnetic hyperfine fields, $P(B_{hf})$, and an average Fe spin direction perpendicular to the film surface. For $x \leqslant 0.50$ the films are paramagnetic and exhibit quadrupole-split doublets that indicate a distribution of electric quadrupole splittings, $P(\Delta E_Q)$, or of electric field gradients. With increasing Tb content, the observed decrease of the Lamb–Mössbauer factor (obtained from the normalized NRIXS spectra) and the overall behavior of the VDOS indicate the tendency to vibrational softening. A plot of $g(E)/E^2$ versus E proves the existence of non-Debye-like vibrational excitations below about 10 meV in the amorphous films that do not exist in crystalline systems. The probability of excitation of these anomalous atomic vibration increases with rising Tb content.

Our investigations demonstrate that the NRIXS technique has evolved into a powerful and presently unique technique for the investigation of lattice dynamics in thin films.

Acknowledgements

We are very grateful to U. v. Hörsten (Duisburg) for the expertise preparation of samples. Work at Duisburg was supported by Deutsche Forschungsgemeinschaft (Ke 273/17-1). Use of the Advanced Photon Source was supported by the US Department of Energy, Basic Energy Sciences, Office of Science, under Contract Nr. W-31-109-Eng-38.

References

1. Seto, M., Yoda, Y., Kikuta, S., Zhang, X. W. and Ando, A., *Phys. Rev. Lett.* **74** (1995), 3828.
2. Sturhahn, W., Toellner, T. S., Alp, E. E., Zhang, X. W., Ando, M., Yoda, Y., Kikuta, S., Seto, M., Kimball, C. W. and Dabrowski, B., *Phys. Rev. Lett.* **74** (1995), 3832.
3. Chumakov, A. I., Rüffer, R., Grünsteudel, H., Grünsteudel, H. F., Grübel, G., Metge, J., Leupold, O. and Goodwin, H. A., *Europhys. Lett.* **30** (1995), 427.
4. Chumakov, A. I. and Sturhahn, W., *Hyp. Interact.* **123/124** (1999), 781.
5. Alp, E. E., Sturhahn, W. and Toellner, T. S., *J. Phys.: Condens. Matter* **13** (2001), 7645.
6. Sturhahn, W., Röhlsberger, R., Alp, E. E., Ruckert, T., Schrör, H. and Keune, W., *J. Magn. Magn. Mater.* **198–199** (1999), 590.
7. Keune, W. and Sturhahn, W., *Hyp. Interact.* **123/124** (1999), 847.

8. Ruckert, T., Keune, W., Sturhahn, W., Hu, M. Y., Sutter, J. P., Toellner, T. S. and Alp, E. E., *Hyp. Interact.* **126** (2000), 363.
9. Roldan Cuenya, B., Keune, W., Sturhahn, W., Toellner, T. S. and Hu, M. Y., *Phys. Rev. B* **64** (2001), 235321.
10. Ruckert, T., Keune, W., Sturhahn, W. and Alp, E. E., *J. Magn. Magn. Mater.* **240** (2002), 562.
11. For a review see Phillips, W. A. (ed), *Amorphous Solids – Low Temperature Properties*, Springer, Berlin, 1981.
12. See the article by J. Jäckle, in [11].
13. Schirmacher, W., Diezemann, G. and Ganter, C., *Phys. Rev. Lett.* **81** (1998), 136, and references therein.
14. Schirmacher, W., Diezemann, G. and Ganter, C., *Physica B* **284–288** (2000), 1147.
15. Suck, H.-B. and Rudin, H., In: H. Beck and H.-J. Güntherodt (eds), *Glassy Metals II, Topics in Applied Physics*, Vol. 53, Springer, Berlin, 1983, p. 217.
16. Toellner, T. S., PhD Dissertation, Northwestern University, Evanston, IL, USA, 1996.
17. Toellner, T. S., *Hyp. Interact.* **125** (2000), 3.
18. Sturhahn, W., *Hyp. Interact.* **125** (2000), 149.
19. Ruckert, T., Tappert, J., Brand, R. A. and Keune, W., *J. Magn. Magn. Mater.* **165** (1997), 411.
20. Röhlsberger, R., Sturhahn, W., Toellner, T. S., Quast, K. W., Hession, P., Hu, M., Sutter, J. and Alp, E. E., *J. Appl. Phys.* **86** (1999), 584.
21. Brand, R. A., *Nucl. Instrum. Methods B* **28** (1987), 398.
22. Heiman, N. and Lee, K., *Phys. Lett. A* **55** (1975), 297.
23. Heiman, N., Lee, K. and Potter, R. J., *J. Appl. Phys.* **47** (1976), 2634; *AIP Conf. Proc.* **29** (1976), 130.
24. Eymery, J. P., Fnidiki, A., Krishnan, R., Tessier, M. and Vitton, J. P., *Phys. Rev. B* **38** (1988), 11931.
25. Unruh, K. M. and Chien, C. L., *Phys. Rev. B* **30** (1984), 4968.
26. Ghafari, M., Gonser, U., Wagner, H. G. and Naka, M., *Nucl. Instrum. Methods* **199** (1982), 197.
27. Ghafari, M., Brand, R. A. and Keune, W., *Hyp. Interact.* **42** (1988), 931.
28. Sturhahn, W. and Kohn, V. G., *Hyp. Interact.* **123/124** (1999), 367.
29. Moffatt, W. G., *The Handbook of Binary Phase Diagrams*, Vol. 3, Genium Publ. Co., Schenectady, New York, 1984.
30. Kress, W., *Phonon Dispersion Curves, One-Phonon Density of States and Impurity Vibrations of Metallic Systems*, Physics Data Vol. 26-1, Fachinformationszentrum Karlsruhe, Karlsruhe, Germany, 1987.

Hyperfine Interactions **144/145:** 77–83, 2002.

Phonons in Confinement and the Boson Peak Using Nuclear Inelastic Absorption

T. ASTHALTER[1], M. BAUER[1], U. VAN BÜRCK[2], I. SERGUEEV[3], H. FRANZ[4] and A. I. CHUMAKOV[5]
[1] *Institut für Physikalische Chemie, Universität Stuttgart, Pfaffenwaldring 55, D-70569 Stuttgart, Germany*
[2] *E13 Experimentalphysik, TU München, James-Franck-Str. 1, D-85747 Garching, Germany*
[3] *E13 Experimentalphysik, TU München, James-Franck-Str. 1, D-85747 Garching, Germany and European Synchrotron Radiation Facility, B.P. 220, F-38043 Grenoble, France*
[4] *Hamburger Synchrotronstrahlungslabor, Deutsches Elektronen Synchrotron, Notkestr. 85, D-22603 Hamburg, Germany*
[5] *European Synchrotron Radiation Facility, B.P. 220, F-38043 Grenoble, France*

Abstract. We have applied nuclear inelastic absorption (NIA) to the molecular glass former dibutylphthalate/ferrocene, both in bulk and in nanoporous matrices having pore sizes of 50 and 25 Å, respectively. The quantity $g(E)/E^2$, where $g(E)$ is the vibrational phonon density of states (VDOS) of the resonant nuclei, exhibits a pronounced maximum at low energies. Confinement in pores leads to a suppression of the VDOS below 1.5 meV, independent of the pore size. Also in the scaled heat capacity $C(T)/T^3$, we observe a decrease of the peak maximum for low temperatures. Our observations are discussed in the light of experimental and theoretical results on nanocrystals and a recent theoretical model for the boson peak.

Key words: Boson peak, inelastic scattering, confinement.

1. Introduction

The low-frequency vibrational properties of glasses and other disordered systems are a topic of long-term interest in condensed matter physics [1]. Especially the microscopic origin of the Boson peak, an excess contribution to the vibrational density of states (VDOS) $g(E)$ that appears as a hump both in the ratio $g(E)/E^2$ and in the scaled specific heat $C(T)/T^3$, is a recent matter of debate. Among the numerous models that aim at an explanation of this phenomenon, we wish to mention the soft-potential model [2, 3] and a model of coupled harmonic oscillators [4]. Whereas the former one assumes the existence of soft localized modes, the latter starts from an ensemble of coupled harmonic oscillators having a distribution of elastic constants induced primarily by disorder, which may, however, be accompanied by anharmonic effects, as discussed in a recent generalized version [5].

Whereas it seems difficult to prove directly the validity of these various models, it might be a rewarding task to attempt an experimental study of the influence of geometrical confinement on the Boson peak for two reasons. First, one might expect to see a cut-off of the VDOS at very low energies/long wavelengths owing to the geometrical confinement. Second, the spectral position of the Boson peak itself may be shifted. Both a hardening originating from the interaction with a matrix having stiff walls and a softening owing to the increase of local free volume around molecules close to the pore wall are possible. Hence, conclusions about the microscopic nature of the softest lattice vibrations in disordered systems may be drawn.

In the present study, we examined the glass former dibutylphtalate/ferrocene at different temperatures under geometrical confinement in porous silica matrices of 25 and 50 Å pore size as well as in the bulk, using nuclear inelastic absorption (NIA) [6–8]. This method is a powerful tool for element-selective studies of the vibrational dynamics of crystalline solids and glasses, giving direct access to the partial VDOS of the resonant atoms. Unlike in inelastic neutron scattering, the interaction process is completely incoherent so that the admixture of coherent cross-sections from any type of atom is exactly zero. Moreover, NIA provides background-free measurements because only the resonant nuclei contribute to the delayed signal. This is especially advantageous in our case, where we introduce the glass former into a matrix whose dynamics are not of immediate interest to us.

2. The experiment

The sample investigated was dibutylphthalate containing 5 mol-% ferrocene enriched to 95% in ^{57}Fe. The preparation of the enriched ferrocene and the loading of the nanoporous matrix – a pellet having 2 mm thickness and 5 mm diameter – are described in [9]. The experiment was carried out at the Nuclear Resonance Beamline ID18 [10] at the ESRF, using a three-bounce high-resolution monochromator (HRM), where the small energy bandwidth is achieved by using two highly asymmetrically cut Si (9 7 5) reflections and where a Ge(3 3 1) symmetrically cut crystal ensures an approximately horizontally outgoing beam. A new set of crystals providing an energy resolution of about 0.5 meV was used. Within this bandwidth, a high flux of $3 \cdot 10^8$ s^{-1} was achieved by collimating the beam in front of the HRM by a compound refractive lens [11]. The sample was mounted into a copper holder sealed with Kapton windows in a closed-cycle cryostat such that the beam entered in 5° grazing incidence from below. Owing to radiative heat transfer from the cryostat window to the sample (distance $\approx$3 mm), the set temperature of the cryostat is not the actual temperature of the sample. The true sample temperature was determined by detailed balance after subtraction of the resolution function from the raw spectra. The HRM was scanned within a range of $\pm$80 meV around the resonance energy in steps of 0.25 meV; for energies in the range $\pm$10 meV, a finer step width of 0.125 meV was used.

3. Results and discussion

The extraction of the one-phonon contribution from the raw NIA spectra and the evaluation of the phonon VDOS was carried out using DOS V2.1 [12]. The broadening of the instrumental function was partly eliminated from the data by deconvolution. Both for the temperature determination and for all further data evaluations, the subtraction of the elastic line from the raw spectra is a key issue, which influences strongly the quality of the data at lowest energies. Usually, the subtraction is done such that the inelastic spectrum has a smooth behaviour. We observed that it was possible to find an optimum percentage of subtraction by comparing the nth moments of the spectrum and the $(n-1)$-th moments of the VDOS, as described in [12].

Figure 1 shows the partial VDOS of the bulk sample at 85 K, where the prominent features are the intermolecular lattice modes ($E < 20$ meV) and the (intramolecular) optical modes of ferrocene at 22.0, 60.1 and 62.6 meV, respectively. The VDOS is normalized to one over the full energy range, thus yielding the probability density per vibrational mode rather than per unit volume. The difference between bulk and nanoporous samples becomes noticeable when inspecting the reduced VDOS $g(E)/E^2$, since low energies are emphasized by the $g(E)/E^2$ plot. This quantity is displayed in Figure 2. We note that for energies below 1.5 meV the VDOS is suppressed in the pores with respect to the bulk, whereas for higher energies the curves are almost identical. The spectral position of the peak in $g(E)/E^2$ remains unaltered.

Qualitatively, the suppression at low energies can be explained by a cut-off of long-wavelength acoustic modes owing to the geometrical confinement. Quantitatively, however, we would expect that the cut-off energy is two times lower in 50 Å pores than in 25 Å pores. In contrast to this simplifying assumption, the position of the cut-off seems to be independent of the pore size within the accuracy of our energy resolution of 0.5 meV. Our finding agrees with neutron time-of-flight results

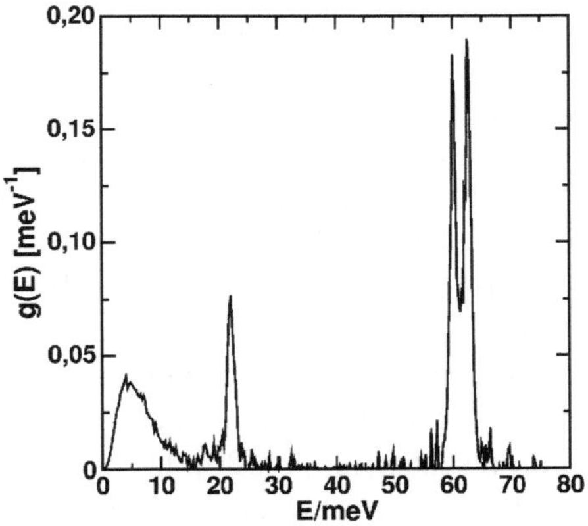

Figure 1. Partial VDOS for bulk ferrocene/dibutylphthalate at $T = 85$ K.

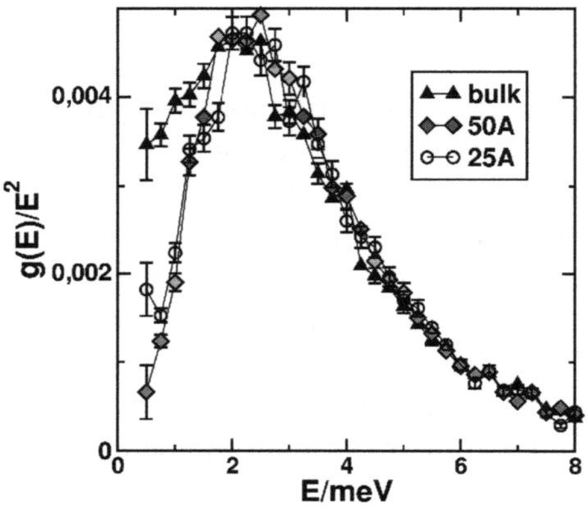

Figure 2. Reduced partial VDOS for ferrocene/dibutylphthalate at $T = 85$ K, bulk versus nanopores. The error bar at 0.5 meV takes the uncertainty of the subtraction of the elastic line into account.

obtained on salol in nanopores [13], where a decrease of $g(E)/E^2$ was observed as well, but the dependence on pore size was also smaller than expected.

In this context, considering the Lamb–Mössbauer factor, which is easily obtained from the VDOS, may be illustrating. In a simple cut-off model, an increase of f_{LM} of up to about 6% for 25 Å pores is expected for $T = 85$ K. We obtain $f_{\mathrm{LM}} = 0.59 \pm 0.02$ for the bulk, 0.58 ± 0.02 for 50 Å and 0.57 ± 0.08 for 25 Å, i.e. no change of the mean-square displacement of the iron atom within the experimental accuracy can be observed.

For the interpretation of our findings, it is important to note that our system consists of a relatively soft solid included in a much harder matrix. This is evident from a comparison of the Debye temperatures: $\Theta_D = 494$ K for vitreous SiO_2 [14] compared to $\Theta_D = 146.2$ K for bulk dibutylphthalate/ferrocene as determined from the VDOS according to

$$(k_B \Theta_D)^2 = \frac{3}{\int_0^\infty \frac{g(E)}{E^2}\,\mathrm{d}E.} \tag{1}$$

The vibrational properties of nanoparticles are the object of intense current research. Lamb's classical theory [15] of a vibrating homogeneous sphere predicts the appearance of discrete acoustic surface breathing and torsional modes, which have been found in numerous Raman scattering experiments (e.g., [16]). On embedding into a matrix these modes do not disappear, but are globally stiffened [17]. However, Lamb's theory does not describe precisely the shape of vibrational spectra. Using Green's function methods on Ag nanocrystals [18], an enhancement of low-frequency modes and a linear scaling of $g(E)$ with E for the surface atoms was predicted in confinement. Experimental evidence for this behaviour was obtained using inelastic neutron scattering [19, 20]. Using NIA on nanocrystalline Fe [21], an enhancement of the VDOS for low energies and a general broadening were observed. An important influence of these effects, however, could be attributed to oxidation effects and grain-boundary vibrational modes, in addition to the pure confinement effects, in a similar study [22]. Both NIA studies did not confirm the linear $g(E)$ scaling, but gave evidence for the conventional $g(E) \propto E^2$ behaviour for low energies.

Our system strongly differs from the investigated nanoparticles in two aspects. First, instead of spherical confinement we have three-dimensionally interconnected channels, which are difficult to treat theoretically. Second, since we observe selectively the iron atoms, which are not directly located close to the pore walls but in the centre of the ferrocene molecules, the observation that the confinement does not introduce any extra surface modes with respect to the bulk is not surprising. In this context, the slight decrease of low-energy modes is consistent with the expectations of [18] where the embedded central atoms have a smaller VDOS at low energies. Hence, our study probes sensitively the inner part of our glass in the pores and avoids oxidation problems, but gives no direct access to surface vibrational properties. Altogether, the most plausible conclusion from our experiment is a slight stiffening of the acoustic non-surface phonons owing to confinement, which depends less critically on pore size than surface acoustic phonons. The deviations from Debye behaviour that are characteristic for a glass, however, are not taken into account in these considerations.

Hence, we have finally made use of the relation

$$C_V(T) = 3k_B \int_0^\infty \frac{g(E)(\beta E)^2 e^{\beta E}}{(e^{\beta E} - 1)^2}\, dE, \tag{2}$$

where $\beta = 1/k_B T$, to obtain the vibrational contribution to the scaled isochoric heat capacity $C_V(T)/T^3$. For energies between 0 and 0.5 meV, the extrapolation $g(E)/E^2 = \text{const.}$, based on the experimental value for $E = 0.5$ meV, was used. The result both for bulk and porous samples is shown in Figure 3. As a consequence of the decrease of the number of very soft acoustic modes in confinement, which

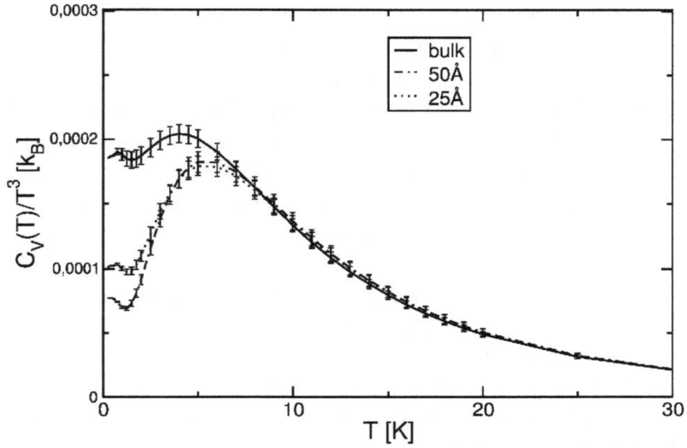

Figure 3. Scaled heat capacity (in units of k_B) for ferrocene/dibutylphthalate, bulk versus nanopores.

dominate the contribution to $C_V(T)$, an overall smaller value for the porous samples is obtained, especially for low temperatures. We have to point out, however, that our calculation relies on the assumption that the shape of the phonon VDOS is essentially temperature-independent below 85 K, which need not be strictly fulfilled in glasses. Results on bulk ferrocene/dibutylphthalate using quasielastic nuclear forward scattering [23], however, indicate a quasi-harmonic behaviour of the Lamb–Mössbauer factor with T, which supports our assumption.

Our finding may be compared to observations [24, 25] where the $C_V(T)/T^3$ maximum characteristic for a glass appeared also in the corresponding crystalline phase, but at higher temperatures and with much lower intensities.

We may also correlate our result to a recent theoretical model [5]. Here, the degree of disorder (represented by a disorder parameter, the width of the distribution of force constants of an ensemble of slightly anharmonic oscillators) correlates with the height and the position of the $C_V(T)/T^3$ maximum. In particular, an increase of the disorder parameter leads to an enhanced maximum and to a shift towards lower temperatures. If and only if we have a simple disappearance of soft modes in confinement, not compensated by any appearance of harder modes, then we would have a more narrow force constant distribution, which would then be consistent with the scenario described in [5].

Acknowledgements

The financial support of the BMBF under contracts 05 SK8 WOB/5 and 05 KS1 WOC/2 is gratefully acknowledged.

References

1. Phillips, W. A. (ed), *Amorphous Solids: Low-Temperature Properties*, Springer, Berlin, 1981.
2. Karpov, V. G., Klinger, M. I. and Ignatiev, F. N., *Sov. Phys. JETP* **57** (1983), 439.
3. Buchenau, U., Galperin, Yu. M., Gurevich, V. L., Parshin, D. A., Ramos, M. A. and Schober, H. R., *Phys. Rev. B* **46** (1992), 2798.
4. Schirmacher, W., Diezemann, G. and Ganter, C., *Phys. Rev. Lett.* **81** (1998), 136.
5. Schirmacher, W., Pöhlmann, M. and Maurer, E., *Phys. Stat. Sol. B* **230** (2002), 31.
6. Seto, M., Yoda, Y., Kikuta, S., Zhang, X. W. and Ando, M., *Phys. Rev. Lett.* **74** (1995), 3828.
7. Sturhahn, W., Toellner, T. S., Alp, E. E., Zhang, X., Ando, M., Yoda, Y., Kikuta, S., Seto, M., Kimball, C. W. and Dabrowski, B., *Phys. Rev. Lett.* **74** (1995), 3832.
8. Chumakov, A. I. and Sturhahn, W., *Hyp. Interact.* **123/124** (1999), 781.
9. Asthalter, T., Sergueev, I., Franz, H., Rüffer, R., Petry, W., Messel, K., Härter, P. and Huwe, A., *Eur. Phys. J. B* **22** (2001), 301.
10. Rüffer, R. and Chumakov, A. I., *Hyp. Interact.* **97/98** (1996), 589.
11. Chumakov, A. I., Rüffer, R., Leupold, O., Barla, A., Thiess, H., Asthalter, T., Doyle, B. P., Snigirev, A. and Baron, A. Q. R., *Appl. Phys. Lett.* **77** (2000), 31.
12. Kohn, V. G. and Chumakov, A. I., *Hyp. Interact.* **125** (2000), 205.
13. Zorn, R., Richter, D., Hartmann, L., Kremer, F. and Frick, B., *J. Physique IV* **10** (2000), Pr7.
14. Stephens, R. B., *Phys. Rev. B* **13** (1976), 852.
15. Lamb, H., *Proc. London Math. Soc.* **13** (1882), 189.

16. Fujii, M., Nagareda, T., Hayashi, S. and Yamamoto, K., *Phys. Rev. B* **44** (1991), 6243.
17. Ovsyuk, N. N. and Novikov, V. N., *Phys. Rev. B* **53** (1996), 3113.
18. Kara, A. and Rahman, T. S., *Phys. Rev. Lett.* **81** (1998), 1453.
19. Frase, H., Fultz, B. and Robertson, J. L., *Phys. Rev. B* **57** (1998), 898.
20. Stuhr, U., Wipf, H., Andersen, K. H. and Hahn, H., *Phys. Rev. Lett.* **81** (1998), 1449.
21. Fultz, B., Ahn, C. C., Alp, E. E., Sturhahn, W. and Toellner, T. S., *Phys. Rev. Lett.* **79** (1997), 937.
22. Pasquini, L., Barla, A., Chumakov, A. I., Leupold, O., Rüffer, R., Deriu, A. and Bonetti, E., *Phys. Rev. B* **66** (2002), 073410.
23. Sergueev, I., Franz, H., Asthalter, T., Petry, W., van Bürck, U. and Smirnov, G. V., submitted to *Phys. Rev. B*.
24. Pohl, R. O. and Swartz, E. T., *J. Non-Cryst. Solids* **76** (1985), 117.
25. Ramos, M. A., Vieira, S., Bermejo, F. J., Davidowski, J., Fischer, H. E., Schober, H., González, M. A., Loong, C. K. and Price, D. L., *Phys. Rev. Lett.* **78** (1997), 82.

Hyperfine Interactions **144/145:** 85–92, 2002.

γ-Ray Holography – Three-Dimensional Imaging of a Local Atomic Structure

P. KORECKI[1] and J. KORECKI[2]
[1]*Institute of Physics, Jagiellonian University, Kraków, Poland; e-mail: korek@surf.if.uj.edu.pl*
[2]*Faculty of Physics and Nuclear Techniques, Academy of Mining and Metallurgy, Kraków, Poland*

Abstract. γ-ray holography is a novel method for three-dimensional (3D) imaging of a local atomic structure, which utilizes the internal reference idea of holography. The principles of this method will be presented with emphasis on the phase information gained by means of complex holograms.

Key words: X-ray holography, Mössbauer effect, resonant nuclear scattering.

1. Introduction

Crystallographic phase problem. In a standard diffraction experiment, an object is illuminated by a plane wave and the diffraction pattern is measured with a far field detector. In this experimental geometry, atomic resolution requires the radiation wavelength which is comparable with inter atomic distances. Then, the diffraction pattern is connected with the atomic structure of the object by Fourier transform. In the X-ray regime, for energy of $E \simeq 10$ keV and $\lambda \simeq 1$ Å the frequency of the scattered waves is about $\nu \simeq 10^{19}$ Hz.

Therefore, only the intensities of the scattered waves can be measured directly and the phase information associated with the scattered waves is lost by time averaging. This makes the direct transformation of the diffraction pattern to the real space impossible. Figure 1 illustrates this difficulty. A two-dimensional object* from Figure 1(a) is illuminated by a monochromatic plane wave. The resulting diffraction image (Fourier transform of the object) is presented in Figure 1(b), where the intensity and phase distributions are shown separately. Figure 1(c) shows the back-transformed images utilizing the whole information, only the intensities and only the phase information, respectively.

In some cases, as it is visible from Figure 1, the information associated with the phase may be crucial for the reconstruction of the object. Generally, the phase gives the information about the absolute positions inside the object. Several methods of physical and computational character have been demonstrated to solve the phase problem [1]. Particularly, it was realized very early that Mössbauer spectroscopy

* The object is the logo of Lufthansa airlines. The Seeheim Workshop on Mössbauer Spectroscopy took place in the Lufthansa training center.

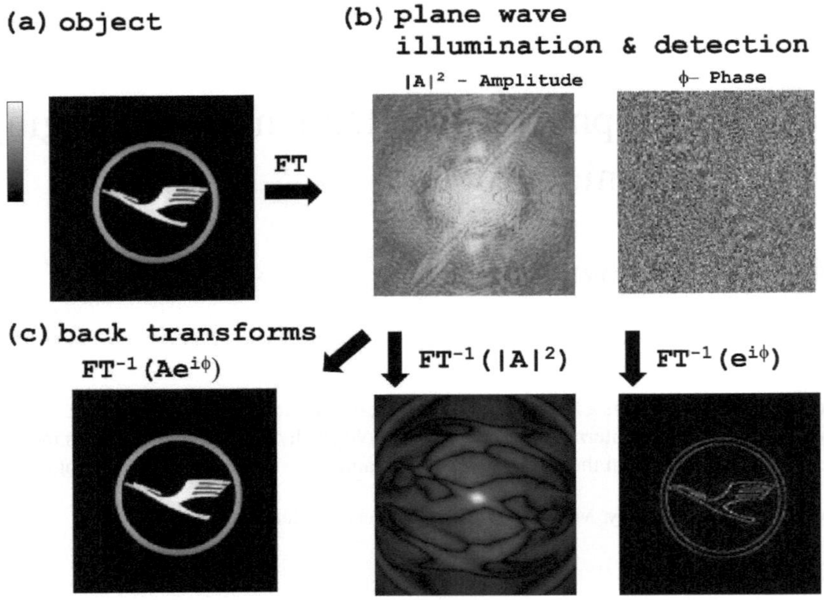

Figure 1. Demonstration of the phase problem. (a) A two-dimensional object. (b) Fourier transform of the object. (c) Back-transforms utilizing the whole complex information (*left*), only the intensity (*middle*) and only the phase information (*right*). Intenstities are shown in an arbitrary color scale. Phase changes from 0 to 2π.

may solve the phase problem by taking advantage of the interference between waves scattered in nuclear and electronic processes [2]. However, diffraction experiments require highly collimated beams, which are difficult to produce with a natural radioactive source. This may change with the development of ultrahigh-intense X-ray free electron lasers.

Holography. Holography was invented by Gabor, as a tool for structure determination at the atomic level [3] free from the phase problem. However, the lack of high-quality lenses in the X-ray regime prevented the progress of the method as a tool for atomic structure investigation. A lens, in the original idea of Gabor, was required to produce a spherical reference wave emerging from a point-focus near the imaged object as shown in Figure 2(a). The hologram, i.e., the interference pattern of the direct and scattered wave could be detected using a screen placed in the far-field and it contained the phase information. The phase information about the relative phase is gained when the reference wave and the object waves are added coherently. The resulting intensity at the detector can be written as:

$$I = |R + O|^2 = |R|^2 + 2|R||O|\cos(\phi_R - \phi_O) + \cdots, \tag{1}$$

where $|R|$, $|O|$, ϕ_R, ϕ_O are the amplitudes and phases of the reference and object waves, respectively.

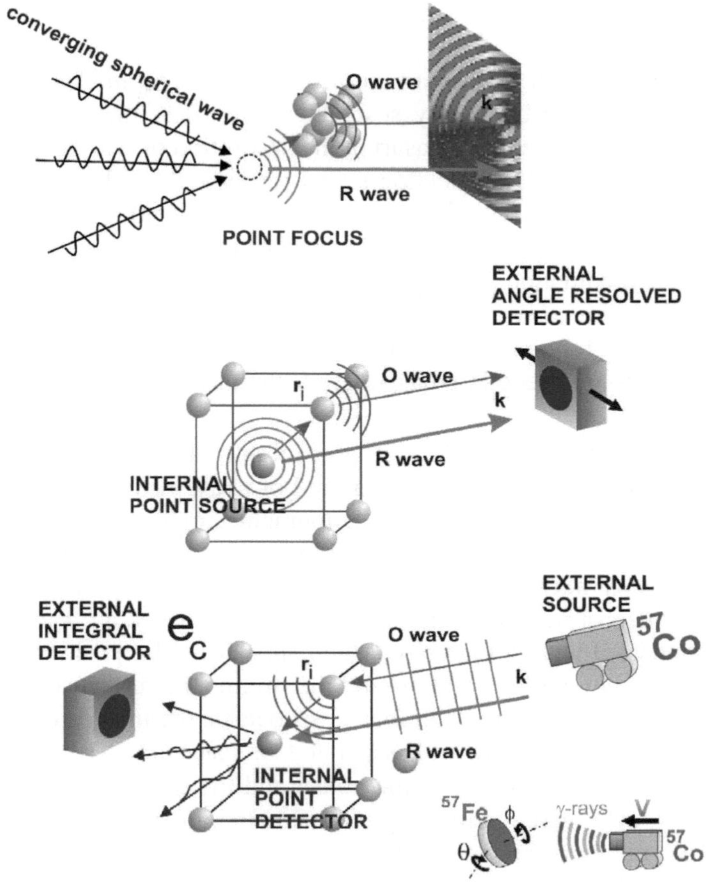

Figure 2. The formation process of the hologram in the Gabor's holographic microscope (*top*), in the Internal Source Holography (*middle*), and in the Internal Detector Holography (*bottom*).

Internal source holography (ISH) [4] and internal detector holography (IDH) [5] are alternative lensless approaches, which were realized experimentally quite recently. They avoid optical elements and thus provide images with atomic resolution by using electrons (photo, Auger, Kikuchi) [6] or X rays [7]. In ISH, atoms *inside* a sample serve as coherent point sources of short-wavelength radiation which is scattered from nearby atoms (Figure 2). This produces an interference pattern corresponding to the local environment around the emitting atoms. Such a pattern, which is collected with an angle resolved detector, may be treated in the same way as an in-line Gabor hologram. The back-reconstruction of the internal source hologram to the real space is performed numerically [8]. The method works only if all emitting sources have similar local environments. However, the long range order is not required as in standard diffraction methods. In fact, internal sources of γ-rays were proposed previously for solving of the phase problem by means of the

Kossel lines analysis [9]. However, such an experiment required a perfect single crystal as the sample.

IDH is connected with ISH by a time-reversal as shown in Figure 2. In X-ray IDH, the interference of elastically scattered waves with the direct unscattered reference wave is monitored by measuring the absorption of particular atoms serving as internal detectors. The detection signal for measuring the local X-ray field intensity is provided by X-ray fluorescence or electron yield and is recorded with an external angle integrating detector. Because of the weak scattering of X rays, the holographic modulation is only in the order of 10^{-4} of the total signal. In addition, for high spatial resolution, holograms have to be recorded over a wide angular range. Therefore, high-brilliance sources and fast two-dimensional energy resolving detectors are needed to make these techniques widely applicable.

Holographic phase problem. Twin images. Initially, ISH was believed to being free from the phase problem. However, a kind of phase still exists [10]. Consider Equation (1), which is fundamental for holographic imaging. From Equation (1) only the absolute value of the phase difference $\phi_R - \phi_O$ between the reference and the object wave may be extracted. As a consequence, twin-images appear in the real-space reconstruction of the hologram. Twin images are spurious images inherently connected with holography, which appear in false positions. For Gabor type holography, they are inverted around the origin (the reference point). In the original Gabor microscope the point focus was outside the imaged object and thus the real and twin images were well separated in space. In ISH and in IDH the internal source/detector is a part of the object and the real and twin images may overlap and add up to zero, completely removing the information about particular scatterers from the reconstructed real-space. Figure 3 illustrates the twin-images problem in ISH. Three holograms of a simple object consisting of a reference atom and eight scatterers placed on rectangular lattice with lattice parameter a, were simulated for different values of a and a constant wavelength λ. The real space reconstruction of the holograms is shown in Figure 3(c). It can be observed that for some values of a hologram, as well as the real-space images, are blind with respect to some inter atomic distances due to real and twin images cancellation in this centrosymetrical system.

Twin images may be eliminated in electron holography and X-ray holography by collecting holograms for many wavelengths [11].

2. Principles of γ-ray holography

IDH can be successfully performed in a table-top experiment utilizing another scattering process: nuclear resonant scattering of γ-rays. For the 14.41 keV transition in ^{57}Fe the corresponding wavelength of 0.86 Å is sufficiently short for atomic resolution. For this kind of radiation the nucleus can be treated as a point scatterer, so the form factor is constant. The angular anisotropy of the reference and object

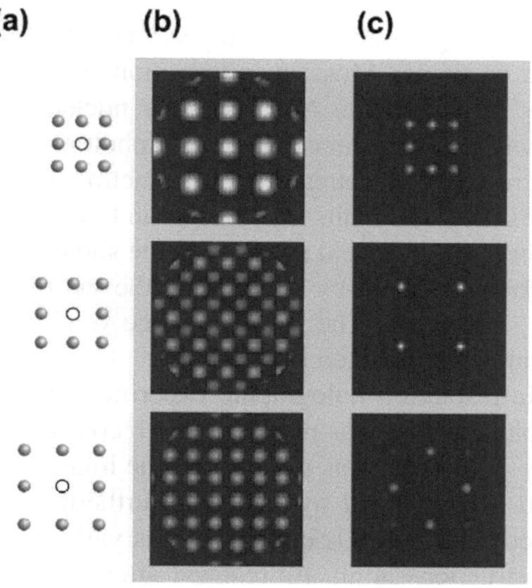

Figure 3. Twin images problem. (a) Objects consisting of a reference atom placed at origin and eight scatterers placed on rectangular lattice with different lattice parameter a. (b) Holograms simulated for objects shown in (a). (c) Real-space reconstructions of the holograms.

waves is caused only by the polarization effects and since the nuclear transitions in ^{57}Fe are M1, the scattering amplitude is a slow varying function of angles. Moreover, the excited state lifetime is long compared to the characteristic lattice vibration times of atoms in the sample and the Lamb–Mössbauer factor, describing the thermal effects for the case of slow scattering, is isotropic. For magnetically ordered systems, the levels are split and the scattering and absorption is influenced by the internal magnetic fields at the nuclei. In consequence, there is a possibility to image only chosen nuclei, with a particular set of the hyperfine interaction parameters, and it is possible to resolve the 3D magnetic structure of the sample. The nuclear resonant scattering amplitude value is very high as compared to the X-ray scattering amplitude given in the forward direction by Zr_0, where Z is the atomic number and r_0 is the classical electron radius. In the exact resonance, the value of the nuclear resonant amplitude can be as high as $440r_0$ for ^{57}Fe and may cause holographic oscillations of several percents in the measured pattern.

In the case of ^{57}Fe the internal conversion coefficient α, responsible for the ratio of the nonradiative to radiative decay probability, is 8.17. This value is almost ideal for the holographic application with the detection through the inelastic channel. The electrons following the conversion process are easy to detect over a full hemisphere above the sample using a simple proportional detector.

γ-ray holography was realized in the IDH geometry [12, 13]. In this version of holographic imaging 14.41 keV γ rays from radioactive source (^{57}Co), illuminate a sample containing Mössbauer isotope (^{57}Fe) (Figure 2). The energy of γ rays is tuned to a particular Mössbauer transition in the sample by changing the Doppler velocity of the source. Any particular nucleus in the sample can be excited in resonant way by either direct Mössbauer photons (acting as the reference wave) or previously resonantly scattered photons (acting as the object wave). The excited nucleus de-excites and emits conversion and Auger electrons, which are then detected to probe the γ-ray field intensity at the site of the absorbing nucleus. A 2D hologram is collected point-by-point while the sample is rotated relative to the direction of the γ-ray beam. The internal source version of γ-ray holography was also discussed but never realized [15].

In the first atomic resolution holographic experiment with γ radiation [12], a hologram of ^{57}Fe epitaxial film was recorded and reconstructed to the real space. Although the quality of the data was reasonable, the images of the next neighbors were very weak and accompanied by artifacts. Furthermore, succeeding experiments [14] performed on a $^{57}Fe_3O_4$ epitaxial films yielded no nuclear images in the real-space reconstructions at all. As realized after some time, this was due to twin-images problem.

3. Complex γ-ray hologram

Multiple wavelength illumination techniques for twin images suppression, cannot be applied in the γ-ray holographic Mössbauer experiment, since the resonance condition fixes the wavelength.

However, very recently, it was demonstrated that twin images can be eliminated by reconstructing a so-called complex hologram [16]. In resonant scattering one is able to directly change the phase of the scattering, and thus of the object waves, by detuning from the resonance. The phase change of $\pi/2$ replaces the cosine function in Equation (1) by the sine function removing thus the sign ambiguity. In addition, since holograms recorded close to the Mössbauer resonance correspond to the same wavelength they can be combined to a single complex hologram. In a holographic experiment the phase change of $\pi/2$ is not directly feasible. Therefore, it was proposed that suitable holograms may be built from linear combinations of two holograms recorded symmetrically bellow and above the exact resonance condition. The sum and difference of symmetrically recorded holograms always corresponds to a hologram with sine and cosine behavior, respectively. Figure 4(a) shows the raw data, i.e., the sine and cosine holograms measured for a $^{57}Fe/MgO(100)$ epitaxial film. Please note, that apart from twin images problem, the difference hologram is ideal for structural imaging since it is background free. The complex hologram was constructed from processed experimental data and its magnitude and phase are shown in Figure 4(b). The high-fidelity real-space reconstruction of the hologram is shown in Figure 4(b) proving the solution of the phase problem. The spatial resolution is better than 0.05 nm.

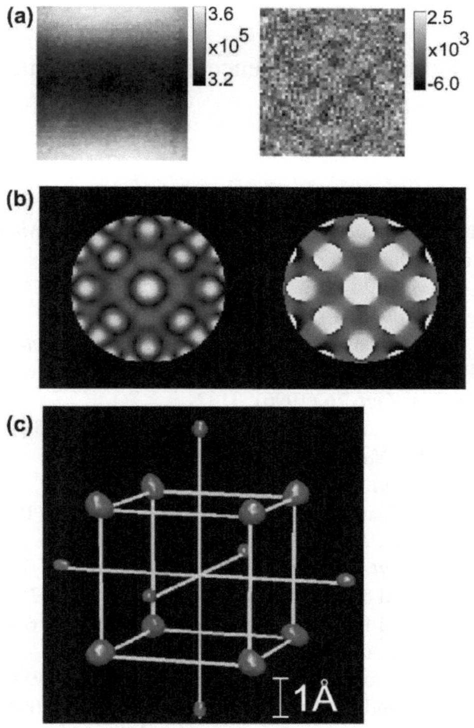

Figure 4. Complex γ-ray holography of $^{57}Fe/MgO(100)$ epitaxial film. (a) Raw data. Sum and difference of the patterns recorded bellow and above the resonance at a half width of the resonance line. The bars show the number of counts. (b) Amplitude (*left*) and phase (*right*) of the complex hologram processed from (a). (c) Real-space reconstruction of bcc local structure in $^{57}Fe/MgO(100)$.

4. Outlook

The most promising feature of γ-ray holography is the ability to record holograms characteristic of ions in different chemical environments and characterized by different hyperfine interaction parameters. Although sensitivity to hyperfine interaction parameters has not been directly shown in holographic experiment yet, such possibility is given by the Mössbauer experiment. Recently, this unique feature of the Mössbauer diffraction that is useful for studies of atomic arrangements in materials was applied to measure diffraction peaks characteristic of spatial arrangements of two nonequivalent Fe sites in a Fe_3Al powder sample [17]. The diffraction profiles were measured with the energy of incident radiation tuned to different hyperfine magnetic field values of both chemical environments. Whereas the structural information gained by γ-ray holography and the Mössbauer diffraction is similar, the fields of applications of these methods are very different. In γ-ray holography the acquisition scheme is based on the detection of conversion and Auger electrons, which can be done very efficiently even for ultrathin films, where the kinematical limit is valid.

Acknowledgements

This work was supported by Polish Science Research Council, Grant No. 5 P03B 069 21.

References

[1] Woolfson, M. M. and Hai, F., *Physical and Non-Physical Methods of Solving Crystal Structures*, Cambridge University Press, 1995.
[2] Parak, F. *et al.*, *Z. Physik A* **244** (1971), 456.
[3] Gabor, D., *Nature (London)* **161** (1948), 777.
[4] Szöke, A. In: D. T. Attwood and J. Bokor (eds), *Short Wavelength Coherent Radiation: Generation and Application, AIP Conf. Proc.* No. 147, AIP, New York, 1986.
[5] Gog, T. *et al.*, *Phys. Rev. Lett.* **76** (1996), 3132.
[6] Harp, G. R., Saldin, D. K. and Tonner, B. P., *Phys. Rev. Lett.* **65** (1990), 1012.
[7] Tegze, M. and Faigel, G., *Nature (London)* **380** (1996), 49.
[8] Barton, J. J., *Phys. Rev. Lett.* **61** (1988), 1356.
[9] Hannon, J. P., Carron, N. J. and Trammel, G. T., *Phys. Rev. B* **9** (1974), 2791; **9** (1974), 2810.
[10] Len, P. M. *et al.*, *Phys. Rev. B* **50** (1994), 11275.
[11] Barton, J. J., *Phys. Rev. Lett* **67** (1991), 3106.
[12] Korecki, P., Korecki, J. and Ślezak, T., *Phys. Rev. Lett.* **79** (1997), 3518.
[13] Korecki, P., Korecki, J. and Karaś, W., *Phys. Rev. B* **59** (1999), 6141.
[14] Korecki, P. and Korecki, J., unpublished.
[15] Odeurs, J., Coussement, R. and L'abbé, C., *Phys. Rev. B* **60** (1999).
[16] Korecki, P., Materlik, G. and Korecki, J., *Phys. Rev. Lett.* **86** (2001), 1534.
[17] Stephens, T. A. and Fultz, B., *Phys. Rev. Lett.* **11** (1997), 336.

Hyperfine Interactions **144/145:** 93–101, 2002.

Diffusion Studies with Synchrotron Radiation and with Neutron Scattering

G. VOGL, B. SEPIOL, M. SLADECEK, L. STADLER and M. KAISERMAYR*
Institut für Materialphysik der Universität Wien, A-1090 Wien, Austria

Abstract. With scattering methods we are able to detect the elementary diffusion jump. This is a report on investigations with methods working in the time domain, i.e., nuclear resonant scattering of synchrotron radiation and neutron spin echo. The accent of this paper is on diffusion in ordered alloys. We finish with an outlook on what will be possible with the upcoming potential of future synchrotron sources.

Key words: diffusion, nuclear resonant scattering, neutron spin-echo.

In the past we have several times reported on results of quasielastic methods, i.e., methods acting in the energy domain, in particular quasielastic Mössbauer spectroscopy and quasielastic neutron scattering (e.g., [1–3]). Here we shall concentrate on methods acting in the time domain. These methods have only recently acquired the potential to study diffusion. This is because only recently third-generation synchrotrons with high brilliance in the X-ray regime and powerful spin echo spectrometers at neutron sources became accessible. The idea of Mössbauer spectroscopy with synchrotron radiation, usually called nuclear resonant scattering of synchrotron radiation (NRS) is schematically shown in Figure 1.

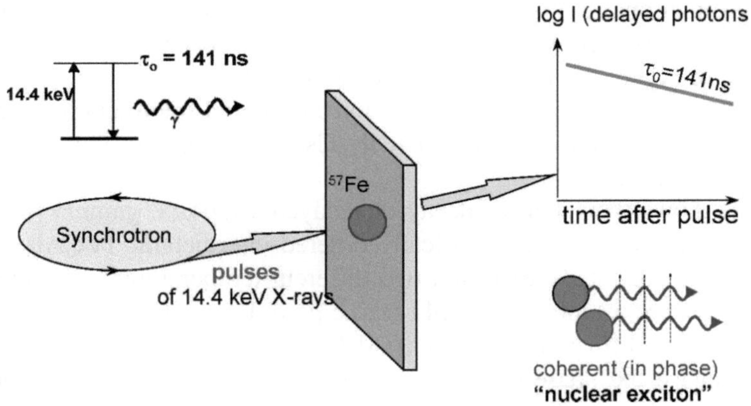

Figure 1. Set-up in investigations of nuclear resonant scattering of synchrotron radiation in forward direction.

* Now: SAP Deutschland AG.

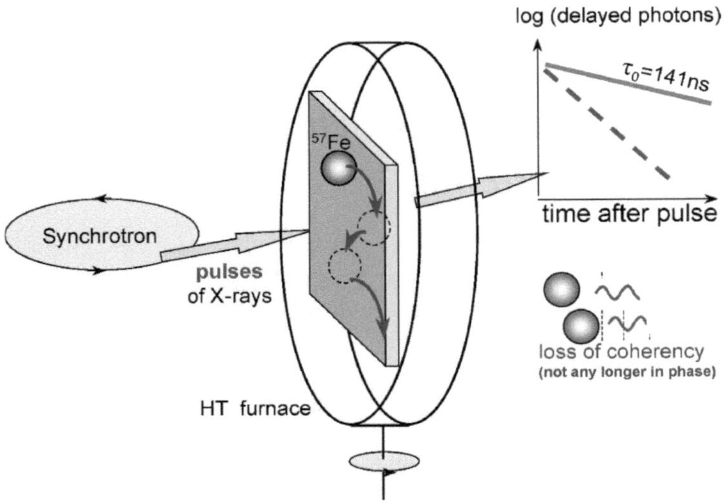

Figure 2. Set-up in diffusion studies by nuclear resonant scattering of synchrotron radiation in forward direction. The arrows symbolise jumping atoms. The sample is a single crystal which can be rotated together with the furnace.

Here the Mössbauer nucleus is ^{57}Fe and the synchrotron pulses (length some tens of picoseconds, highly monochromatic around 14.4 keV, energy width meV) follow each other after about 170 ns. In forward scattering (and in Bragg directions) the 14.4 keV radiation re-emitted from all ^{57}Fe nuclei is in phase, i.e., there is maximum intensity. The re-emitted photon intensity from an ideally thin sample would decay with a half life of 141 ns (Figure 1).

If, however, the ^{57}Fe atoms start diffusing, i.e. in a crystalline lattice to jump from lattice site to lattice site, then the phase relation in forward direction [4] (and into Bragg directions [5]) is lost and with increasing time more and more of the intensity goes incoherently into 4π. The intensity will now decay more rapidly (dashed line in Figure 2). For an early review see [6].

In this paper we concentrate on diffusion in ordered intermetallic alloys with a particular simple cubic structure, namely CsCl structure (in materials science named B2 structure).

Figure 3 shows the intensity decay of delayed 14.4 keV gamma radiation in forward direction for a stoichiometrically ordered intermetallic $Fe_{50}Al_{50}$ alloy (in the following abbreviated FeAl) and two different temperatures. At 900°C a diffusion jump occurs seldom within 141 ns, i.e., the decay of the delayed radiation hardly notices the diffusion at all. At 1030°C, however, the decay is strongly accelerated because the phase relation between re-emitted parts of the radiation is strongly disturbed by sufficiently fast site changes of the re-emitting atoms, and with increasing time less and less intensity goes coherently into forward direction. That is why the intensity decay in forward direction is considerably faster than at 900°C. For more details see [7].

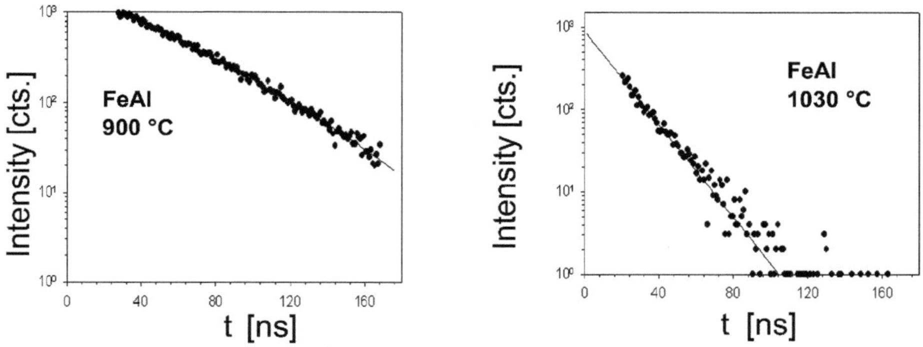

Figure 3. Intensity decay is accelerated at 1030°C (right). FeAl crystal, synchrotron beam parallel to [1 1 0] crystal direction.

To be more exact: the intensity in forward direction I_{FS} is essentially the square of the momentum–time correlation function of the diffusing atoms

$$I_{FS} \propto \exp \frac{-t}{\tau} \left| S(\mathbf{Q}, t) \right|^2,$$

where the momentum–time correlation function in the simplest case of jump diffusion is (see [6])

$$S(\mathbf{Q}, t) = \exp\left[\frac{-t}{\tau} \sum_{i=1}^{N} N^{-1} \{ 1 - \exp(-i\mathbf{Q}\boldsymbol{l}_i) \} \right].$$

The decay with time t allows one to determine the jump rate $1/\tau$, and the phase factor $\exp(-i\mathbf{Q}\boldsymbol{l}_i)$ – with $\mathbf{Q}$ the wave vector of the synchrotron radiation – makes it possible to determine the jump vectors $\boldsymbol{l}_i$ just by playing with specimen orientation. $\boldsymbol{l}_i$ denotes the variety of jump vectors numbered from 1 to N.

This orientation dependence is shown in Figure 4 where the time integrated intensity of the delayed 14.4 keV radiation (see inset, shadowed area) is plotted versus sample orientation.

Comparison with the above formula leads to an unambiguous model for the diffusion jumps (Figure 5).

The majority of the Fe atoms in $Fe_{50}Al_{50}$ ordered alloy jumps preferentially to third nearest neighbour sites (thicker arrow in Figure 5), but a minority jumps to next (second) nearest neighbour sites (thinner arrow). Some atoms perhaps even jump to fourth nearest neighbours. Sounds a little surprising; are there really no jumps to the antistructure site, i.e. the nearest neighbour site (bold arrows)?

That was the reason why we searched for other systems with B2 structure in order to support or falsify the model. In particular a system with a lower tendency for ordering is highly desirable for comparison. In such a system we expect that jumps of the transition element atoms to the anti-structure sites will indeed occur, even though we expect the occupation time τ of these sites to be smaller. We will

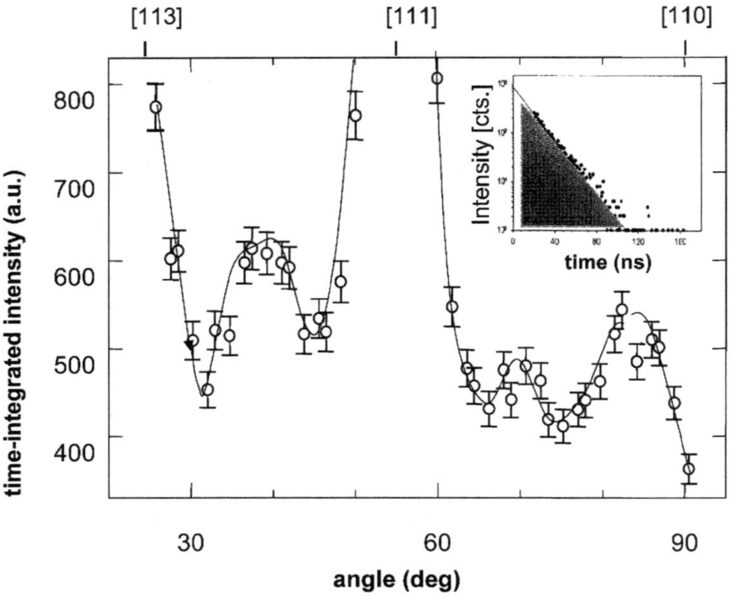

Figure 4. Orientation dependence of time integrated intensity (see inset) for nearly 40 orientations of an FeAl single crystal with its (110) plane horizontal, i.e., synchrotron beam in this plane, parallel to directions from [1 1 3] through [1 1 1] to [1 1 0]. Angle relative to [0 0 1]. Sample temperature 1030°C. Data taken from [7]. The solid line is a simulation according to the described jump model (see text).

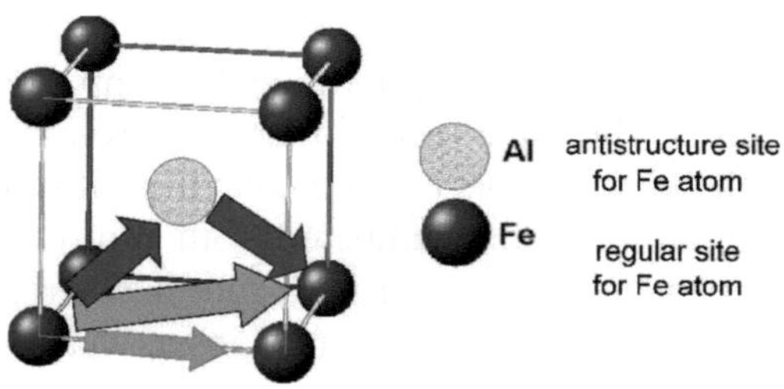

Figure 5. Jumps of Fe atoms in $Fe_{50}Al_{50}$.

then face a more complicated situation from the experimental point of view: there will be several jumps with different lengths and different frequencies, i.e., we will have to solve a jump matrix.

Unfortunately there exists no other intermetallic ordered Fe alloy with B2 (CsCl) structure. We are faced with the decisive drawback of Mössbauer spectroscopy and its synchrotron variant to be easily feasible only with ^{57}Fe. We fortunately found a number of B2 alloys containing Ni or Co which are accessible for neutron methods able to probe atom dynamics. Here – as declared above – we dwell on methods working in the time domain, i.e. we shall not report results of quasielastic neutron

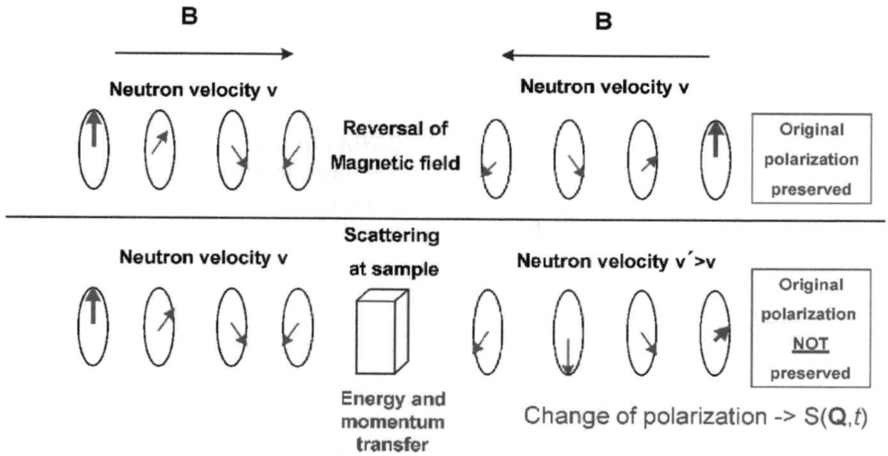

Figure 6. Principle of neutron spin-echo spectroscopy.

scattering that may be found in [8]. By way of example we report on neutron spin echo results on NiGa.

Figure 6 shows the principle of neutron spin-echo spectroscopy. The idea of the experiment is due to F. Mezei (see, e.g., [9]) and is as follows: polarized neutrons (i.e., neutrons with their spins aligned) are Larmor precessing during their flight through a magnetic coil. After they have passed a certain distance they fly along the same distance through coil with magnetic field with exactly opposite direction, i.e., the neutrons reverse their sense of precession. If the second magnetic coil has exactly the same length as the first one and if the neutron velocity has not changed, the neutrons will arrive with full polarization as at the beginning of the game. That is controlled by an analyzer. If, however, e.g., by scattering with energy transfer at a sample, the velocity of the neutron has changed during the game, then the polarization will not be preserved. Again the momentum–time correlation function $S(\mathbf{Q}, t)$ is measured.

We have chosen an intermetallic alloy with B2 structure (NiGa) where diffusion of the transition element (Ni) is sufficiently fast and where again practically only its diffusion will be visible for our method. Here the trick is: the incoherent scattering cross section for Ga atoms is negligibly small compared to that of Ni atoms (coherent scattering is avoided by measuring off the Bragg reflections). That implies that Ga is practically invisible for the neutron spin echo method and we will probe only Ni diffusion.

In NiGa indeed the situation is more complicated than for Fe jumps in ordered FeAl. A fit with jumps to second nearest neighbours made no sense (dashed line in Figure 7) and the comparison with theory yields jumps to the anti-structure sites (full line) and back to regular sites (Figure 8).

This finding implies that in other B2 systems than FeAl – here for NiGa – the residence time of the transition element atoms on anti-structure sites may be

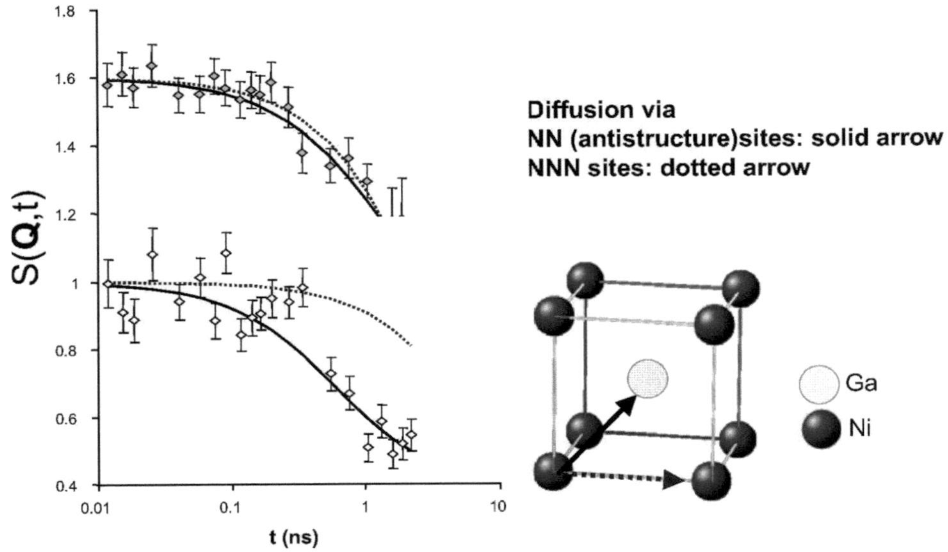

Figure 7. Neutron spin echo at SPAN (Hahn-Meitner Institut Berlin). Momentum–time correlation function $S(\mathbf{Q}, t)$ for Ni diffusion in a single crystal of nearly stoichiometric NiGa B2 ordered intermetallic alloy. Temperature 1130°C, direction of **Q** close to [1 1 0], length of $\mathbf{Q} = 1.45$ Å^{-1} (*top*) and 1.82 Å^{-1} (*bottom*). The lines represent the momentum–time correlation functions according to diffusion via nearest-neighbour (NN) jumps into anti-structure sites (full lines) and next nearest-neighbour jumps (dashed lines). It is evident that jumps to anti-structure sites are the better fit for $\mathbf{Q} = 1.82$ Å^{-1}, whereas for $\mathbf{Q} = 1.45$ Å^{-1} no distinction is possible. From [10].

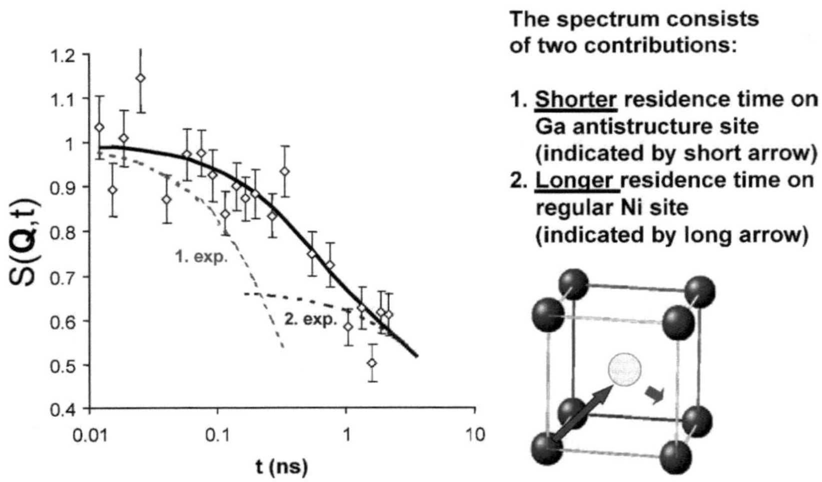

Figure 8. Neutron spin echo on NiGa as in Figure 7. Temperature 1130°C, direction of **Q** close to [1 1 0], length of $\mathbf{Q} = 1.82$ Å^{-1}. The solid line is a sum of two exponentials (dashed lines) with different relaxation times. The weight of first exponential is for clarity normalised to one, the weight of the second exponential is about 0.65. From [10].

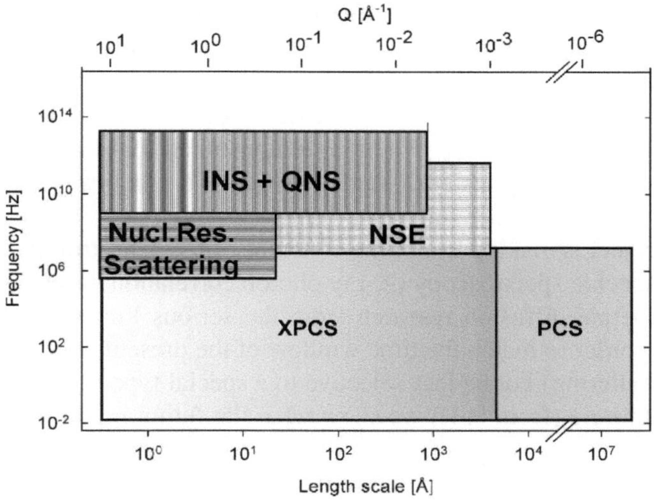

Figure 9. Comparison of methods able to determine jump frequency and length scale of diffusion. In the atomistic range (length scale in the range of Å) we have the neutron methods (inelastic and quasielastic neutron scattering INS and QNS and neutron spin echo NSE) and nuclear resonant scattering including classical Mössbauer spectroscopy. For much larger length scale (far beyond the atomistic range) photon correlation spectroscopy (PCS) of coherent laser light is used. We expect that the development and application of this method extended to the X-ray range (XPCS) will be more and more successful with increasing coherency of X-ray sources.

considerable. The results are fully confirmed, though with weaker resolution – by results from QNS [10]. This supports our conjecture that also in FeAl the Fe jump is via anti-structure sites. However, they are occupied for a much shorter residence time than the regular sites, so that the jumps of the Fe atoms are – even though they are executed via anti-structure sites – "effectively" to regular second or third nearest neighbour sites.

Putting all information on diffusion of the transition element atoms in B2 alloys together we end up with the following picture of diffusion in B2 alloys:

- Diffusion is always *via jumps into a vacancy on anti-structure nearest neighbour (NN)* sites. This is true for stoichiometric (or nearly stoichiometric) alloys, but also for non-stoichiometric alloys. For the latter we refer to [11]. Ab initio calculations by Fähnle *et al.* [12] confirm and explain these findings.
- From investigations with the quasielastic methods [10, 11], on which we could not report here due to lack of space, we further conclude that for the transition element atoms the *residence time on anti-structure sites increases with increasing deviation from stoichiometry.*
- From comparison with tracer data [13] we conclude that in B2 alloys the *diffusivity of the transition element atoms increases with deviation from stoichiometry.*

We are presently working on developing further methods with the aim to fully exploit the possibilities given by the present and future highly brilliant synchrotron light sources.

One project is working with grazing incidence of 14.4 keV radiation in order to scan diffusion in surface layers. We have demonstrated that the high brilliance of synchrotron radiation enables such studies which had practically been unfeasible with conventional Mössbauer sources.

Another project is making use of the coherence of synchrotron radiation for dynamic X-ray speckle spectroscopy (X-ray photon correlation spectroscopy XPCS). This would liberate diffusion research from the serious limitation to rather high diffusivities in order to match the time window of the present methods (Mössbauer and neutron scattering) but be less selective to a special type of atoms.

In the final figure (from [14]) we show what the future might hold for research on the elementary diffusion jump in addition to nuclear resonant scattering of X-rays, neutron spin echo spectroscopy (QNS) quasielastic and inelastic neutron spectroscopy (QNS and INS).

Acknowledgements

We wish to thank R. Rüffer, O. Leupold and G. Grübel (all ESRF Grenoble), C. Pappas (BENSC, HMI Berlin), J. Combet and B. Frick (ILL Grenoble) and H. Thiess (now Zeiss, Wetzlar) for fruitful discussions and help with measurements. The authors are grateful to H. Schicketanz (Universität Wien) for supplying NiGa samples. Investigations were supported by Austrian Fonds zur Förderung der wissenschaftlichen Forschung (FWF), contract No. P15421 and F&E Vorhaben, Hahn-Meitner Institut Berlin (project Ofenbau).

References

1. Vogl, G., In: G. J. Long and F. Grandjean (eds), *Mössbauer Spectroscopy Applied to Magnetism and Materials Science*, Vol. 2, Plenum Press, New York, 1996, p. 85.
2. Kaisermayr, M., Sepiol, B., Combet, J., Rüffer, R., Pappas, C. and Vogl, G., *J. Synchrotron Rad.* **9** (2002), 210.
3. Vogl, G. and Feldwisch, R., In: J. Kärger, P. Heitjans and R. Haberlandt (eds), *Diffusion in Condensed Matter*, Vieweg, 1998, p. 40; Vogl, G. and Sepiol, B., In: P. Heitjans and J. Kärger (eds), *Diffusion in Condensed Matter*, 2nd edn, Springer, 2003, to appear.
4. Sepiol, B., Meyer, A., Vogl, G., Franz, H. and Rüffer, R., *Phys. Rev. B* **57** (1998), 10433; Sepiol, B., Meyer, A., Vogl, G., Rüffer, R., Chumakov, A. I. and Baron, A. Q. R., *Phys. Rev. Lett.* **76** (1996), 3220.
5. Thiess, H., Kaisermayr, M., Sepiol, B., Sladecek, M., Rüffer, R. and Vogl, G., *Phys. Rev. B* **64** (2001), 104305.
6. Vogl, G. and Sepiol, B., *Hyp. Interact.* **123/124** (1999), 595.
7. Sepiol, B., Czihak, C., Meyer, A., Vogl, G., Metge, J. and Rüffer, R., *Hyp. Interact.* **113** (1998), 449.

8. Kaisermayr, M., Combet, J., Ipser, H., Schicketanz, H., Sepiol, B. and Vogl, G., *Phys. Rev. B* **61** (2000), 12038; Kaisermayr, M., Combet, J., Ipser, H., Schicketanz, H., Sepiol, B. and Vogl, G., *Phys. Rev. B* **63** (2001), 054303.
9. Mezei, F., *Int. J. Modern Phys. B* **7** (1993), 2885.
10. Kaisermayr, M., Pappas, C., Sepiol, B. and Vogl, G., *Phys. Rev. Lett.* **87** (2001), 175901; Kaisermayr, M., Rennhofer, M., Vogl, G., Pappas, C. and Longeville, S., *Phys. Rev. B* **66** (2002), 024302.
11. Feldwisch, R., Sepiol, B. and Vogl, G., *Acta Metall. Mater.* **43** (1995), 2033.
12. Fähnle, M., Meyer, B., Mayer, J., Oehrens, J. S. and Bester, G., *Mat. Res. Soc. Symp. Proc.* **527** (1998), 23; Vogl, G., Sepiol, B., Czihak, C., Rüffer, R., Weinkamer, R., Fratzl, P., Fähnle, M. and Meyer, B., *Mat. Res. Soc. Symp. Proc.* **527** (1998), 197.
13. Vogl, G., Sepiol, B., Kaisermayr, M., Sladecek, M., Thiess, H. and Rüffer, R., In: R. Dieckmann and C. T. Liu (eds), *Nonstoichiometric Ceramics and Intermetallics*, UEF, New York, 2001, unpublished.
14. Grübel, G. and Vogl., G., *Synchr. Rad. News* **15** (2002), 14.

Hyperfine Interactions **144/145:** 103–117, 2002.

High-Pressure Studies of Magnetism and Lattice Dynamics by Nuclear Resonant Scattering of Synchrotron Radiation

GERHARD WORTMANN, KIRSTEN RUPPRECHT and HUBERTUS GIEFERS
Physics Department, University of Paderborn, D-33095 Paderborn, Germany

Abstract. A review is given on recent high-pressure experiments using nuclear resonant scattering of synchrotron radiation. We shortly introduce the methodological aspects connected with high-pressure experiments applying nuclear forward scattering and nuclear inelastic scattering of synchrotron radiation. Selected examples for the study of magnetism are given for the Laves phases $LuFe_2$ and $ScFe_2$, where we studied the variation of the magnetic ordering temperature and of the Fe band moment as a function of pressure. An actual example for the study of lattice dynamics is a recent investigation of the phonon density-of-states in metallic iron with special emphasis on hcp ε-Fe, where the pressure-induced texture is used, in conjunction with theoretical calculations, to extract density of phonon states as seen parallel and perpendicular to the hexagonal c-axis.

Key words: high-pressure, magnetism in RFe_2 Laves phases, phonons in hcp iron.

1. Introduction

Immediately with the first successful nuclear resonant scattering studies with synchrotron radiation (SR) on polycrystalline samples [1, 2] it was realised that these experiments are extremely suited for high-pressure experiments. This is due to the fact that SR exhibits already almost laser-like properties and that its collimation can be enhanced, with only minor loss of intensity, by optical elements like mirrors and X-ray lenses, to spot sizes in the 10 μm range. Samples of this size allow, with diamond-anvil cells, pressures in 100 GPa (1 Mbar) range [3]. Applying the method of coherent (elastic) nuclear forward scattering (NFS) of SR [1] for the investigation of magnetic properties of RFe_2 Laves phases, model systems for Fe band magnetism, high-pressure experiments above 100 GPa were performed at room temperatures as well as at low temperatures [4, 5]. Even higher pressures were reached with the method of incoherent nuclear inelastic scattering (NIS) of SR [2], where the phonon density-of-states (DOS) of iron was studied at pressures up to 153 GPa [6, 7].

In this review we report on recent high-pressure studies performed with ^{57}Fe-NFS to study magnetic properties of RFe_2 Laves phases, which demanded for high pressure and, *simultaneously*, for high temperature. We also present recent high-pressure ^{57}Fe-NIS studies of lattice dynamics, which concentrate on a new aspect, namely the use of pressure-induced structural textures in an initially polycrystalline sample to study the dependence of the phonon-DOS on the crystallographic directions. Such an information is normally only obtainable by the use of single-crystalline samples.

In this review, we introduce in Chapter 2 to the methodological aspects connected with high-pressure NFS and NIS experiments, here the demands on the beamline instrumentation and on the diamond-anvil cells. In Chapter 3 we present recent high-pressure and high-temperature studies of $LuFe_2$ and $ScFe_2$ and discuss, together with previous results on YFe_2, the dependence of the magnetic ordering temperatures and hyperfine fields on the lattice spacings using theoretical predictions for magnetic band moment systems. Chapter 4 is devoted to NIS studies of iron with special emphasis on the phonon DOS in ε-Fe, the hexagonal high-pressure phase, which is of special geophysical interest, since this phase constitutes the solid core of the Earth's inner core. Using the texture of the pressurized sample, we are able to determine the phonon DOS in ε-Fe as seen parallel and perpendicular to the hexagonal c-axis. In this respect the derived sound velocities, and a possible anisotropy with respect to the hexagonal c-axis, are of special geophysical interest. In the final chapter we will give some conclusions and an outlook for future developments.

2. Experimental

The NFS and NIS experiments reported here were performed at beamline ID22N of ESRF (Grenoble) employing the 14.413 keV gamma resonance of ^{57}Fe. The principal components of ID22N are very similar and compatible with those at ID18, the other "Mössbauer beamline" at ESRF [8]. A sketch of the set-up of this beamline for high-pressure experiments is given in Figure 1. The SR beam, coming from a specially adjusted undulator, is passing at the front window of the optical hutch through a compound refractive lens (CRL), which further reduces the (small) opening angle of the SR beam towards parallelism (which is especially useful for the performance of the high resolution monochromator (HRM)), then impinging on the high-heatload pre-monochromator (PM) with a pair of Si(1,1,1) crystals (cooled down to about 120 K) with an energy resolution of about 2 eV. The SR beam is then entering the "heart" of the beamline instrumentation, the HRM. For the NFS experiments, the HRM consists of two pairs of ("nested") Si(4,2,2) and Si(9,7,5) crystals with a bandwidth of about 4.5 meV. For the present NIS studies two pairs of Si(3,3,3) and Si(9,7,5) crystals provided an energy resolution of 3.5 meV. The next elements, a focusing monochromator and (optional) a further CRL, reduce the initial beam size of about 1.2 mm $\times$ 0.8 mm ($h \times v$) to the size (mostly somewhat smaller) of the sample in the DAC. For the present NFS studies, the beam was collimated to about 100 μm $\times$ 100 μm, for the present NIS experiments we used a 90 μm $\times$ 60 μm spot size with 3 $\times$ 10^8 photons/s at 90 mA

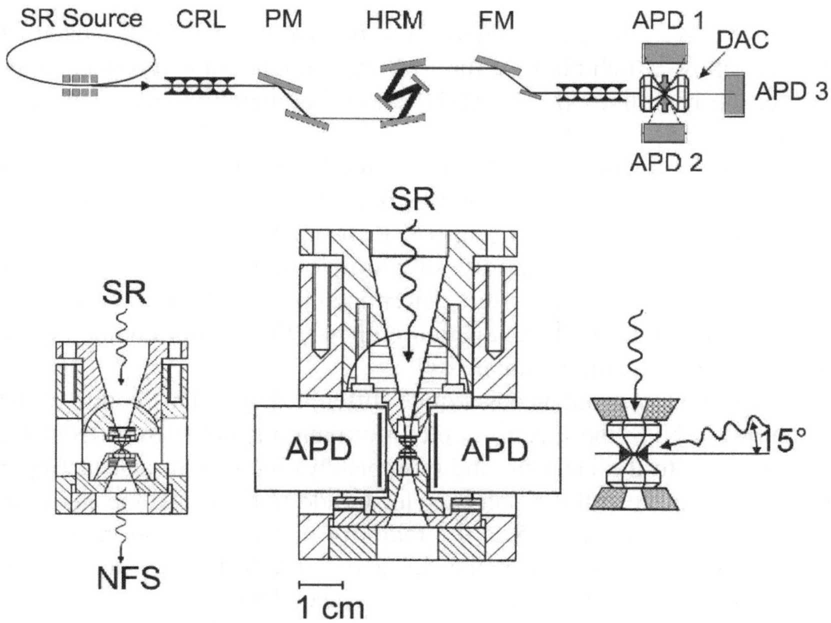

Figure 1. Upper panel: sketch of the experimental set-up at the beamline ID22N of ESRF for h.p. NFS and NIS studies with CRL for compound refractive lens, PM for pre-monochromator, HRM for high-resolution monochromator, FM for focusing monochromator, DAC for diamond-anvil cell and APD 1, 2, 3 for avalanche photo diode detectors used for NIS (1 + 2) and NFS (3) studies. Lower panel: sketches of the diamond-anvil cells used *(left)* for NFS and *(middle)* for NIS studies. On the right is a sketch of the diamond anvils in the DAC used for NIS studies with two different directions of the SR beam with respect to the diamond anvil axis (see text).

storage ring current. In recent NIS experiments the use of Kirkpatrick–Baez (K–B) multilayers [9] instead of a bent Si(1,1,1) crystal provided a beam size of 15 μm $\times$ 10 μm, comparable to a 5 μm $\times$ 5 μm size obtained with K–B mirrors in the above mentioned NIS experiment at 153 GPa [7].

We turn now to the "heart" of the high-pressure experiment, the diamond anvil cell (DAC). The technology for generation of high pressure (and also reliably monitoring its value) with a DAC is well developed, also for normal Mössbauer experiments [3, 10]. Here we describe just the special needs for NFS and NIS experiments. As shown schematically in Figure 1 upper panel, NFS is a "straight-forward" method, the monochromatized beam is entering the cell via the first diamond, passing through the sample and leaving the cell via the second diamond. The photoabsorption in the diamonds is, for the 14.4 keV of ^{57}Fe, about 75% for a 5 mm path through the diamonds. One should note that for the 21.5 keV resonance of ^{151}Eu, also successfully applied for NFS experiments with a DAC [4, 5, 11], the absorption in the diamonds amounts to only 40%. For ^{57}Fe- and ^{151}Eu-NFS experiments, a $Ta_{90}W_{10}$ alloy was used as gasket material, which allows also for a good collimation of the SR radiation as well as for classical Mössbauer experiments [5]. We developed previously a relatively small DAC, made from a special CuBe alloy, which allowed for NFS experiments in a cryomagnet system (available at the beamlines ID18/ID22N) at low temperatures and, with external magnetic fields [4, 5].

In the present high-temperature/high-pressure NFS experiments, we used a new DAC with the same dimensions as above, but now made from a special Ti-alloy with a small thermal expansion. Up to 670 K, heating was possible in air without corrosion of the cell, enabling the use of a simple mini-oven without vacuum or an inert gas atmosphere. The pressure, determined by ruby fluorescence, remained unchanged when increasing the temperature up to 670 K. After cooling, there was a small decrease of pressure by about 10%.

For NIS experiments, one needs a quite different DAC. This is due to the fact that, in contrast to NFS, the spectroscopic information is not transmitted "straight-forward", but, due to the inelastic and (completely) incoherent and isotropic reemission process, in the full 4π solid angle. In addition, due to the high internal conversion coefficient of $\alpha = 8.2$, the majority of the information is now contained in the delayed emission of the 6.4 and 7.0 keV $K_{\alpha,\beta}$ X-rays of Fe, and only 11% in the re-emitted 14.4 keV γ-rays. The cell has to be modified to allow for a detection of the re-emitted radiation mainly perpendicular to the incoming beam [6, 7, 12]. For this purpose, at first, beryllium, which allows for a sufficient transmission of the $K_{\alpha,\beta}$ X-rays of Fe, was used instead of the $Ta_{90}W_{10}$ alloy as gasket material. Secondly, the dimension of the cell was enlarged to allow for large opening in the cylindrical mantle, in which two avalanche photo diodes (APD) detectors were placed as near as possible to the pressurized sample. Thirdly, the diamonds were mounted in a free configuration without a supporting belt to enable the largest possible solid angle for the re-emitted radiation [12]. Fourthly, we had, in addition, modified our h.p. cells for the present experiments by two additional slits in the mantle, perpendicular to the openings for the APDs, allowing now also NIS studies with the SR beam almost perpendicular (at an angle of 70–90°) to the diamond anvil axis (see Figure 1 lower right panel). This modification of the h.p. cell enabled us to study the impact of a preferred orientation on the phonon-DOS in ε-Fe [13]. In the (almost) perpendicular configuration, the absorption of the incident 14.4 keV SR radiation by the first diamond anvil is strongly reduced, resulting in a considerable increase of the inelastic count rate.

3. NFS studies of magnetism at high pressures and high temperatures

In continuation of our previous NFS and Mössbauer studies of RFe_2 Laves phases (Y, Gd, Sc) at pressures up to 105 GPa and temperatures between 15 and 300 K [4, 5], we performed now also NFS studies at high pressures *and* high temperatures. Here we studied the magnetic properties of $LuFe_2$ (with cubic C15 structure) at room temperature and pressures up to 80 GPa. In addition we performed NFS studies of $LuFe_2$ at various pressures up to 39 GPa and temperatures up to 600 K.

We measured for $ScFe_2$ (with hexagonal C14 structure) NFS spectra at pressures up to 16 GPa and temperatures up to 650 K, in continuation of previous NFS and Mössbauer studies [4, 5, 14]. Our present interest was focused on the variation of the magnetic ordering temperatures T_C in $LuFe_2$ and $ScFe_2$ with pressure. Interestingly, one observed for YFe_2, isostructural to $LuFe_2$, first an increase of T_C (monitored by a magneto-elastic anomaly seen in h.p. XRD measurements) and then, at higher pressures, by our NFS and Mössbauer studies, a continuous decrease of T_C with pressure [4].

Figure 2(a) shows NFS spectra of $LuFe_2$ at various pressures up to 80 GPa. The NFS spectra were fitted, according to the properties of C15 Laves phases [4, 5] with two magnetically different sites with (slightly) different hyperfine fields B_{hf} and different angles with respect to main axis of the quadrupole interaction tensor, ΔE_Q. Details of these fits, which did not take into account effects of magnetic texture and pressure distributions, can be found in [15, 16]. The spectra reveal, with increasing pressure, a decrease of B_{hf} and an increase of ΔE_Q, from $\Delta E_Q =$ $(-)0.33$ mm/s at 0 GPa to $(-)0.65$ mm/s at 80 GPa, where the NFS spectrum exhibits a pure quadrupole interaction. From the variation of B_{hf} with pressure we derived a magnetic ordering temperature of 295 K around 75 GPa [15].

Figure 2(b) shows typical NFS spectra of $LuFe_2$ at 18 GPa and various (increasing) temperatures from 295 to 585 K. The sample thickness was larger than in the pressure series (Figure 2(a)); because of this and the focusing optics, NFS spectra of good statistics could be recorded within 10 minutes. The disappearance of the magnetic structures and the occurrence of a "sharp" Bessel beat (due to thickness effects) [17] in the NFS spectra at the highest temperatures indicate the absence of magnetic order. From a series of additional spectra between 300 and 585 K, taken during heating and cooling of the h.p. cell, we derived a magnetic ordering temperature of $T_C = 573(5)$ K at 18 GPa. Figure 3(a) presents a plot of the (inverse) NFS counting rate around the Bessel minimum from 48 to 64 ns (normalized with the overall NFS count rate in the time interval 20 to 160 ns (see Figure 3(a)). The magnetic ordering temperature T_C is attributed to the maximum of the inverse count rate in Figure 3(a), the decrease at higher temperatures is due to the shift of the Bessel minimum to larger times because of a concomitant decrease of the f-factor with increasing temperatures [17].

All magnetic ordering temperatures T_M obtained for $LuFe_2$ and $ScFe_2$ as well as for YFe_2 are plotted in Figure 3(b) as a function of the Fe–Fe distance d_{Fe-Fe}. Such a presentation allows for the direct comparison of the C15 ($LuFe_2$, YFe_2) and C14 ($ScFe_2$) structures, since the local surroundings of the Fe sites, numbers of neighbours and coordination, are identical or very similar for the two structures [4]. The pressure dependence of the lattice parameters of the investigated RFe_2 systems was determined either by angle-dispersive X-ray diffraction (XRD) studies performed with the same DACs as used in the NFS experiments at beamline ID30 of ESRF or by energy-dispersive XRD studies at beamline F3 of HASYLAB (DESY, Hamburg) [16, 18].

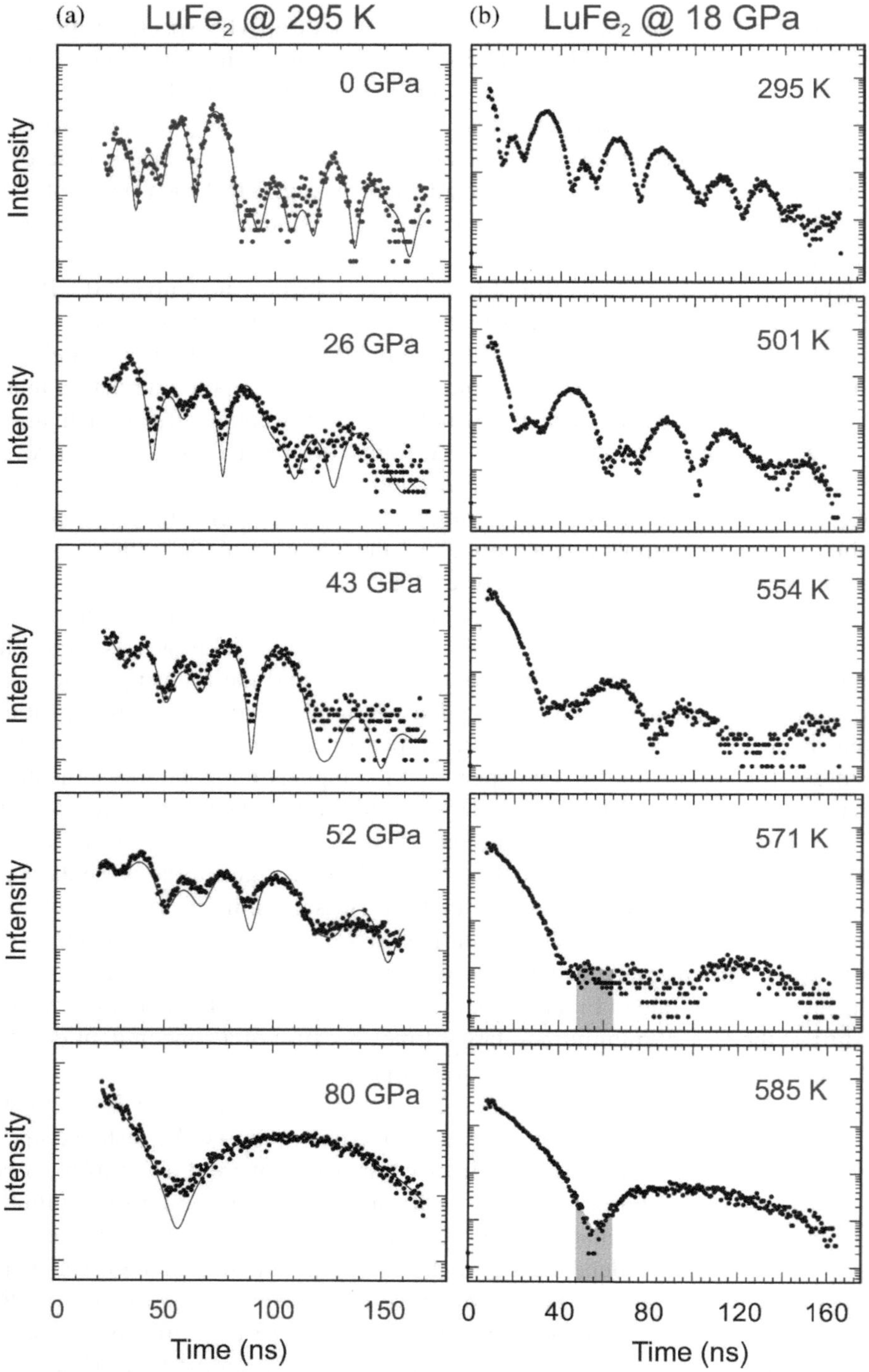

Figure 2. (a) NFS spectra of $LuFe_2$ at 295 K and various pressures. (b) NFS spectra of $LuFe_2$ at 18 GPa and various temperatures.

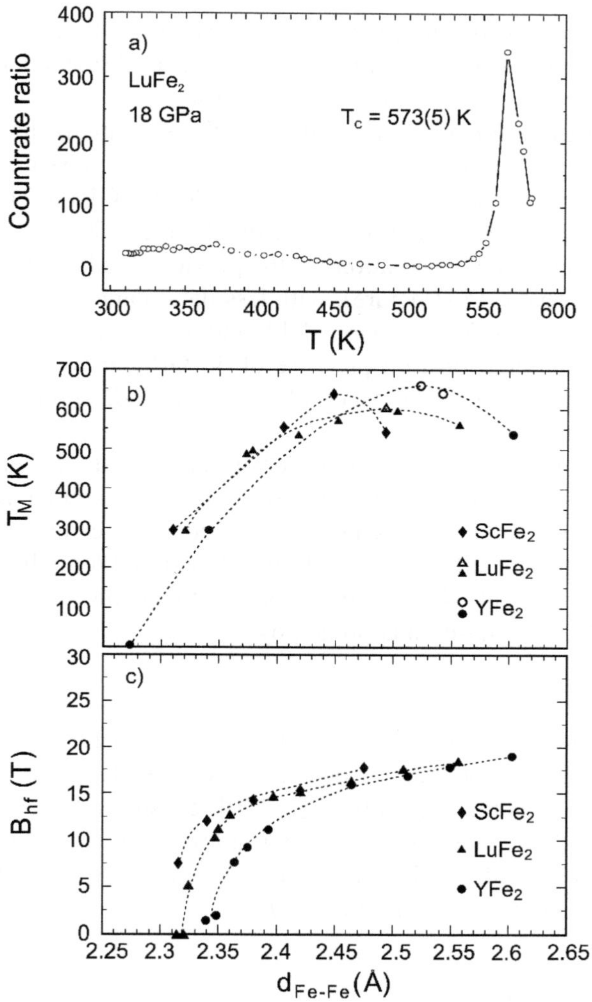

Figure 3. (a) Normalized inverse NFS countrate at the Bessel minimum observed for $LuFe_2$ at 18 GPa above the ferromagnetic ordering temperature T_C (see shaded area in Figure 2(b)). (b) Magnetic ordering temperatures T_M, obtained from NFS (full symbols) and from XRD [15] (open symbols) studies, and (c) magnetic hyperfine fields B_{hf} in $LuFe_2$, $ScFe_2$ and YFe_2 at room temperature and various pressures as a function of the Fe–Fe distances d_{Fe-Fe}. The data of YFe_2 and, in part, of $ScFe_2$ are from [4, 5]. The Fe–Fe distances d_{Fe-Fe} were derived from XRD studies [16, 18]. The dotted lines in (b) and (c) are guides to the eyes; for $ScFe_2$, the fm–afm transition at 35 GPa is indicated by a double bar.

Figure 3(c) exhibits the magnetic hyperfine fields B_{hf} of $LuFe_2$, $ScFe_2$, and YFe_2, measured at various pressures at room temperature, plotted also versus d_{Fe-Fe}. The continuous decrease of B_{hf} down to $d_{Fe-Fe} = \sim 2.37$ Å (corresponding for $ScFe_2$ to ~30 GPa and for $LuFe_2$ to ~40 GPa) as well as for YFe_2 to ~2.40 Å (corresponding to ~50 GPa) indicates, when compared with the variation of T_C

values and extrapolated to saturation values $B_{\mathrm{hf}}(p, T = 0\ \mathrm{K})$, also to a continuous decrease of the Fe band moment, assuming, as discussed in [4], a direct proportionality between B_{hf} and the Fe moment μ_{Fe}.

The most interesting magnetic property for all 3 systems is the initial increase of the ferro-magnetic (fm) ordering temperatures T_{C} with pressures (Figure 3(b)). We observe for $LuFe_2$ first an increase from 562 K at 0 GPa to 605 K at 10 GPa and then a decrease to 490 K at 30 GPa and finally to 295 K at about 75 GPa (see Figure 2(a)), similar to the observations in the previous studies of YFe_2 [4, 5]. For $ScFe_2$ (with hexagonal C14 structure), with less data points, a similar behavior of T_{C} is indicated. We know from previous NFS and Mössbauer studies that $ScFe_2$ is ferromagnetically ordered up to a pressure of 35 GPa, above which a transition to antiferromagnetic (afm) order takes place, evidenced, similar to $TiFe_2$, by a nonmagnetic 2a site [5, 14, 19].

The variation of the ferromagnetic ordering temperature T_{C} as well as of the magnetic moments (via the hyperfine fields B_{hf}) is very similar for the 3 investigated RFe_2 systems. It can be discussed in a model for ferromagnetism with $3d$ band moments in intermetallic compounds, developed by Wohlfahrt [20]. Applying this model for the present RFe_2 systems, the ferromagnetic ordering temperature T_{C} is determined in a mean-field approach by

$$T_{\mathrm{C}} = A_{\mathrm{Fe\text{–}Fe}} \mu_{\mathrm{Fe}}^2, \tag{1}$$

where $A_{\mathrm{Fe\text{–}Fe}}$ accounts for the ferromagnetic exchange between the Fe band moments μ_{Fe}. These band moments are depending sensitively on the $d_{\mathrm{Fe\text{–}Fe}}$ distance, the $3d$ bandwidth is increasing with $d_{\mathrm{Fe\text{–}Fe}}^{-5}$ [20], causing a concomitant decrease of the $3d$ electron density-of-states at the Fermi level, $D(E_{\mathrm{F}})$. Taking the variation of the band moment with the lattice spacing into account, Wohlfahrt *et al.* [20] derived the following pressure variation of T_{C}:

$$\frac{\mathrm{d}T_{\mathrm{C}}}{\mathrm{d}P} = \frac{5\kappa T_{\mathrm{C}}}{3} - \frac{\kappa\alpha}{T_{\mathrm{C}}}, \tag{2}$$

where κ is the compressibility and α is a term which accounts for the variation of the band moment with the lattice spacing. The observed initial increase of T_{C} with pressure points to a dominance of the first term of Equation (2), indicating a strong increase of the ferromagnetic exchange via a RKKY mechanism, overcompensating the concomitant decrease the band moment, the latter contained in the second term of Equation (2). At higher pressures, the second term becomes dominant and $\mathrm{d}T_{\mathrm{C}}/\mathrm{d}p$ changes its sign to a negative value, as experimentally observed. This proves that Equation (2) can qualitatively explain the observed variation in all three systems at pressures up to about 30 GPa, since T_{C} and μ_{Fe} as well as the Fe–Fe distances are very similar. At higher pressures and smaller moments the behaviour changes more drastically, possibly also due to afm interactions and, finally, due to a complete loss of the band moment. This loss of magnetism was observed in YFe_2 by additional measurements at low temperature around 90 GPa, corresponding to $d_{Fe–Fe}$ φ 2.3 Å [4, 5]. From the very similar variation of B_{hf} and T_M in $LuFe_2$ and YFe_2 with C15 structure, a similar critical value of $d_{Fe–Fe}$ can be expected for $LuFe_2$. $ScFe_2$ with C14 structure may behave different from YFe_2 and $LuFe_2$, especially above the fm–afm transition around 35 GPa, then similar to $TiFe_2$ [19]. A more detailed account will be given elsewhere [21].

4. NIS high-pressure study of phonon density-of-states in hcp iron

The new technique of nuclear inelastic scattering (NIS) introduced in 1995 [2] is especially suited to measure phonon density-of-states (DOS) of Mössbauer nuclei. The first high-pressure NIS studies with the 14.4 keV resonance were devoted to metallic iron, where the phonon-DOS of cubic (bcc, α-Fe) and, after the pressure induced phase transition around 13 GPa, of the hexagonal (hcp, ε-Fe) phase was studied first at the ESRF up to 42 GPa [6] and then at the APS at pressures up to 153 GPa [7]. The phonon DOS of ε-Fe and the derived thermodynamic parameters, e.g., the sound velocities, are of actual geophysical interest, since this iron phase is the main component of the Earth's solid inner core. Here we report on recent NIS experiments performed at the ESRF, where the phonon DOS in α-Fe and ε-Fe was measured up to 40 GPa with a spectral resolution of 3.5 meV [13], an considerable improvement in comparison to ~6.5 meV of the first study [6]. Here we put our emphasis on ε-Fe and use the texture, often produced in a high-pressure cell, to extract information on the phonon DOS as seen parallel and perpendicular to the hexagonal c-axis of ε-Fe. It is known that ε-Fe orients, after the bcc-hcp transition around 13 GPa, with the c-axis preferentially parallel to the load axis of the high-pressure cell, which is the diamond anvil axis in the DAC (see Figure 1 lower right panel).

The experimental set-up, in particular the special DAC used for ^{57}Fe-NIS experiments is described in Chapter 2. Typical NIS spectra recorded for α-Fe (bcc) at ambient pressure and for ε-Fe (hcp) at 40 GPa are shown in Figure 4(a). The spectra exhibit drastic differences due to (i) the different structures and (ii) the different binding strength, the latter also reflected by the intensity of the (inelastic) side wings, which scale with $1 - f_{LM}$, where f_{LM} is the Lamb–Mössbauer factor for the elastic (recoil-free) scattering of the gamma radiation [2].

The corresponding phonon DOS of α-Fe (at 0 GPa) and ε-Fe (at 40 GPa), derived from the measured NIS spectra by subtraction of multi-phonon excitations and by the elastic line [2], are shown in Figure 4(b). They exhibit characteristic spectral features (peaks) for the α- and ε-phase and, as already visible in the NIS spectra, a clear shift of all spectral features to higher energy. Additional high-pressure NIS spectra of α-Fe may be found in [7, 13]. Here we concentrate on ε-Fe, where one observes systematic differences in the phonon DOS for spectra taken at 28 and 40 GPa with two different orientations, 0° and 75°, of the h.p. cell with respect to the beam (see Figure 5). This indicates the existence of a preferred orientation in the ε-Fe sample with implications on the phonon spectra.

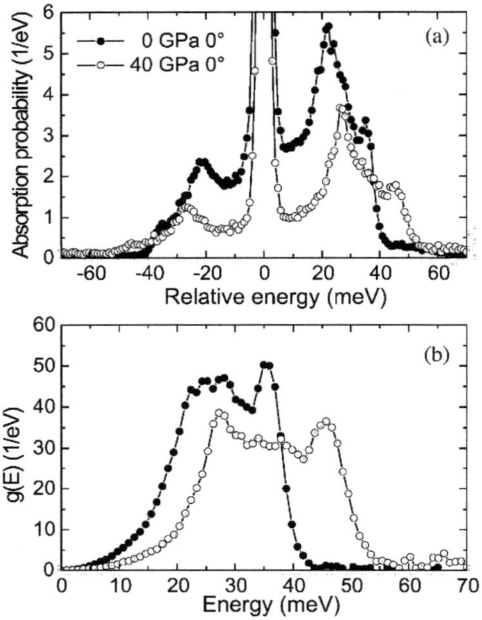

Figure 4. (a) NIS spectra in α-Fe (0 GPa, full dots) and in ε-Fe (40 GPa, open dots). The NIS spectra are shown after normalisation according to the procedure in [2]. The intensity of the inelastic part (sidebands) decreases by about 50% between 0 and 40 GPa, reflecting the change in f_{LM} from 0.80 to 0.88. (b) Phonon-DOS $g(E)$ of α-Fe (0 GPa, full dots) and ε-Fe (40 GPa, open dots) derived from the NIS spectra in (a).

In order to explain the dependence of the observed phonon DOS on the orientation of the h.p. cell, theoretical calculations of the phonon dispersion relations and the derived phonon DOS were performed for ε-Fe with an *ab initio* method described in [7, 13, 22]. Figure 6 exhibits the dispersion relations for different directions in the Brillouin zone of the hcp lattice calculated with the lattice parameters of ε-Fe at 40 GPa [23]. The dispersion relations were used to derive the integral phonon DOS $g(E)$ for a polycrystalline sample. The theoretical DOS is shown in Figure 7(a) and compared in Figure 7(d) with the experimental DOS, obtained, in order to randomize texture effects, by adding the normalized DOS of the NIS spectra measured with 0° and at 75° orientations.

Figures 7(b) and (c) exhibit the calculated phonon DOS as seen parallel and perpendicular to the hexagonal c-axis. The projected phonon DOS along the c-axis exhibits two sharp features of the two optical phonon branches at the Γ point of the Brillouin zone; the projected DOS perpendicular the c-axis, averaged over the (a, b) plane, exhibits a multi-peaked structure with a maximum around 38 meV.

In order to explain the derived experimental phonon DOS shown in Figures 7(e, f), one has to remind that the probability of nuclear inelastic scattering varies with $\cos^2\vartheta$, where ϑ is the angle between the k-vector of the incident synchrotron radiation and the polarization vector of the excited or annihilated phonons, as proved by

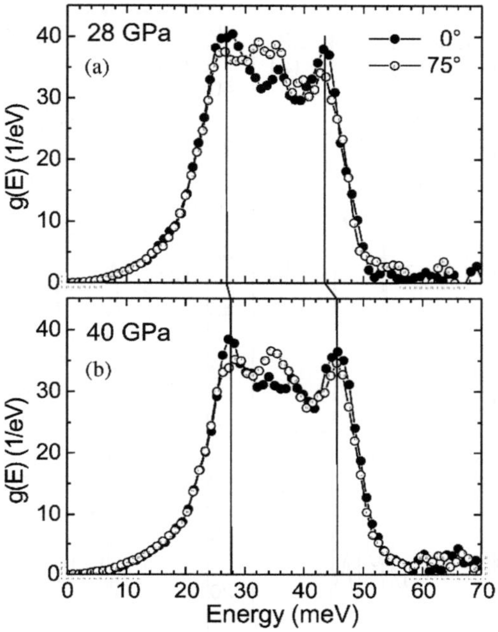

Figure 5. Phonon DOS, $g(E)$, of ε-Fe (a) at 28 GPa and (b) at 40 GPa. Full dots correspond to measurements performed at an angle of 0° between the SR beam and the axis of the diamonds, open dots to measurements performed at an angle of 75°.

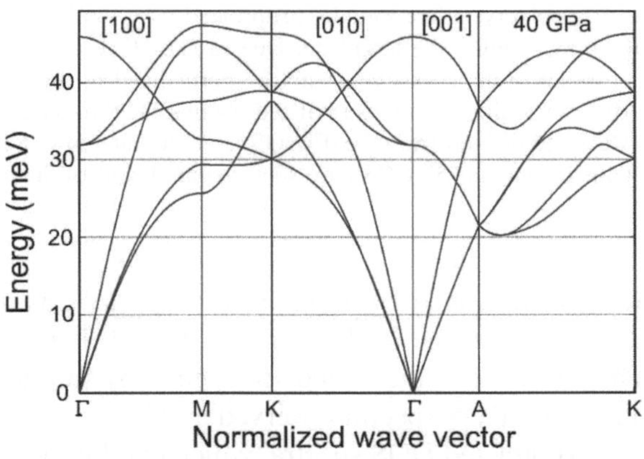

Figure 6. Dispersion relations of ε-Fe at 40 GPa calculated by an *ab initio* method [13, 23] with the standard assignments of the directions in the first Brillouin zone.

NIS studies of single crystalline samples [24, 25]. Assuming a preferred orientation of the c-axis of ε-Fe along the diamond axis, then the theoretical DOS as displayed in Figure 7(b) should be enhanced and the one perpendicular to the c-axis shown in Figure 7(c) should be diminished in the experimental 0°-NIS spectra. When measured at 75° with respect to the diamond axes, just the opposite should be the case. Other crystallographic directions, e.g., the (1,0,1) direction, should vary much

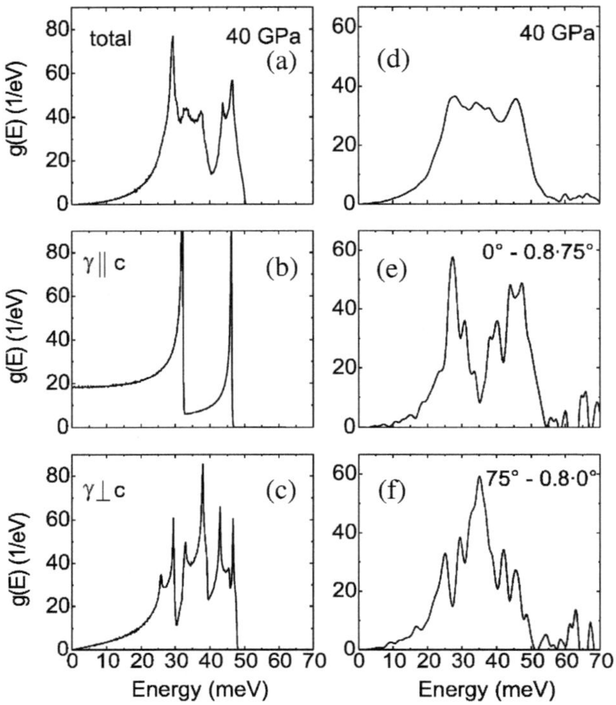

Figure 7. (a)–(c): Theoretical phonon DOS of ε-Fe derived from the dispersion relations of Figure 6: (a) total phonon DOS for a polycrystalline sample, (b) projection of the DOS along the c-axis and (c) perpendicular to the c-axis within the a,b-plane. Due to van Hove singularities of the optical modes, the DOS seen parallel to the c-axis could not be normalized in (b). (d)–(f): Experimental phonon DOS of ε-Fe (d) from addition of the normalized phonon DOS measured at 0° and 75° (see Figure 3), (e) phonon DOS as obtained from the 0° DOS after subtraction of 80% of the 75° DOS, (f) phonon DOS as obtained from the 75° DOS after subtraction of 80% of the 0° DOS.

less, since they are less influenced by a texture of the c-axis as directly reflected by the corresponding peaks in the XRD spectra [26, 27]. It is then obvious that texture effects can be enhanced in the 0° and 75° spectra by subtracting appropriate amounts of the normalized DOS spectra observed at 75° and 0°, respectively. Figures 7(e, f) show such subtracted DOS spectra, which exhibit the most pronounced differences when 80% of the spectra of the other direction were subtracted. Considering that the experimental spectral resolution of 3.5 meV is not contained in the theoretical spectra, then there is very good agreement in the characteristic spectral features between the theoretical total DOS and the averaged experimental DOS in Figures 7(a) and (d) as well as between the projected DOS and the subtracted experimental DOS in Figures 7(b, c) and (e, f).

With this information we attribute the peaks at 27 (28) meV and 44 (46) meV in the 28 (40) GPa DOS spectra to the optical modes seen along the c-axis. This interpretation is supported by recent Raman studies of ε-Fe [28, 29], where the

Raman active E_{2g} mode was observed at frequencies of 215 cm^{-1} (28 GPa) and 225 cm^{-1} (40 GPa), which corresponds exactly to the maxima at lower energy in the measured phonon DOS.

The derived thermodynamic and elastic parameters of ε-Fe can be also used to look for a possible anisotropy with respect to the c-axis. It may be reminded that the observed anisotropy in the sound velocities of the Earth's inner core were speculatively attributed to a highly textured ε-Fe core [30, 31]. The present values for the average sound velocities v_D, derived from the slope of the phonon DOS in the low energy range from 0 to 14 meV [13], with $v_D = 4.47(6)$ and 4.43(6) km/s at 28 GPa and $v_D = 4.70(6)$ and 4.49(6) km/s for 40 GPa at 0° and 75°, respectively, indicate, still within the error bars, slightly higher v_D values along the c-direction. The difference is more pronounced for the derived Debye temperatures with $\Theta_D =$ 507(5) K and 499(5) at 28 GPa and $\Theta_D = 550(5)$ K and 527(5) K at 40 GPa at 0° and 75°, respectively. Future NIS studies on highly-textured ε-Fe samples, performed at higher pressures and higher statistical accuracy, may allow to answer the question about anisotropic elastic properties of ε-Fe with implications for the Earth's core.

5. Summary

High-pressure experiments applying both NFS and NIS have made a big progress in the last few years; studies of extremely small samples (in the μg and even sub-μg range) are possible at pressures well above 1 Mbar, where other methods have difficult or no access. The option of high or low temperatures has been already widely applied for NFS. First NIS studies at high pressure and high temperature were just recently performed at the APS [32].

NFS h.p. studies of the RFe_2 Laves phases deliver detailed information on the variation of the magnetic properties as function of pressure and temperature, together with accompanying XRD h.p. studies, a magnetic phase diagram has been established.

NIS h.p. studies enable in an unprecedented way the study of the phonon DOS in iron well above 1 Mbar with access to important elastic and thermodynamic parameters of geophysical interest. Texture can be used for a *mode-selective analysis* of the DOS and for a search of an anisotropy in the sound velocities of ε-Fe. Similar ^{57}Fe-NIS h.p. studies are presently applied to Invar alloys as well as to Fe(2+) systems with high-spin/low-spin transitions [33]. Also the ^{119}Sn-resonance has been successfully used for high pressure NIS studies [34].

Acknowledgements

We want to thank Bryan Doyle, Olaf Leupold, and Alexander Chumakov for their expert help with the NFS and NIS experiments as well as for friendly cooperation. The enriched RFe_2 samples were prepared by Jan Zukrowski. In particular we want to mention the various contributions of our former members of the Paderborn group, Tobias Friedmann and Rainer Lübbers to the present work. Finally we want to thank Dario Alfè for his calculations and for useful discussions.

Work supported by the BMBF (05 KS1PPB/3 and 05 KS1PPA/5).

References

1. Hastings, J. B., Siddons, D. P., van Bürck, U., Hollatz, R. and Bergmann, U., *Phys. Rev. Lett.* **66** (1991), 770.
2. Seto, M. *et al.*, *Phys. Rev. Lett.* **74** (1995), 3828; Sturhahn, W. *et al.*, *Phys. Rev. Lett.* **74** (1995), 3832.
3. Jayarama, A., *Rev. Mod. Phys.* **55** (1983) 65; Mao, H.-K. and Hemley, R. J. (eds), *Ultrahigh Pressure Mineralogy; Physics and Chemistry of the Earth's Deep Interior*, Mineralogical Society of America, Washington, DC, 1998.
4. Lübbers, R., Wortmann, G. and Grünsteudel, H. G., *Hyp. Interact.* **123/124** (1999), 529.
5. Lübbers, R., Rupprecht, K. and Wortmann, G., *Hyp. Interact.* **128** (2000), 115.
6. Lübbers, R., Grünsteudel, H. F., Chumakov, A. I. and Wortmann, G., *Science* **287** (2000), 1250.
7. Mao, H.-K. *et al.*, *Science* **292** (2001), 914.
8. Rüffer, R. and Chumakov, A. I., *Hyp. Interact.* **97–98** (1996), 589; see also: http://www.esrf.fr/exp_facilities/ID18/index.htlm with a link to ID22N.
9. ESRF Highlights 2001, p. 113; see also: http://www.esrf.fr./info/highlights/2001/methods/METH3.html.
10. Pasternak, M. P. and Taylor, D. E., In: G. Long and F. Grandjean (eds), *Mössbauer Spectroscopy Applied to Magnetism and Material Science*, Vol. 2, Plenum, New York, 1996, p. 167.
11. Pleines, M. *et al.*, *Hyp. Interact.* **120/121** (1999), 181.
12. Lübbers, R., PhD thesis, Paderborn, 2000.
13. Giefers, H., Lübbers, R., Rupprecht, K., Wortmann, G., Alfe, D. and Chumakov, A. I., *High Press. Res.* **22** (2002), 451; see also: Lübbers, R. *et al.*, *ESRF Highlights*, 2000, p. 48.
14. Rupprecht, K., Lu, J., Wortmann, G., Prezewoznik, J. and Zukrowski, J., *Acta Physica Polonica A* **100** (2001), 789.
15. Rupprecht, K. *et al.*, *High Press. Res.* **22** (2002), 189.
16. Friedmann, T., Diploma thesis, Paderborn, 2001.
17. van Bürck, U., Siddons, D. P., Hastings, J. B., Bergmann, U. and Hollatz, R., *Phys. Rev. B* **46** (1992), 6207.
18. Reiß, G., PhD thesis, Paderborn, 2000.
19. Lu, J., Hesse, H. J., Zukrowski, J., Przewoznik, J. and Wortmann, G., In: Ortalli, I. (ed), *Conf. Proc. "ICAME-95"*, Vol. 50, SIF, Bologna 1996, p. 243.
20. Wohlfahrt, E. P., In: *Physics of Solids under High Pressure*, North-Holland, 1981, p. 175; Wagner, D. and Wohlfahrt, E., *J. Phys. F* **11** (1981), 2417.
21. Rupprecht, K., to be published.
22. Alfè, D. *et al.*, *Nature* **401** (1999), 462.
23. Alfè, D., private communication.
24. Chumakov, A. I. *et al.*, *Phys. Rev. B* **56** (1997), 10578.
25. Paulsen, H. *et al.*, *Phys. Rev. Lett.* **86** (2001), 1351.
26. Wenk, H.-R. *et al.*, *Nature* **405** (2000), 1044.
27. Dubrovinsky, L. S. *et al.*, *Phys. Rev. Lett.* **84** (2000), 1720.
28. Merkel, S. *et al.*, *Science* **288** (2000), 1626.
29. Olijnyk, H., Jephcoat A. P. and Refson, K., *Europhys. Lett.* **53** (2001), 504.
30. Stixrude, L. and Cohen, R. E., *Science* **267** (1995), 1972.
31. Creager, K. C., *Nature* **356** (1992), 309 and *Science* **278** (1997), 1284.

32. Sturhahn, W., private communication.
33. Giefers, H. *et al.*, unpublished results.
34. Hu, M. Y., PhD thesis, Northwestern University, Chicago, 1999; Giefers, H. *et al.*, unpublished results.

Hyperfine Interactions **144/145:** 119–127, 2002.

High-Pressure Mössbauer Spectroscopy of Perovskite Iron Oxides

S. NASU[1], T. KAWAKAMI[1], S. KAWASAKI[2] and M. TAKANO[2]
[1] *Division of Materials Physics, Department of Physical Science, Graduate School of Engineering Science, Osaka University, Toyonaka, Osaka 560-8531, Japan*
[2] *Institute for Chemical Research, Kyoto University, Uji, Kyoto-fu 611-0011, Japan*

Abstract. Using a diamond anvil cell and a low-temperature cryostat with a superconducting solenoid, high-pressure ^{57}Fe Mössbauer spectroscopy with a radioactive source at 4.5 K under external magnetic fields up to 7 T has been developed. Pressure–temperature magnetic phase-diagrams for perovskite iron oxides, $SrFeO_3$, $CaFeO_3$ and $Sr_2LaFe_3O_9$ are presented up to about 70 GPa. High-pressure Mössbauer measurements under external magnetic fields proved that the electronic ground states of these oxides switch to uniform-charge and ferromagnetic (and most probably metallic) states under extremely high pressure.

Key words: DAC, high pressure, perovskite, $CaFeO_3$, $SrFeO_3$, $Sr_2LaFe_3O_9$, magnetic phase diagram.

1. Introduction

Various types of metal–insulator transitions and related phenomena in strongly correlated electron systems such as 3*d* transition metal oxides have been extensively studied [1]. Perovskites containing Fe–O octahedra, in which the iron has an unusually high valence state of Fe^{4+} (d^4), provide us with a unique and interesting class of compounds. Measurements of photoelectron spectra and the Hall effect for $SrFeO_3$ have suggested that the electric and magnetic properties are dominated by oxygen hole character [2, 3]. Fe^{4+} is expressed as $Fe^{3+}\underline{L}$, where $\underline{L}$ stands for a hole in the oxygen octahedron. The dynamics of holes leads to a variety of ground states such as metallic antiferromagnetism with a screw spin structure for $SrFeO_3$ [4, 5], a metal ($T > 290$ K, orthorhombic) to semiconductor ($T < 290$ K, monoclinic) transition coupled to the breathing phonon mode for $CaFeO_3$ [6–9]. Furthermore, $Sr_2LaFe_3O_9$, with a formal valence of $Fe^{11/3+}$, may be considered as $SrFeO_3$ doped with electrons so that the hole concentration is decreased to 2/3. Although the O $2p$ holes remain delocalized in $SrFeO_3$, those in $Sr_2LaFe_3O_9$ become confined in two-thirds of the oxygen (111) layers, or the doped electrons become confined in one-third of these layers, so that two kinds of Fe sites are generated in such a way that $3(Fe^{3+}\underline{L}^{2/3})$ (paramagnetic metal, $T > 200$ K) converts to $2Fe^{3+} + Fe^{3+}\underline{L}^2$ (antiferromagnetic (AF) insulator, $T < 207$ K) [10–

12]. The accompanying structural change is known to be much smaller [10–13] than for $CaFeO_3$.

Fe^{4+} oxides generally change their electronic properties remarkably under high-pressure [14–17]. We have shown that the transition in $CaFeO_3$ occurred at an almost constant temperature of 290 K below 17 GPa but disappears above 20 GPa [17] and pressure induces in $Sr_2LaFe_3O_9$ the switching from the charge-disproportionate antiferromagnetic state to the charge-uniform (most probably metallic) ferromagnetic state [18, 19].

In this paper, we report on the pressure–temperature magnetic phase-diagrams of $SrFeO_3$, $CaFeO_3$ and $Sr_2LaFe_3O_9$ determined by using high-pressure ^{57}Fe Mössbauer spectroscopy in a diamond anvil cell (DAC) and a low-temperature cryostat with a superconducting solenoid. Especially the effects of the external magnetic field on the Mössbauer spectra under high pressure will be discussed. We present some direct evidence of the electronic ground states switching to the charge-uniform ferromagnetic states under extremely high pressure.

2. Experimental

The high-pressure ^{57}Fe Mössbauer measurements were performed in transmission geometry using a DAC of Basset type [20]. ^{57}Fe-enriched (usually 50%) sample powder and fine ruby powder were enclosed in the hole of a Re gasket with a diameter in the range 200 to 600 μm. The pressure–transmitting medium was a 4 : 1 methanol : ethanol solution. Ruby manometry was used to determine the pressure at 300 K and 77 K. The pressure at 4.5 K is assumed to be the same as the pressure at 77 K. A high-density γ-ray source of 370 MBq ^{57}Co in Rh (0.5 mm × 0.5 mm in size) and a common γ-ray source of 975 MBq ^{57}Co in Rh (4 mm in diameter) were used. The velocity scale of the spectrum was relative to α-Fe at room temperature. A high-pressure Mössbauer cryostat with superconducting solenoids and the Mössbauer velocity transducer was used. For the detection of the γ-rays a Xe-filled proportional counter was used. The direction of the magnetic field is parallel to the direction of the γ-rays. High magnetic fields were applied to the absorber specimen and also inevitably to the γ-ray source. Since the Fe in Rh matrix has a local magnetic moment, although its spin glass behavior requires rather large concentration of magnetic impurity [21], one can expect induced hyperfine fields at the ^{57}Fe nuclei in Rh. In this investigation we determined the line-shapes of the ^{57}Fe Mössbauer spectra rising from the ^{57}Co in Rh source (975 MBq, 4 mm × 4 mm × 6 μm) experimentally at 4.5 K under external magnetic fields up to 7 T, as shown in Figure 1. The induced hyperfine fields at ^{57}Fe nuclei in the Rh source at 4.2 K were observed and determined as a function of H_{ext} from the measurements with a Pd-2%Fe single line absorber placed outside the cryostat. In these measurements, the induced hyperfine fields at ^{57}Fe nuclei in Rh have nearly the same magnitude but the opposite direction with respect to what was reported previously [22]. As shown in Figure 1, the observed total hyperfine fields are cancelled with each other

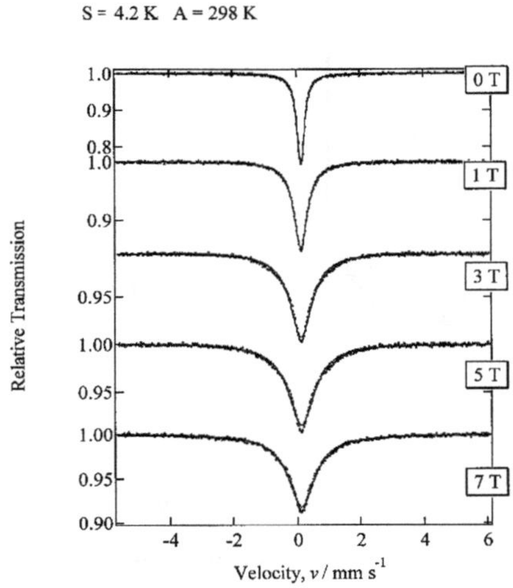

Figure 1. Typical ^{57}Fe Mössbauer spectra obtained from the combination of ^{57}Co in Rh (4 mm × 4 mm × 6 μm) source under high magnetic fields up to 7 T at 4.5 K placed inside of the high pressure Mössbauer cryostat with superconducting solenoids and Pd-2%Fe absorber under no magnetic field at 300 K placed outside of the cryostat. Velocity scale is relative to α-Fe at 300 K.

and rather single-line like Mössbauer absorption spectra have been observed for the Pd-2%Fe absorber placed outside the cryostat. The spectral shape under high magnetic field is rather complicated. However, in this investigation we focused our attention on the intensity change of the $\Delta m = 0$ (2nd and 5th) absorption lines in Mössbauer spectra induced by the longitudinally applied external magnetic fields.

3. Results and discussion

The high-pressure ^{57}Fe Mössbauer spectra of $SrFeO_3$ have been reported already in [23]. At ambient pressure and 298 K, the spectrum is a paramagnetic singlet typical (non-charge disproportionation paramagnetic phase; NCD P) for the valence state of Fe^{4+} (or $Fe^{3+}\underline{L}$). With a decrease in temperature, magnetic ordering (non-charge disproportionation magnetic order phase; NCD MO) occurs below 134 K. At 4.5 K, the spectrum consists of a magnetically split sextet of Fe^{4+} (or $Fe^{3+}\underline{L}$) and its hyperfine field is 33.1 T corresponding to Fe^{4+} ($t_{2g}^3 e_g^1$) in a high-spin configuration. Essentially the same behavior is preserved up to 15 GPa, although the magnetic ordering temperature increases to 210 K at 15 GPa. By measuring the count-rate at zero Doppler velocity as a function of temperature (Mössbauer thermal scanning method), the pressure-dependent magnetic ordering temperatures have been determined. In the pressure range between 15 and 17 GPa, the Mössbauer spectra change markedly and the magnetic ordering temperature suddenly increases to more than 400 K. These results have been summarized in a pressure–temperature (P–T) magnetic phase diagram as shown in Figure 2. No change in the crystal structure has been observed up to 56 GPa in a high-pressure X-ray diffraction study. In order

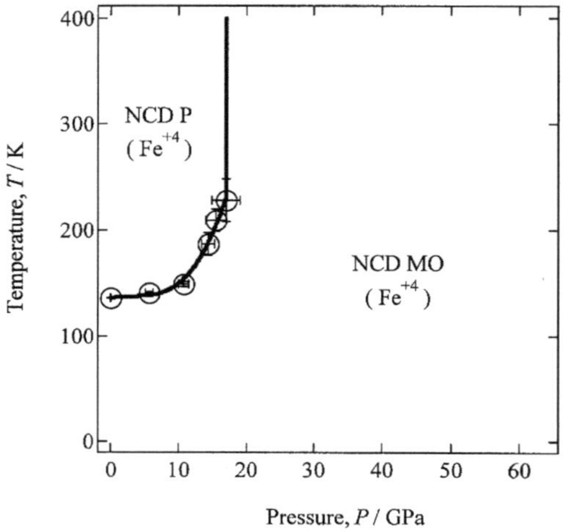

Figure 2. Pressure–temperature (P–T) magnetic phase diagram for $SrFeO_3$ determined by the high-pressure ^{57}Fe Mössbauer spectroscopy. Non-charge disproportionation paramagnetic phase is denoted as NCD P. Non-charge disproportionation magnetic ordered phase is denoted as NCD MO.

to determine the magnetic structure under high pressure, the Mössbauer spectra at 13 GPa and 4.5 K have been measured under longitudinally applied external magnetic fields. $SrFeO_3$ shows a helical spin structure at ambient pressure and the Mössbauer spectrum under external high magnetic fields shows an enhancement of the intensities of the $\Delta m = 0$ (2nd and 5th) absorption lines due to the spin-canting [23], as shown in the bottom spectrum of Figure 3. However, the spectral shapes at 4.5 K and 13 GPa are quite different from the shape in the spectrum measured at ambient pressure as shown in Figure 3. Mössbauer spectra under external magnetic fields show the nearly complete vanish of the 2nd and 5th absorption lines, which suggests that the magnetic moments of $SrFeO_3$ at 13 GPa and 4.5 K align in the direction of the external magnetic field. Even at 1.1 T almost all magnetic moments align in the direction of the external magnetic field suggesting a ferromagnetic coupling of the Fe magnetic moments at 13 GPa and 4.5 K. The remarkable line broadening observed under high magnetic fields is due to the line broadening of the ^{57}Co source in Rh, but the nearly complete disappearance of the 2nd and 5th absorption lines is clearly seen. From the above results, we can conclude that the spin–structure of $SrFeO_3$ shows a switching from a helical to a ferromagnetic one at a pressure of 13 GPa. The critical pressure at which the transition from helix to ferromagnetic occurs has not yet been determined precisely, although the preliminary results obtained show that the helical structure still survives at 5 GPa and 4.5 K.

The crystal structure of $CaFeO_3$ is of the $GdFeO_3$-type [6–8] and the Fe–O–Fe bond angle is about 160° at ambient pressure, lower than the 180° bonding angle found in $SrFeO_3$. Therefore the p–d hybridization effect in $CaFeO_3$ is weaker than in the case of $SrFeO_3$. $CaFeO_3$ shows a bad conductivity around room temperature in comparison with $SrFeO_3$ and the metal–insulator transition occurs at 290 K.

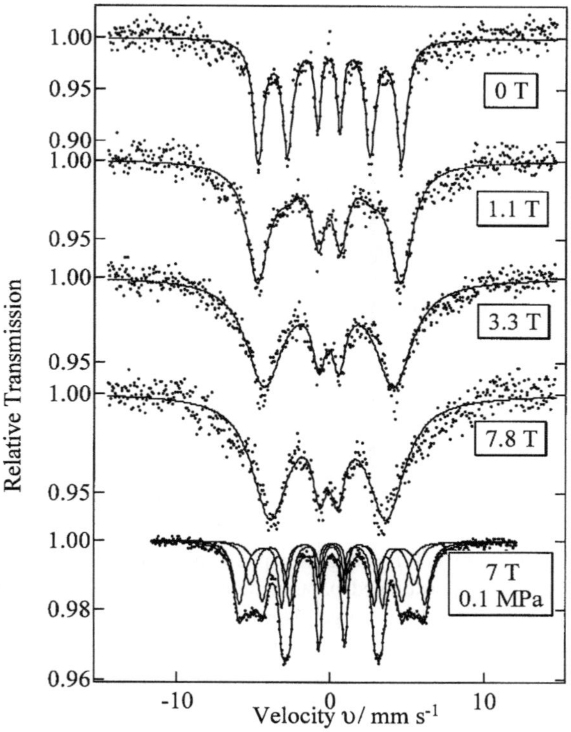

Figure 3. Typical ^{57}Fe Mössbauer spectra obtained from $SrFeO_3$ at 13 GPa and 4.5 K with and without longitudinally applied external magnetic fields of 1.1, 3.3 and 7.8 T. The high-magnetic-field Mössbauer spectrum at 4.5 K and 7 T under 0.1 MPa is shown in the lowest spectrum. The spectrum can be analyzed by using 4 sublattice model and shows clearly the enhancement of the intensities for the $\Delta m = 0$ absorption lines. The velocity scale is relative to α-Fe at 300 K.

At the same temperature the Mössbauer spectrum becomes split gradually into two absorption lines suggesting a charge disproportionation (charge separation: $2Fe^{4+}$ (or $Fe^{3+}\underline{L}$) $\leftrightarrow$ $Fe^{3+} + Fe^{5+}$ (or $Fe^{3+}\underline{L}^2$)) with two different charge states with decreasing temperature in a second-order transition. $CaFeO_3$ below 125 K as well as $SrFeO_3$ below 134 K show an antiferromagnetic long-range order with a helical spin structure. Pressure effects on the charge disproportionation and the magnetic ordering of $CaFeO_3$ have been investigated up to 65 GPa and summarized in the pressure–temperature phase diagram shown in Figure 4 [17]. The magnetic ordering temperature increases with increasing pressure, but the charge disproportionation temperature does not show any significant change up to 17 GPa and is suddenly suppressed above 20 GPa. Below 17 GPa the localization of charge at the Fe sites occurs and the resultant Mössbauer spectra consist of two components rising from Fe^{3+} and Fe^{5+} (or $Fe^{3+}\underline{L}^2$) below 290 K. Above 20 GPa, the spectra show a single component with a uniform charge state of Fe^{4+} (or $Fe^{3+}\underline{L}$,). Similarly to the case of $SrFeO_3$, the Mössbauer spectra under a longitudinally applied

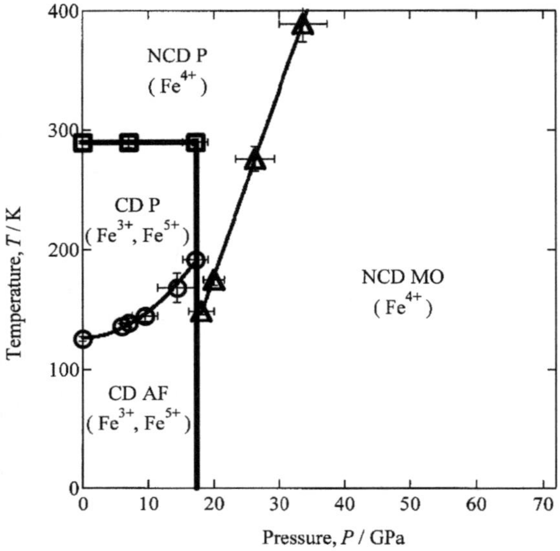

Figure 4. Pressure–temperature (P–T) magnetic phase diagram for $CaFeO_3$. The low-pressure phases are a non-charge disproportionated paramagnetic phase (NCD P), a charge disproportionated paramagnetic phase (CD P) and a charge disproportionated antiferromagnetic phase (CD AF) depending on the temperature and the pressure. The high-pressure phase is a non-charge disproportionated, magnetically ordered phase (NCD MO). Each boundary determined by the experiments is shown as squares, triangles and circles.

external magnetic field have been measured for $CaFeO_3$ at 39 GPa and 4.5 K. The Mössbauer spectrum for $CaFeO_3$ at 39 GPa and 4.5 K under an external magnetic field of 7 T applied longitudinally to the γ-ray propagation direction clearly shows the disappearance of the $\Delta m = 0$ lines, suggesting a ferromagnetic coupling of Fe in this compound. However, the response to the external magnetic fields of the magnetic moments in $CaFeO_3$ differs slightly from the case of $SrFeO_3$. Even at 3 T the Fe magnetic moment in $CaFeO_3$ does not align to the direction of the external magnetic field, although an external magnetic field of 1.1 T is enough to align completely the Fe magnetic moment in $SrFeO_3$. Therefore, the electronic ground states of $SrFeO_3$ and $CaFeO_3$ under high pressure are ferromagnetic but the detailed electronic structure might be different in the two compounds. Ferromagnetic $CaFeO_3$ under high pressure has a rather large magnetic anisotropy energy as compared to $SrFeO_3$.

$Sr_2LaFe_3O_9$ has the same cubic perovskite structure as $SrFeO_3$ and may be considered as $SrFeO_3$ doped with electrons. At ambient pressure this compound shows a metal–insulator transition at 207 K. At the same temperature a magnetic transition occurs from a paramagnetic metal to an antiferromagnetic insulator. At ambient pressure a paramagnetic singlet corresponding to the charge non-disproportionated state was observed above 207 K, while the low temperature spectra consist of two magnetically split components. The change around 200 K is recognized as

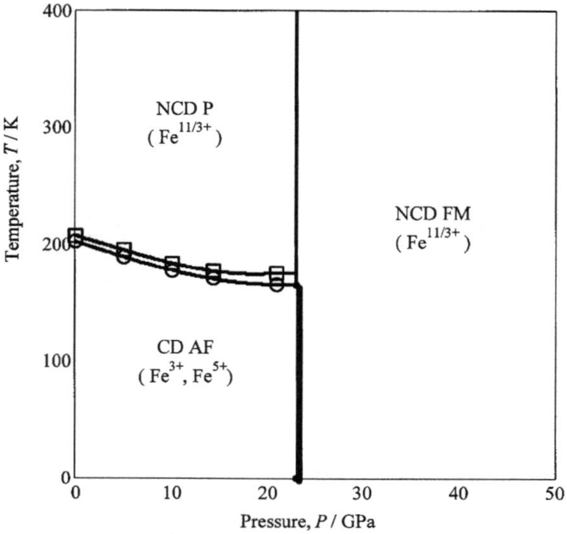

Figure 5. Pressure–temperature (P–T) magnetic phase diagram for $Sr_2LaFe_3O_9$ determined by the high-pressure ^{57}Fe Mössbauer spectroscopy. NCD P: Non-charge disproportionation paramagnetic phase. CD AF: Charge disproportionation antiferromagnetic phase. NCD FM: Non-charge disproportionation ferromagnetic phase. Boundaries between NCD P and CD AF were determined by using Mössbauer thermal scanning method and the transition between these phases is first-order indicating a slight hysteresis of the transition temperatures. Squares and circles show maximum and minimum transition temperatures at each pressure. Vertical dashed lines show the boundary between charge disproportionated antiferromagnetic phase and the charge non-disproportionated ferromagnetic phase, although the boundaries at 4.5 K and above 300 K were determined experimentally.

a first-order transition. Essentially the same behavior was observed up to 21 GPa, except for the decrease of the transition temperature. These pressure effects on $Sr_2LaFe_3O_9$ are summarized in the pressure–temperature phase diagram shown in Figure 5. In the pressure range 21 to 25 GPa the charge disproportionation is suppressed but the magnetic transition temperature is suddenly increased to more than 400 K. Figure 6 shows typical Mössbauer spectra obtained for $Sr_2LaFe_3O_9$ at 42 GPa and 4.5 K with and without longitudinally applied external magnetic fields. The intensities of the 2nd and 5th lines in the spectra nearly vanish when the external magnetic fields exceed 2 T, indicating that the magnetic moment of Fe in this compound is aligned parallel to the direction of the external magnetic fields. From these results we can conclude again that pressure switches the electronic ground state from the antiferromagnetic and charge disproportionated state to a ferromagnetic and charge non-disproportionated state. In the Mössbauer spectrum measured under an external magnetic field of 1 T, the 2nd and 5th lines are slightly visible as shown in Figure 6, indicating that the magnetic anisotropy of ferromagnetic $Sr_2LaFe_3O_9$ under pressure is stronger than in $SrFeO_3$ but weaker than in $CaFeO_3$.

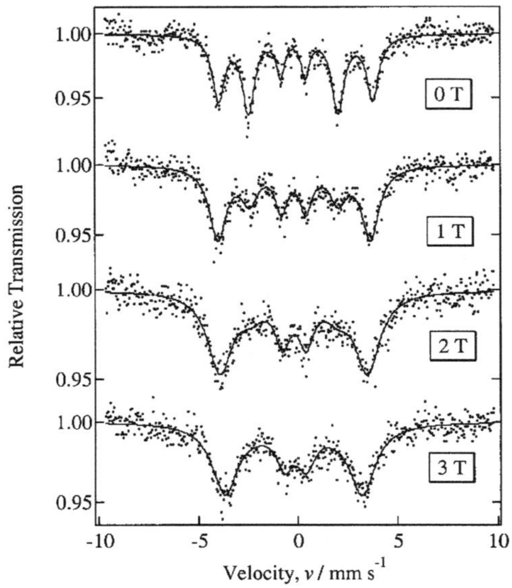

Figure 6. Typical ^{57}Fe Mössbauer spectra obtained for $Sr_2LaFe_3O_9$ at 42 GPa and 4.5 K with and without longitudinally applied external magnetic fields of 1, 2 and 3 T. The velocity scale is relative to α-Fe at 300 K.

4. Summary

The pressure–temperature magnetic phase-diagrams of $SrFeO_3$, $CaFeO_3$ and $Sr_2LaFe_3O_9$ have been determined by applying high-pressure ^{57}Fe Mössbauer spectroscopy using a DAC and a low-temperature cryostat with a superconducting solenoid. The effects of the external magnetic field on the Mössbauer spectra under high-pressure have been discussed in detail. We found direct evidence for the electronic ground states switching to charge-uniform ferromagnetic states from antiferromagnetic (with helical spin–structure) metal ($SrFeO_3$), antiferromagnetic (with helical spin–structure) insulator ($CaFeO_3$) and antiferromagnetic (with collinear spin–structure) insulator ($Sr_2LaFe_3O_9$) under extremely high pressure.

Acknowledgements

This work has been supported by Core Research for Evolutional Science and Technology (CREST) of the Japan Science and Technology Corporation (JST) and a Grant-in-Aid for COE Research (10CE2004) and for Scientific Research (A, 12304018) from the Ministry of Education, Culture, Sports, Science and Technology.

The authors are grateful to A. Fujimori, S. Endo, S. Morimoto, T. Sasaki, K. Kuzushita and T. Yamada for their respective contributions to this work.

References

1. Imada, M., Fujimori, A. and Tokura, Y., *Rev. Mod. Phys.* **70** (1998), 1211.
2. Bocqest, A. E., Fujimori, A., Mizokawa, T., Saitoh, T., Na-matame, H., Suga, S., Kimizuka, N., Takeda, Y. and Takano, M., *Phys. Rev. B* **45** (1992), 1561.
3. Hayashi, N., Terashima, T. and Takano, M., *J. Mater. Chem.* (to be published).
4. MacChesney, J. B., Sherwood, R. C. and Potter, J. F., *J. Chem. Phys.* **43** (1965), 1907.
5. Takeda, T., Yamaguchi, Y. and Watanabe, H., *J. Phys. Soc. Jpn.* **33** (1972), 967.
6. Takano, M., Nakanishi, N., Takeda, Y., Naka, S. and Takada, T., *Mater. Res. Bull.* **12** (1977), 923.
7. Morimoto, S., Yamanaka, T. and Tanaka, M., *Physica (Amsterdam)* **237B–238B** (1997), 66.
8. Woodward, P. M., Cox, D. E., Moshopoulou, E., Sleight, A. W. and Morimoto, S., *Phys. Rev. B* **62** (2000), 844.
9. Takeda, T. *et al.*, *Solid State Sci.* **2** (2001), 673.
10. Battle, P. D., Gibb, T. C. and Lightfoot, P., *J. Solid State Chem.* **84** (1990), 271.
11. Ishikawa, T., Park, S. K., Katsufuji, T., Arima, T. and Tokura, Y., *Phys. Rev. B* **58** (1998), R13 326.
12. Li, J. Q., Matsui, Y., Park, S. K. and Tokura, Y., *Phys. Rev. Lett.* **79** (1997), 297.
13. Matsuno, J., Mizokawa, T., Fujimori, A., Mamiya, K., Takeda, Y., Kawasaki, S. and Takano, M., *Phys. Rev. B* **60** (1999), 4605.
14. Takano, M., Nasu, S., Abe, T., Yamamoto, K., Endo, S., Takeda, Y. and Goodenough, J. B., *Phys. Rev. Lett.* **67** (1991), 3267.
15. Nasu, S., Abe, T., Yamamoto, K., Endo, S., Takano, M. and Takeda, Y., *Hyp. Interact.* **67** (1991), 529.
16. Rosenberg, G. Kh., Milner, A. P., Pasternak, M. P., Hearne, G. R. and Taylor, R. D., *Phys. Rev. B* **58** (1998), 10 283.
17. Kawakami, T., Nasu, S., Sasaki, T., Morimoto, S., Endo, S., Kawasaki, S. and Takano, M., *J. Phys. Soc. Jpn.* **70** (2001), 5701.
18. Kawakami, T., Nasu, S., Sasaki, T., Kuzushita, K., Morimoto, S., Endo, S., Yamada, T., Kawasaki S. and Takano, T., *Phys. Rev. Lett.* **88** (2002), 037602.
19. Kawakami, T., Nasu, S., Sasaki, T., Kuzushita, K., Morimoto, S., Endo, S., Kawasaki S. and Takano, M., *J. Phys. Condens. Matter* **14** (2002), 10 713.
20. Bassett, W. A., Takahashi, T. and Stook, P. W., *Rev. Sci. Instrum.* **38** (1967), 37.
21. Rusby, R. L., *J. Phys. F* **4** (1974), 1265.
22. Nasu, S., Sasaki, T., Kawakami, T., Tsutsui T. and Endo, S., *J. Phys. Condens. Matter.* **14** (2002), 11 167.
23. Nasu, S., *Hyp. Interact.* **128** (2000), 101.

Hyperfine Interactions **144/145:** 129–139, 2002.

Thin Films Made Fast and Modified Fast

A Mössbauer Illumination of Photon- and Ion-Beam Induced Processes

PETER SCHAAF*, ALEXANDER MÜLLER, ETTORE CARPENE and MICHAEL KAHLE
Universität Göttingen, Zweites Physikalisches Institut, Bunsenstrasse 7/9, D-37073 Göttingen, Germany; e-mail: pschaaf@uni-goettingen.de

Abstract. Thin films are playing a more and more important role for technological applications and there are many aspects of materials surface processing and thin film production, ranging from simple heat treatments to ion implantation or laser surface treatments. These methods are often very complicated, involving many basic processes and they have to be optimized for the desired application. Nuclear methods, especially Mössbauer spectroscopy, can be successfully applied for this task and some examples will be presented for laser-beam and ion-beam based processes.

Key words: laser, ion-beams, magnetic texture, iron carbides, CEMS, CXMS, MOMS, spin orientation.

1. Introduction

Coating formation and surface processing is becoming a key technology in materials science. Thin films are becoming more and more important in the application of smart materials. This covers the improvement of surface properties such as hardness, corrosion resistance, wear resistance, friction, electrical, thermal, magnetic, and optoelectronic properties for specific applications [1, 2]. Many techniques evolved for the preparation of thin films and coatings. Here, we present a method involving short pulsed laser irradiation for the direct synthesis of surface layers in reactive atmospheres, another advanced method of surface treatment. Irradiation of surfaces with short laser pulses of high intensity in a reactive atmosphere can result in a direct coating formation if the treatment parameters are properly adjusted [3–5]. Pulsed-laser nitriding and laser carburizing are attractive methods to improve metal surface properties, such as hardness, wear and corrosion resistance, with the advantage of simple experimental set-up, rapid treatment and precise process control.

In the last few years ion beam irradiation of ferromagnetic thin films, tens of nanometers thick, was found to be a very efficient way to induce magnetic anisotropy due to inverse magnetostriction effects [6, 7]. In particular, studies on magnetic texturing associated with noble-gas ion irradiation have been established

* Author for correspondence.

in nickel, iron and permalloy films [8–13], and the dependence on important parameters such as ion energy, ion fluence, substrate temperature, applied magnetic field and mechanical strain has been investigated. Also different mechanisms have been found as shown for Co/Pt multilayers [14, 15]. In most previous investigations, the magnetic properties were analyzed using methods probing the near-surface part of the films, such as magneto-optical methods or magnetic force microscopy [13–15]. Angle-dependent Mössbauer spectroscopy of ^{57}Fe marker-layers, placed in different depths of ^{nat}Fe films offers a very elegant way to gather depth-sensitive information on anisotropic magnetic hyperfine fields and spin orientations [16]. Up to now, this technique has mainly been applied to characterize structural and electronic properties of single crystals [17, 18]. A promising approach to gain information about the spin distributions in different sample layers by using depth-sensitive and angle-dependent Conversion Electron Mössbauer Spectroscopy is presented here in combination with magneto-optical Kerr effect, Rutherford backscattering spectrometry and X-ray diffraction investigations. Three 77 nm thick Fe films containing a 15 nm ^{57}Fe marker layer at different depths were deposited on Si(100) substrates and irradiated under well-defined mechanical stress with 200 keV Xe ions. The orientation distribution of the spins within the sample plane was analyzed by Mössbauer spectroscopy, tilting the sample with respect to the incident γ-beam and observing the angle dependent relative line intensity variation of the sextet lines. For the sample with the sensitive marker layer in the middle of the ferromagnetic film, all spins were found to be oriented exactly along the mechanical stress axis. This agrees with the angular properties of the macroscopic hysteresis curves measured by magneto-optic Kerr effect. However, in most of the previous investigations the magnetic properties were analyzed using methods probing either the near-surface part of the films, like magneto-optical methods or magnetic force microscopy [13–15], or they measure the magnetization of the whole film, like vibrating sample magnetometer or SQUID. Only recently, Nuclear Resonance Scattering (NRS) has been involved in determining spin structures [19, 20].

2. Experimental

Pure iron was used as starting material for the laser carburizing treatments, the samples having 1–1.5 mm thickness and were mechanically polished with SiC grinding paper followed by 1 μm diamond paste. A Siemens XP2020 XeCl excimer laser with $\lambda = 308$ nm, 55 ns (FWHM) pulse duration, and 8 Hz repetition rate was used for the excimer laser treatments. The laser fluence was set to the desired fluence (0.5 to 5 J/cm^2) with a spot size of 5×5 mm^2 employing a focusing beam homogenizer and a variable beam attenuator [3]. For the irradiations, the samples were mounted in a chamber, that was first evacuated to less than 10^{-3} Pa and then filled with nitrogen gas (99.999%) or methane (99.5%) at a pressure of 0.01–0.15 MPa prior to the treatment. Larger sample areas were treated by meandering (i.e. scanning) the surfaces relative to the laser beam by a motorized and computer-controlled sample stage [3]. The shift of each laser spot was set according to the desired overlap and the total number of laser shots.

The carbon profiles have been obtained by Rutherford Backscattering Spectrometry (RBS) with a 900 keV He^{++} beam and a backscattering angle of 165°. The data analysis was performed with the RUMP code [21]. Phase analysis was done by X-ray diffraction in Bragg–Brentano geometry and under 2–5° grazing incidence (GIXRD) employing Cu-K_α radiation. The iron samples have been analyzed by Conversion Electron and Conversion X-ray Mössbauer Spectroscopy (CEMS and CXMS) [3].

For the magnetic texturing experiments, three special films of 10×5 mm^2 size have been deposited on ultrasonic-cleaned, 0.5 mm thick Si(100) substrates, each 40×15 mm^2 in size. The natural iron films were deposited by electron-beam evaporation and the 15 nm thick ^{57}Fe marker-layers by an effusion cell. The marker-layer was placed either directly at the Fe/Si interface (sample I), in the middle of the ^{nat}Fe film (sample M) or at the surface of the film (sample S). The total iron film thickness was 77(3) nm for all three samples. The depositions were performed with the substrate held at room temperature in the same vacuum chamber with a base-pressure of 4×10^{-6} Pa without breaking the vacuum. These as deposited samples have been characterized by RBS, CEMS, XRD, and in addition by Magneto-Optical Kerr Effect (MOKE). Before irradiation the samples were mounted on a special target holder allowing the defined bending of the Si-substrate to a bending radius $R = 1$ m (curvature $1/R = 1$ m^{-1}) as controlled by a Dektak surface profiler. According to Stoney [22] this corresponds to a strain of 108(11) GPa in the iron film. Then, the samples were irradiated at room temperature with 200 keV Xe-ions to a fluence of 1×10^{16} cm^{-2}, provided by the Göttingen ion implanter IONAS at a pressure of about 10^{-5} Pa in the implantation chamber. This implantation energy corresponds to a mean implantation depth of 34 nm (SRIM2000 code [23]) and was found to induce the largest change in magnetization [12]. By using an electrostatic X–Y sweeping system ($f_x = 120$ Hz, $f_y = 5$ Hz) the ions were homogenously distributed over the irradiated area of 10×10 mm^2. The ion beam current was kept at about 0.3 μA, in order to avoid beam heating of the samples. After the irradiation the strain was relaxed by releasing the bending and the samples were again measured by RBS, CEMS, XRD, and MOKE, again in this sequence. Due to the limited thickness of the iron films and the high enrichment of the ^{57}Fe marker-layers (96%), almost all of the information for Mössbauer spectroscopy is encountered from the marker-layers. The MOKE-set-up is a PCSA-ellipsometer using a photo-elastic modulator and online field measurement [24], with a maximum applied field of about 1500 Oe provided by a Helmholtz coil.

3. Magnetic Orientation Mössbauer Spectroscopy – MOMS

Traditionally, conventional Mössbauer spectroscopy is performed with perpendicular (normal) incidence of the γ-beam. For magnetic sextets, the relative line intensities are depending on the angle θ between the γ-direction and direction of the hyperfine field B, the latter being co-linear with the spin direction ($I_{1,6}/I_{3,4}$ is always 3) [25–28]:

$$\frac{I_{2,5}}{I_{3,4}} = \frac{\frac{1}{2}(1-\cos^2\theta)}{\frac{1}{8}(1+\cos^2\theta)} = 4\frac{(1-\cos^2\theta)}{(1+\cos^2\theta)}. \tag{1}$$

This results in the line intensity ratios 3 : 2 : 1 for a random spin orientation (in space), 3 : 4 : 1 for all spin aligned in the film plane and thus being perpendicular to the γ-beam and 3 : 0 : 1 when all spins are pointing out of the film plane, i.e. line 2 and 5 of the sextet disappear. Now, knowing that for thin films and foils the spins are almost completely aligned in the film plane, one might ask if we can resolve the spin distribution within the plane by means of Mössbauer spectroscopy. This becomes possible when we tilt the sample normal with respect against the γ-direction by an angle α and then rotate the sample around its normal n by the angle φ. This is sketched in Figure 1.

The variation of the line intensities with the angle φ gives the information about the hyperfine field orientation in the film plane (i.e. in the ^{57}Fe marker-layers). This variation can be simulated by a linear combination of n magnetic orientations ψ_i:

$$\frac{I_2}{I_3}(\varphi) = 4\sum_{i=1}^{n} c_i \frac{1-\sin^2(\alpha)\sin^2(\varphi-\psi_i)}{1+\sin^2(\alpha)\sin^2(\varphi-\psi_i)} + 4\left(1-\sum_{i=1}^{n} c_i\right)\frac{1-\cos^2(\alpha)}{1+\cos^2(\alpha)}, \tag{2}$$

where c_i are the fractions of the spins pointing in the direction ψ_i in the film plane and the remaining spins are pointing out of plane (along the plane normal). This measurement of the sample for all φ with non-normal incidence is named

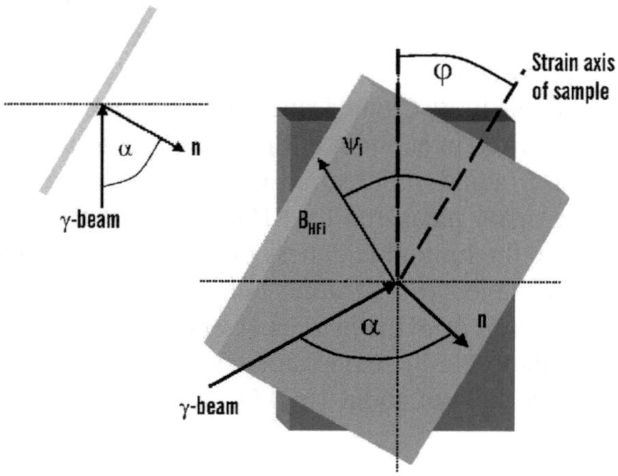

Figure 1. Schematic view of the angle-dependent Mössbauer setup (MOMS). ψ_i indicates the in-plane vector of the magnetization direction B_{HF}, α is the tilt-angle of the sample in the X–Y plane (angle between γ-direction and surface normal n and φ is the rotation angle of the sample around its normal).

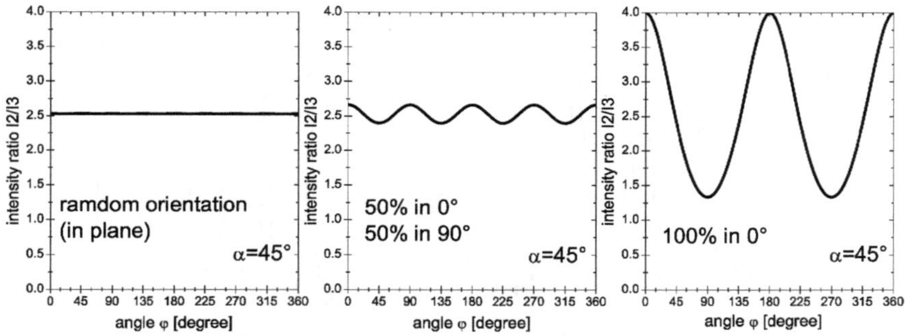

Figure 2. Characteristic patterns for MOMS of thin iron films with an incidence angle of $\alpha = 45°$ assuming all spins lying in the plane.

Magnetic Orientation Mössbauer Spectroscopy (MOMS), allowing to resolve the spin distribution in the sample.

Considering an intensity ratio $I_2/I_3 = 4$ for the α-iron from the normal incidence spectrum and thus a 100% in-plane alignment of the hyperfine field-vectors, Equation (2) can be simplified to:

$$\frac{I_2}{I_3}(\varphi) = 4\sum_{i=1}^{n} c_i \frac{1 - \sin^2(\alpha)\sin^2(\varphi - \Psi_i)}{1 + \sin^2(\alpha)\sin^2(\varphi - \Psi_i)}, \tag{3}$$

which gives some characteristic variations of the line intensities as shown in Figure 2 for $\alpha = 45°$. There it becomes obvious that MOMS can really distinguish different spin distributions and their directions in iron films.

This method was employed to investigate the spin distribution in thin iron films on silicon before and after Xe-ion irradiation. The results are presented below.

4. Results and discussion

4.1. LASER CARBURIZATION OF IRON

By irradiating pure iron substrates in methane atmosphere with a pulsed excimer laser, a rather thick and polycrystalline cementite layer was obtained [4]. Cementite (θ-Fe_3C) is of great technological importance for the mechanical properties of steels and iron alloys, however, it is hard to obtain as a single phase since it is typically embedded in the steel matrix. Only few investigations on the Fe_3C electronic structure [29], its mechanical [30] and thermodynamical [31] properties can be found in the literature, due to the difficulty in obtaining it in its pure state. Recently, a single-phase cementite film has been prepared by a special Physical Vapor Deposition technique [30], and no other method has been successful in synthesizing pure cementite.

The Mössbauer spectra and their analyses are shown in Figure 3 for an iron substrate being excimer laser treated with 4 J/cm^2 in 2 bar CH_4 and 33×33 pulses.

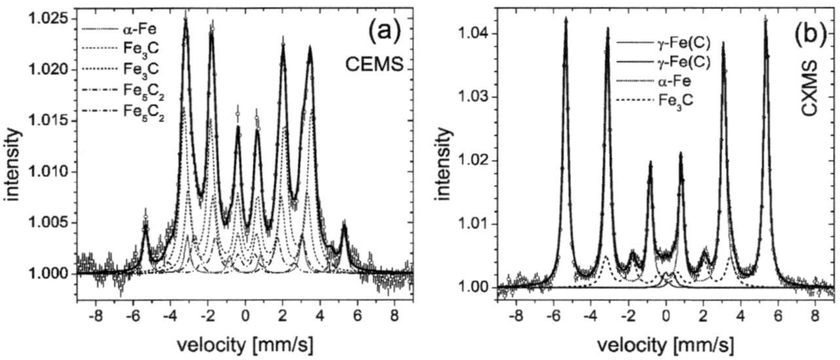

Figure 3. Mössbauer spectra of excimer laser carburized iron, left CEM spectrum, right CXM spectrum (4 J/cm^2, 33 × 33 pulses, 2 bar CH_4).

The CEM spectrum, with an information depth of about 150 nm, shows 75(2)% cementite (Fe_3C) and 16(2)% Hägg carbide (Fe_5C_2), while the remaining part is non-reacted α-Fe. We believe the latter contribution mostly originates from the border of the treated area where the spot overlap is incomplete, but some inhomogeneity of the layer cannot be excluded. The CXMS spectrum, which gives information on a depth of the order of 20 μm, revealed only 15% Fe_3C in addition to 3% austenite, indicating that the Fe_3C layer is confined to a depth of the order of 3 μm. All hyperfine parameters agree well with those reported in the literature [32–34]. This phase analysis and composition coincides with the XRD analysis. The RBS spectra of the original Fe and the sample irradiated with 256 pulses show after the laser treatment in methane a reduction of the Fe yield exactly to the cementite stoichiometry (25 at.% C and 75 at.% Fe) to a depth of at least 650 nm.

In conclusion, a method of obtaining a rather thick, polycrystalline cementite layer has been presented. For more results on laser nitriding and laser carburizing of iron, steel, aluminum, and silicon please refer to the references [3–5, 35, 36].

4.2. MAGNETIC TEXTURING BY ION-IMPLANTATION

Figure 4 shows the RBS spectra of the three samples I, M, and S before and after ion irradiation. One distinguishes the ^{57}Fe, ^{nat}Fe and Si components of the film and the implanted Xe ions (scaled by factor 5), whose concentration peaks at 30(2) nm, in agreement with the theoretical prediction mentioned before. A small sputtering effect of about 2 nm is visible, e.g., the shift of the Si edge. The broadening of the ^{57}Fe marker-layers upon irradiation is not very prominent, thus preserving the depth information of the marker layers. In addition to a weak ⟨110⟩ crystallographic texturing effect of the deposited and irradiated films, the XRD spectra show a small lattice expansion, which is due to the implanted Xe stored in the iron lattice.

The Mössbauer spectra taken with normal incidence were fitted using mainly the α-iron sextet, before implantation only sample I showed some interface contri-

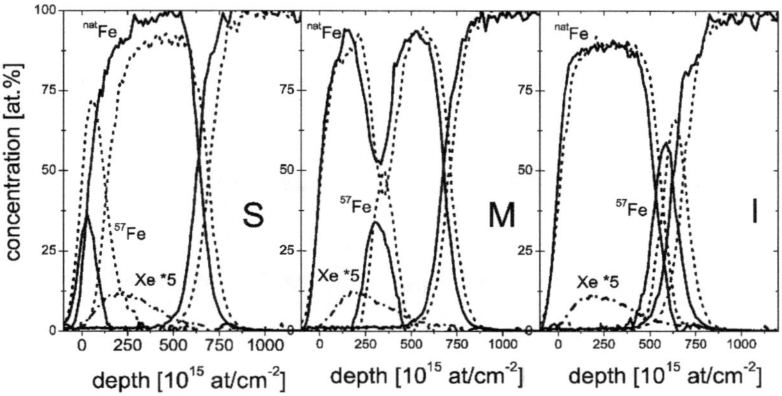

Figure 4. Depth profiles as RBS results for the three samples I, M, and S before (dotted) and after the Xe implantation (solid). The Xe concentration is scaled by a factor of 5.

Table I. Results from MOMS (maxima of I_2/I_3) and MOKE (easy axis ε) of ^{57}Fe/Fe samples I, M, and S

Sample	As deposited			Irradiated		
	MOMS		MOKE	MOMS		MOKE
	c_i [%]	ψ_i [°]	ε [°]	c_i [%]	ψ_i [°]	ε [°]
I	68(5)	70(8)	–	86(3)	87(4)	63(3)
	32(5)	160(8)	–	14(3)	153(3)	157(3)
M	–	–	–	95(2)	0(3)	1(3)
	–	–	–	5(2)	90(3)	91(3)
S	26(2)	46(4)	–	80(2)	71(3)	105(4)
	74(2)	136(4)	–	20(2)	161(3)	20(4)

butions (17% of iron silicide) in addition to the pure iron. At the interface to the substrate (sample I), we added a second sextet and a doublet of together about 31% fraction, indicating the increasing silicon fraction in the iron after implantation. After the irradiation we observed also the formation of a paramagnetic iron silicide (FeSi) with parameters in good agreement with those reported in [37, 38]. At the surface (sample S) we observed the formation of some FeO after implantation, while for this sample S and sample M the implanted Xe is visible as a small additional sextet with 8(2)% relative area and $B_{\mathrm{HF}} = 26(2)$ T representing iron atoms having a Xe atom as nearest neighbor. All these analyses of the Mössbauer spectra taken with normal incidence of the γ-beam indicate a complete in-plane orientation of the magnetization direction in the samples before and after irradiation, the values are given in Table I. Only sample M shows a smaller percentage of in plane alignment, but still within the error limit. Maybe in this case some spins also point out of the plane.

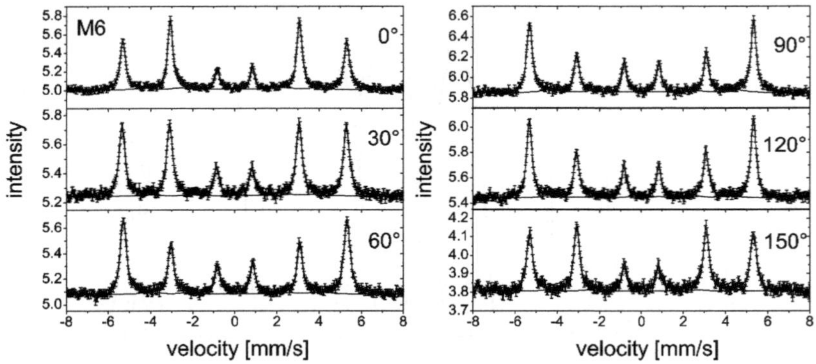

Figure 5. MOMS series of sample M for $\alpha = 45°$ and various angles φ.

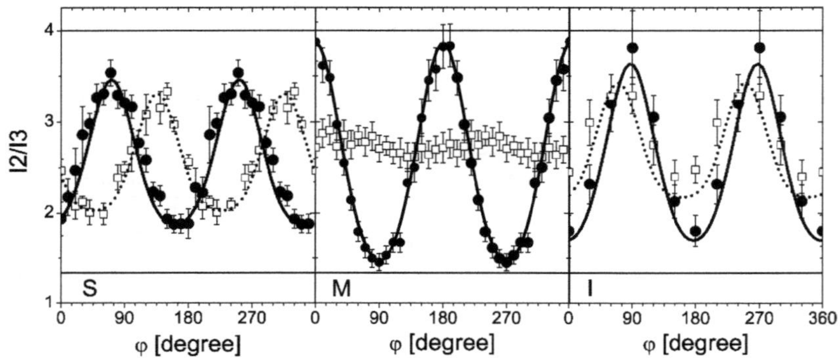

Figure 6. Intensity variations for the three samples as obtained by MOMS. The solid lines are fits with Equation (3) and parameters given in Table I. The theoretical minimum (1.3333) and maximum (4) is indicated.

For the CEMS measurements a proportional-counter was used, that surrounded the sample and could be tilted with respect to the γ-beam ($\alpha = 0$–$80°$) and also could be rotated around the sample-normal ($j = 0$–$360°$) as indicated already in Figure 1. With that, the samples have been measured with $\alpha = 45°$ and for angles φ ranging from 0° to 360°. Some of these spectra for sample M are shown in Figure 5.

It is clearly seen, that there is a nice variation of the relative line intensities with the angle φ. After analyzing the spectra and taking the line intensity for the α-Fe subspectrum, the variations for all samples were achieved as displayed in Figure 6.

Figure 6 presents the results of the angle-dependent Mössbauer analysis (MOMS) on the α-Fe fraction of the three samples. There, the 0°-direction is the direction of the external strain during implantation. The lines are the results of fits with Equation (3) for as deposited and irradiated samples and considering two in-plane terms of the hyperfine field direction. The parameters are summarized in Table I.

Before implantation (open symbols), the three samples show a different behavior. At the Fe/Si-interface (sample I) one can clearly observe an oscillation of the

I_2/I_3 ratio, indicating a preferred spin direction (68% at 70°). Possible interpretations are a small tilt of the sample during deposition or small stray fields which are known to affect the anisotropy. Another reason for the alignment of the field might be the lattice mismatch of Si and Fe which leads to permanent inverse magnetostriction effects [6]. The substrate orientation was observed by Laue-images and found to be ⟨011⟩ in 0°-direction. In the middle of the film (sample M) with the ^{57}Fe embedded in the ^{nat}Fe, no preferred hyperfine field direction was significant before implantation. The only disturbance of the film structure resulting in strains could occur due to the different deposition methods of the neighboring chemically and magnetically identical layers, but the spin orientation was nearly isotropic. The behavior at the surface (sample S) was similar to that at the interface (sample I). It showed a preferred hyperfine field direction with a rather strong intensity. The preferred direction was at 136(4)° with a fraction of 74(2)%.

The situation after irradiation is indicated by the filled symbols in Figure 6. Sample I shows the smallest effect, as the intensity and the direction of the oscillation changes only slightly. Since the penetration depth of the Xe was only 34 nm, less than 8% of the Xe-ions deposited their energy in this part of the film and also widened the lattice. The most obvious change of the hyperfine field parameters took place in the middle of the film (M). The φ-dependence changed from a nearly isotropic distribution to a nearly perfect (95(2)%) alignment in 0° direction of the sample. In this depth we also had the highest Xe-content after irradiation, indicating the strong dependence of the texturing effect on this parameter. Combined with the relaxation of the internal strains of the as-deposited film during irradiation, the Xe-ions induced a small widening of the lattice, as confirmed by the XRD-analysis, resulting in a strong inverse magnetostriction effect. The widening is much larger in the middle of the film than at the Fe/Si interface and it induced a net compressive strain, resulting in the 0° alignment of the B_{HF}-direction [13]. The behavior observed at the surface (S) was similar to that at the interface: the amplitude and the direction of the φ-dependence changed, but here the change of the angle was much larger than in the case of sample I. All ions pass this surface layer, resulting in the highest energy deposition. But hardly any Xe remained in this part, resulting again in lower lattice deformations. All the results of the MOMS experiments are summarized in Table I.

To compare the results with the macroscopic magnetic properties of the samples we made an angle dependent longitudinal MOKE-analysis. The measurements are shown in Figure 7 and the results also given in Table I. Before implantation, the magnetic remanence, as obtained from the hysteresis curves, are nearly isotropic. After implantation we obtain a rather strong magnetic anisotropy with a fourfold and an uniaxial contribution as can be seen in the polar plots. Since MOKE probes the first 20–30 nm of the film, at least for sample S the information range of CEMS and MOKE are the same. One would expect a similar preference direction of the magnetization, since we would expect the hyperfine field and the easy axis to coincide. It is not yet clear, why the methods give different results, but MOMS gives

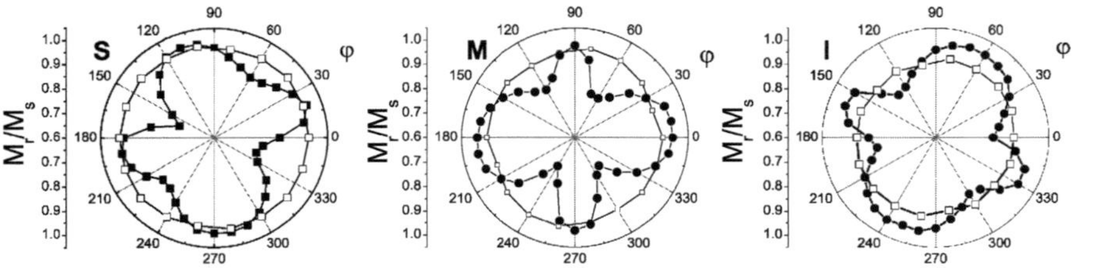

Figure 7. Polar plots of the film remanence as measured by MOKE (sample S, M, and I, from left).

an instantaneous ("non-destructive") picture of the spin distribution, the MOKE analysis demands the application of a strong external field and thus immanently changes actively the spin distribution. Further investigations are under way in this direction. Nevertheless, for the middle of the film (sample M), MOMS and MOKE results are in perfect agreement, although the MOKE signal from this film-depth is very small. At the interface I, the measurements are not directly comparable, since they probe non-overlapping parts of the film. As can be seen in Table I, the uniaxial part of the anisotropy is not equal in the three samples.

This behavior is not understood so far and further experiments have to be carried out in that direction.

5. Summary

The efficiency of laser nitriding and laser carburizing in Fe, Al, Ti, Si has been investigated for different types of pulsed lasers. Mössbauer spectroscopy significantly contributes to understanding the details of the processing and it can help to optimize the treatmens. On the other hand, MOMS seems to be a very interesting method to study magnetic anisotropies – with the help of ^{57}Fe marker-layers also depth dependent – and to resolve the spin distribution in a non-destructive way. A complete uniaxial spin orientation was found by that in the middle of the iron film, coinciding exactly with the external strain axis.

Acknowledgements

This work was partly supported by the Deutsche Forschungsgemeinschaft (DFG) under grant Scha 632/3 and by the DFG within the SFB 602 in Göttingen under grant B4.

References

1. Bäuerle, D., *Laser Processing and Chemistry*, Springer Verlag, Berlin, Heidelberg, 2000.
2. Cahn, R. W. and Haasen, P., *Physical Metallurgy*, North-Holland Physics Publishing, Amsterdam, 1983.
3. Schaaf, P., *Progress in Materials Science* **47**(1) (2002), 1.
4. Carpene, E. and Schaaf, P., *Appl. Phys. Lett.* **80** (2002), 891.

5. Schaaf, P., Han, M., Lieb, K.-P. and Carpene, E., *Appl. Phys. Lett.* **80** (2001), 1091.
6. Sander, D., *Rep. Progr. Phys.* **62** (1999), 809.
7. Chikazumi, S., *Physics of Ferromagnetism*, 2nd edn, Clarendon Press, Oxford, 1997.
8. Lewis, W. A., Farle, M., Clemes, B. M. and White, R. L., *J. Appl. Phys.* **75** (1994), 5644.
9. Farle, M., Saffari, H., Lewis, W. A., Kay, E. and Hagstrom, S. B., *IEEE Trans. MAG* **28** (1992), 2940.
10. Neubauer, M., Reineke, N., Uhrmacher, M., Lieb, K.-P., Münzenberg, M. and Felsch, W., *Nucl. Instrum. and Methods B* **148** (1998), 332.
11. Gupta, R., Lieb, K.-P., Luo, Y., Müller, G. A., Schaaf, P., Uhrmacher, M. and Zhang, K., *J. Magn. Magn. Mater.* (2002), submitted.
12. Lieb, K.-P., Zhang, K., Schaaf, M. G. A., P., Uhrmacher, M., Felsch, W. and Münzenberg, M., *Acta Physica Polonica A* **100** (2002), 751.
13. Zhang, K., Lieb, K.-P., Schaaf, P., Uhrmacher, M., Felsch, W. and Münzenberg, M., *Nucl. Instrum. and Methods B* **161–163** (2000), 1016.
14. Devolder, T., Chappert, C., Chen, Y., Cambril, E., Bernas, H. P. J. J. and Ferre, J., *Appl. Phys. Lett.* **74** (1999), 3383.
15. Bernas, H., Devolder, T., Chappert, C., Ferre, J., Kottler, V., Chen, Y., Vieu, C., Jamet, J. P., Mathet, V., Cambril, E., Lemerle, K. O. S., Rousseaux, F. and Launois. H., *Nucl. Instrum. and Methods B* **148** (1999), 872.
16. Zemcik, T., Kraus, L. and Zaveta, K., *Hyp. Interact.* **51** (1989), 1051.
17. Woike, T., Imlau, M., Angelov, V. S. J. and Delley, B., *Phys. Rev. B* **61** (2000), 12249.
18. Reuther, H., Behr, G. and Teresiak, A., *J. Phys. Condens. Matter* **13** (2001), L225.
19. Niesen, L., Murgarza, A., Rosu, M. F., Goehoom, R., Jungblut, R. M., Roozeboom, F., Baron, A. Q. R., Chumakov, A. I. and Rüffer, R., *Phys. Rev. B* **58** (1998), 8590.
20. Röhlsberger, R., Bansmann, J., Senz, V., Jonas, K. L., Bettac, A., Leupold, O., Rüffer, R., Burkel, E. and Meiwes-Broer, K. H., *Phys. Rev. Lett.* **86** (2001), 5597.
21. Doolittle, L. R., *Nucl. Instrum. and Methods B* **15** (1986), 227.
22. Stoney, G. G., *Proc. Roy. Soc. (London) A* **82** (1909), 172.
23. Ziegler, J. F. and Manoyan, J. M., *Nucl. Instrum. and Methods B* **35** (1988), 215.
24. Müller, A., Magentic texturing of iron films by ion implantation, PhD thesis, Universität Göttingen, Göttingen, 2003.
25. Gonser, U. (ed), *Mössbauer Spectroscopy*, Vol. 5, Springer-Verlag, Berlin, 1975.
26. *Ibid*, chapter 1, p. 1.
27. Wertheim, G. K., *Mössbauer Effect: Principles and Applications*, Academic Press, New York and London, 1964.
28. Wegener, H., *Der Mössbauer Effekt und seine Anwendungen in Physik und Chemie*, Bibliographisches Institut, Hochschultaschenbücher-Verlag, Mannheim, 1966.
29. Häglung, J. H., Grimvall, G. and Jarlborg. T., *Phys. Rev. B* **44** (1991), 2914.
30. Li, S. J., Ishihara, M., Yumoto, H., Aizawa, T. and Shimotomai, M., *Thin Solid Films* **316**, (1998), 100.
31. Kunze, J., *Nitrogen and Carbon in Iron and Steel*, Akademie Verlag, Berlin, 1990.
32. Ron, M., In: R. L. Cohen (ed), *Applications of Mössbauer Spectroscopy II*, Academic Press, New York, 1980, p. 329.
33. Schaaf, P., Wiesen, S. and Gonser, U., *Acta Metall.* **40** (1992), 373.
34. Schaaf, P., Krämer, A., Wiesen, S. and Gonser, U., *Acta Metall.* **42** (1994), 3077.
35. Carpene, E. and Schaaf, P., *Phys. Rev. B* **65** (2002), 224111.
36. Carpene, E., Flank, A.-M., Traverse, A. and Schaaf, P., *J. Phys. D* **35** (2002), 1428.
37. Fanciulli, M., Rosenblad, C., Weyer, G., Svane, A., Christensen, N. E., von Känel, H. and Rodriguez, C. O., *J. Phys. Condens. Matter* **9** (1997), 1619.
38. Desimoni, J. and Sanchez, F. H., *Hyp. Interact.* **122** (1999), 277.

Hyperfine Interactions **144/145:** 141–149, 2002.

Influence of the Interfaces on Magnetic Properties of Fe/Ag and Fe/Cu Multilayers Prepared by Sputtering

A. P. KUPRIN*, L. CHENG, Z. ALTOUNIAN and D. H. RYAN
Physics Department and Centre for the Physics of Materials, McGill University, 3600 University Street, Montreal, Quebec H3A 2T8, Canada

Abstract. Dependence of magnetic behavior of ultrathin Fe layers on the width of the interfaces was investigated by using dc-magnetron sputtering to grow (Fe(110)/Ag(111)) multilayers (ML) with thin interfaces and (Fe(111)/Cu(111)) multilayer with wide interfaces. Thicknesses of non-magnetic layers in all cases were 25 Å. The Fe/Ag MLs were prepared with thickness of Fe layers t_{Fe} = 8.8 Å and 2.4 Å, while Cu/Fe ML had the largest t_{Fe} = 11.0 Å together with widest interfaces. In-field ^{57}Fe conversion electron Mössbauer spectroscopy (CEMS) down to 50 K was employed to show that Fe layers are not uniform, to characterize interfaces, and to distinguish between static and relaxation behavior. This allowed unambiguous interpretation of T-dependencies of susceptibility and spontaneous magnetisation upon crossover to island structure of the Fe layers.

Key words: low-temperature conversion electron Mössbauer spectroscopy, magnetic multilayers, interfaces, temperature decay of magnetization.

1. Introduction and methods

Data storage and information transport technologies at present involve magnetic structures with lateral sizes less than a micrometer with a continuing tendency to approach a nanometer scale where high stability to thermal magnetization decay is a key issue. Though magnetism of thin Fe films was extensively studied experimentally, including Mössbauer spectroscopy, in connection with the 3D–2D crossover [1] and a variety of magnetic states of fcc γ-Fe that depend on the atomic volume [2], the finite-size effects associated with a decrease of lateral size of the layers received less attention. In most of the works Fe films were grown epitaxially in UHV that resulted in high monocrystalline quality and practically atomically flat interfaces. Previously, we investigated how significant interface roughness and intermixing in Fe/Cu multilayers prepared by sputtering may result in break up of the layers into isolated islands [3]. As sputtering provides a convenient way to vary the width of the interfaces we used this method to prepare two different kind of multilayers. Two Fe/Ag multilayers were grown to investigate transition

---

* Present address: Physics Department, University of California, San Diego, 9500 Gilman Drive, La Jolla, CA 92093-0319, USA.

Table I. Structural parameters and growth conditions of Fe/Ag and Fe/Cu multilayers. N is the number of bilayer repetitions, t_{Fe}, t_{nm} are the thicknesses of Fe and non-magnetic elemental layers, $\sigma_{nm/Fe}$ and $\sigma_{Fe/nm}$ are the roughnesses of the interfaces (NM/Fe means Fe on top of non-magnetic metal), P_{Ar} is an Ar pressure during deposition, and R_{nm} and R_{Fe} are the deposition rates

Name	Spacer	N	t_{Fe} (Å)	t_{nm} (Å)	$\sigma_{nm/Fe}$ (Å)	$\sigma_{Fe/nm}$ (Å)	P_{Ar} (mTorr)	R_{nm} (Å/s)	R_{Fe} (Å/s)
ML8.8	Ag	25	8.8	25.3	6.6	13.0	10	3.0	1.0
ML2.4	Ag	25	2.4	28.3	7.7	9.8	10	3.0	1.0
ML11.0	Cu	36	11.0	25.1	12.5	34.0	20	1.9	0.8

to island morphology of the Fe layers in the case when interfaces are sharp due to a strong de-mixing tendency and small band hybridization between Fe and Ag. One Fe/Cu multilayer was also prepared to represent the opposite case of continuous layers with wide interfaces. In all cases, the thickness of non-magnetic layers t_{nm} was much larger than thickness of the Fe layers t_{Fe} to escape interlayer coupling. Fe(110)/Ag(111) and Fe(111)/Cu(111) orientations were found by high-angle XRD. Parameters of the multilayers were determined from fits to small-angle X-ray reflectivity data and are summarized in Table I. The crucial details about magnetic and structural properties on local scale both in remanent state as well as with in-plane external field were obtained by ^{57}Fe low-temperature conversion electron Mössbauer spectroscopy (LT-CEMS) down to 50 K. We built two gas proportional counters based on the prototype designed at Mainz [4] and optimized for low-temperatures in respect of S/N ratio and energy resolution. Important features of the counters are: thin sensitive volume of 3 mm in thickness only (1 mm anode-sample distance) to compensate for increased gas density, spring support for the anode wire to accommodate shrinkage of the dimensions, double mylar windows in the detector's body and its cap with inter-window volume connected to the sensitive volume to keep walls of the sensitive chamber always flat, and electrical isolation of the detector body from the inner parts of the cryostats. Operating voltage was optimized for each working temperature in respect of reducing measurement time [5]. Gas-flow He + 4%CH_4 counter was used in the temperature range 320–90 K inside standard liquid N_2 cryostat, while gas-filled He counter was used down to 50 K in the in-house made cryostat filled with He-gas for heat exchange and cooled by the vibration-isolated closed cycle He refrigerator. Both counters allowed simultaneous acquisition of spectra in backscattering CEMS and transmission geometries, and showed stable long-term operation and proved to be very effective for our thin samples (with natural isotopic content). Magnetic properties in the bulk were characterized by initial susceptibility χ_{ac} measured with the ac excitation field of 70 Hz and amplitude of 5 Oe and by spontaneous magnetization M_s extrapolated to zero field from measurements in the external field of 4 T strong enough to saturate our multilayers at any orientation.

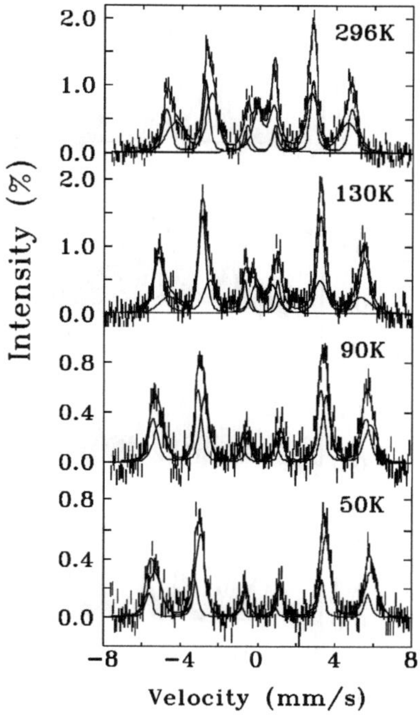

Figure 1. Conversion electron Mössbauer spectra for the (Fe 8.8 Å/Ag 25.3 Å)×25 multilayer in the range 50 K–room temperature. Strong Zeeman splitting persist to RT. Note a weak signal that appears at around zero velocity at the increased temperatures only. Details of decomposition are described in the text.

2. Results

Selected CEMS spectra of the Fe/Ag multilayers ML8.8 and ML2.4 are shown in Figure 1 and Figure 2 respectively. At 50 K spectra represent a ferromagnetic (FM) Zeeman splitting. Asymmetrical broadening of lines suggests at least two components that were modeled by two sextets with symmetrical Gaussian distribution of hyperfine fields (P(B_{hf})) for each of them. This contrasts with MBE-grown multilayers when Fe(110)/Ag(111) orientation exhibits a single sextet only, but two or three sextets are necessary to fit spectra for Fe(100)/Ag(100) orientation and may be ascribed to the different physical nature of deposition processes used. Considering that top and bottom interfaces are distinctly different as reflected in Table I, and similar to the case of sputtered Fe(111)/Cu(111) multilayers [3], it is logical to ascribe the sextet (S_{bi}) with larger average field $\langle B_{hf} \rangle$ having narrower P(B_{hf}) and smaller isomer shift (we used α-Fe at RT as a reference) to the bottom interface together with the inner regions, and another sextet (S_t) to the top interface, following results on site-probed Fe(100)/Ag(100) multilayers ([6] and references therein). Widths of both distributions are much broader for ML2.4, especially for S_{bi} and partial contribution of S_t is strongly decreased. Partial contributions of S_{bi}

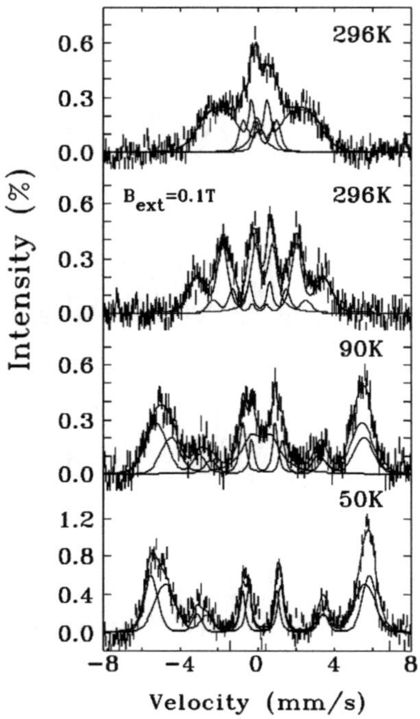

Figure 2. Conversion electron Mössbauer spectra for the (Fe 2.4 Å/Ag 28.3 Å)×25 multilayer in the range 50 K–room temperature. Note that lines are much wider than for the multilayer shown in Figure 1. Zeeman splitting collapses at around room temperature to spin-glass-like pattern. External in-plane magnetic field of 0.1 T increases hyperfine splitting and resolves constituent lines; details are in the text.

and S_t for both samples roughly correspond to the ratio of interface roughnesses in Table I, suggesting that the inner regions of pure Fe are rather small. At 50 K the average field of S_{bi} is essentially the same for both samples $\langle B_{hf}^{bi}\rangle = (35.4 \pm 0.2)$ T, while splitting of S_t is notably less for thinner Fe layers with $\langle B_{hf}^{t}\rangle = (35.0 \pm 0.1)$ T and (32.5 ± 0.3) T for ML8.8 and ML2.4, respectively.

On raising temperature, signal around zero velocity starts to appear pointing to superparamagnetizm of a part of the Fe layers. Its parameters practically do not change with temperature (isomer shift $\delta = +0.2$–0.4 mm/s, quadrupole splitting $\Delta = 0.8$–0.9 mm/s) and are in the range for small Fe clusters in Ag seen previously even in MBE-grown Fe/Ag system [7] and references therein. But its partial contribution to the spectrum increases first at the expense of the S_t, in agreement with expectations of a weaker magnetic interactions in more rough and alloyed top interface. For ML8.8 it is reliably resolved at 130 K and corresponds to a doublet (D). Nevertheless, its partial contribution is relatively small being 17% in the RT spectrum and external in-plane magnetic field $B_{ext} = 0.1$ T does not increase its contribution and results in the identical spectrum. Dominant contribution of the FM signal and completely in-plane alignment of magnetization for both sextets in the whole temperature range studied shows that the 8.8 Å thick Fe layers are quite uniform and at least several atomic layers (AL) thick on macroscopic scale (1 AL Fe(110) = 2.03 Å) with Curie temperature (T_c) far above RT. For ML2.4 a doublet appears at lower temperatures and constitutes 20% of the signal already at 90 K. Further increase of temperature leads to a dramatic spectral changes, but affects S_t more drastically. At RT as compared to 50 K, $\langle B_{hf}^{bi} \rangle$ drops a little more than twice to (15.3 ± 0.4) T, while $\langle B_{hf}^{t} \rangle$ drops almost six times to (5.6 ± 0.3) T. Also, width of the P(B_{hf}) for S_{bi} increases 2.4 times that makes spectrum look like that of an amorphous Fe spin-glass alloy. In addition, a single line appears at about zero velocity. At the same time, direction of magnetization shows no temperature dependence remaining (33 ± 1)° tilted from the normal. Small in-plane B_{ext} = 0.1 T strongly increases $\langle B_{hf}^{t} \rangle$ to (15.0 ± 0.6) T, effect typical for SP behavior [1]. $\langle B_{hf}^{bi} \rangle$ increases only about 30% to (20.6 ± 0.3) T, but P($\langle B_{hf}^{bi} \rangle$) decreases 1.8 times such that Zeeman lines become nicely resolved. Both of these effects suggest more collinear magnetic structure for S_{bi} considering that 1st near-neighbors contribute around 50% to B_{hf} on ^{57}Fe [8]. For both sextets magnetization now lies totally in-plane and single line component disappears suggesting that SP relaxation is effectively suppressed. Hence, one can estimate that at least 33% of the Fe atoms are SP at RT and already at 320 K most of the Fe atoms are engaged in fluctuations as the spectrum (not shown) exhibits a broad single line at zero velocity with typical long relaxation wings [9].

CEMS spectra of the Fe/Cu multilayer ML11.0 are shown in Figure 3. Lines of the Zeeman pattern are much broader than for Fe/Ag multilayers in agreement with a somewhat better solubility of Fe in Cu. Also the paramagnetic (PM) signal in the central part of the spectrum is still present at the lowest temperature. Main features of the spectral shape and its evolution with temperature follows those of Fe/Cu multilayers studied in our group previously ([3] and references therein). Hence, we used the same model to fit present spectra. Upon warming the thinnest FM parts of the Fe layers become PM that is reflected in population of the strong doublet with smaller splitting at the expense of the Zeeman split component. The latter one is modeled at 50 K by two Gaussian distributions. The one with larger field, $\langle B_{inn} \rangle$ = (32.6 ± 1.4) T represents inner FM regions, while the lower field one, $\langle B_{int} \rangle$ = (17.8 ± 2.0) T, – alloyed Fe-rich FM top interfaces. These fields are smaller than $\langle B_{hf}^{bi} \rangle$ and $\langle B_{hf}^{t} \rangle$, respectively, at each temperature of the experiment. This exposes a weaker magnetic exchange in a more intermixed Fe/Cu system notwithstanding that on average ML11.0 has the thickest Fe layers among samples studied of 5.3 AL (1 AL Fe(110) = 2.07 Å). Also magnetization is (62 ± 5)° tilted from the normal, that is about twice more than for ML2.4 with sharpest interfaces, and points to a weaker perpendicular magnetic anisotropy (PMA) due to less sharp interfaces. As in the Fe/Ag multilayers the bottom interfaces are much sharper, but are strikingly different, remaining PM in the whole T-range studied and are modeled by a quadrupole doublet with larger splitting. Its partial contribution to the spectrum is around 10% and independent of temperature. Disappearance of

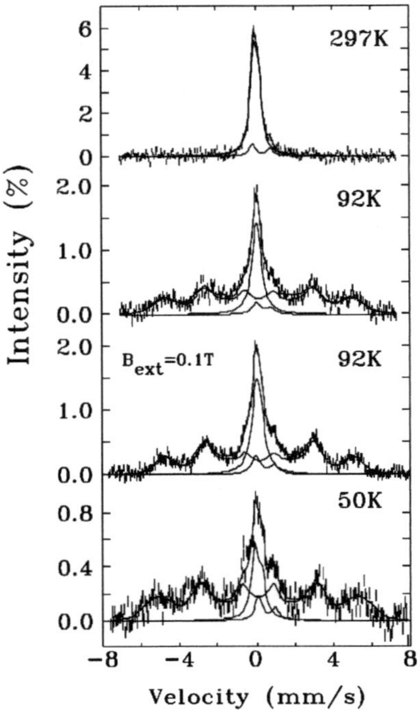

Figure 3. Conversion electron Mössbauer spectra for the (Fe 11.0 Å/Cu 25.0 Å)×36 multilayer in the range 50 K–room temperature. Wide distribution of hyperfine fields is present at low temperatures, which completely collapses on warming before reaching room temperature. External in-plane magnetic field of 0.1 T does not increase hyperfine splitting.

FM contribution happens at $T_c = 250$ K, that is lower than for Fe/Ag multilayers. In-plane $B_{ext} = 0.1$ T applied in the range 90–180 K does not change either visible Zeeman splitting or any of the parameters of the model except orientation of magnetization, rotating it completely into the plane of the multilayer. This shows the absence of any SP relaxation.

Measurements of susceptibility complement ^{57}Fe Mössbauer spectroscopy due to sensitivity not only to critical phenomena and magnetic orientation, but also to domain formation and motion [10] and due to a five orders of magnitude larger characteristic time scale. For ML8.8 film a steady increase of $\chi_{\parallel}$ on warming from 5 K to a very large value without saturation at RT presented in Figure 4 in principal agrees with conclusion of T_c to be much higher RT obtained from Mössbauer data. And the absence of any pronounced features may be explained by the smearing effect of small but measurable SP islands with large size distribution (due to finite roughness of the interfaces) also seen to be present in Mössbauer spectra. SP behavior is much stronger in ML2.4 due to smaller sizes of isolated Fe islands expected for thinner layers with the blocking temperature (T_B) estimated from Mössbauer data to be slightly above RT. Now $\chi_{\parallel}$ shows a pronounced hump at a temperature around 150 K, i.e., about twice lower than T_B, situation similar for

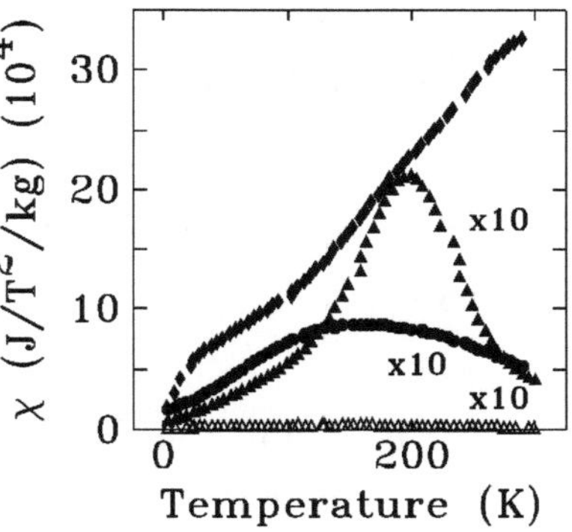

Figure 4. Temperature dependence of the real part of the initial susceptibility: diamonds – for the (Fe 8.8 Å/Ag 25.3 Å)×25 multilayer, circles – for the (Fe 2.4 Å/Ag 28.3 Å)×25 multilayer, triangles – for the (Fe 11.0 Å/Cu 25.1 Å)×36 multilayer. Filled symbols represent susceptibility measured with in-plane excitation field, $\chi_{\|}$, open – measured in perpendicular orientation, $\chi_{\perp}$.

nm-size α-Fe particles [9], except for a much broader width of the hump for our sample due to large size distribution of the Fe islands expected for thin sputtered layers. This broad hump cannot be due to formation and motion of domains as no spin-reorientation is detected at these temperatures in Mössbauer spectra [10]. ML11.0 also exhibits a similarly strong signal in $\chi_{\|}$ below T_c. But its several times smaller width immediately suggests another explanation than SP relaxation as this multilayer has the roughest interfaces that would result in the largest size-distribution of the Fe islands. Support for the rejection of SP behavior obtained from Mössbauer data comes from T-dependence of $\chi_{\perp}$ that shows no features at all. We note that with excitation the field oriented perpendicular to the film and almost in-plane magnetization at 50 K $\chi_{\perp}$ is insensitive to order–disorder transition and formation of perpendicular domains is excluded [10]. Thus, the peak in $\chi_{\|}$ is attributed to the in-plane rotational response of magnetization.

One of the important parameters for practical applications is thermal decay of spontaneous magnetisation, M_s. At low temperatures this process is governed by thermal excitations of spin waves and experimental data in most of the works are fit to a power law [11]. The crossover from 3D to 2D behavior is marked by the change in the exponent from $n = 3/2$ (Bloch-like) to $n = 1$ (quasi-linear) if the energy gap at zero wave vector, whose origin is magnetic anisotropy, is not vanishingly small [12]. For high-quality Fe(110) thin films this happens at thickness around 3 AL [1]. M_s for ML8.8 decreases with temperature most slowly but clearly non-linearly as shown in Figure 5. The fitted to $T^{3/2}$-law spin-wave parameter $B_{\|} = (4.3 \pm 0.8)\ 10^{-5}\ K^{-3/2}$ is 5.4 times larger than for the similar multilayer grown by MBE [1]. This agrees with the more strongly reduced exchange coupling expected inside interface areas in sputtered films [13]. For ML2.4 $B_{\|}$ experiences an additional increase of 2.0. Though remanent magnetization has a substantial normal component detected in Mössbauer spectra (Figure 2), this does

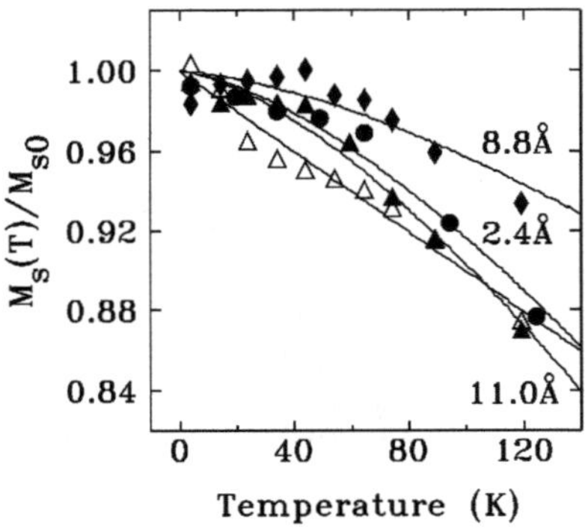

Figure 5. Temperature dependence of reduced spontaneous magnetization obtained from extrapolation of the in-field data up to 4 T to zero field for the (Fe 8.8 Å/Ag 25.3 Å)×25 multilayer (diamonds), (Fe 2.4 Å/Ag 28.3 Å)×25 multilayer (circles), (Fe 11.0 Å/Cu 25.0 Å)×36 multilayer (triangles). Filled symbols represent measurements with the in-plane field, open – in perpendicular orientation.

not necessarily imply that interface magnetic anisotropy became stronger than for ML8.8 as it may prevail merely due to weaker dipolar shape anisotropy as 1–2 AL thick Fe layers are unavoidably broken up into isolated SP patches. This is in contrast with clearly linear M_svT for sputtered Fe(111)/Cu(111) multilayer with discontinuous Fe layers [3] and may be explained by a stronger interlayer coupling in this system that results in a stronger PMA as there are 6 NN in the adjacent atomic planes for Fe(111) orientation, while only 2 NN for Fe(110), even when Fe is more susceptible to intermixing with Cu than Ag. Excessively wide interfaces of ML11.0 should strongly decrease both PMA and in-plane shape anisotropy thus suppressing linear dependence of M_svT. In fact, the decrease of M_s is the fastest with $B_{\parallel} = (9.7 \pm 0.6)\ 10^{-5}\ \mathrm{K}^{-3/2}$ as illustrated in Figure 5. Interestingly, M_svT measured in perpendicular orientation cannot be fitted with $T^{3/2}$-law, contrary to linear fit that gives $B_{\perp} = (101 \pm 7)\ 10^{-5}\ \mathrm{K}^{-1}$. In this orientation external field is applied at a very large angle to the remanent magnetization and should decouple the very outer regions of the Fe layers with small numbers of Fe NN from more uniform inner ones, thus effectively thinning and driving them into 2D regime.

3. Conclusions

Sputtering allows preparation of ultrathin Fe layers with either narrow interfaces or wide and alloyed interfaces. In both Fe/Ag and Fe/Cu multilayers top and bottom interfaces are distinctly different. Top interfaces are much wider and are ferromagnetic in both systems. Bottom interfaces are less alloyed, they are ferromagnetic in Fe/Ag multilayers, but non-magnetic in Fe/Cu multilayers. Widening of the interfaces suppresses perpendicular magnetic anisotropy, drastically increases the decay rate of spontaneous magnetization with temperature, and induces crossover from 3D to 2D magnetism by aligning magnetization normal to the easy plane.

Acknowledgements

This work was supported by grants from the Natural Sciences and Engineering Research Council of Canada and Fonds pour la formation de chercheurs et l'aide à la recherche, Québec.

References

1. Gutierrez, C. J., Qiu, Z. Q., Tang, H., Wieczorek, M. D., Mayer, S. H. and Walker, J. C., *Phys. Rev. B.* **44** (1991), 2190.
2. Keune, W., Schatz, A., Ellerbrock, R. D., Fuest, A., Wilmers, K. and Brand, R. A., *J. Appl. Phys.* **79** (1996), 4265.
3. Kuprin, A. P., Hanžel, D., Ensling, J., Spiering, H. and Gütlich, P., unpublished.
4. Kuprin, A. P. and Novakova, A. A., *Nucl. Instr. Meth. B* **62** (1992), 493.
5. Kuprin, A. P., Cheng, L., Altounian, Z. and Ryan, D. H., *J. Appl. Phys.* **87** (2000), 6591.
6. Keavney, D. J., Wieczorek, M. D., Storm, D. F. and Walker, J. C., *J. Magn. Magn. Mater.* **121** (1993), 49.
7. Kuprin, A., Wiarda, D. and Ryan, D. H., *Phys. Rev. B* **61** (2000), 1267.
8. Volkening, F. A., Jonker, B. T., Krebs, J. J., Koon, N. C. and Prinz, G. A., *J. Appl. Phys.* **63** (1988), 3869.
9. Bødker, F., Mørup, S., Pedersen, M. S., Svedlindh, P., Jonsson, G. T., Garcia-Palacios, J. L. and Lazaro, F. J., *J. Magn. Magn. Mater.* **177–181** (1998), 925.
10. Arnold, C. S., Johnston, H. L. and Venus, D., *Phys. Rev. B* **56** (1997), 8169.
11. Bayreuther, G., *Hyp. Interact.* **47** (1989), 237.
12. Erickson, R. P. and Mills, D. L., *Phys. Rev. B* **44** (1991), 11825.
13. Mathon, J. and Ahmad, S. B., *Phys. Rev. Lett.* **37** (1988), 660.

Hyperfine Interactions **144/145:** 151–160, 2002.

Magnetic Phases in Alloys and Nanostructured Systems

J.-M. GRENECHE
Laboratoire de Physique de l'Etat Condensé, UMR CNRS 6087, Université du Maine, F72085 Le Mans Cedex 9, France; e-mail: greneche@univ-lemans.fr

Abstract. The structure of nanocrystalline and nanostructured iron-containing systems is discussed on the basis of the hyperfine interactions observed by ^{57}Fe Mössbauer spectrometry. In the case of nanocrystalline alloys, the structure, topology and composition of both the crystalline grains and the intergranular phase can be approached by Mössbauer spectrometry, provided a series of spectra is collected as a function of temperature and as a function of the volumetric crystalline fraction. Moreover, several examples of nanostructured powders are reviewed in order to illustrate the contribution of the Mössbauer spectrometry to the study of the structural and chemical orders associated to the atomic intermixing process during the milling procedure. Finally, the magnetic properties are discussed as a function of the thickness of grain boundaries.

Key words: nanostructure powders 81.07.Wx, magnetic properties of nanostructures 75.75.+a, Mössbauer spectroscopy of solids 76.80.+y.

1. Introduction

During the last decade, much attention was devoted to the investigation of the structure of nanostructured materials which exhibit unconventional physical properties [1, 2]. Indeed, high density of interfaces up to 10^{25} m^{-3} for grains of 10 nm diameter are observed in nanostructured materials whereas up to 50–60% of atoms can be located in grain boundaries when their width is significantly thick, with grains of about 5 nm diameter. Consequently, in contrast to microstructured materials, the structural modelling of nanostructured materials thus requires that of the grain boundaries, the contribution of which is extremely high because of the confinement effect. Review papers were recently devoted to the structure of intercrystalline interfaces [3] and grain boundaries in nanocrystalline materials [4].

It is important to mention that the structure of grain boundaries in metallic systems is controversially debated: Gleiter assumed that nanocrystalline materials consist of perfect crystalline grains and completely disordered grain boundaries with low density [5]. Such a concept of a "gaslike" structure [6, 7] was not supported by further diffraction experiments. Transmission Electronic Microscopies, small-angle scattering technique and EXAFS allowed to observe the presence of grain boundaries but it remains difficult to separate both contributions in order to characterize the atomic disorder. As a consequence of its atomic scale sensitivity,

^{57}Fe Mössbauer spectometry is especially efficient for modelling the structure of grain boundaries in iron-containing alloys because the hyperfine structures of crystallites and grain boundaries can be "a priori" separated. Nevertheless, the very complex hyperfine structure does not usually allow an easy description into two components because the fitting procedure leads to different solutions. But their respective evolution versus temperature might bring an unambiguous attribution and a quantitative estimation of both proportions.

In the case of ionic systems, it has recently been shown that Mössbauer spectrometry is an excellent tool for characterisation of the atomic topology of grain boundaries as well as their magnetic structure and for estimating accurately the atomic proportions, in the case of nanostructured antiferromagnetic ferric fluorides (FeF_3) prepared by high-energy ball milling [8, 9]. In addition, the atomic layer interface located between the grain boundaries and the nanocrystalline grain has been clearly evidenced. The mutual analysis of both zero-field and in-field Mössbauer spectra allow to conclude that the crystalline grain and the grain boundaries resulting from a cubic and a random packing of FeF_6 octahedral units behave as an antiferromagnet and a speromagnet, respectively (as is schematically illustrated in Figure 1). The thermal evolution of the total magnetic properties is also strongly dependent on the nanostructural properties because the coupling between single domain grains is progressively established when the magnetic correlation length becomes smaller than the thickness of grain boundaries [10]. In this case, the magnetic coupling is due to the superexchange term whose correlation is roughly less than 1 nm. This study clearly demonstrated the usefulness of ^{57}Fe Mössbauer spectrometry, providing the hyperfine structure of a series of spectra recorded versus temperature and/or versus external magnetic field can be modelled.

Let us note that in the case of nanoparticle of iron oxide (γ-Fe_2O_3 maghemite [11, 12]), the surface layer [13] and its magnetic canting which can be also well

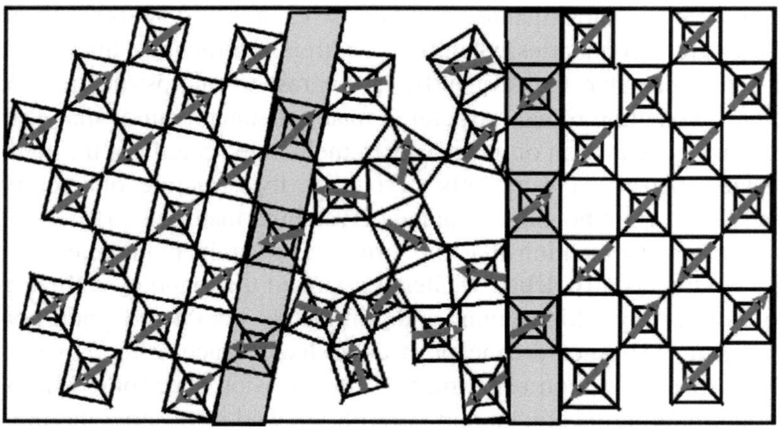

Figure 1. 2D schematic representation of grain boundary located between two crystalline grains showing the atomic topology and the magnetic structures, see [8].

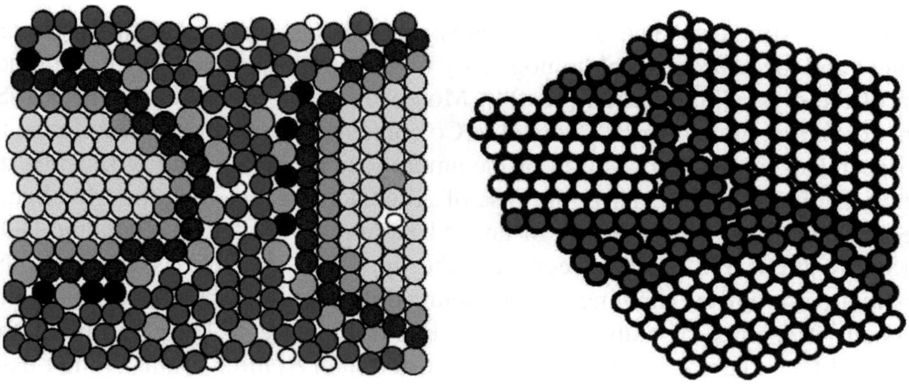

Figure 2. 2D schematic representation of the atomic structure nanocrystalline alloys (left) and nanostructured powders (right).

characterized and quantified from sets of Mössbauer data give rise to unconventional properties in addition to their fundamental interest.

In the case of metallic systems, the magnetic correlation length is expected to range from a few nanometers up to ten nanometers. When it overcomes the thickness of grain boundaries, nanocrystalline grains are thus magnetically coupled, favoring the presence of penetrating fields, i.e., polarization effects, in the presence of grain boundaries displaying smaller magnetic ordering temperatures. Such features might originate some disagreements about the expected hyperfine structure and consequently misinterpretations of Mössbauer spectra.

We illustrate such phenomena in the case of nanostructured and nanocrystalline alloys. Their atomic representation is schematized in Figure 2. In the next sections, we will discuss the structure of grain boundaries, their role on magnetic properties and also the consequences on the hyperfine structure.

2. Nanocrystalline alloys

The nanocrystalline alloys correspond to nanocrystalline grains embedded within an amorphous matrix, obtained from a subsequent annealing of the amorphous precursor which thus exhibit two stages of crystallization. The pioneering alloy was discovered in 1988 by Yoshizawa *et al.* [14], and labelled as Finemet ($Fe_{73.5}Cu_1Nb_3B_9Si_{13.5}$). Then new systems were found: Nanoperm family alloys (FeMB(Cu), M = Nd, Zr, Hf, ...) [15] and more recently Hitperm ((Fe,Co)MBCu with M = Zr, Hf, Nb, ...) [16]. These two-phase structural and soft magnetic alloys were widely investigated by a combination of several techniques (diffraction, electron and atom probe microscopies, magnetic measurements and Mössbauer spectrometry).

The contribution of Mössbauer spectrometry is quite important [17]. All the data allow to characterize the structure and the chemical composition of both the intergranular phase (or the amorphous remainder) and the nanocrystalline grains. In the case of Finemet alloys, the hyperfine structure is very complex because of the DO3 structural type of FeSi grains: nevertheless one can estimate the Si content and the non-chemical homogeneity of the amorphous remainder as a function of the crystalline fraction [18–20]. Moreover, the composition of (Fe, Co)Si nanocrystalline grains which emerged in Co containing Finemet has been found to be dependent on the composition of the amorphous precursor but not sensitive to the crystalline fraction [21]. In the case of Al containing Finemet, the Mössbauer spectra clear show that most of Si and Al atoms are located in the cubic-DO3 Fe(Si,Al) grains [22]. On the contrary, the hyperfine structure is less complex for Nanoperm alloys, allowing a more detailed description of both structural and magnetic properties of both components. But in the case of Hitperm alloys, the structural type and content of Fe and Co located in crystalline grains are not easy to determine, even by combining Mössbauer with X-ray diffraction data: numerous studies are in progress [23–25]. The presence of impurities which remains difficult to be clearly identified from the hyperfine structure, has been controversially debated [26–29] as well as the chemical composition and the magnetic nature of the interface located between the crystalline grains and the amorphous remainder [30, 31]. It is important to note that such a concept fairly supports Atom Probe Field Ion Microscopy, EXAFS and Ferromagnetic resonance [32].

All the results allow thus to propose an atomic scale modelling representation which is illustrated in Figure 2(left) [30]. Spectra have to be recorded at different temperatures on nanocrystalline alloys with different crystalline volumetric fraction while a fitting model has to be developed in order to describe simultaneously all spectra [33]. The validity of the model requires a check of the physical dependence of each hyperfine parameter.

In addition to the estimate of the Fe content in both the amorphous remainder and the crystalline grain, the distribution of the hyperfine field due to the amorphous remainder can be interpreted in terms of local environments, i.e. the chemical composition [34], while its evolution versus temperature gives information on the magnetic coupling between the crystalline grains and the amorphous remainder. When the crystalline volumetric fraction is low, the amorphous remainder is weakly Fe-depleted and the crystalline grains are strongly coupled to the amorphous matrix at low temperature while they behave as single magnetic domain particle with superparamagnetic behaviour at high temperature (see Figure 3a). This can be clearly evidenced through the significant reduction of the hyperfine field due to the crystalline grains, which increases when the temperature increases. Indeed the distance between grains is larger than the ferromagnetic correlation length, preventing thus magnetic coupling through the paramagnetic amorphous matrix. On the contrary, when the volumetric crystalline fraction increases, the intergranular phase becomes chemically heterogeneous as a consequence of the atomic diffusion mechanism, giving rise to Fe- and B-rich zones and Fe–M-rich zones (M = Nd, Zr, ...). A structural modelling approach requires decomposition of the hyperfine distribution into different (Gaussian) components and comparison of their positions with data of literature; it is also important to note that the tem-

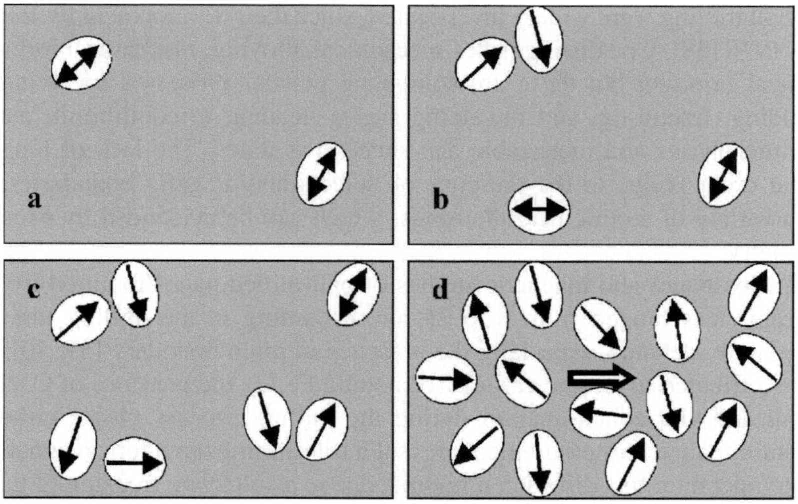

Figure 3. Schematic representation of magnetic behaviours at high temperature as a function of the increasing volumetric crystalline fraction (a–d). Single arrays represent blocked magnetization (static regime) while double arrays symbolize the presence of magnetization fluctuations in single domain nanocrystalline grains, i.e. the superparamagnetic behaviour (see text).

perature evolution of the hyperfine field distribution of the amorphous confirms the existence of different local orders. Nevertheless, the hyperfine structure of the intergranular phase at high temperature is disturbed because the magnetic grains might induce some polarization effects. Indeed, the prevailing contribution of the nanostructure implies short distances between crystalline grains, which become smaller than the ferromagnetic exchange length: it favours a progressive magnetic coupling between grains (see Figure 3), preventing thus the emergence of superparamagnetic fluctuations. It is consistent with the occurrence of non-zero values of hyperfine fields at temperatures much higher than the "Curie temperature" of the amorphous matrix. Such a feature is usually explained by the presence of penetrating fields [35, 36] and/or dipolar contribution [37].

The structural modelling of the intergranular phase remains consequently a delicate task because the total hyperfine structure is not well resolved at low temperature preventing an easy decomposition while that of the amorphous component is disturbed at high temperature. However, it is suggested that the first atomic layer in close contact with the crystalline grains mainly consists of M atoms, preventing the diffusion of atoms towards grains and thus their growth. Moreover, further information evidenced by complementary techniques is needed.

3. Nanostructured powders

Because of their applied potentiality due to easy tailoring as well as their non-conventional physical properties, nanostructured powders usually prepared by high-energy ball milling were widely investigated, since their development by Benjamin *et al.* in 1976 [38]. One distinguishes mechanical alloying, mechanical milling and mechanical grinding but these are solid-state powder processes involving iterative welding, fracturing, and rewelding stages, leading to equilibrium and non-equilibrium phases and metastable and successive states. The lack of long-range structural order is due to the presence of defects and/or grain boundaries, originating a variety of atomic neighbourings, which can be evidenced by Mössbauer spectrometry.

Both Mössbauer and magnetic studies of ball-milled nanostructured iron powders (grain sizes ranged from 8 to 25 nm, according to the milling time) were controversially debated, especially the presence of grain boundary [39, 40], which was then attributed to an iron carbide (cementite Fe_3C), the presence of C resulting from ball and vial contamination during the milling process [41]. Further low-temperature magnetic measurements reveal a transition from a ferromagnetic state at high temperature to a disordered regime, due to a collective freezing of the magnetic moments of crystalline grains at low temperature. The authors suggest that the grain boundaries thus display a spin-glass-like behaviour [42], as previously concluded in the case of nanocrystalline FeRh alloys [43].

Beside ground iron powders, one has to mention small iron clusters produced by inert gas condensation technique and then consolidated together [44]. The Mössbauer study clearly reveals two almost equiprobable components: a 33 T magnetic sextet assigned to the crystalline bcc-Fe grains, the presence of a second magnetic component unambiguously attributed to the grain boundaries. Indeed the main characteristics were the following: higher isomer shift, higher hyperfine field at low temperature, larger linewidth and smaller Debye temperature. Those data are consistent with completely disordered iron grain boundaries as suggested by the gaslike concept [44].

During the last decade, much attention was devoted to some binary ferrous milled powders such as Fe–Mn, Fe–Ni, Fe–Cr, Fe–Ti, . . . , synthesized by mechanical attrition in a high-energy ball mill, which were studied by a combination of techniques: X-ray diffractometry, transmission electron microscopy and Mössbauer spectrometry. Grain size, crystalline constraints were estimated from Bragg peaks broadening while the crystalline grains and the grain boundary component were quantified from Mössbauer spectra, especially the hyperfine field distribution. The thickness of grain boundaries was estimated at about 0.5 nm (2 atomic layers) in fcc-type alloys (Fe–Mn, Fe–Ni), but larger than 1nm in the case of bcc-alloys (Fe–Cr, Fe–Mo and Fe–Ti [45]).

The chemical order and the nature of grain boundaries were studied during the alloying process of elementary elements induced during the mechanical procedure in the case of several binary systems. Indeed the intermixing process is of particular interest in the presence of non-miscible elements. The short-range and long-range ordering were respectively investigated by Mössbauer spectrometry and X-ray or neutron diffraction, revealing thus metastable, transient, stationary states.

The shape of hyperfine field distribution allows to get a quantitative estimation of anti-site defects in the case of milled Fe–V alloys with different compositions, as a function of milling time and subsequent annealing treatment [46], while stationary hyperfine field distributions Fe–M alloys (M = Cr, W) are interpreted as due to irregular interfaces between nanometer-sized Fe-rich zones and M-rich zones [47, 48].

In the case of equiatomic $Fe_{50}Re_{50}$ and $Fe_{58}Ta_{42}$ alloy, an "intermediate" phase is clearly identified during the solid-state reaction from the Mössbauer spectra [48, 49]. It is suggested that this "intermediate" phase is due to highly convoluted fractal interfaces between the powder grains, as also suggested by neutron small-angle-scattering [50, 51].

Several groups studied the bcc-$Fe_{50}Cu_{50}$ solid solution prepared by mechanical alloying of elemental Fe and Cu powders. An increasing density of interfaces during further milling is observed, resulting in an interdiffusion of Cu and Fe, as suggested from the shape of hyperfine field distributions which clearly demonstrates that the Fe and Cu were alloyed on an atomic level [52].

Moreover, it has been recently shown that the contaminant atoms introduced during the milling process are preferentially located at grain boundaries, chemically bonded with iron forming some extra phase.

Those different examples clearly demonstrate how the structure of nanostructured properties can be decomposed into two components and how the hyperfine structure can provide a quantitative estimate and a qualitative description of both crystalline grains and grain boundaries in the case of iron-containing powders. It is important to emphasize that the analysis of the hyperfine structure in terms of hyperfine field distribution requires close attention. The lack of resolution of the hyperfine structure is probably the most difficult point for getting accurate values of relative absorptions, leading then to the relative proportions of iron atoms located in these two zones. It is thus necessary to assume either similar recoil-free factor values or different recoil-free factor values but both composition and structure of grains and grain boundaries have to be well known. Studies of amorphous systems allowed to conclude that the values of recoil-free factor in amorphous alloys are up to 10% smaller than that of the analogous crystalline alloy [53]. The final stage is to estimate the relative volume but this requires the knowledge of the density and the chemical composition of each component.

Another feature which might influence the hyperfine structure is due to the occurrence of polarization effects originating from interactions between grains. Indeed one expects that the crystalline grain and the grain boundary are characterized by two different Curie temperatures, the highest assigned to the crystalline grains (T_c^G) and the lowest to the grain boundaries (T_c^{GB}), assuming the same composition for both components. When the measuring temperature is ranged between T_c^G and T_c^{GB}, the magnetic behaviour results from single domain magnetic grains (because of their size) and paramagnetic grain boundaries (as intergranular phase). Such a description is schematised in Figure 4.

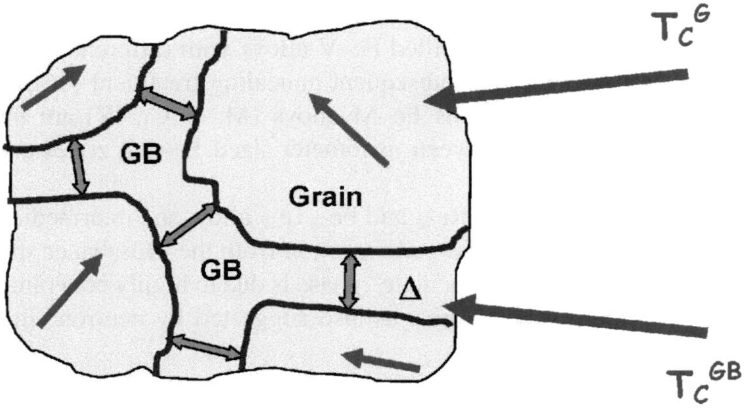

Figure 4. Schematic description of a nanostructured powder composed of grains and grain boundaries (thickness Δ), both characterized by their Curie temperatures.

The thickness of grain boundaries Δ has thus to be compared to the ferromagnetic correlation length which is estimated at about ten nm in the case of metallic systems. It is clear that the presence of narrow grain boundaries favours the magnetic coupling between crystalline grains, probably originating the occurrence of polarization effects, modifying thus the hyperfine structure of the grain boundaries. Consequently, the fitting model has to be checked at various temperatures in order to identify the pure hyperfine structure to be attributed to each zone. In contrast, an unusual phenomenon was recently observed with paramagnetic nanocrystalline grains separated by ferromagnetic disordered grain boundaries [54]. Moreover, it is also important to emphasize recent studies on the role of grain boundaries on the magnetic and transport properties in islanded Fe(110) thin films [55].

Acknowledgements

Acknowledgements are due to A. Slawska-Waniewska, J. Borrego, N. Randrianantoandro, M. Miglierini, I. Skorvanek and N. Cowlam for fruitful discussions.

References

1. Gleiter, H., *Acta Mater.* **48** (2000), 1.
2. Suryanarayana, C. and Koch, C. C., *Hyp. Interact.* **130** (2000), 5.
3. Lojkowski, W. and Fecht, H. J., *Progress in Materials Science* **45** (2000), 339.
4. Fultz, B. and Frase, H. N., *Hyp. Interact.* **130** (2000), 81.
5. Gleiter, H., In: M. Nastasi, D. M. Parkin and H. Gleiter (eds.), *Mechanical Properties and Deformation Behavior of Materials Having Ultra-Fine Microstructures*, Kluwer Academic, Dordrecht, 1993, p. 3.
6. Zhou, X., Birringer, R., Herr, U. and Gleiter, H., *Phys. Rev. B* **35** (1987), 9085.
7. Birringer, R., Gleiter, H., Klein, H. P. and Marquardt, *Phys. Lett. A* **102** (1984), 365.
8. Guérault, H. and Greneche, J. M., *J. Phys. Condens. Matter* **12** (2000), 4791.

9. Guérault, H., Tamine, M. and Greneche, J. M., *J. Phys. Condens. Matter* **12** (2000), 9497.
10. Greneche, J. M., *Czech. Journal of Physics* **52** (2002), 139.
11. Dormann, J. L., Fiorani, D. and Tronc, E., *Adv. Chem. Phys.* **98** (1997), 283.
12. Tronc, E., Ezzir, A., Cherkaoui, R., Chanéac, C., Noguès, M., Kachkachi, H., Fiorani, D., Testa, A. M., Greneche, J. M. and Jolivet, J. P., *J. Magn. Magn. Mater.* **221** (2000), 63.
13. Coey, J. M. D., *Phys. Rev. Lett.* **27** (1971), 1140.
14. Yoshizawa, Y., Oguma, S. and Yamauchi, K., *J. Appl. Phys.* **64** (1988), 6044.
15. Suzuki, K., Kataoka, N., Inoue, A., Makino, A. and Masumoto, T., *Mat. Trans. JIM* **31** (1990), 743.
16. Willard, M. A., Laughlin, D. E., McHenry, M. E., Thoma, D., Sickafus, K., Cross, J. O. and Harris, V. G., *J. Appl. Phys.* **84** (1998), 6773.
17. Greneche, J. M., Miglierini, M. and Slawska-Waniewska, A., *Hyp. Interact.* **126** (2000), 27.
18. Hampel, G., Pundt, A. and Hesse, J., *J. Phys. Condens. Matter* **4** (1992), 3195.
19. Rixecker, G., Schaaf, P. and Gonser, U., *J. Phys. Condens. Matter* **4** (1992), 10295.
20. Greneche, J. M., *Hyp. Interact.* **111** (1997), 261 and references therein.
21. Borrego, J. M., Conde, C. F., Conde, A. and Greneche, J. M., *J. Non-Cryst. Solids* **287** (2001), 120.
22. Borrego, J. M., Conde, A., Todd, I., Frost, M., Davies, H. A., Gibbs, M. R. J., Garitaonandia, J. S., Barandiaran, J. M. and Greneche, J. M., *J. Non-Cryst. Solids* **287** (2001), 125.
23. Kemény, T., Kaptas, D., Kiss, L. F., Balogh, J., Bujdoso, L., Gubicza, J., Ungar, T. and Vincze, I., *Appl. Phys. Lett.* **76** (2000), 2110.
24. Kopcewicz, M., Grabias, A., Willard, M. A., Laughlin, D. E. and McHenry, M. E., *IEEE Trans. Magn.* **37** (2001), 2226.
25. Blazquez, J. S., Conde, A. and Greneche, J. M., *Appl. Phys. Lett.* **81** (2002), 1612.
26. Kemény, T., Varga, L. K., Kiss, L. F., Balogh, J., Pusztai, T., Toth, L. and Vincze, I., *Mater. Sci. Forum* **269–272** (1998), 419.
27. Kemény, T., Kaptas, D., Balogh, J., Kiss, L. F., Pusztai, T. and Vincze, I., *J. Phys. Condens. Matter* **11** (1999), 2841.
28. Vincze, I., Kemény, T., Kaptas, D., Kiss, L. F. and Balogh, J., *Hyp. Interact.* **113** (1998), 123.
29. Balogh, J., Bujdoso, L., Kaptas, D., Kemény, T., Vincze, I., Szabo, S. and Beke, D. L., *Phys. Rev. B* **61** (2000), 4109.
30. Greneche, J. M. and Slawska-Waniewska, A., *J. Magn. Magn. Mater.* **215-216** (2000), 264.
31. Slawska-Waniewska, A. and Greneche, J. M., *Phys. Rev. B* **56** (1997), R8491.
32. Schmool, D. S., Garitaonandia, J. S. and Barandiaran, J. M., *Phys. Rev. B* **58** (1998), 12159.
33. Miglierini, M. and Greneche, J. M., *J. Phys. Condens. Matter* **9** (1997), 2303; **9** (1997), 2321.
34. Borrego, J. M., Conde, C. F., Conde, A., Peña-Rodríguez, V. A. and Greneche, J. M., *J. Phys. Condens. Matter* **12** (2000), 8089.
35. Hernando, A. and Kulik, T., *Phys. Rev. B* **49** (1994), 7064.
36. Suzuki, I. and Cadogan, J. M., *Phys. Rev. B* **58** (1998), 2730.
37. Franco, V., Conde, C. F., Conde, A., Kiss, L. F., Kaptàs, D., Kemény, T. and Vincze, I., *J. Appl. Phys.* **90** (2001), 1558.
38. Benjamin, J. S. and Volin, T. E., *Metall. Trans.* **5** (1974), 1929.
39. Del Bianco, L., Ballesteros, C., Rojo, J. M. and Hernando, A., *Phys. Rev. Lett.* **81** (1998), 4500.
40. Del Bianco, L., Hernando, A., Bonetti, E. and Navarro, E., *Phys. Rev. B* **56** (1997), 8894.
41. Balogh, J., Kaptas, D., Kemeny, T., Vincze, I. and Radnoczi, G., *Phys. Rev. Lett.* **82** (1999), 4150.
42. Bonetti, E., Del Bianco, L., Fiorani, D., Rinaldi, D., Caciuffo, R. and Hernando, A., *Phys. Rev. Lett.* **83** (1999), 2829.
43. Hernando, A., Navarro, E., Multigner, M., Yavari, A. R., Fiorani, D., Rosenberg, M., Filoti, G. and Caciuffo, R., *Phys. Rev. B* **58** (1998), 5181.
44. Herr, U., Jing, J., Birringer, R., Gonser, U. and Gleiter, H., *Appl. Phys. Lett.* **50** (1987), 472.

45. Fultz, B., Kuwano, H. and Ouyang, H., *J. Appl. Phys.* **77** (1995), 3458.
46. Ziller, T., Le Caer, G., Isnard, O., Cénédèse, P. and Fultz, B., *Phys. Rev. B* **65** (2002), 024204.
47. Le Caer, G., Delcroix, P., Shen, T. D. and Malaman, B., *Phys. Rev. B* **54** (1996), 12775.
48. Cooper, R. J., Randrianantoandro, N., Cowlam, N. and Greneche, J. M., *J. Phys. Condens. Matter* **9** (1997), 1425.
49. Randrianantoandro, N., Cooper, R. J., Greneche, J. M. and Cowlam, N., *J. Phys. Condens. Matter* **14** (2002), 9713.
50. Cooper, R. J., Randrianantoandro, N., Cowlam, N. and Greneche, J. M., *Mat. Sci. Eng. A* **226**(8) (1997), 84.
51. Cocco, G., Cowlam, N. and Enzo, S., *Mat. Sci. Eng. A* **178** (1994), 29.
52. Jiang, J. Z., Gonser, U., Gente, C. and Bormann, R., *Appl. Phys. Lett.* **63** (1993), 2768.
53. Schurer, P. J. and Morrish, A. H., *Phys Rev. B* **23** (1981), 1006.
54. Amils, X., Nogues, J., Surinach, S., Munoz, J. S., Baro, M. D., Hernando, A. and Morniroli, J. P., *Phys. Rev. B* **63** (2001), 052402.
55. Sefrioui, Z., Menendez, J. L., Navarro, E., Cebollada, A., Briones, F., Crespo, P. and Hernando, A., *Phys. Rev. B* **64** (2001), 224431.

Hyperfine Interactions **144/145:** 161–173, 2002.

Magnetic and Structural Features of Rare Earth Intermetallic Compounds as Seen by ^{57}Fe Mössbauer Probe

E. M. BAGGIO-SAITOVITCH, D. R. SÁNCHEZ, M. B. FONTES, S. N. MEDEIROS, E. PASSAMANI and H. MICKLITZ*
Centro Brasileiro de Pesquisas Fisicas, Rua Dr. Xavier Sigaud 150, Urca, 22290-180 Rio de Janeiro, Brazil

Abstract. ^{57}Fe Mössbauer spectroscopy has been used to study the magnetic structure of some intermetallic compounds having strongly correlated electronic properties including some magnetic superconducting compounds. The common characteristic of the presently studied systems is the fact that they have a crystalline structure closely related to the $ThCr_2Si_2$- and $BaNiSn_3$-type of structures, where the transition metal carries no magnetic moment. This specific property allows using ^{57}Fe as Mössbauer probe located at the transition metal site to detect very small transferred hyperfine fields (B_{thf}) from the neighboring rare earth sites, and in many cases, to determine the angle θ between B_{thf} and the z-axis component of the electric field gradient (EFG).

Key words: borocarbides, heavy fermion systems, magnetic structure, Mössbauer spectroscopy, superconductivity.

1. Introduction

Strongly correlated electron systems (SCES) present a variety of physical properties that arise from the competition between Kondo effect and RKKY interactions [1]. The approach to the quantum critical point (QCP) in the Doniach phase diagram is followed by a decrease in the magnetic ordering temperature (usually of AF type). The variation of J/W can be induced by chemical doping and hydrostatic pressure in order to approach the QCP. The understanding of the anomalous aspects of these systems naturally involves the study of their magnetic properties. Magnetism, superconductivity or the coexistence of both are the main features of the RNi_2B_2C and RNiBC (RS = rare earth or Y) compounds [2]. A better understanding of the variety of magnetic structures existing in these compounds and their connection with the superconducting state is necessary.

The ^{57}Fe Mössbauer spectroscopy is not a very useful technique for studying the superconducting transition at T_c, however it is a powerful tool for the study of

---

* Permanent address: II. Physikalisches Institut, Universität zu Köln, Zülpicherstr. 77, 50937 Köln, Germany.

magnetism by means of magnetic hyperfine interactions at the ^{57}Fe nucleus. This technique is used here to study the magnetic properties and structural properties by means of electrical hyperfine interactions, of the $Ce(Ru_{1-x}Fe_x)_2Ge_2$, $Ce(Co_{1-x}Fe_x)Ge_3$, RNi_2B_2C and RNiBC (R = rare earth or Y) series. A common characteristic of these compounds is that the ^{57}Fe probe carries no magnetic moment, which is ideal to study magnetic structures through the transferred magnetic hyperfine field (B_{thf}) at the ^{57}Fe nucleus from the neighboring magnetic atoms [3–5]. These systems crystallize in the tetragonal structure of the $ThCr_2Si_2$ [6], $LuNi_2B_2C$ [7] (both I4/mmm), LuNiBC [7] (P4/nmm) and $BaNiSn_3$ [8] (I4/mm), respectively. The last two symmetry groups are variants of the group I4/mmm of the $ThCr_2Si_2$-type of structure. The magnetism of these compounds is exclusively due to the f moments of the R atoms and can be studied by the B_{thf} (via RKKY interaction or dipolar field) at the ^{57}Fe nucleus [3–5]. The magnetic and quadrupole interaction are, in general, of the same order of magnitude in these kinds of systems, and the transferred hyperfine fields are of the order of $\sim$1 T. Therefore, to observe such small hyperfine fields it is necessary that the ^{57}Fe probe carries no magnetic moment, as it occurs in the compounds presented in this work.

2. Results and discussion

2.1. $Ce(Ru_{1-x}Fe_x)_2Ge_2$ AND $Ce(Co_{1-x}Fe_x)Ge_3$ SYSTEMS

The $Ce(Ru_{1-x}Fe_x)_2Ge_2$ ($0 \geqslant x$-Fe $\geqslant 1$) series was the first system we have investigated using ^{57}Fe Mössbauer spectroscopy, in which the Fe atoms carries no magnetic moment. The samples are formed in the $ThCr_2Si_2$-type of crystalline structure in the complete range of concentrations. The $CeRu_2Ge_2$ orders ferromagnetically below $T_c = 7.5$ K, with the Ce-f moments aligned along the c-axis [9], while the $CeFe_2Ge_2$ is a non-magnetic moderate heavy fermion (HF) system with Sommerfeld coefficient $\gamma \sim 400$ mJ/mol K^2 [10].

Mössbauer measurements were taken at two fixed temperatures, 4.2 K and 300 K, shown in Figure 1. A clear asymmetric line shape of the spectra is seen for the Fe concentrations in the magnetic region, previously established by DC magnetization measurements. The Mössbauer spectra at 4.2 K were analyzed using the diagonalization of the complete Hamiltonian, with quadrupole and magnetic interactions of similar magnitude and accounting for the angle θ between B_{thf} and the z-axis component of the EFG. Specifically for this series, a double set of fitting parameters could be found, with no regarding for the correct physical picture. One of the possible sets of hyperfine parameters is shown in Figure 2. In any case, one can observe a rapid change of the direction of the magnetic moment of the Ce with x-Fe. This behavior is consistent with the DC magnetization measurements, which suggest a change from ferromagnetism to a spin glass type of behavior with Fe concentration [11]. Above $x = 0.6$ it is no longer possible to define θ and B_{thf} from the fits. As shown by the bulk magnetic measurements, at a critical concentration

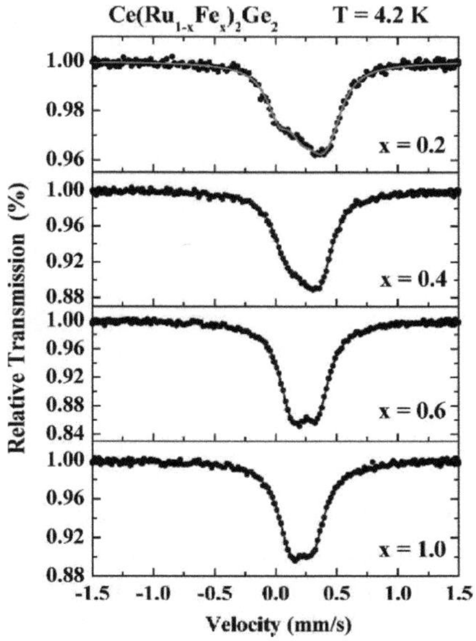

Figure 1. ^{57}Fe Mössbauer spectra of the $Ce(Ru_{1-x}Fe_x)_2Ge_2$ series at $T = 4.2$ K.

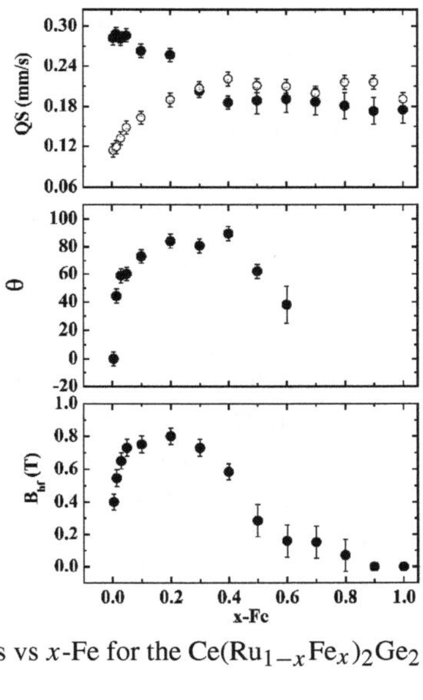

Figure 2. Hyperfine parameters vs x-Fe for the $Ce(Ru_{1-x}Fe_x)_2Ge_2$ series: (•) 4.2 K and (○) 300 K.

$x_c \sim 0.7$ occurs the break down of magnetism of the system [11] and this behavior is reflected in the Mössbauer data.

There is a discrepancy in the values of QS above the magnetic transition and 4.2 K in the low Fe concentrations of these compounds where the system is magnetically ordered. A detailed discussion of this anomalous behavior will be published elsewhere.

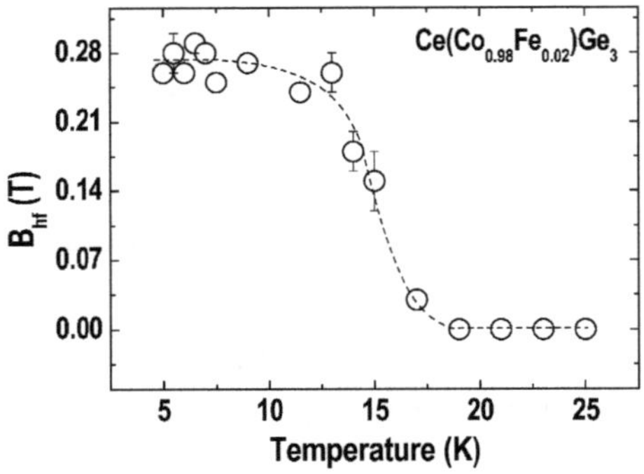

Figure 3. Temperature dependence of B_{thf} at the ^{57}Fe nucleus in $Ce(Co_{0.98}{}^{57}Fe_{0.02})Ge_3$.

We are also currently studying the $Ce(Co_{1-x}Fe_x)Ge_3$ series [12, 13]. The $CeFeGe_3$ is a non-magnetic moderate heavy fermion system ($\gamma = 150$ mJ/mol K^2) [12], while $CeCoGe_3$ is a magnetic HF with a complex magnetic phase diagram presenting two transitions at ~21 K and ~18 K [13], possibly of antiferromagnetic (AF) type. The magnetic transition temperature decreases with the Fe doping, probably followed by a change in the kind of magnetic order.

Assuming an AF spin arrangement, one could expect, in general, a zero transferred field at the Fe site. The Fe probe is placed inside, and close to the base, of a pyramid formed by the Ce atoms in the $BaNiSn_3$ crystalline structure. Consequently, B_{thf} does not cancel completely at the Fe site, even for a collinear antiferromagnetic order, leading to this small measured field (Figure 3). Any other kind of magnetic arrangement would also result in an observable transferred field, possibly with higher value of B_{thf}. Just for sake of clarity, in the $ThCr_2Si_2$-type of structure the Fe atoms sits inside a symmetric tetrahedra formed by R atoms, giving a zero B_{thf} for a AF spin arrangement of the R magnetic moments.

Therefore, the nature of the magnetism for x-Fe $= 0.02$ is still not completely clear, since the position of Fe atoms in the lattice favors a non-zero transferred field for all kind of spin arrangement. From Mössbauer measurements the magnetic moments are probably aligned along the c-direction in an AF configuration. However, more studies are needed in order to completely understand the modification of the magnetic properties of $CeCoGe_3$ system introduced by Fe-doping.

2.2. RNi_2B_2C AND RNiBC

The crystalline structure of the RNi_2B_2C series can be viewed as a stacking of Ni_2B_2 layers spatially separated by R–C planes. The detailed knowledge of their magnetic structure, which seems to be responsible for the suppression of superconductivity in some compounds of this series, is important for the understanding of the interplay between superconductivity and magnetism.

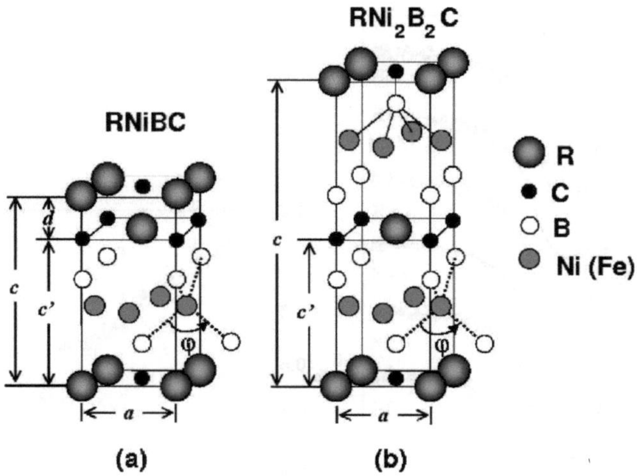

Figure 4. Crystal structures of (a) RNiBC and (b) RNi_2B_2C compounds. The relevant lattice parameters a and c' used in the text are indicated.

The similarities between the RNi_2B_2C and the RNiBC structure are illustrated in Figures 4(a) and (b): while in RNi_2B_2C [$\equiv(RC)_1(NiB)_2$] there are alternating layers of R–C and Ni_2–B_2, an additional R–C layer exists between the Ni_2–B_2 layers in the RNiBC [$\equiv(RC)_2(NiB)_2$] series. Thus, in both classes of compounds the Ni_2–B_2 layers, which are responsible for superconductivity, are sandwiched between two R–C layers. These Ni_2–B_2 layers are built from NiB_4 tetrahedra with a B–Ni–B bonding angle φ as indicated in Figures 4(a) and (b). In RNiBC series the lattice parameters c and d (see Figure 1(a)) turned out to be essentially independent of R with $c = 7.55$ Å and $d = 2.41$ Å [14, 15]. In RNi_2B_2C (see Figure 4(b)) $(c/a) = 2\,(c'/a)$, with c' being the distance of the R–C layers between which the Ni_2–B_2 layers are sandwiched and a being the lattice parameter in the basal plane. In the RNiBC compounds c' is given by $c' = c - d$ (Figure 4(a)).

2.2.1. *Magnetism in RNi_2B_2C and RNiBC series*

The magnetic structure of the RNi_2B_2C and RNiBC were investigated by Mössbauer spectroscopy through the transferred magnetic hyperfine field (B_{hf}), mainly resulting from the four nearest neighbor magnetic R atoms (two from the R layer above and two from the R layer below the Ni plane), at the Ni site.

$TbNi_2B_2C$ develops an AF order at $T_N \sim 15$ K and a small ferromagnetic component below about 8 K has been observed [16]. However, more studies were necessary to know their specific ferromagnetic arrangement. The magnetic order at 15 K becomes evident from the low-temperature ^{57}Fe Mössbauer spectra shown in Figure 5(a), a magnetic broadening appears below 15 K.

The temperature dependence of the transferred magnetic hyperfine field (B_{hf}) is shown in Figure 5(b). B_{hf} increases as the temperature decreases below T_N,

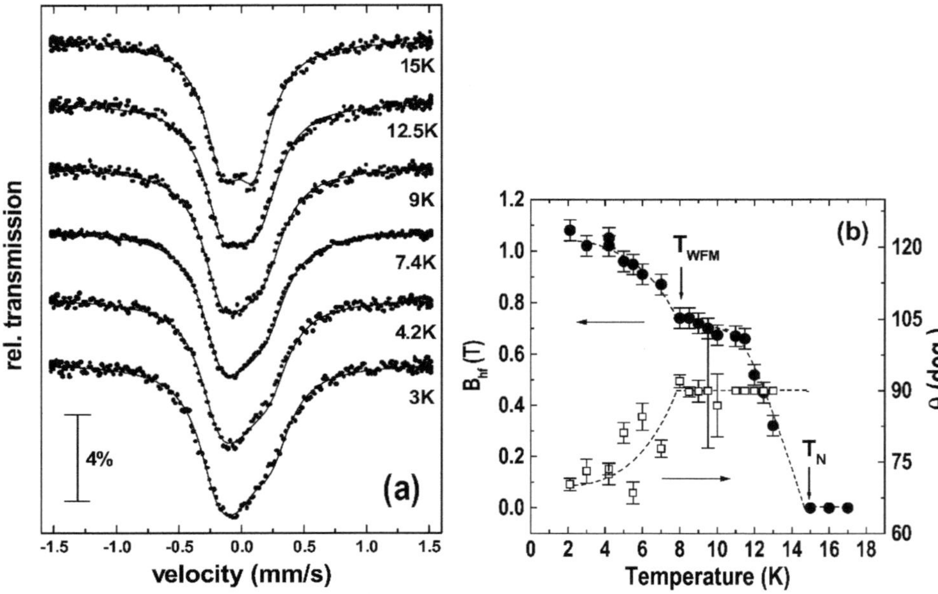

Figure 5. (a) ^{57}Fe Mössbauer absorption spectra and (b) temperature dependence of B_{hf} of ^{57}Fe and of the angle θ between B_{hf} and V_{zz} in $Tb(Ni_{0.99}{}^{57}Fe_{0.01})_2B_2C$.

reaching a value of about 0.8 T at 8 K (see Figure 2(b)). Between 8 and 5 K, a small, but significant increase in B_{hf} is observed. Simultaneously below 8 K, the angle θ between B_{hf} and the main component of the electric field gradient V_{zz} (in *c* direction) decreases to about 70° (Figure 2(b)). These results are consistent with the picture that one spin component of Tb remains fluctuating down to about 8 K and is locked below this temperature, a fact that we relate to the weak ferromagnetism [4].

$GdNi_2B_2C$ is known to have AF order below $T_N \sim 20$ K, and with a rearrangement of spins below $T_R \sim 14$ K [17]. Lowering the temperature below T_N a symmetric broadening appears in the spectrum which further increases for lower temperatures (see Figure 6(a)). The temperature dependence of B_{hf} is shown in Figure 6(b). In order to explain the symmetric $GdNi_2B_2C$ spectra with a combined magnetic and electrical quadrupole interaction the angle θ between B_{hf} and V_{zz} either is just the so-called "magic angle" ($\theta \sim 55°$) or is not well defined, i.e., all angles between $0 \leqslant \theta \leqslant 90°$ occur with equal probability. These results are in agreement with a spiral-like spin structure model proposed in [9] for the spin structure below T_R [5].

Neutron diffraction studies [18–20] reveal a ferromagnetic (FM) coupling between neighboring R–C planes (separation $d = 2.41$ Å). The coupling between the $(R–C)_2$ double layers, however, turned out to be different along the RNiBC series, typical for systems wherein RKKY interaction is dominant: for Er [18], Dy and Tb [19, 20] it is FM, while it is simple collinear AF for Ho [18]. Such an AF coupling also was proposed for Gd [21].

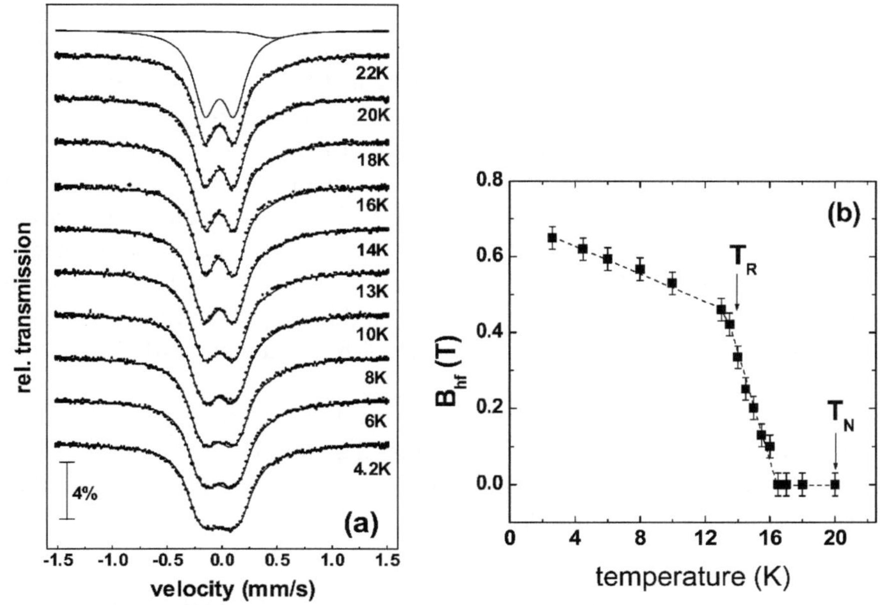

Figure 6. (a) ^{57}Fe Mössbauer spectra of $Gd(Ni_{0.99}Fe_{0.01})_2B_2C$ in 22 K $\geqslant T \geqslant$ 4.2 K. (b) Temperature dependence of B_{hf} at the ^{57}Fe nucleus in $Gd(Ni_{0.99}Fe_{0.01})_2B_2C$.

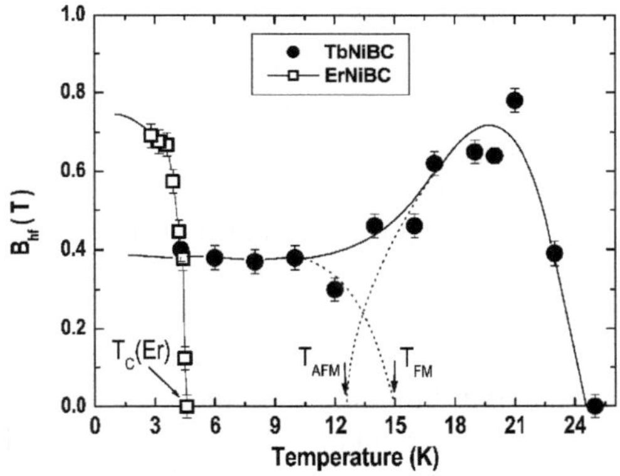

Figure 7. Temperature dependence of the B_{thf} at the ^{57}Fe nucleus for the $Er(Ni_{0.99}{}^{57}Fe_{0.01})BC$ (open squares) and for the $Tb(Ni_{0.99}{}^{57}Fe_{0.01})BC$ (full circles). The indicated temperatures are the ferromagnetic T_{FM} and the antiferromagnetic T_{AM} transitions observed by preliminary neutron diffraction measurements in TbNiBC [20].

From Mössbauer experiments a transferred B_{hf}, at the Fe (Ni) site was observed for the FM Er and Tb but not for the AF Ho and Gd. The temperature dependence of B_{hf} for Er nicely follows the magnetization curve for $T \leqslant T_M \approx 4.7$ K (Figure 4). In the case of Tb, however, a magnetic broadening of the quadrupole doublet already starts below 24 K.

The temperature dependence of B_{hf} for Tb is shown in Figure 7, where the onset temperatures T_{FM} and T_{AF} for FM and AF order, respectively, as obtained from neutron diffraction data [19, 20] are indicated by arrows. To understand the $B_{hf}(T)$ as shown in Figure 4, we need to look at the neutron diffraction results: TbNiBC displays a 2-dimensional magnetic ordering (two closest layers coupled ferromagneticaly, with their moments in the basal plane, and without magnetic coupling between the double layers) at around 25 K, a 3-dimensional ferromagnetic ordering at 15 K and below 12.5 K a combined ferro- and antiferro-magnetic ordering. Therefore, the B_{hf} observed between $15 \leqslant T \leqslant 25$ K may indicate a short range 2-dimensional magnetic ordering and below ~14 K the ^{57}Fe nucleus senses a transferred B_{hf} due to the 3-dimensional ferromagnetic ordering of the Tb moments.

No magnetic hyperfine field, on the other hand, was observed for Ho, Dy and Gd. This is in agreement with the experimental data for Ho and Gd. However, this result is not clear for the Dy compound whose magnetic structure is very similar to that found for the Tb by neutron diffraction.

2.2.2. *Superconductivity and magnetism in RNi_2B_2C: Pair breaking field*

$HoNi_2B_2C$ orders AF at $T_N = 8.5$ K and becomes a superconductor at 8 K. An incommensurate modulated magnetic structure appears at ~6 K (see Figure 8(a)), which seems to be responsible for the reentrant behavior, i.e., the suppression of superconductivity between 4.7 K $< T <$ 6 K [2] (Figure 5(b)).

The loss of superconductivity in this range of temperature is attributed to the appearance of a pair breaking field at the Ni layer, where ^{57}Fe probe is located. The transferred B_{hf} at the Ni(^{57}Fe) site, resulting from the fourth nearest neighbors of Ho atoms (two from the Ho layer above and two from the Ho layer below the Ni plane), will not cancel in the interval 4.7 K $< T <$ 6 K, due to the incommensurate AF structure of $HoNi_2B_2C$, as shown in Figure 8. The transferred magnetic hyperfine field (B_{hf}) shown as a function of the temperature (Figure 8(c)), reveals clearly that such a magnetic pair breaking field indeed exists in the reentrant region for the $HoNi_2B_2C$ compound. Such a kind of pair breaking field is not observed in case of $DyNi_2B_2C$ and $ErNi_2B_2C$, where superconductivity coexists with magnetic order [3].

2.2.3. *Superconductivity and structural features*

^{57}Fe ME studies of RNi_2B_2C compounds [22, 23] above their magnetic ordering temperatures reveal one quadrupole doublet whose splitting $|\Delta E_Q|$ increases from Er to Gd. The $|\Delta E_Q|$ values for RNiBC (R = Er, Ho, Dy, Tb, Gd) and for RNi_2B_2C (R = Er, Ho, Dy, Tb, Gd, Nd, Pr) are plotted in Figure 9 as a function of the structural parameter (c'/a) which is a measure for the B–Ni–B bonding angle φ (see Figures 4(a) and (b)).

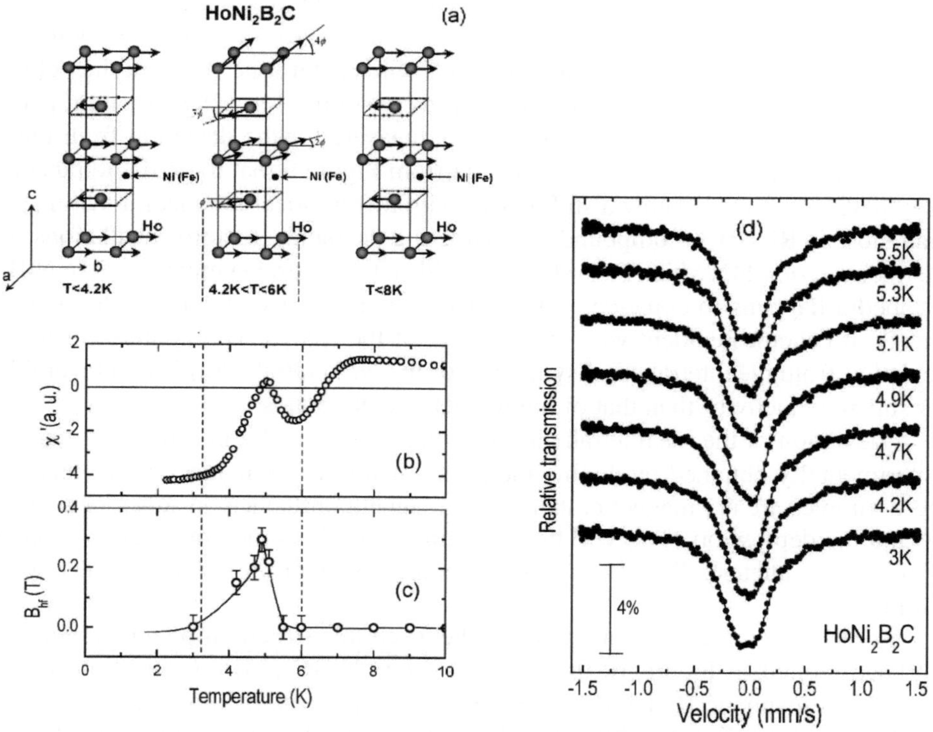

Figure 8. (a) Magnetic structure of the $HoNi_2B_2C$ for three different temperature regions, (b) χ_{ac}, (c) B_{thf} at the Ni site, and (d) temperature dependence of ^{57}Fe Mössbauer spectra for the $Ho(Ni_{0.99}{}^{57}Fe_{0.01})_2B_2C$ compound.

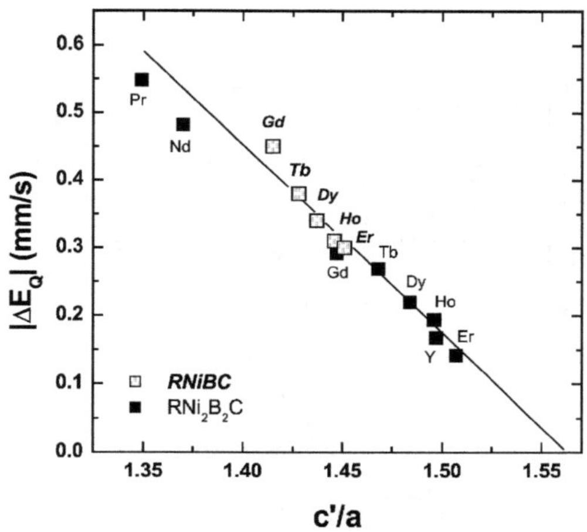

Figure 9. Room temperature quadrupole splitting $|\Delta E_Q|$ observed in RNiBC and RNi_2B_2C compounds as a function of the (c'/a)-ratio.

Figure 9 shows that *all* compounds, RNi_2B_2C as well as RNiBC, have the *same* linear correlation between (c'/a) and $|\Delta E_Q|$. This clearly indicates that both the crystallographic and the electronic structure (which determine ΔE_Q) at the Fe(Ni) site in RNiBC and RNi_2B_2C compounds are very similar for equal structural parameter (c'/a). Furthermore, since $|\Delta E_Q|$ is a measure for the deviation from ideal tetrahedral symmetry, we have to conclude from Figure 9 that the NiB_4 tetrahedra in RNiBC compounds show a stronger deviation from ideal tetrahedral symmetry than those in RNi_2B_2C compounds. Considering the band structure calculations by Mattheiss *et al.* [24], which have shown that ideal tetrahedral symmetry of the NiB_4 tetrahedra happens to coincide with a relative high density of states at the Fermi energy, it becomes evident why the structure of the RNiBC compounds (stronger deviation from ideal tetrahedral symmetry of the NiB_4 tetrahedra) is less favorable for superconductivity than that of the RNi_2B_2C compounds.

A deviation of the ideal tetrahedral symmetry of the NiB_4 in $LuNi_2B_2C$ can be experimentally obtained by doping the Lu site with the nonmagnetic La. Because these samples are nonmagnetic, there is no magnetic influence on superconductivity and the depression of T_c will be determined only by structural effects (change of (c'/a) parameter). The above arguments are also valid for YNi_2B_2C if doped with La.

In order to see if there is any relation between this structural parameter and T_c we have plotted in Figure 10 the T_c value for LuNiBC [25] together with those for the non-magnetic RNi_2B_2C (R = Lu, Sc) compounds, the non-magnetic mixtures $(Lu_{1-x}La_x)Ni_2B_2C$ [26], $(Y_{1-x}La_x)Ni_2B_2C$ [27] and the magnetically ordered RNi_2B_2C (R = Tm, Er, Ho, Dy) compounds as a function of the (c'/a) parameter. The non-magnetic $LuNi_2B_2C$, $ScNi_2B_2C$, $(Lu_{1-x}La_x)Ni_2B_2C$ and LuNiBC show a linear relation between (c'/a) and T_c (solid line in Figure 10). The non-magnetic mixtures $(Y_{1-x}La_x)Ni_2B_2C$ also show a linear relationship between

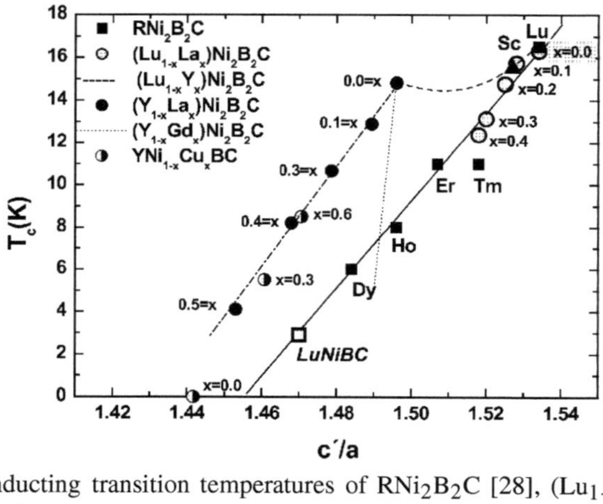

Figure 10. Superconducting transition temperatures of RNi_2B_2C [28], $(Lu_{1-x}La_x)Ni_2B_2C$ [26], $(Y_{1-x}La_x)Ni_2B_2C$ [27], LuNiBC [25], and $YNi_{1-x}Cu_xBC$ [29].

(c'/a) and T_c (dashed line in Figure 10) with a slope which is identical to that for the other non-magnetic compounds. This slope of T_c vs (c'/a), is thus a measure for the decrease of T_c with decreasing (c'/a).

The T_c values found for $(Y_{1-x}La_x)Ni_2B_2C$ are higher (for the same (c'/a) value) than those of RNi_2B_2C family of borocarbides. This feature is likely related to the fact that the Y does not behave as a rare earth in these series of compounds. Furthermore, the YNiBC becomes superconducting when doped with Cu [29], and their T_c vs c'/a plot follows quite the same trend observed for the $Y_{1-x}La_xNi_2B_2C$ series (Figure 10).

Quite surprisingly the magnetically ordered RNi_2B_2C (R = Dy, Ho, Er) compounds with T_N close to T_c are on the same line as the non-magnetic compounds. Only T_c of $TmNi_2B_2C$ with $T_N \ll T_c$ is slightly below the line (see Figure 10), indicating that there is a clear paramagnetic pair-breaking effect due to exchange scattering in the framework of Abrikosov–Gorkov theory for this compound. Thus, the pair-breaking effect of the R magnetic moment seems to be very small for Dy, Ho, Er and it is $\Delta T_c \approx -2$ K for Tm. This finding is quite in contrast to the generally accepted opinion that T_c in all magnetically ordered RNi_2B_2C compounds scales with the de Gennes factor [30, 31]. However, it is in perfect agreement with very recent theoretical calculations [32] which have shown that the Ni(3*d*), electrons which are mainly responsible for superconductivity, do not feel the R magnetic moments in the case that the R magnetic moments are AF ordered, i.e., for Dy and Ho.

We want to point out that the mixtures $Y_{1-x}Gd_xNi_2B_2C$ [33] (dotted line in Figure 10) and $Lu_{1-x}Gd_xNi_2B_2C$ [34] cannot be considered in our linear relation of T_c with (c'/a), because in both cases, the pair breaking effect (T_c will decrease according to Abrikosov–Gorkov theory) is more effective in reduction of T_c than the structural effect.

Taking into account what was discussed above, the (c'/a) parameter (a measure of the distortion of the NiB_4 tetrahedra) seems to be a relevant parameter which determines T_c in *all* non-magnetic and antiferromagnetically ordered RNi_2B_2C and RNiBC compounds. On the other hand, the T_c scale with (c'/a) ratio (Figure 7) and the de Gennes factor [20] in a similar way for the heavy rare earth RNi_2B_2C. However, in the Abrikosov–Gorkov framework it is not possible to explain the differences in T_c of the nonmagnetic $LuNi_2B_2C$, $ScNi_2B_2C$, YNi_2B_2C, $LaNi_2B_2C$, and the R–C double layer LuNiBC compound.

The superconducting properties of the RNi_2B_2C phases are attributed to an electron–phonon mechanism [24]. Although the Fermi electrons in these materials have predominant Ni(3*d*) character, the superconductivity occurs only when a special *s–p* band is optimally aligned relative to the Fermi level (i.e., when the NiB_4 tetrahedra angles are nearly "ideal") and which exhibits strong electron–phonon coupling [24].

Since we expect that the structure by itself cannot determine T_c, more work is necessary to elucidate the role of the structural effects on the electron–phonon coupling, which seems to be essential to describe the nature of the depression of T_c in the rare earth-nickel-borocarbides series of compounds.

3. Conclusion

We have shown that ^{57}Fe Mössbauer spectroscopy is a valuable technique to study the R spins arrangement in compounds where the Mössbauer probe carries no magnetic moment. The SCES we have studied belongs to a class of materials with a crystalline structure satisfying this condition, and it was possible to detect very small transferred hyperfine fields, as well as the direction of the R magnetic moments. Furthermore, Mössbauer spectroscopy clearly indicated that the crystallographic and the electronic structure of RNiBC and RNi_2B_2C are very similar for the same structural parameter (c'/a). This was the common characteristic that allowed to find the existence of a linear correlation between T_c and c'/a, for both series of compounds.

Acknowledgements

We thank the financial support from FAPERJ/RJ, VW Foundation, Capes/Brazil, CNPq/Brazil and DAAD/Germany. D. R. S. thanks Latin American Center for Physics (CLAF) for his Post-doc fellowship. This work was supported by PRONEX-MCT.

References

1. For reviews see: Stewart, G. R., *Rev. Mod. Phys.* **56** (1984), 755 and Grewe, N. and Steglich, F., In: K. A. Gschneidner, Jr. and L. L. Eyring (eds), *Handbook on the Physics and Chemistry of Rare Earths*, Vol. 14, Elsevier, Amsterdam, 1991, p. 343.
2. Müller, K.-H. and Narozhnyi, V. N., *Rep. Prog. Phys.* **64** (2001), 943.
3. Sánchez, D. R., Micklitz, H., Fontes, M. B., Bud'ko, S. L. and Baggio-Saitovitch, E. M., *Phys. Rev. Lett.* **76** (1996), 507.
4. Sánchez, D. R., de Melo, M. A. C., Fontes, M. B., Bud'ko, S. L., Baggio-Saitovitch, E. M., Hillberg, M., Wagener, W., Klauß, H.-H., Walf, G. H. and Litterst, F. J., *Phys. Rev. B* **57** (1998), 10268.
5. Sánchez, D. R., Micklitz, H. and Baggio-Saitovitch, E. M., *J. Phys.: Condens. Matter* **9** (1997), L299.
6. Parthé, E. and Chabot, B., In: K. A. Gschneidner and L. Eyring (eds), *Handbook on the Physics and Chemistry of Rare Earths*, Vol. 6, North-Holland, Amsterdam, 1984, Ch. 48, p. 113.
7. Siegrist, T., Zandbergen, H. W., Cava, R. J., Krajewski, J. J. and Peck, Jr., W. F., *Nature (London)* **367** (1994), 254.
8. Dörrscheidt, W. and Schäfer, H., *J. Less-Common Met.* **58** (1978), 209.
9. Loidl, A., Knorr, K., Knopp, G., Krimmel, A., Caspary, R., Böhm, A., Sparn, G., Geibel, C., Steglich, F. and Murani, A. P., *Phys. Rev. B* **46** (1992), 9341.
10. Ebihara, T., Motoki, K., Toshima, H., Takashita, M., Kimura, N., Sugawara, H., Ichihashi, K., Settai, R., Onuki, Y., Aoki, Y. and Sato, H., *Phys. Rev. B.* **206/207** (1995), 219.

11. Fontes, M. B., Bud'ko, S. L., Continentino, M. A., El-Massalami, M., Sampaio, L. C., Hundley, M. F., Lacerda, A., Guimarães, A. P. and Baggio-Saitovitch, E. M., *Phys. Rev. B* **53** (1996), 11678.
12. Yamamoto, H., Ishikawa, M., Hasegawa, K. and Sakurai, J., *Phys. Lett. A* **196** (1994), 83.
13. Pecharsky, V. K., Hyun, O.-B. and Gschneidner, Jr., K. A., *Phys. Rev. B* **47** (1993), 11839.
14. Huang, Q., Lynn, J. W., Santoro, A., Chakoumakos, B. C., Cava, R. J., Krajewski, J. J. and Peck, Jr., W. F., *Physica C* **271** (1996), 311.
15. Fontes, M. B., Trochez, J. C., Continentino, M. A., Giordanengo, B., Bud'ko, S. L., Sánchez, D. R. and Baggio-Saitovitch, E. M., *Phys. Rev. B* **60** (1999), 6781.
16. Cho, B. K., Canfield, P. C. and Johnston, D. C., *Phys. Rev. B* **53** (1996), 8499.
17. Canfield, P. C., Cho, B. K. and Dennis, K. W., *Physica B* **215** (1995), 337.
18. Chang, L. J., Tomy, C. V., McK Paul, D., Andersen, N. H. and Yethiraj, M., *J. Phys.: Condens. Matter* **8** (1996), 2119.
19. Skanthakumar, S. and Lynn, J. W., *Physica B* **259–261** (1999), 576.
20. Bourdarot, F., Lynn, J. W., Baggio-Saitovitch, E., Huang, Q., Sanchez, D. R. and Skanthakumar, S., *Phys. Rev. B* **64** (2001), 104410.
21. El Massalami, M., Giordanengo, B., Mondragon, J., Baggio-Saitovitch, E. M., Takeuchi, A., Voiron, J. and Sulpice, A., *J. Phys.: Condens. Matter* **7** (1995), 10015.
22. Zeng, Z., Sánchez, D. R., Guenzburger, D., Ellis, D. E., Baggio-Saitovitch, E. M. and Micklitz, H., *Phys. Rev. B* **55** (1997), 3087.
23. Sánchez, D. R., Micklitz, H. and Baggio-Saitovitch, E. M., *J. Phys.: Condens. Matter* **12** (2000), L177.
24. Mattheiss, L. F., Siegrist, T. and Cava, R. J. , *Solid State Commun.* **91** (1994), 587.
25. Gao, L., Qui, X. D., Cao, Y., Meng, R. L., Sun, Y. Y., Xue, Y. Y. and Chu, C. W., *Phys. Rev. B* **50** (1994), 9445.
26. Sánchez, D. R., Micklitz, H. and Baggio-Saitovitch, E. M., to be published.
27. Sánchez, D. R., Bud'ko, S. L. and Baggio-Saitovitch, E. M., *J. Phys.: Condens. Matter* **12** (2000), 9941.
28. Lynn, J. W., Skanthakumar, S., Huang, Q., Sinha, S. K., Hossain, Z., Gupta, L. C., Nagarajan, R. and Godart, C., *Phys. Rev. B* **55** (1997), 6584.
29. Freudenberger, J., Drechsler, S.-L., Fuchs, G., Kreyssig, A., Nenkov, K., Shulga, S. V., Müller, K.-H. and Schultz, L., *Physica C* **306** (1998), 1.
30. Cho, B. K., Canfield, P. C. and Johnston, D. C., *Phys. Rev. B* **52** (1995), R3844.
31. Canfield, P. C., Cho, B. K. and Dennis, K. W., *Physica B* **215** (1995), 337.
32. Hyeonjin, D., Sigrist, M., Cho, B. K. and Lee, S.-I., *Phys. Rev. Lett.* **83** (1999), 5350.
33. El Massalami, M., Bud'ko, S. L., Giordanengo, B., Fontes, M. B., Mondragon, J. C., Baggio-Saitovitch, E. M. and Sulpice, A., *Phys. Stat. Sol. (b)* **189** (1995), 489.
34. Cho, B. K., Canfield, P. C. and Johnston, D. C., *Phys. Rev. Lett.* **77** (1996), 163.

Hyperfine Interactions **144/145:** 175–181, 2002.

Magnetism in $FeCl_2$ at High Pressures

W. M. XU and M. P. PASTERNAK
School of Physics and Astronomy, Tel Aviv University, 69978 Tel Aviv, Israel

Abstract. Anhydrous $FeCl_2$ has been studied by ^{57}Fe Mössbauer effect up to 61 GPa and at $5 < T < 300$ K using diamond-anvil-cells. In the 3.5–31 GPa range an average hyperfine field of ~1.2 T is observed, attributed to mutual cancellation of *Fermi-contact* and *orbital* terms. At $P \geqslant 34$ GPa the orbital term is abruptly quenched resulting in enhanced hyperfine field. At $P > 42$ GPa the onset of a diamagnetic phase is observed, signifying a total collapse of magnetism at $P > 61$ GPa.

Key words: high-pressure, Mössbauer spectroscopy, ferrous halide, magnetism, spin crossover, Mott transition.

1. Introduction

For many decades anhydrous $FeCl_2$ has attracted the interest of Mössbauer spectroscopists. This compound which orders antiferromagnetically at rather low temperature (~24 K) exhibits the unusual spin-flop phenomenon (matamagnetism) [1]. Despite the Fe^{2+} large magnetic moment of 4.5 μ_B as obtained from neutron diffraction studies [2] a surprisingly small hyperfine magnetic field, H_{hf}, of 0.45 T has been measured at $T \ll T_N$ [3]. This has been explained in terms of a *nearly* cancellation of the *Fermi-contact* and the *orbital* terms contributions to the hyperfine field. Pressure studies to 0.5 GPa by Vettier and Yelon [4] reported a structural phase transition at about 0.4 GPa, from a rhombohedral r-$FeCl_2$ phase to a closer packed, hexagonal, h-$FeCl_2$ phase with a CdI_2 structure-like, isomorphous to ferrous bromide and iodide at ambient pressure.

The ferrous halides are *Mott insulators*, i.e., apart from ordering antiferromagnetically they are also good insulators. The reason being the strong electron–electron correlation within the d-bands resulting in (i) narrow d-bands, therefore in relatively large moments, and in a large d–d (U-gap) or d–p energy gap (Δ-gap). External pressure being an excellent non-evading tool has been chosen for tuning this band gap, bearing in mind that with pressure the bandwidth increases and eventually the gap will close, leading to delocalization of the electronic wave function, hence metallization. This will concur with the collapse of the magnetic moment [5]. With the analogous FeI_2 [6] a pressure induced orbital momentum quenching has been observed at 18 GPa. Above 28 GPa and down to 5 K no magnetic interaction has been detected, leading to the conclusion that a paramagnetic–diamagnetic transition took place. Resistivity studies asserted the onset of a metallic state.

In the present work high-pressure (HP) magnetic and structural properties of $FeCl_2$ were studied for the first time using ^{57}Fe MS up to 61 GPa in the temperature range of 5–300 K. The main purpose of this study was to monitor the progressive transformation of the magnetic state with decreasing interatomic distances and for possible structural changes near the Fe region.

2. Experimental

The sample was prepared by metal-vapor reaction of the elements sealed in an evacuated quartz tube. $PdCl_2$ and metallic iron enriched to 25% in ^{57}Fe were heated to 700°C. Dry and pure Cl_2 vapor obtained from the decomposition $PdCl_2$ reacted with Fe to finally produce microcrystals of the transparent anhydrous $FeCl_2$. Because of to the extreme delinquency of the material, the sample had to be treated and loaded into the diamond anvil cell (DAC) under exceptionally dry conditions using a special glove box. The size of the anvils culet was 400-μm in diameter and Re gasket with initial sample cavity dimensions of 100-μm diameter by 40-μm height was used. Liquid argon was used as a pressure transmitting medium and the pressure gradient was ~5% as measured by the ruby fluorescence method.

Mössbauer spectra were recorded up to 61 GPa and in the 5–300 K temperature ranges using a top-loaded, variable-temperature, LHe cryostat. The source used was a commercial 10 mCi ^{57}Co(Rh) *point source* (0.5 mm × 0.5 mm). Spectra recorded at $T < T_N$ and up to 34 GPa were analyzed using a general program that combines magnetic and electric field gradient (*efg*) interactions of comparable magnitude [7]. Due to texture effects caused by lack of perfect hydrostatic conditions in this anisotropic layered compound, partial polarization of the γ-rays with respect to main axis of the *efg* was present and was taken into account during spectra fitting procedures. For spectra recorded in the 34–50 GPa range at $T < T_N$, the following *Hamiltonian*, applicable for magnetic interaction larger than electric interaction, was used:

$$H = \mu \cdot H_{hf}\frac{I_z}{I} + \frac{e^2 q_{zz} Q(3\cos^2\theta - 1)}{8I(2I - 1)}\left(3I_z^2 - I(I - 1)\right), \tag{1}$$

where μ is the nuclear moment, I_z and I_z^2 are spin-operators, H_{hf} is the hyperfine field, $e^2 q_{zz} Q$ is the quadrupole coupling, twice the value of QS, the *quadrupole splitting*, and θ is angle between H_{hf} and q_{zz}.

3. Results

Typical Mössbauer spectra recorded in the 3.5–61 GPa range at 300 K (paramagnetic phase) and 5 K are shown in Figures 1 and 2, respectively. In Figure 3 we summarize the pressure dependence of the various hyperfine parameters and the variation of the Néel temperature with pressure is shown in Figure 4. The latter was determined from the $H_{hf}(T, P)$ data. Based on the distinctive changes in the spectra

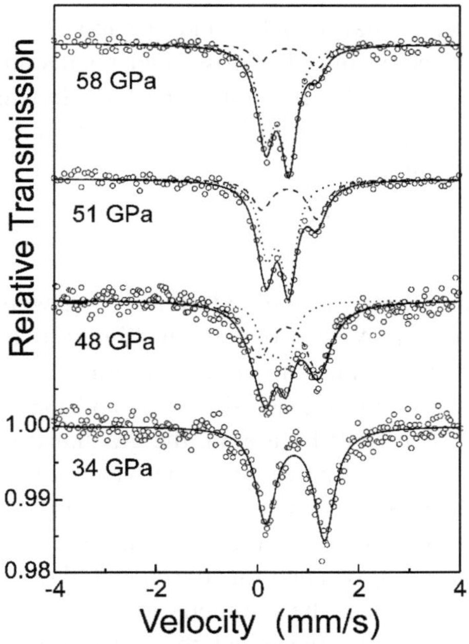

Figure 1. Mössbauer spectra recorded at 300 K and at various pressures. The solid line is the theoretical fit to the experimental points. The dotted lines in the 48–58 GPa spectra represent the new diamagnetic phase, dashed lines represent the remnant of the previous phase.

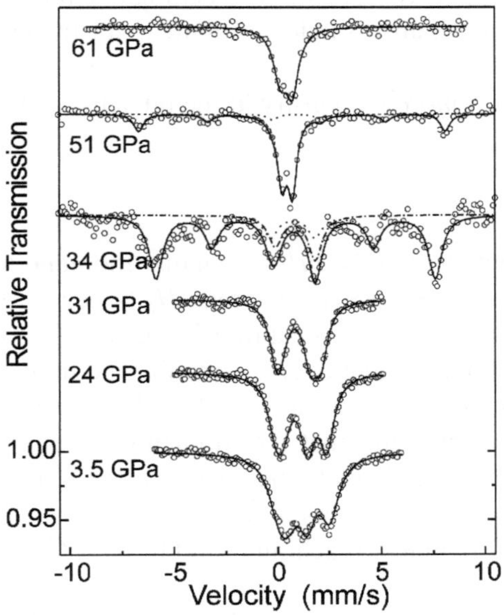

Figure 2. Mössbauer spectra recorded at 5 K ($T \ll T_N$). The solid line is a theoretical spectrum taking into account texture effects (see text). At 34 GPa one observes the onset of a large H_{hf} coexisting with that of the previous phase (dashed-dot line). At 51 GPa a coexistence of the magnetic (dotted line) and a non-magnetic (dashed line) is observed and at 61 GPa the magnetic interaction is quenched.

features caused by either crystallographic and/or band-structure phase transitions occurring with pressure increasing, the Mössbauer results are classified into three pressure ranges: 0–34 GPa, 34–42 GPa and 42–61 GPa. In what follows we discuss and propose possible interpretations of the drastic differences pertinent to these three pressure ranges regimes.

4. Discussion

0–34 GPa In this range spectra recorded at $T > T_N$ were fitted with a single quadrupole-split component. As can be seen in Figure 1, because of texture effects the intensities of the two lines at 34 GPa are unequal. This is due to texture effects. The large values of the isomer shift (IS) and QS are typical of ferrous compounds. As can be seen from Figure 3a, the gradual decrease in IS, which in ^{57}Fe is proportional to the *negative* value of the s-electron density at the nucleus ($\rho_s(0)$) suggests a gradual increase in density, of about 15% in this pressure range. One also observes a gradual increase in *efg* (Figure 3b) explained as a pressure-induced distortion of the $FeCl_6$ octahedron consequential of a relative elongation of Cl–Fe–Cl length along the z-axis, the symmetry axis of q_{zz}. This trend in distortion reaches a maximum at ~15 GPa after which the equatorial $FeCl_6$ distances start to increase. At ~35 GPa QS reaches its ambient pressure value. Measurements carried out at $T \ll T_N$ reveals an unusual small H_{hf} (see Figure 3c). In this pressure range, H_{hf} varies between 0.4, at 0 GPa, to 1.5 T, reaching 5 T maxima at ~15 GPa.

The two important mechanisms that contribute to the hyperfine field of $Fe^{2+}(d^6)$ are the *Fermi-contact* term H_c which is proportional to $(\rho_s(\uparrow) - \rho_s(\downarrow))$, the difference in *spin-up* and *spin-down* density at the nucleus, and the orbital term, H_{orb} which can be expressed as:

$$H_{orb} = -2\mu_B\langle r^{-3}\rangle\langle l_z\rangle, \tag{2}$$

where $\langle l_z\rangle$ is the z-component of the *orbital* angular momentum and, μ_B is the Bohr magneton. Generally H_{orb} is positive whereas H_c is negative [3]. In the h-$FeCl_2$ phase $|H_{orb}|$ is comparable to $|H_c|$, resulting in the small H_{hf}. The spatial dependence of H_{orb} in r^{-3}, similar to that of *efg*, is manifested in the pressure variation of H_{hf} (Figure 3c) which follows the trend of the QS. Also, from the fitting parameters we find that $\theta \sim 0°$, suggesting that the moments are aligned along the c-axis, which suggests the same orbital level scheme as in r-$FeCl_2$ in which the component of the orbital angular momentum lies along the trigonal axis [8]. In this pressure range T_N increases substantially, reaching ~200 K at 32 GPa (see Figure 4).

34–42 GPa A drastic increase in H_{hf} is observed at 34 GPa (see Figure 1). With some remnants of the h-$FeCl_2$ phase left, the spectrum consists of a symmetric sextet characterized by $H_{hf} = 36$ T, ten times that of its maximum value in the 0–34 GPa range. The hyperfine parameters were obtained from least-squares fitting using the *Spin-Hamiltonian* described in (1). A value of $e^2qQ(3\cos^2\theta - 1) = 0$

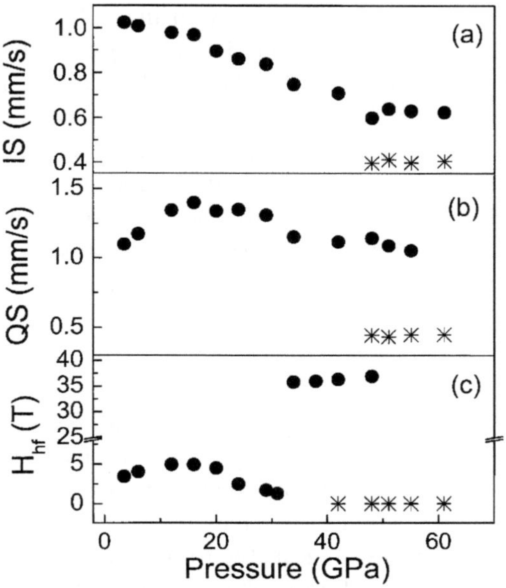

Figure 3. Pressure dependence of the hyperfine field parameters. In the 0–48 GPa range the IS(P) decreases gradually indicating a monotonous increase in the *s*-density. In the 47–61 GPa a diamagnetic phase (*) evolves with considerably smaller IS. Values of IS are with respect to Co(Rh) at 300 K.

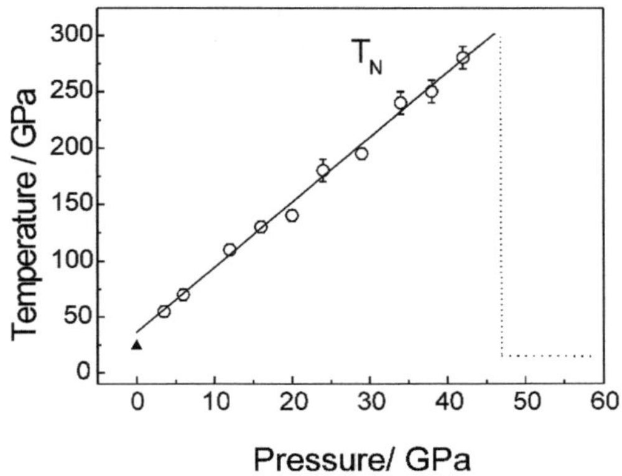

Figure 4. The progressive increase in T_N as function of pressure. The value at 0 GPa was taken from [4]. At $P > 45$ GPa the magnetic interaction collapses. The straight line through the experimental points is a guide to the eyes.

is deduced causing the symmetric magnetic spitting pattern. Since QS as measured at $T > T_N$ is definitely not zero (see Figures 1 and 3) one must conclude that the angle θ between the principal axis of the *efg* and that of the magnetic moment is $\approx 55°$. The same θ has been deduced in NiI_2 [9] and CoI_2 [10], isostructural layered compounds. At 40 GPa, and at $T < T_N$, no remnants of the low-pressure phase is observed; only a magnetic-split spectrum is present.

Although drastic changes occur in the hyperfine field, no observable changes occur in the IS and QS (see Figures 1 and 3). This suggests that no abrupt volume change and no distortion of the Fe–Cl octahedra take place. This jump in H_{hf} is explained as due to orbital-term quenching caused by the significant increase in the crystal field splitting within the $^5T_{2g}$ low-lying levels manifold. The results show that once $10Dq$ reaches a critical value exceeding the *exchange energy* a *discontinuous* quenching occurs. This will also result in a change in the direction of the *efg* principal axis with respect to H_{hf}. In this pressure range T_N increases further, reaching a value close to 275 K at 42 GPa (see Figure 4). QS at $T > T_N$, is independent of pressure in contrast to its behavior at the previous pressure-range regime.

42–61 GPa At $P > 45$ GPa drastic changes occur. Room-temperature Mössbauer spectra in this pressure range clearly show the onset of a new doublet (see Figure 1) characterized by a considerably smaller IS (0.4 mm/s) and |QS| (0.4 mm/s). The sharp drop in IS implies a first-order phase transition accompanied by a volume decrease and the decrease in QS suggests the formation of a more symmetric Fe–Cl octahedron. No magnetic interaction is observed down to 5 K (see Figure 2); the new high-pressure phase is *diamagnetic* and its abundance increase reaches 100% at 60 GPa. This diamagnetic phase can be accounted for either due to a high to low-spin (HS $\rightarrow$ LS) transition of Fe^{2+} or to a *Mott* transition in which the magnetic moment collapses but it should concur with metallization. The first mechanism is a result of crystal field splitting increase with pressure whereas the second mechanism is created by *d–d* or *p–d* bands closure. Both phenomena have been observed in high-pressure studies of similar binary ferrous compounds. The *Mott* transition is responsible for the collapse of magnetism in FeI_2 [6], whereas in the case of FeO [11] a *spin-crossover* is responsible. In both mechanisms one would expect a substantial diminishing in Fe^{2+} radius, which explains the sudden decrease in the IS, e.g., increase in s-density.

5. Summary and conclusions

In the present work high-pressure hyperfine interaction studies of $FeCl_2$ phases induced by pressure have been carried out using ^{57}Fe MS. In the h-$FeCl_2$ low pressure phase, within the 3.5–31 GPa range, very small H_{hf} values, 1–5 T, have been measured. The phenomenon is explained on the basis of mutual cancellation of the *Fermi-contact* and *orbital* terms. At ~34 GPa, a sudden jump of H_{hf} to 36 T is observed concurrent with the disappearance of the quadrupole shift and no change in IS. This is explained as due to crystal field increase. As a result the orbital term is quenched and the magnetic moment, previously parallel to the q_{zz} symmetry axis, is now tilted by ~55°.

At $P > 42$ GPa, the onset of a nonmagnetic phase takes place, coexisting with the previous phase up to 61 GPa when magnetism in $FeCl_2$ collapses completely.

Acknowledgement

This work was supported by ISF grant 90/04.

References

1. Starr, C., Bitter, F. and Kaufman, A. R., *Phys. Rev.* **58** (1940), 977.
2. Wilkinson, M. K., Cable, J. W., Wollan, E. O. and Koehler, W. C., *Phys. Rev.* **113** (1959), 497.
3. Ono, K., Ito, A. and Fujita, T., *J. Phys. Soc. Japan* **19** (1964), 2119.
4. Vettier, C. and Yelon, W. B., *Phys. Rev. B* **11** (1974), 4700.
5. Pasternak, M. P. and Taylor, R. D., In: G. J. Long and F. Grandjean (eds), *Mössbauer Spectroscopy Applied to Magnetism and Materials Science*, Plenum Press, New York, 1996, p. 167.
6. Pasternak, M. P., Xu, W. M., Rozenberg, G. Kh., Taylor, R. D., Hearne, G. R. and Sterer E., *Phys. Rev. B* **65** (2002), 035106.
7. Kundig, W., *Nucl. Instr. Meth.* **48** (1966), 219.
8. Simkin, D. J., *Phys. Rev.* **177** (1968), 1008.
9. Pasternak, M. P., Taylor, R. D., Chen, A., Meade, C., Falicov, L. M., Giesekus, A., Jeanloz, R. and Yu, P. Y., *Phys. Rev. Lett.* **65** (1990), 790.
10. Pasternak, M. P. and Taylor, R. D., *J. Appl. Phys.* **70** (1992), 5956.
11. Pasternak, M. P., Taylor, R. D., Jeanloz, R., Li, X., Nguyen, J. H. and McCammon, C. A., *Phys. Rev. Lett.* **79** (1997), 5046.

Hyperfine Interactions **144/145:** 183–198, 2002.

Electronic and Molecular Structure Studies of a (Bio)Inorganic Complex – A Multi-Technique Approach

ECKHARD BILL, UDO BECKMANN and KARL WIEGHARDT
Max-Planck-Institut für Strahlenchemie, Stiftstrasse 34-36, D-45470 Mülheim, Germany

Abstract. The electronic structure of the dimeric iron complex $[Fe_2(bbpo)Cl_4]^{n-}$ with the tetranionic ligand 1,2-bis(3,5-di-*tert*-butyl-2-hydroxyphenyl)oxamide was studied in two different oxidation states. Combined Mössbauer and EPR investigations in conjunction with magnetic susceptibility measurements are used to demonstrate the complementarity of the techniques. The dianion $[Fe_2(bbpo)Cl_4]^{2-}$(**2**) is a high-spin iron(III) compound with magnetic properties resulting from competing exchange and zero-field interactions of the iron sites. Unusual integer-spin EPR spectra could be recorded from an excited spin septet, $S_t = 3$, of the coupled system. Electrochemical oxidation yields the monoanion $[Fe_2(bbpo^{ox1})Cl_4]^-$ ($\mathbf{2^{ox}}$) which also is an iron(III) compound; the ligand is redox *non-innocent* and forms a π radical that is bound to one of the iron centers and the total spin $S_t = 1/2$ ground states of the molecule originates from exchange coupling of the three paramagnetic centers. A simplified spin coupling scheme assuming infinitively strong radical-iron(III) interaction yields a reasonable and rational interpretation of the magnetic and electronic properties of $\mathbf{2^{ox}}$.

Key words: bioinorganic chemistry, molecular magnetism, non-innocent ligand, exchange interaction, integer spin EPR, applied field Mössbauer.

1. Introduction

Measurements of the electronic structure are fundamental for the understanding of the physical and chemical properties of transition metal complexes, and hyperfine interactions are effective probes of the distribution of valence electrons. In this program ^{57}Fe Mössbauer spectroscopy is a selective and valuable technique which is independent of spin- and valence states for iron-containing complexes, in contrast to general magnetic resonance techniques. While Mössbauer spectroscopy can well stand on its own also in (bio)inorganic applications, its power is substantially increased when it is combined with EPR and magnetic susceptibility measurements. The approach will be demonstrated here for an inorganic molecule with unusual oxidation states and interesting redox behavior including oxidations of non-innocent ligands. The experimental techniques applied are controlled-potential electrochemistry, UVvis-, EPR- and ^{57}Fe-Mössbauer spectroscopy and magnetic susceptibility measurements at variable temperatures and magnetic fields. From the interpretation of the complementary data a consistent description of the electronic structures of the different redox derivatives will be derived.

2. Materials

The tetranionic ligand 1,2-bis(3,5-di-*tert*-butyl-2-hydroxyphenyl)oxamide, H_4(bbpo), coordinates trivalent transition metal ions and forms dimeric complexes. The di-iron(III) compound was obtained as the green crystalline salt $[N(n\text{-}Bu)_4]_2$ $[Fe_2(bbpo)Cl_4]$ and its molecular structure was determined from single-crystal X-ray diffraction measurements. A schematic view of its dianion, $[Fe_2(bbpo)Cl_4]^{2-}$ (compound **2**), is shown in Scheme 1. The iron sites are bridged by an oxamide group and the iron distance is 5.54 Å. Two o-aminophenolate units in the molecule provide potentially redox-active ligand groups that bind to the central ions. The cluster can be taken as spectroscopic model for enzymes with redox active organic ligands in the vicinity a transition metal center. The synthesis, details of molecular structure, electrochemical and spectroscopic properties were described recently [1] as well as the experimental methods [2, 3]. Here we describe the systemat-

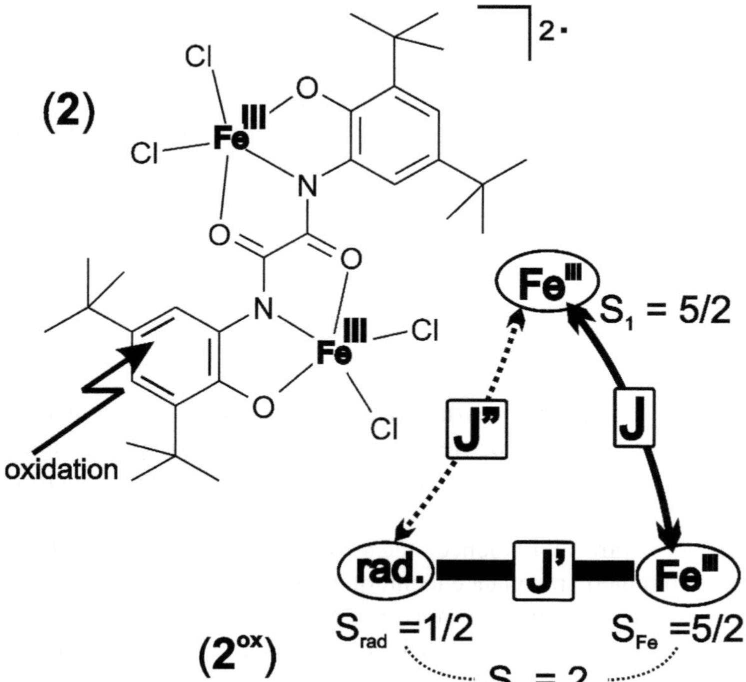

Scheme 1. Schematic view of the molecular structure of $[Fe_2^{III}(bbpo)Cl_4]^{2-}$ (**2**) and spin coupling scheme of the oxidized molecule ($\mathbf{2^{ox}}$) with two ferric ions and a ligand radical. The zick-zack arrow indicates the part of the molecule (aminophenolate group) that is attacked by electrochemical oxidation at the formation of $\mathbf{2^{ox}}$.

ics of the spectroscopic investigations by using applied-field Mössbauer and EPR spectroscopy and magnetic susceptibility measurements.

Solutions of the di-anionic complex **2** in CH_2Cl_2 can be reversibly oxidized by electrochemical methods at −0.5 V (vs. ferrocenium/ferrocene couple) and coulometric conversions yield quantitative amounts of the one-electron oxidized mono-anion $\mathbf{2}^{ox}$ [1]. Since the solvent dichloromethane is opaque for the 14 keV ^{57}Fe Mössbauer radiation and the product is much less stabile in other solvents, the solvent had to be removed for preparations of Mössbauer samples. To this end a 1-ml aliquot of oxidized complex $\mathbf{2}^{ox}$ generated in CH_2Cl_2 solution (0.10 M [N(*n*-Bu)$_4$]PF_6) at 248 K was transferred to a Mössbauer cup and the solvent CH_2Cl_2 was evaporated under reduced pressure at 243 K within 2 h. Subsequently, the sample was rapidly cooled to 80 K. In a parallel experiment the remaining solution was monitored spectrophotometrically in order to check the redox stability of the monoanion. No decomposition was detected within a 2 h time span.

3. Results

3.1. IRON VALENCE AND MOLECULAR OXIDATION STATES

Zero-field Mössbauer spectra of solid **2** and $\mathbf{2}^{ox}$ were recorded at 4.2 K. Under these conditions both compounds show slightly asymmetric quadrupole doublets with virtually identical isomer shifts of 0.45 ± 0.03 mm s^{-1} and very similar quadrupole splittings of 0.98 ± 0.02 mm s^{-1} and 0.92 ± 0.08 mm s^{-1} (Figure 1). The values are typical for high-spin ferric ions. The quadrupole splittings are large in view of the S-state of the ferric ions which have no valence contribution to the electric field gradient (efg) tensor (at least in a ligand-field description). The observed large efg discloses considerable *asymmetry* in covalency of the iron $3d^5$ configuration which must be induced by differences in chloride and phenolate ligand bonds. Similar values are found for other asymmetrically coordinated high-spin ferric ions. Iron(III) oxo groups are typical examples [4]. The presence of almost identical electric Mössbauer parameters clearly shows that the one-electron oxidation of **2** is a *ligand centered* process and a ligand *radical* is formed in $\mathbf{2}^{ox}$ instead of the formation of an high-valent Fe(IV) ion. Hence, H_4(bbpo) is a redox-active, *non-innocent* ligand.

The line width of the zero-field spectrum of $\mathbf{2}^{ox}$ is apparently much larger than that of the non-oxidized starting complex **2** (and therefore the Mössbauer parameters are less accurate). The broadening, however, is not related to structural inhomogeneity of the sample, but is of magnetic origin. The feature is typical for compounds with odd number of unpaired electrons and half-integer spin. In contrast to integer-spin complexes like **2**, Kramers systems like $\mathbf{2}^{ox}$ ($S = 1/2$, see below) show sizable spin expectation already at the very weak earth-field. Because of the quenched orbital momentum of the S-state ion, the spin relaxation rates are often in intermediate range or close to limit of slow rates. Hence, residual magnetic moments induce hfs and line broadening. Complex **2**, on the other hand, has narrow Mössbauer lines in zero field because exchange interaction of the iron

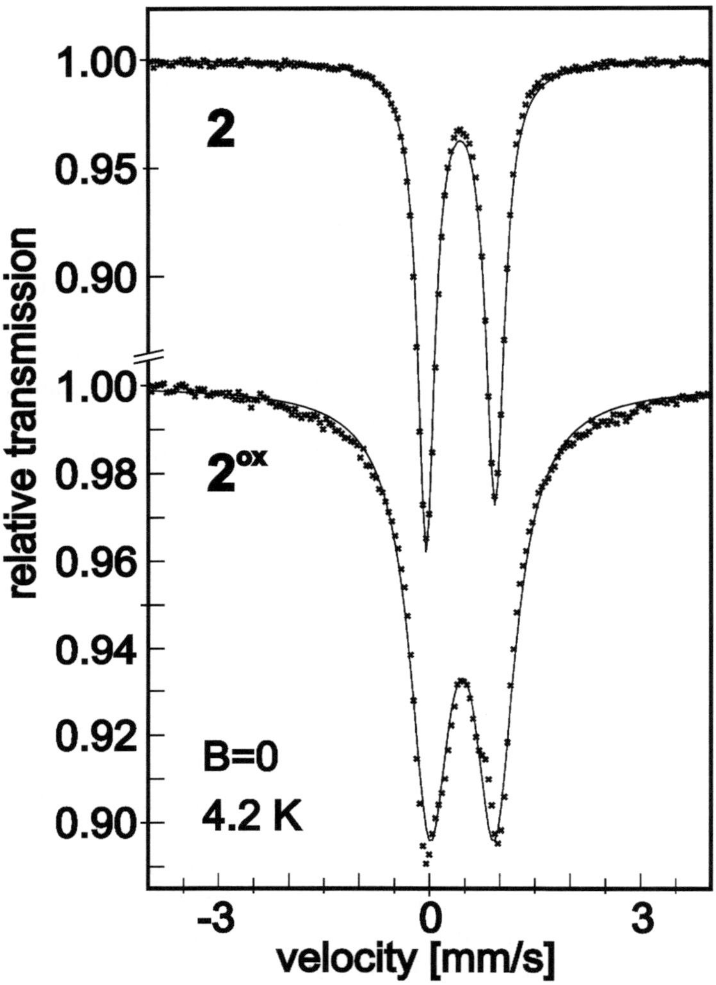

Figure 1. Zero-field Mössbauer spectra of $[Fe_2^{III}(bbpo)Cl_4]^{2-}$ (**2**) and $[Fe_2^{III}(bbpo^{ox1})Cl_4]^{-}$ ($\mathbf{2^{ox}}$).

sites apparently leads to integer total spin states, which do not show sizable spin expectation values without moderate polarizing fields.

3.2. MAGNETIC PROPERTIES OF THE NON-OXIDIZED DIANON $[Fe_2^{III}(bbpo)Cl_4]^{2-}$(**2**)

Magnetic properties of dimeric iron complexes are usually dominated by exchange interaction of the iron sites. Experimental data obtained from static magnetic susceptibility measurements, EPR, and applied-field Mössbauer spectra can be described within the spin the Hamiltonian formalism that parameterizes electronic exchange interaction, single-ion zero-field splitting (zfs) and Zeeman interactions by using:

$$H = -2J\mathbf{S}_1 \cdot \mathbf{S}_2 + \sum_{i=1,2} \{D_i[S_{z,i}^2 - \tfrac{1}{3}S_i(S_i+1) + E/D(S_{x,i}^2 - S_{y,i}^2)] + \mu_B \mathbf{S}_i \cdot g_i \cdot \mathbf{B}\}. \quad (1)$$

For the diiron compound **2**, the spins are $S_1 = 5/2$, $S_2 = 5/2$. J is the (isotropically taken) exchange coupling constant, and D and E/D are the axial and rhombic zfs parameters. After diagonalization of the Hamiltonian , EPR resonances can be derived from the eigenvalues E_i of the spin system, magnetic moments correspond to the derivatives of the eigenvalues, $\partial E_i/\partial B$, and EPR transition probabilities and the spin expectation values are obtained from the eigenfunctions. Electric and magnetic hyperfine splittings of the ^{57}Fe nuclear levels are described by the usual nuclear Hamiltonian [5].

Figure 2 displays the magnetization M_{mol} of solid **2** measured in the temperature range 2–300 K using a SQUID magnetometer at fields of 1, 4 and 7 T. The data were sampled on a $1/T$ scale and the molar susceptibilities were corrected for underlying diamagnetism by use of tabulated Pascal constants ($\chi_{dia} = -825 \cdot 10^{-6}$ cm^3 mol^{-1}). Apparently, the spins of the two ferric ions in **2** are weakly antiferromagnetically (AF) coupled; the complex possesses a total spin

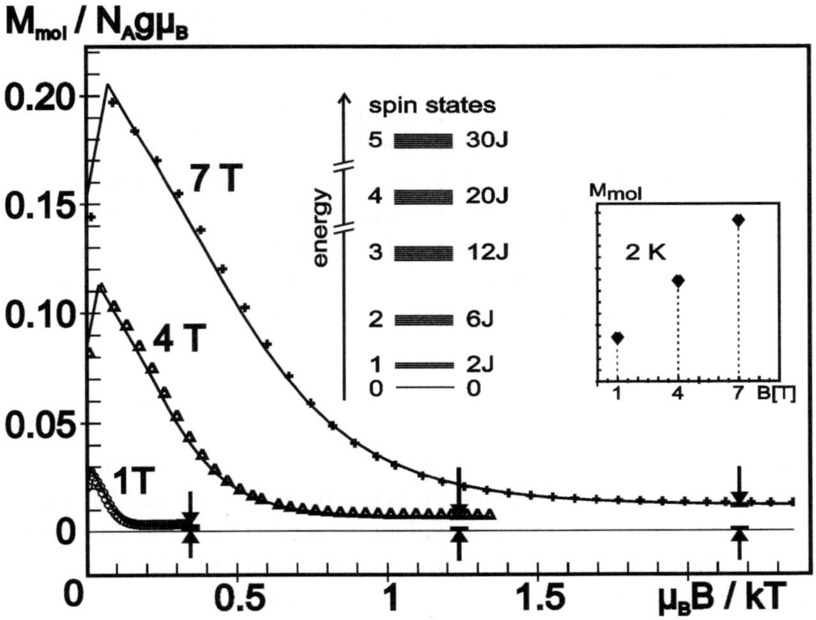

Figure 2. Temperature dependence of the molar magnetization of $[Fe_2^{III}(bbpo)Cl_4]^{2-}$ (**2**) at fields of 1, 4 and 7 T. The lines are spin Hamiltonian simulations using Equation (1) with $S_1 = 5/2$, $S_2 = 5/2$. The insets show the sequence of coupled spin manifolds and an amplified view of the magnetization at base temperature (2 K).

$S_t = 0$ ground state, as shown by the almost vanishing molar magnetization at low temperatures. Increasing temperature (to the left) leads to Boltzmann population of the excited $S_t = 1, 2$ states and increasing magnetization. The slope of that curves is a measure of the exchange splitting of the multiplets. The decline of M_{mol} below $\mu_B B/kT = 0.1$ represents the usual $1/T$ behavior at high temperatures due to equal population of the magnetic M_s-sublevels. The corresponding sequence of the underlying spin manifolds and their exchange splitting is shown as an inset in Figure 2 (zero-field splittings are omitted). In detail, however, the total spin states are energetically not so well isolated as anticipated in this graph. This is best observed for the $S_t = 0$ ground state in the multi-field variable-temperature measurement, because even at base temperature some residual magnetization is observed (vertical arrows). An amplified view of the 2-K magnetic data is plotted in the right inset of Figure 2 as function of the applied field. Apparently, the fields are able to induce a weak magnetic moment for the singlet, which is not expected for an isolated spin singlet state. Such field-induced multiplet-mixing can be enabled only by competing exchange and single-ion zero-field splitting (ZFS). There are a few cases for which alternatively antisymmetric exchange [6] has been identified as the origin of level mixing (the interaction responsible for canted antiferromagnetism in solids), but this does not occur for centrosymmetric molecules [7–9]. Simulations of the multiple-field magnetic data yield a sensitive and unique measure of both, the exchange coupling constant $J = -9.8 \pm 0.5$ cm^{-1} and of the ZFS parameter $|D| = 1.9 \pm 0.3$ cm^{-1}, and reproduce almost perfectly the experimental result. The rhombicity parameter can hardly be determined from the 'macroscopic' static susceptibility measurement of powder samples; the value $E/D = 0.18$ was taken from an X-band EPR investigation of **2** described below.

We mention that the low value found for the exchange coupling constant is not unusal for an di-iron(III) complex without an oxo bridging ligand. A similar but even weaker AF coupling has been reported for the μ-oxalato bridged dinuclear species $[Fe_2^{III}(\mu\text{-ox})(ox)_4]^{4-}$ where a coupling constant J of -3.3 cm^{-1} has been established [10]. On the other hand, an hydroxo- or a monodentate alkoxo bridge mediates exchange couplings of the order of -30 cm^{-1} [11, 12].

3.3. MAGNETIC MÖSSBAUER SPECTRA OF $[Fe_2^{III}(bbpo)Cl_4]^{2-}$ (**2**)

The presence of spin mixing and induced magnetization of the $S_t = 0$ ground state in **2** could also be demonstrated by magnetically perturbed Mössbauer spectra recorded at 1.8 K with 7 T field applied perpendicular to γ. At a first glance the spectrum shown in Figure 3 (bottom) appears to be a diamagnetic pattern. A simulation for $S = 0$, however, reveals a small but significant deviation of the magnetic splitting from the diamagnetic limit (dotted lines). A weak internal field is present that is oriented opposite to the applied field. The magnetic spectrum can be reasonably well simulated with the appropriate spin Hamiltonian (Equation (1)) and with the D, E/D and J parameters from above (which also hold for 1 and 3 T

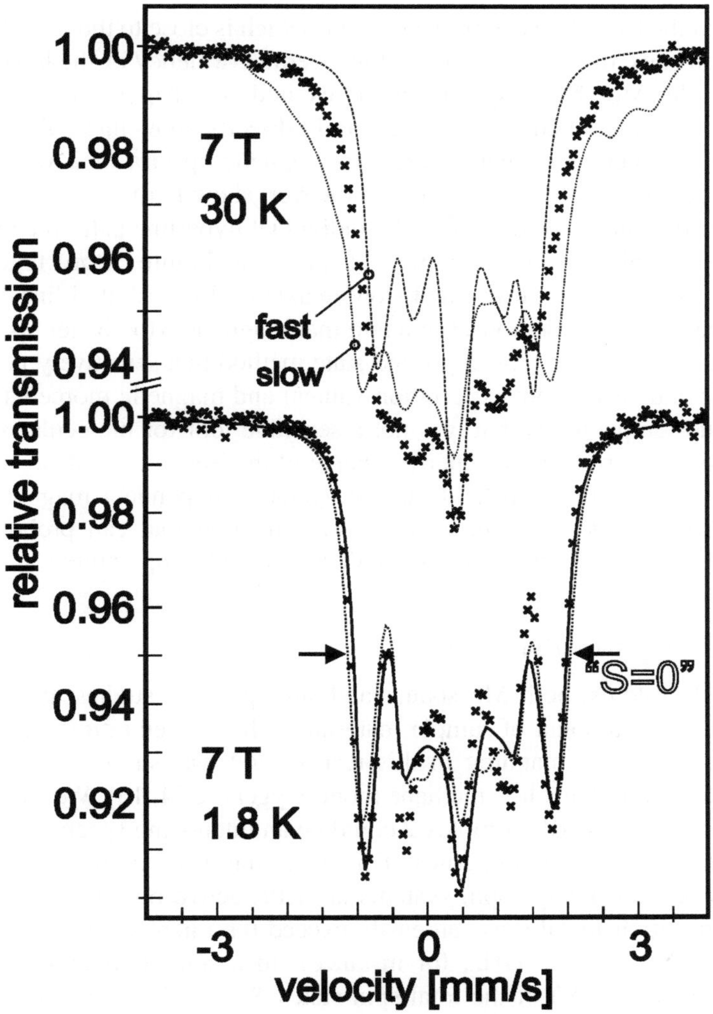

Figure 3. Applied-field Mössbauer spectra of $[Fe_2^{III}(bbpo)Cl_4]^{2-}$ (**2**) at 7 T and 1.8 K and 30 K. The solid line and dashed lines are spin Hamiltonian simulations using Equation (1) with $S_1 = 5/2$, $S_2 = 5/2$, or $S = 0$ (bottom, dotted line) (see text).

applied field spectra, not shown). The symmetry of the hyperfine pattern reveals large asymmetry of the electric field gradient (efs) tensor, $\eta = 0.9 \pm 0.1$, in accord with the low molecular symmetry of the sites. We note that the hyperfine coupling constant, $A/g_N\mu_N = -21.85$ T, cannot be accurately determined from small internal field, its value was rather taken from the simulation of the oxidized analog given below which has a ground state $S_t = 1/2$ with large magnetic moment and it can be assumed that the oxidation affect mostly one of the iron sites. The relatively high value of the hyperfine coupling constant, which is close to the −22 T expected for ionic iron sites, discloses relatively low overall covalency of the bonds.

In principle, Mössbauer spectra of the excited spin triplet and quintet manifolds measured at elevated temperatures could directly show internal fields due to a first-order magnetic moment. Therefore, magnetic spectra of **2** were recorded also at 30, 60 and 120 K. Their shapes, however, show hardly any resolution because enhanced spin relaxation blurs the individual hyperfine pattern coming from the electronic sublevels. An example is depicted in Figure 3(top) together with preliminary 'simulations' in fast and slow relaxation limit (dotted lines) for comparison. The deviations demonstrate again that magnetic Mössbauer spectroscopy of molecular systems is a typical ground state method that sensitively probes local electronic spin density (spin expectation values) and magnetic moments mainly at liquid helium temperatures, but it is not a sensitive tool for molecular spin states and magnetic properties if AF super-exchange of the iron sites leads to spin singlet ground states, $S_t = 0$. Multi-field- and temperature dependent magnetic susceptibility measurements represent an alternative method that can probe exchange splittings of the spin system of the order kT at ambient temperature.

3.4. EPR OF $[Fe_2^{III}(bbpo)Cl_4]^{2-}$(2)

In biological systems, both Mössbauer and susceptibility studies are often hampered by the large amount of sample material or the concentration required. EPR spectroscopy is less demanding in this respect and was successfully applied in studies of di-iron proteins like methane monooxygenase [4, 13]. We supposed that the homovalent di-iron(III) complex **2** could be an interesting spectroscopic model to study integer-spin EPR properties. This is an unusal program at usual X-band frequencies because integer-spin systems are EPR active only in few cases since the zfs of transition metal ions can easily exceed the microwave quantum energy (0.3 cm^{-1} at X-band, 9.5 GHz, for instance). In a spin-coupled system like **2**, the zfs parameters $D(S_t)$ of the total spin states $S_t = 1$–5 are determined by the single-ion zfs parameter D_i of the iron sites. In the limit of dominating isotropic exchange interaction ($|J_0| \gg D_i$) the $D(S_t)$ value can be derived from the local one by using spin projection technique for $S_i = 5/2$ which is an application of the Wigner–Eckart theorem [9, 14].

$$D(S_t) = 2 \cdot d_t \cdot D_i \tag{2}$$

with $d_t = -16/5, -10/21, -1/45, 1/7, 2/9$ for $S_t = 1, 2, 3, 4$, respectively. Here, D_i is the diagonal 3×3 matrix of the Fe(III) sites with components $D_{xx} = E_i - 1/3D_i$, $D_{yy} = -E_i - 1/3D_i$, $D_{zz} = 2/3D_i$, $D(S_t)$ is the corresponding zfs matrix of the coupled states. To avoid the introduction of additional parameters the anisotropic spin coupling term is ignored in these equations, which might be reasonable because the dipolar interactions of the widely separated Fe(III) ions in **2** should be at least an order of magnitude smaller than the microwave quantum [9]; and antisymmetric exchange [7–9] does not occur for centrosymmetric molecules.

The relations (2) show that the septet state, $S_t = 3$, is distinguished among the spin manifolds, because it has almost vanishing contribution form single-ion zfs. Hence, EPR transitions can be expected to occur between the magnetic sublevels, whereas for $S_t = 1, 2, 4$ large ZFS ($D_i \approx 2$ cm^{-1}) should prevent EPR transitions or spread the resonances in the powder spectra in extremely wide field ranges. A realistic picture of the septet splittings and the X-band EPR transitions is shown in the right inset of Figure 4 for field orientations at the z- and the y-axes. The diagram is calculated by using the spin Hamiltonian (Equation (1)) with $D_i = 1.9$ cm^{-1}, $E/D_i = 0.18$, which are the values obtained from the SQUID magnetic measurement. With $J = -12$ cm^{-1} the spin septet is 114 cm^{-1} ($E/k \approx 208$ K) above the ground state and EPR signals should be measurable above ≈30–50 K (which is the case). The corresponding Boltzmann population of the $S_t = 3$ manifold is shown in the left inset in Figure 4.

The experimental X-band EPR spectrum measured from solutions of **2** in dichloromethane at 90 K is shown in Figure 4 (bold line). The absorption derivative trace is a typical integer-spin pattern with derivative lines spread in the field range 0–700 mT. The wide distribution resembles the distribution of transitions indicated in the level diagrams (Figure 4, right insets). The maximum of the absorption in the powder (frozen solution) spectrum appears to be close to $g = 2$ as indicated by the quasi 'zero-crossing' at $B \approx 340$ mT, which is typical for zfs of moderate strength ($D < h\nu$). For comparison a simulated absorption pattern underlying the spectrum is shown as dashed trace above the derivative data. A preliminary simulation of the EPR spectrum of **2** is shown as dotted line in Figure 4. We use an effective Hamiltonian for $S = 3$ (Equation (1) with $S_2 = 0$ and the term $-2JS_1S_2$ being omitted) to describe exclusively septet transitions. Possible contributions from other multiplets as well as the level mixing described above are completely neglected. In view of this crude approximation the simulation result fits reasonably well which supports the interpretation of zfs given above. The obtained zfs parameters are $D(S = 3) = -0.052 \pm 0.008$ cm^{-1}, $E/D(S = 3) = 0.18 \pm 0.05$, which yields $D_i = 1.2 \pm 0.4$ cm^{-1}, $E/D_i = 0.18 \pm 0.04$ when it is converted to local values for the iron sites by using the above relation (2). The numbers are in the range of the magnetization result, the deviation indicate either influence from the neglected dipolar contribution or from rotations of the local g-tensors which also were ignored. We note that the determination of the rhombicity parameter is fairly sensitive, although the fit of the low-field lines and outer wings at high field certainly could be improved by introducing higher order terms (S^4) in the spin Hamiltonian, as was done for other integer-spin systems [2]. For simplicity we did not introduce this ad-hoc term here. A future study of the multiple frequencies will explore the details of the EPR properties of **2**.

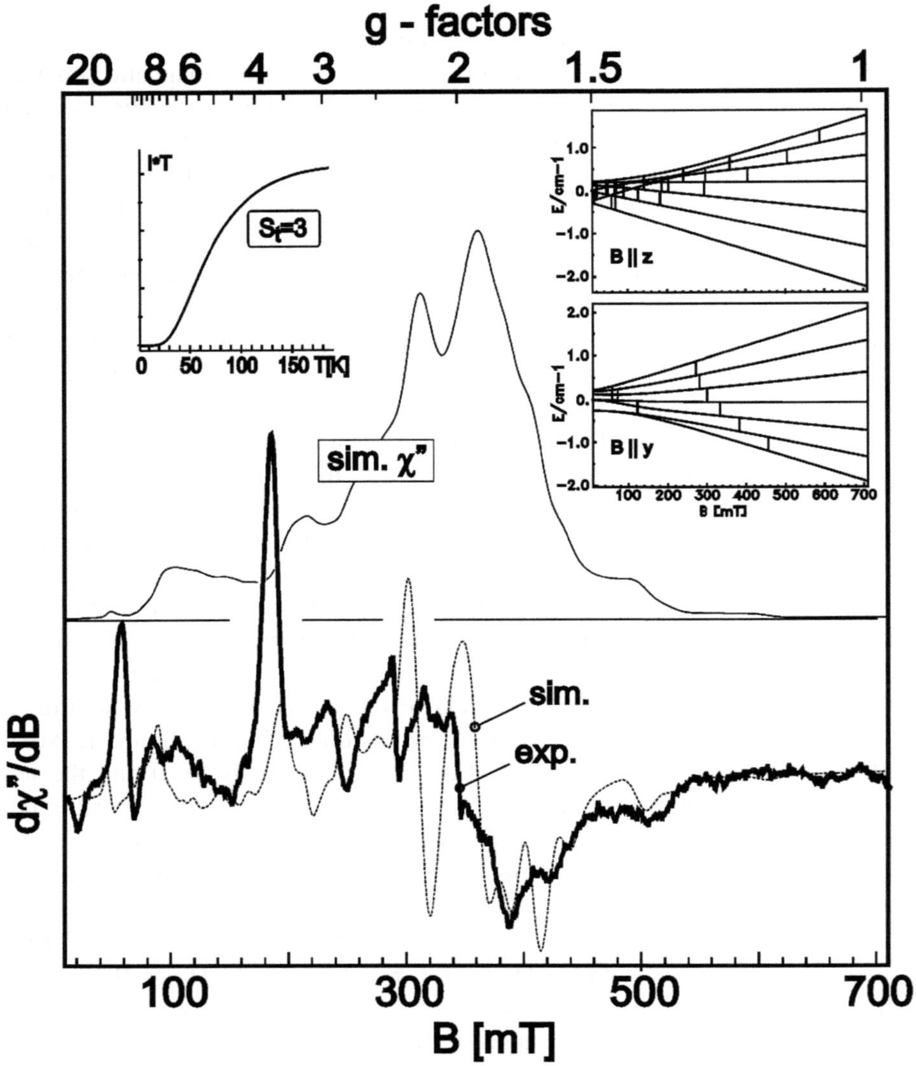

Figure 4. X-band EPR spectrum of $[Fe_2^{III}(bbpo)Cl_4]^{2-}$ (**2**) in frozen CH_2Cl_2(0.10 M $[N(n\text{-}Bu)_4]PF_6$) solution (bold line, bottom); and preliminary spin Hamiltonian simulations for the $S_t = 3$ spin manifold in absorption derivative mode (bottom, dashed line) and absorption mode (top). Experimental conditions: microwave frequency 9.6353 GHz, power 20 mW, modulation 1 mT/100 kHz, temperature 90 K. The right inset shows the Zeeman and zf splittings of the $S_t = 3$ manifold for fields applied in z- and y-directions and the corresponding EPR transitions at 9.635 GHz. The left inset shows the Boltzmann population of the $S_t = 3$ manifold.

3.5. EPR OF THE OXIDIZED MONOANION $[Fe_2^{III}(bbpo^{ox1})Cl_4]^-$ ($\mathbf{2^{ox}}$)

The oxidized monoanion $\mathbf{2^{ox}}$ could not be investigated by magnetic susceptibility measurements because the complex could not be properly separated from the excess of $[N(n\text{-}Bu)_4]PF_6$ salt used for the electrochemical generation and without exact knowledge of sample mass and concentration the magnetic measurements cannot be accurately evaluated. The magnetic properties of $\mathbf{2^{ox}}$ were derived from complementary EPR and Mössbauer data. The X-band EPR spectrum of $\mathbf{2^{ox}}$ was recorded at 10 K from the solution used in the electrochemical cell. It shows a rhombic powder pattern with g values $g_x = 1.998$, $g_y = 1.944$, and $g_z = 1.825$, which clearly establishes that $[Fe_2^{III}(bbpo^{ox1})Cl_4]^-$ possess an $S_t = 1/2$ ground state.

The g values of oxidized $\mathbf{2^{ox}}$ deviate significantly from $g = 2.0$, in view of the ^6S state of high-spin Fe(III) and a ligand radical which also do not contribute significant spin–orbit coupling. This does *not* indicate formation of Fe(IV); the large g-shifts rather owe their origin to level mixing of the ground state doublet with excited quartet due to competing exchange and ZFS, similarly to what was found for **2**. We have used the spin coupling scheme depicted in Scheme 1 for a quantitative interpretation of the g values. In this picture the $S_t = 1/2$ ground state is attained by a *strong* coupling between one high spin ferric ion and the ligand π radical yielding a fictitious 'local' spin $S_2 = 2$ and, in addition, a *weak* AF coupling of this state with the second high spin ferric ion. Based on results from similar systems [15] it is safe to assume that the coupling constant, J', between the radical and the adjacent ferric ion is an order of magnitude stronger than the long-range iron–iron coupling, J, which should be similar to what is found for the non-oxidized starting complex **2**. The coupling J'' between the second ferric ion and the ligand radical is assumed to be negligible small or zero. Situations similar to this simplified model are encountered in the bioinorganic literature of high-spin Fe(III)/Fe(II) dimers ($S_1 = 5/2$, $S_2 = 2$) which are found in the OH-bridged diiron cores of purple-acid phosphatases, methane monooxygenase, and related model compounds [4, 13, 16], or also the dimeric iron-sulfur proteins [17]. These dimers are more or less strongly coupled and spin–orbit interaction at the ferrous ions gives rise to substantial zfs. The cluster ground state properties are modulated by level mixing and can yield g values that span the range $g = 1.5$–2. The situation for $\mathbf{2^{ox}}$ is slightly different in that the 'local' zfs of both 'local' spins should be fairly similar and g values hardly deviate from $g = 2$. Nevertheless, we achieved a quantitative interpretation for the properties of $\mathbf{2^{ox}}$ from a corresponding spin-Hamiltonian which is based on the simplified coupling scheme outlined above by using Equation (1) with $S_1 = 5/2$ for an Fe(III) ion and the fictitious spin $S_2 = 2$ for the other (oxidized site) of the molecule. The approach allows us to probe the effect of ligand oxidation on the electronic properties of the respective Fe(III) site even if their spin coupling itself is unmeasurably strong.

A second order perturbation treatment of the spin coupling Hamiltonian (Equation (1)) yields the following g matrix for the $S_t = 1/2$ ground state of the $S_1 = 5/2/S_2 = 2$ pair [14, 16, 17]

$$g_t = \frac{7}{3}g_1 - \frac{4}{3}g_2 + \frac{-28/45(g_1 - g_2)(8D_1 + 3D_2)}{J} \tag{3}$$

(dipolar coupling and antisymmetric exchange again ignored). The relation visualizes the different influences of g_t. The first terms are the usual strong coupling limits; and the correction is typically small because of the factor $(g_1 - g_2)$. Particularly in this case of ferric ions on both sites, g_1 and g_2 should not differ much from $g = 2$.

It is *a priori* not clear which of the parameters of S_2 is mostly affected by the ligand oxidation. There is a 'physical' influence coming just from the strong 'local' spin coupling and a possible 'chemical' influence on the electronic structure of the iron(III). We used the full coupling Hamiltonian to derive theoretical g values and probe the situation. The value for D_1 was taken from the 'unperturbed' zfs parameters of the homovalent dimer described above and that for D_2 was converted from the same starting value by using spin projection in the strong coupling limit for the *iron(III)-radical* pair, which yields $D_2 = 4/3\ D_{Fe}$. The (isotropic) g value for S_1 was fixed to $g_1 = 2$ and, arbitrarily chosen, g_2 was treated as a fit parameter, as well as the coupling constant J. Possible Euler rotations of the D and g matrices were not considered. With this simple approach we obtained a perfect fit of the experimental g values with optimized variables $g_2 = (2.028, 1.99, 1.988)$ and $J = -12\ \text{cm}^{-1}$. The values are convincingly close to what would be obtained for chemically *unperturbed* iron sites: when $J = -9.8\ \text{cm}^{-1}$ of non-oxidized **2** is converted for the oxidized system with $S_2 = 2$ (strong 'local' coupling), one obtains $J = (7/6)J\ (S_1 = 5/2, S_2 = 5/2) = -11.5\ \text{cm}^{-1}$, and shifts of 0.02 of from $g = 2$ are usual for the electronic g value of iron(III). Hence, the unusual spectroscopic EPR g values of dimeric $\mathbf{2^{ox}}$ are fully consistent with the concept of ligand oxidation and the simplified coupling scheme with virtually infinitively strong Fe(III)-radical coupling for one site of $\mathbf{2^{ox}}$ with only little mutual influence on the electronic structures.

3.6. MAGNETIC MÖSSBAUER SPECTRA OF OXIDIZED $[Fe_2^{III}(bbpo^{ox1})Cl_4]^-$ ($\mathbf{2^{ox}}$)

Magnetic Mössbauer measurements of the monoanion $\mathbf{2^{ox}}$ display very broad spectra with absorptions spread over a large velocity range of $\sim$10 mm s^{-1} contrasting the analogous spectra of **2** which cover a range of only $\sim$3 mm s^{-1}. For the simulations we have used the spin coupling scheme outlined above and the same electronic Hamiltonian (Equation (1)) was applied as for the EPR interpretation for simulations of the spectra. The ZFS parameters were readily taken from the EPR result; isomer shift, quadrupole splitting, asymmetry parameter and (isotropic) A constant of the ferric site could be taken identical with those of the di-ferric starting

Table I. Mössbauer and spin Hamiltonian parameters for **2** and $\mathbf{2^{ox}}$ at 4.2 and 1.8 K

	$[Fe_2^{III}(bbpo)Cl_4]^{2-}$ (**2**)	$[Fe_2^{III}(bbpo^{ox1})Cl_4]^{1-}$ ($\mathbf{2^{ox}}$)	
		Fe_1	Fe_2
S_{Fe} [a]	5/2	5/2	2
S_t [b]	0	1/2	
δ [mm s^{-1}][c]	0.45	0.45	0.49
ΔE_Q [mm s^{-1}][d]	0.98	0.98	−1.15
η [e]	0.9	0.9	0.8
α, β, γ [deg][f]	0; 0; 0	10; 6; 0	11; 65; 0
Γ [mm s^{-1}][g]	0.28	0.28	0.28
$A_{exp}/g_N\mu_N[T]$ [h]	−21.85	−21.85	−21.05
$A_{local}/g_N\mu_N[T]$ [h′]	−21.85	−21.85	−18.04[m]
D [cm^{-1}][i]	1.9	1.9	2.5
E/D [j]	0.18	0.18	0.18
g [k]	2	2	(2.028, 1.99, 1.988)
J [cm^{-1}][l]	−9.8	−12	

[a] local spin at a single Fe-site; [b] ground state; [c] isomer shift vs. α-Fe at 298 K; [d] quadrupole splitting; [e] asymmetry parameter of the efg; [f] Euler angles, rotate efg in p.a.s of D matrix; [g] line width (hwfm); [h] isotropic Mössbauer hyperfine coupling constants; [h′] local hyperfine coupling constants for $S_{Fe} = 5/2$; [i] zero-field splitting parameter for S_i; [j] rhombicity parameter; [k] electronic g value for S_i; [l] J Heisenberg coupling constant; [m] $A_{local} = (6/7)A_{exp}$.

complex **2**. Only those of the oxidized site with spin $S_2 = 2$ were independently optimized and their results are summarized in Table I.

The values of quadrupole splitting and isomer shift of the ferric site with $S_2 = 2$ (next to the ligand radical) are found to be slightly increased with respect to those of the starting compound or the other ferric site of oxidized **2**, whereas the local hyperfine coupling constant, $A_{local}/g_N\mu_N = -18$ T, (with respect to local iron spin) is significantly decreased (−21.8 T before oxidation). While the variations of the quadrupole splitting and isomer shift reflect an attenuated π-donor capability of the oxidized ligand, which is perfectly in accord with ‘chemical intuition’, the reduction of the local hyperfine coupling of the ^{57}Fe nucleus at the ‘oxidized’ iron site seems at first view to indicate the opposite trend. However, it does not originate from variations of overall covalency but from covalent delocalization of *radical* β spin density into the α spin carrying metal d-orbitals. The same effect is observed in complexes with *diamagnetic* central ions for which relative strong super-hyperfine coupling of the (‘pure’) radical spin with the metal nuclei is observed in the EPR spectra. That can only be explained by delocalization of radical spin density onto the central metal ion [1, 15]. The delocalization mechanism is completely analogous to the origin of the “*transferred*” or “*super-transferred hy-*

perfine fields" described in solid-state applications of Mössbauer spectroscopy and magnetic materials [18].

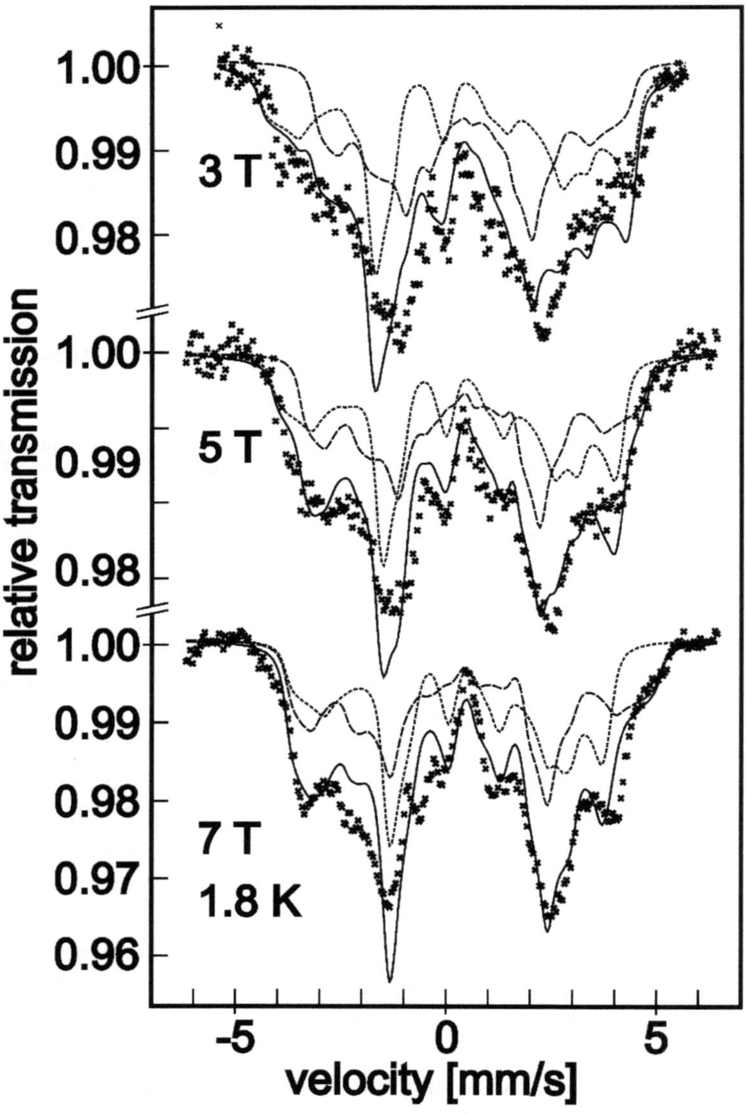

Figure 5. Applied-field Mössbauer spectra of the oxidized complex $[Fe_2^{III}(bbpo^{ox1})Cl_4]^-$ ($\mathbf{2^{ox}}$) at 3, 5, 7 T and 1.8 K. The lines are spin Hamiltonian simulations for the two different iron sites Fe(1), $S_1 = 5/2$ (dotted line) and Fe(2), $S_2 = 2$ (dashed line) and their superposition (solid line). The experimental data are corrected for residual contamination of non-oxidized material by subtracting 10% of spectra of **2**, as shown in Figure 3 for instance.

4. Conclusions

The study presents a detailed description of the electronic and magnetic properties of di-iron complex $[Fe_2^{III}(bbpo)Cl_4]^{2-}$ (**2**) with the ligand 1,2-bis(3,5-di-*tert*-butyl-2-hydroxyphenyl)oxamide and its electrochemically generated one-electron oxidation product $[Fe_2^{III}(bbpo^{ox1})Cl_4]^{-}$ (**2ox**). The spectroscopic data show that the ligand is oxidized forming the trianon $(bbpo^{ox1})^{3-}$ which is a π radical ($S = 1/2$) of the *o*-iminobenzosemiquinonate type. Formation of high-valent iron(IV) species was not observed. In this sense the ligand is *non-innocent* and controls the high-protenial redox properties of the complex.

The di-iron(III) complex $[Fe_2^{III}(bbpo)Cl_4]^{2-}$ (**2**) represents an integer spin system that is characterized by AF exchange interaction and weaker but competing zero-field splitting of the iron sites. Mössbauer, magnetization and EPR studies were used to derive a consistent description of the magnetic and electronic properties.

The three-spin system of $[Fe_2^{III}(bbpo^{ox1})Cl_4]^{-}$ (**2ox**) can be described by strongly AF exchange coupling of the ligand radical and the adjacent high spin ferric ion yielding a fictitious $S^* = 2$ state which in turn couples weakly to the second high spin ferric ion yielding the observed $S_t = 1/2$ ground state. The spin structure provides interesting magnetic properties of long-range coupled systems with competing interactions.

References

1. Beckmann, U., Bill, E., Weyhermüller, T. and Wieghardt, K., *Eur. J. Inorg. Chem.* (2002), submitted.
2. Burdinski, D., Bill, E., Birkelbach, F., Wieghardt, K. and Chaudhuri, P., *Inorg. Chem.* **40** (2001), 1160.
3. Meyer, K., Bill, E., Mienert, B., Weyhermüller, T. and Wieghardt, K., *J. Am. Chem. Soc.* **121** (1999), 4859.
4. Rodriguez, J. H., Ok, H., Xia, Y.-M., Debrunner, P. G., Hinrichs, B. E., Meyer, T. and Packard, N. H., *J. Phys. Chem.* **100** (1996), 6849.
5. Trautwein, A. X., Bill, E., Bominaar, E. L. and Winkler, H., *Structure and Bonding* **78** (1991), 1.
6. Kauffmann, K. E., Popescu, C. V., Dong, Y., Lipscomb, J. D., Que, L. and Münck, E., *J. Am. Chem. Soc.* **120** (1998), 8739.
7. Moriya, T., *Phys. Rev.* **120** (1960), 91.
8. Dzyaloshinsky, I., *J. Phys. Chem. Solids* **4** (1958), 241.
9. Bencini, A. and Gatteschi, D., *EPR of Exchange Coupled Systems*, Springer, Berlin, 1990.
10. Armentano, D., Demunno, G., Faus, J., Lloret, F. and Julve, M., *Inorg. Chem.* **40** (2001), 655.
11. Dewitt, J. G., Bentsen, J. G., Rosenzweig, A. C., Hedman, B., Green, J., Pilkington, S., Papaefthymiou, G. C., Dalton, H., Hodgson, K. O. and Lippard, S. J., *J. Am. Chem. Soc.* **113** (1991), 9219.
12. Tolman, W. B., Liu, S. C., Bentsen, J. G. and Lippard, S. J., *J. Am. Chem. Soc.* **113** (1991), 152.
13. Bossek, U., Hummel, H., Weyhermüller, T., Bill, E. and Wieghardt, K., *Angew. Chem.* **107** (1995), 2885.
14. Scaringe, R. P., Hodgson, D. J. and Hatfield, W. E., *Molecular Phys.* **35** (1978), 701.

15. Chaudhuri, P., Verani, C. N., Bill, E., Bothe, E., Weyhermüller, T. and Wieghardt, K., *J. Am. Chem. Soc.* **123** (2001), 2213.
16. Sage, J. T., Xia, Y. M., Debrunner, P. G., Keough, D. T., Dejersey, J. and Zerner, B., *J. Am. Chem. Soc.* **111** (1989), 7239.
17. Gibson, J. F., Hall, D. O., Thornley, H. M. and Whatley, F. R., *Proc. Nat. Acad. Sci. USA* **56** (1966), 987.
18. Grandjean, F., In: G. J. Long (ed.), *Mössbauer Spectroscopy Applied to Inorganic Chemistry*, Vol. 2, Plenum, New York, 1987, p. 241.

Hyperfine Interactions **144/145:** 199–207, 2002.

Myoglobin and Related Biomolecular Systems Studied with Mössbauer Spectroscopy and Nuclear Forward Scattering

H. WINKLER
Institut für Physik, Universität zu Lübeck, Germany

Abstract. Lineshape analysis of conventional Mössbauer spectra of oxymyoglobin indicated the presence of a dynamic disorder of the FeO_2 moiety. Complementary studies in the time domain by means of nuclear forward scattering (NFS) of synchrotron radiation have confirmed this observation and allowed a more detailed description of the underlying mechanism. Also in the time domain the so-called spectral dimension of deoxymyoglobin has been determined and is compared with the result known from studies in the frequency domain.

Key words: nuclear resonant forward scattering, oxy- and deoxymyoglobin, dynamic disorder, spectral dimension.

1. Introduction

The history of Mössbauer studies of hemo- and myoglobin is nearly as old as the history of Mössbauer spectroscopy itself. The reason is obvious: both proteins are easily available, they contain iron and are respiratory proteins and thus important biomolecules which are vital for all vertebrates. Hemoglobin, however, is a tetramer which poses sometimes problems to the spectroscopist because the four irons need not be necessarily in the same state. Therefore interest concentrated more on myoglobin (Figure 1) which has almost the same structure as one of the subunits of hemoglobin.

2. Dynamic disorder of the FeO_2 moiety in oxymyoglobin

The task of myo- and hemoglobin is either to transport or to store oxygen. To fulfil this function requires the ability to reversibly bind molecular oxygen. Pauling has proven by susceptibility measurements that in the deoxy form the iron is in its ferrous high-spin state. This is obviously the essential point for the reversibility of the oxygen binding. The first question is, therefore, how the iron is prevented from being oxydized irreversibly to iron(III). The main cause of irreversible oxidation of inorganic iron containing molecules are collisions with other molecules which finally give rise to the formation of a ferric iron dimer coupled by an oxygen bridge.

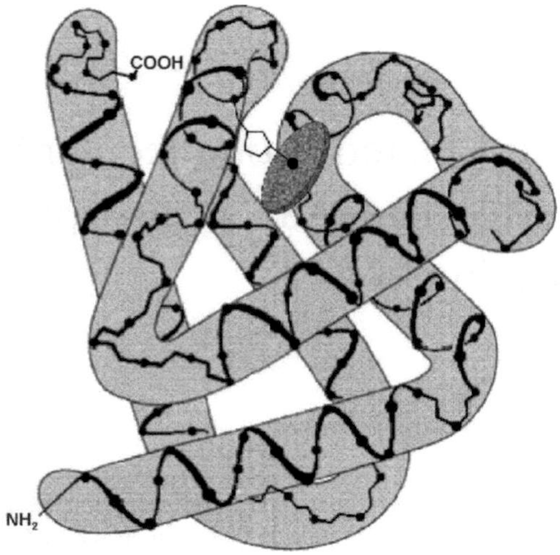

Figure 1. Sketch of the structure of oxymyoglobin. The disk models the heme prosthetic group (i.e. the porphyrin with Fe in its center).

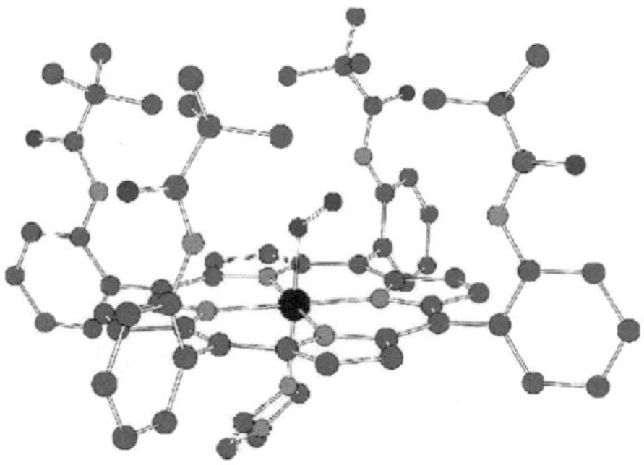

Figure 2. Structure of the so-called picket-fence porphyrin. Located in the center is Fe which has oxygen and imidazole as axial ligands.

In the globins the heme pocket protects the iron. Collman and coworkers [1] have shown that in the so-called picket-fence porphyrin (Figure 2) the pickets simulate perfectly the protective power of the heme pocket and thus form a functional model for myo- and hemoglobin. In the oxygenated form the iron sites of both, the native proteins and the model, are proven to be diamagnetic, i.e. their total spin is zero. Three different models for the FeO_2 moiety have been suggested: the bent end-on model of Pauling, the equidistance model of Griffith and the superoxide model of Weiss (Figure 3). The Griffith model had been discarded soon due to the results of X-ray diffraction studies. The distinguishing feature of the two other models is that the iron is either Fe(II) or alternatively Fe(III). As it seems more probable that the iron would remain in the ligated state still iron(II) but now in the low-spin state the Pauling model has been mostly favored. Mössbauer spectroscopy had been expected to resolve the problem but the spectra opened more questions than they

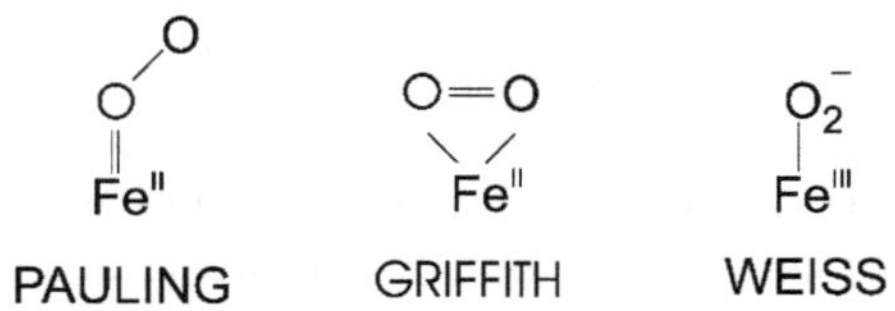

Figure 3. Various models of the FeO_2 bond.

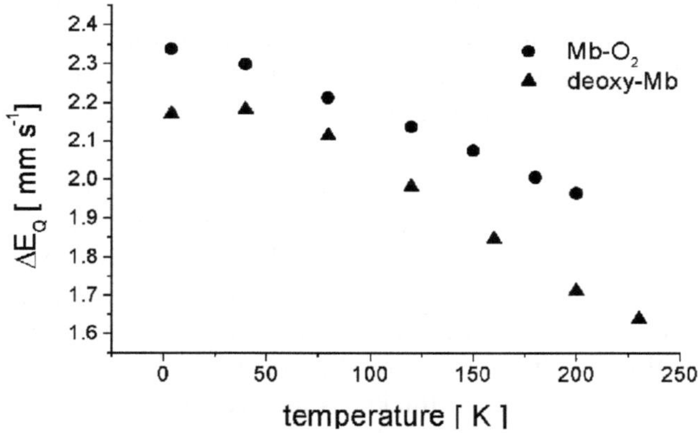

Figure 4. The temperature dependence of the quadrupole splitting in deoxy- and in oxymyoglobin.

answered and the precise nature of the FeO_2 bond remained a matter of controversy for a long time.

What has to be explained by any model is first of all the rather large quadrupole splitting [2]. Since in the low-spin ferrous state all three t_{2g} orbitals would be doubly occupied the electron shell should be almost spherical and the components of the electric field gradient (efg) tensor small. Furtheron, it has been observed that oxymyoglobin exhibits a strong temperature dependence of the quadrupole splitting like deoxymyoglobin (Figure 4). For the latter an explanation can be given in terms of a Boltzmann population of excited orbital states. In oxymyoglobin, however, the electron shell ought to be very stable so that excited states are far away in energy. Extensive MO calculations by Herman and Loew [3] showed for the bent end-on model that an asymmetric delocalization of a 3*d* electron due to covalency is present so that a formal Fe^{III}–O_2^- description seemed appropriate and the large quadrupole splitting and even the temperature dependence could be explained this way. However, for molecular orbitals only the total spin is a good quantum number and this was found to be zero. If this spin is artificially separated in a spin $1/2$ at the iron coupled antiferromagnetically to another spin $1/2$ at the ligand the resulting state lies according to Herman and Loew more than 13,000 cm^{-1} above the ground state. So it is a matter of taste whether one prefers to speak of a Fe^{II}–O_2 or a Fe^{III}–O_2^- moiety in the oxygenated globins.

What has to be explained in addition is that the two lines broaden asymmetrically at intermediate temperatures. For the picket-fence porphyrin, where this behavior is also very pronounced, a satisfactory explanation was given by Spartalian *et al.* [4] by invoking a dynamic process of the kind that the terminal oxygen jumps between four different positions, two and two of which are equivalent with respect to the efg tensor they produce. The spectra could be fitted very well by this model. A corresponding explanation for myoglobin seemed to be impossible because the X-ray structure of oxymyoglobin indicated that a rotation of the oxygen molecule in the heme pocket is sterically hindered. However, after refining the structure Phillips [5] suggested that a structural disorder of the oxygen might be possible with a secondary position of the terminal oxygen rotated by about 40°. Nevertheless, no rigorous mathematical treatment of the protein data has been performed since then.

After the advent of nuclear resonant forward scattering (NFS) as an additional tool for the study of hyperfine interactions we decided to investigate the problem again, because we knew that NFS is much less sensitive to external disturbances like mechanical vibrations which can broaden Mössbauer resonance lines. Our NFS data show apart from an accelerated decay of the delayed counts no spectacular variations with temperature (Figure 5). If the spectra are fitted with quadrupole splitting and effective thickness as free parameters using the progam package SYNFOS [6], we get also a decrease of the splitting but an increase of the effective thickness, which is physically unreasonable and obviously due to the faster decay with time. Of course, there exists a correspondence between frequency domain and time domain of the kind that line broadening means faster decay. In the present case, however, the faster decay is obviously caused by the dephasing which results from a dynamic disorder. Since it was known from single crystal studies [7] that the largest component V_{zz} of the efg tensor lies within the heme plane, we have constructed a model which consists of two electric field gradients with both efg components $V_{zz,1}$ and $V_{zz,2}$ lying in this plane but separated by an unknown angle of rotation β around the heme normal. The dynamics is introduced by a jump frequency and the Boltzmann population of the secondary position determined by the energy difference ΔE. The results received by an appropriately adapted version of SYNFOS [8] are $\beta = 40°$ and $\Delta E = 111\ \mathrm{cm}^{-1}$ (Figure 6).

The temperature dependence of the transition rate w_{12} from the ground state 1 to the excited state 2 (Figure 7) can be fitted by the expression

$$w_{12}(T) = w_\mathrm{T} + w_0 \cdot \mathrm{e}^{-E_\mathrm{A}/k_\mathrm{B}T} \tag{1}$$

with $w_\mathrm{T} = 10^7\ \mathrm{s}^{-1}$, $w_0 = 10^9\ \mathrm{s}^{-1}$ and $E_A = (260 \pm 35)\ \mathrm{cm}^{-1}$, which indicates that thermal activation plus a low-temperature tunneling process are responsible for the dynamics of the FeO_2 moiety [9].

3. Spin–lattice relaxation in deoxymyoglobin

Another kind of dynamics can be observed in deoxymyoglobin, namely spin–lattice relaxation. For this purpose an external magnetic field has to be applied

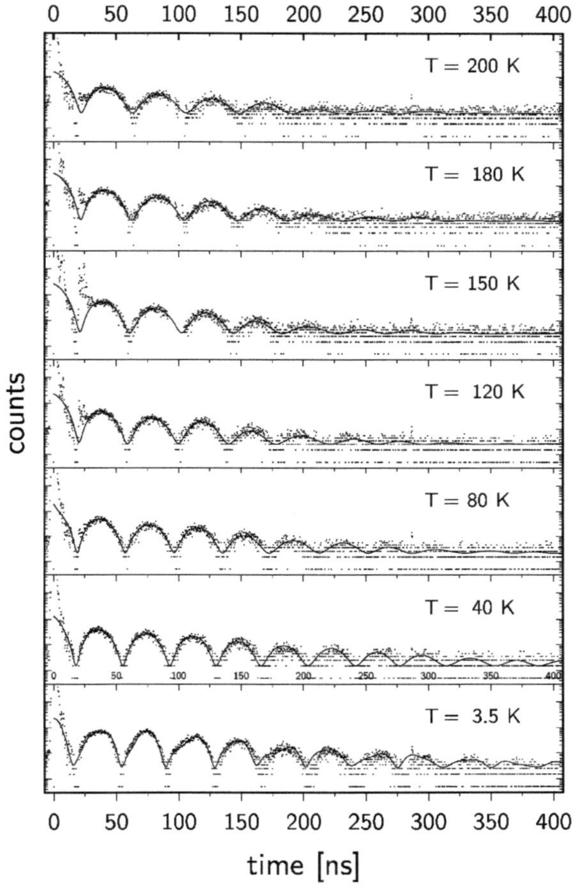

Figure 5. NFS of oxymyoglobin at various temperatures. The solid lines are the results of fits assuming dynamic disorder of the terminal oxygen as described in the text. The richer structure visible at 3.5 K is due to an impurity of ferric high-spin metmyoglobin which has been taken into account in the fit.

in order to produce finite spin expectation values in the various magnetic substates. Spin–lattice relaxation induces jumps between the magnetic substates of the ferrous high-spin state. Stapleton and coworkers [10] have observed by means of EPR measurements on several iron proteins in the ferric state that in many cases the relaxation rates do not follow a T^9 law as theoretically predicted for the Raman process in Kramers systems but with an appreciably lower power than 9. At the time when fractals became popular they have attributed this behavior to a purported fractional dimensionality of the proteins. This value can be also approximately obtained by counting the α-carbons of a protein backbone within a certain radius and determining the slope of the straight line which results from a double-logarithmic plot. For many proteins a number of about 1.6 is found [11].

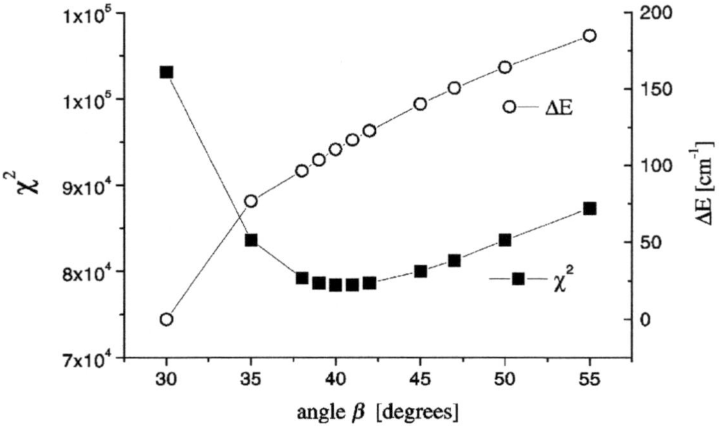

Figure 6. Goodness-of-the-fit χ^2 and energy distance ΔE of the secondary position of the terminal oxygen as a function of the rotation angle β.

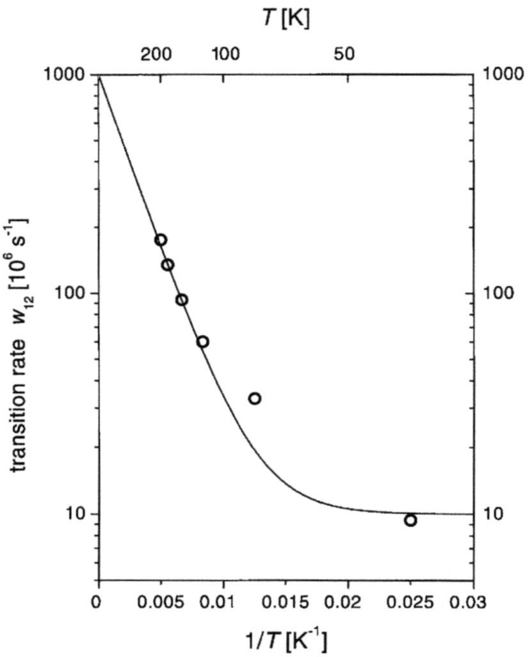

Figure 7. Transition rate of the terminal oxygen between the two conformations in oxymyoglobin as a function of temperature. The solid line is the result of the fit with Equation (1).

Conventional Mössbauer spectroscopy of deoxymyoglobin at variant temperatures in external fields of 2 and 6 T yielded also a value of 1.6 [12], if the probability for a jump from a lower state i up to state j which is by Δ_{ij} higher in energy is assumed to vary with temperature according to

$$w_{ij} = w_0 \frac{\Delta_{ij}^{d_S}}{\exp(\Delta_{ij}/kT) - 1}, \tag{2}$$

where w_0 is a scaling factor. The inverse process is determined by the principle of detailed balance. The temperature-dependent term in (2) corresponds to the energy density of phonon states with $\hbar\omega = \Delta_{ij}$, which is given by the three factors $1/\{\exp(\Delta_{ij}/kT) - 1\}$ for the phonon occupation number, $\Delta_{ij}^{d_S - 1}$, with $d_S = 3$ for the density of states in a Debye solid, plus an additional factor Δ_{ij}. If one takes the exponent d_S as free fit parameter one gets as result the fractional dimension characteristic for a particular protein.

The concept of Stapleton has, however, evoked much criticism because it turned out that the fractional dimension is not well defined but depends among others on the solvent. Solid-state theoreticians as, e.g., Krummhansl [13] pointed out that the unusual phonon spectra can be explained more simply by the assumption of an anisotropic medium as phonon carrier or by non-central molecular forces. So the fractal concept of proteins is nowadays regarded as an overinterpretation.

Spin–lattice relaxation can be observed not only in the frequency domain by conventional Mössbauer spectroscopy but even better in the time domain by NFS. Fitting our data by means of the appropriate SYNFOS routine (Figure 8) we got the much higher value of 2.26 for d_S as compared to the above-mentioned value of 1.6. The reason for the discrepancy might be explained by the different kind of sample material: microcrystals in the conventional measurements and frozen solution in the NFS study. In a frozen solution the anisotropy of the phonon carrier may be less pronounced. It follows that the spectral fractional dimension is not a well-defined property of proteins but only a phenomenological parameter. Nevertheless, it is of great value to know that one can vary this parameter when fitting relaxation spectra, because one finds often relaxation phenomena in systems like biological macromolecules when investigated in applied fields. In order to arrive at a consistent picture it is mandatory to interpret these spectra theoretically by including dynamics of some kind.

4. Summary

In summary, one may say that it is often worthwhile to repeat former Mössbauer studies by means of NFS. Although it is generally correct that in principle the information obtainable by NFS is the same as by conventional Mössbauer spectroscopy, the results from NFS can nevertheless provide new insights because many sources of distorsions are absent in NFS. The time spectra

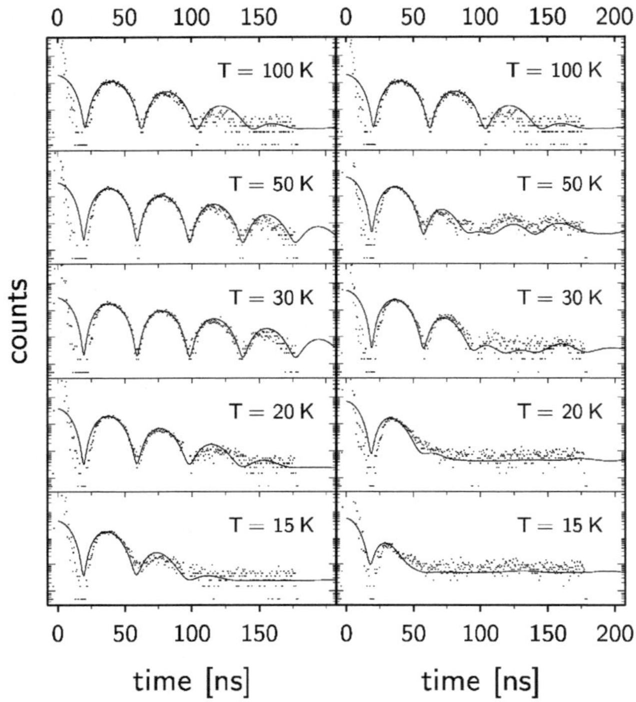

Figure 8. NFS of deoxymyoglobin in an applied field of 2 T (left column) and 6 T (right column) at various temperatures.

- are almost free of background noise,
- insensitive to mechanical vibrations, and
- have a precise and stable time scale.

Making use of these advantages the investigations of oxymyoglobin by means of NFS have put the hypothesis of dynamic disorder on firmer grounds. In deoxymyoglobin the presence of spin–lattice relaxation and its interpretation in terms of a spectroscopic dimension <3 has been confirmed.

Acknowledgements

The NFS data have been taken together with V. Schünemann, R. Benda, Ch. Herta and A. X. Trautwein from Lübeck at HASYLAB in Hamburg with support from O. Leupold, Yu. Shvyd'ko, H. D. Rüter and E. Gerdau. We thank F. Parak and coworkers in his lab in Munich for providing us with excellent Mb samples over the years.

Support from the German Federal Ministry of Education and Research (BMBF) under grant no. 05 SK8FLA8 is gratefully acknowledged.

References

1. Collman, J. P., Gagné, R. R., Reed, C. A., Halbert, T. R., Lang, G. and Robinson, W. T., *Am. Chem. Soc.* **97** (1975), 1427.
2. Bade, D. and Parak, F., *Z. Naturforsch.* **33c** (1978), 488.
3. Herman, Z. S. and Loew, G. H., *J. Am. Chem. Soc.* **102** (1980), 1815.
4. Spartalian, K., Lang, G., Collman, J. P., Gagne, R. R. and Reed, C. A., *J. Chem. Phys.* **63** (1975), 5375.
5. Phillips, S. E. V., *J. Mol. Biol.* **142** (1980), 531.
6. Haas, M., Realo, E., Winkler, H., Meyer-Klaucke, W., Trautwein, A. X., Leupold, O. and Rüter, H. D., *Phys. Rev. B* **56** (1997), 14082.
7. Maeda, Y., Harami, T., Morita, Y., Trautwein, A. and Gonser, U., *J. Chem. Phys.* **75** (1981), 36.
8. Haas, M., Realo, E., Winkler, H., Meyer-Klaucke, W., Trautwein, A. X. and Leupold, O., *Phys. Rev. B* **61** (2000), 4155.
9. Herta, C., Winkler, H., Benda, R., Haas, M. and Trautwein, A. X., *Eur. Biophys. J.* **31** (2002), 478.
10. Stapleton, H., Allen, J. P., Flynn, C. P., Stinson, D. G. and Kurtz, S., *Phys. Rev. Lett.* **45** (1980), 1456.
11. Allen, J. P., Colvin, J. T., Stinson, D. G., Flynn, C. P. and Stapleton, H. J., *Biophys. J.* **38** (1982), 299.
12. Winkler, H., Ding, X.-Q., Burkardt, M., Trautwein, A. X. and Parak, F., *Hyp. Interact.* **91** (1994), 875.
13. Krummhansl, J. A., *Phys. Rev. Lett.* **56** (1986), 2696.

Hyperfine Interactions **144/145:** 209–222, 2002.

Mössbauer Spectroscopy in the Energy and in the Time Domain, a Crucial Tool for the Investigation of Protein Dynamics

ALEXANDRA HUENGES, KLAUS ACHTERHOLD and FRITZ G. PARAK*
Physik-Department E17, TUM, 85747 Garching, Germany

Abstract. New experiments on MbCO crystals and MbCO in a glycerol/water solution are described. Phonon-assisted Mössbauer effect reveals solid state like harmonic vibrations and is used to determine the phonon density of states spectrum at the iron center of myoglobin. Above 200 K, a new type of fluctuations arises which is essential for protein function. The experiments can be explained by a two state model. In the "rigid" state, harmonic vibrations are present in the full temperature regime. Above 200 K a "flexible" state can be reached. The quasi-diffusive motions in this state can be described by an overdamped Brownian oscillator with only one frequency. The Boltzmann probability to reach the flexible state is used. The influence of the solvent surrounding the protein is investigated. The dynamical transition temperature of the protein is compared with a glass transition.

Key words: myoglobin, density of phonon states, Brownian oscillator, conformational transition.

1. Introduction

Proteins are the bricks of life. Their structure with different degrees of flexibility as well as their structural changes are decisive for the biological function. One of the best studied proteins is the myoglobin. It is a globular molecule with a diameter of about 30 Å and a molecular mass of 17,800 Dalton. It is built up by a sequence of 153 amino acids arranged mainly in 8 α-helices. The active center is formed by the iron ion of a heme group. The well-known function is the storage of an O_2 molecule [1]. Recently it was found that myoglobin also plays an important role in the regulation of the NO content of the cells [2, 3]. Even this relatively simple protein adopts two different conformations (which means slightly different structures), the so-called "relaxed" r-conformation where the heme iron is ligated, for instance by an O_2 or a CO molecule, or the "tensed" t-conformation, where the 6th coordination of the iron is empty. X-ray structure analysis is the most important method to determine the 3-dimensional architecture of protein molecules. Myoglobin was the first protein were the structure has been solved [4, 5]. Nevertheless, its structure is still under investigation [6–9]. Myoglobin became a paradigm in the study of protein dynamics [10]. From the functional properties there are two

* Author for correspondence.

most important questions: how can a ligand like O_2 enter the heme pocket and bind to the iron although there is no channel from the surface into the pocket which is large enough. The answer comes from the investigation of structural distributions and structural fluctuations in proteins. Another problem is the understanding of the physical mechanism of the conformational changes between r and t by structural relaxation.

The X-ray structure analysis of the L29W mutant of myoglobin gave deep insight into several aspects of protein dynamics [11]. In this point mutant the residue lysine (L) at position 29 in the sequence of amino acids (counted from the N-terminus) is replaced by a tryptophan (W) (see Figure 1). This bulky residue reduces the velocity of ligand binding by two orders of magnitude. Therefore, intermediate structures between ligated and unligated can be stabilized at low temperatures. For the discussion we compare three different X-ray structures on the CO ligated mutant performed at 105 K. The structure measured in the dark shows the ligand CO bound to the iron. If the crystal is illuminated by a laser during the structure determination, the molecule structure remains identical, a structural relaxation from the ligated to the unligated conformation does not take place on the time scale of several days. However, the bond to the CO ligand is broken and the CO is found in the heme pocket on the distal side near Trp29. The situation

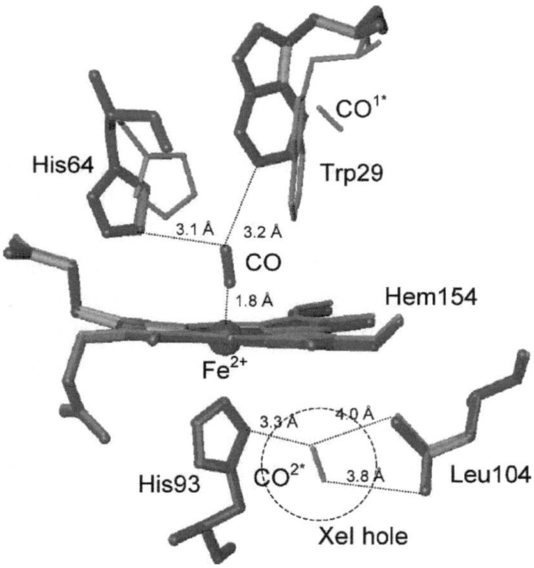

Figure 1. The structure of the active site of L29W MbCO at 105 K. For details see [11]. Thick lines: MbCO dark structure. The iron is situated within the heme plane and the CO binds at a distance of 1.8 Å to the iron. By photodissociation at temperatures below 180 K the structure remains the same but the CO molecule is now found below the Trp29 (CO^{1*}). After extended illumination above 180 K and cooling to 105 K during the illumination the CO molecule is found in the xenon hole (CO^{2*}). Moreover, the structure has strongly changed (thin lines, note the movements of His64 and Trp29).

changes completely if the laser illumination is performed at a temperature slightly above 180 K and the crystal is cooled during illumination to 105 K. Again the CO bond is broken. But now the CO molecule is found behind the heme group at the so-called proximal side. Moreover, on the distal side the structure has significantly changed. Above 180 K a conformational relaxation takes place. One can distinguish two temperature regimes separated by a characteristic temperature, T_c, called the dynamical transition temperature. Below T_c, in the "low temperature regime" the molecules are frozen in a conformation: only above the dynamical transition temperature can they adopt their structure to the ligation state if it is changed. Since this is a necessary condition for the protein function the temperature regime above T_c is called the "physiological regime". In the following it will be shown that the Mössbauer spectroscopy can contribute significantly to the understanding of the physics of protein dynamics in both regimes.

2. Early investigation of protein dynamics by Mössbauer spectroscopy

An anomalous temperature dependence of the mean square displacement, $\langle x^2 \rangle$, of the heme iron in myoglobin was first described by [12]. The Lamb–Mössbauer factor of metmyoglobin crystals was determined between 4.2 K and room temperature. Above 200 K, the $\langle x^2 \rangle$ values increase dramatically with temperature. This behavior was later approved and investigated in detail on different biomolecules [13–22]. It is obvious to correlate the mean square displacements below 200 K with the low temperature regime and those above with the physiological temperature regime. This implicates that the strong increase of the $\langle x^2 \rangle$ values above T_c may be necessary for protein function. Above T_c, the Mössbauer absorption changes the spectral shape. In addition to the narrow Lorentzians, broad lines become visible [18, 19, 21, 23–25]. It is an exciting issue to investigate these phenomena in more detail and to correlate it with the biological function.

3. Investigations in the time domain; Mössbauer effect with synchrotron radiation

The feasibility of Mössbauer experiments with synchrotron radiation was shown by Gerdau and co-workers [26]. Nuclear forward scattering by a Mössbauer nucleus is the most direct way of Mössbauer investigations with synchrotron radiation. The principles of this method are explained in a number of publications [27–29]. An application to myoglobin is described in [30]. In principle, nuclear forward scattering has clear advantages for the studying of protein dynamics. As mentioned before, at physiological temperatures proteins show broad lines in addition to the narrow Mössbauer absorption lines. In absorption experiments, the broad lines reveal themselves at high Doppler velocities where their contribution to the absorption is only small. It is difficult to analyze them with sufficient statistics. In the time domain, they reveal themselves at early times, where the counting rate is high.

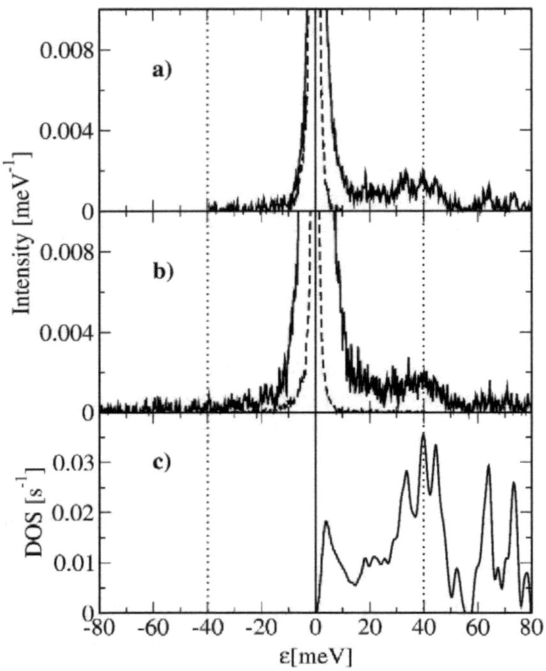

Figure 2. Result of phonon-assisted Mössbauer effect measured on a sample containing a huge number of MbCO crystals. (a) $T = 40$ K; (b) $T = 235$ K. Solid line: energy dependence of the phonon-assisted Mössbauer absorption measured as the intensity of delayed radiation. Stokes processes appear at positive energies (energy is deposited in the thermal bath), anti-Stokes processes at negative energies (energy is taken from the thermal bath). At 40 K anti-Stoke processes are practically absent. Dashed line: energy resolution of the experimental equipment (for details see [33]). (c) Density of phonon states. The dotted line indicates a temperature independent vibration.

Therefore, nuclear resonant forward scattering is in principle well suited to study those lines. Unfortunately, at present the dead time of the counting electronics is still rather long (about 10 ns). Up to now, no experiment was performed on proteins in the most interesting very short time regime.

Decisive information on the dynamics of proteins came from "phonon-assisted Mössbauer effect". Synchrotron radiation with an energy width of about 1 meV is scattered by molecules, containing the Mössbauer isotope ^{57}Fe. If the energy of the synchrotron radiation does not match the Mössbauer transition, delayed nuclear resonance scattering can nevertheless occur. The energy balance is ensured by the absorption or emission of phonons with the energy ε. Varying the energy difference ε between the Mössbauer excitation energy and the synchrotron radiation, the density spectrum of phonon states coupling to the iron is obtained [31–33].

Figure 2 shows experiments on small myoglobin crystals, ligated with CO. They were randomly oriented in the synchrotron beam. The measurements at 40 K (a) and 235 K (b) reveal Stokes and anti-Stokes processes. The creation of phonons can happen at all temperatures, whereas the anti-Stokes annihilation process is nearly

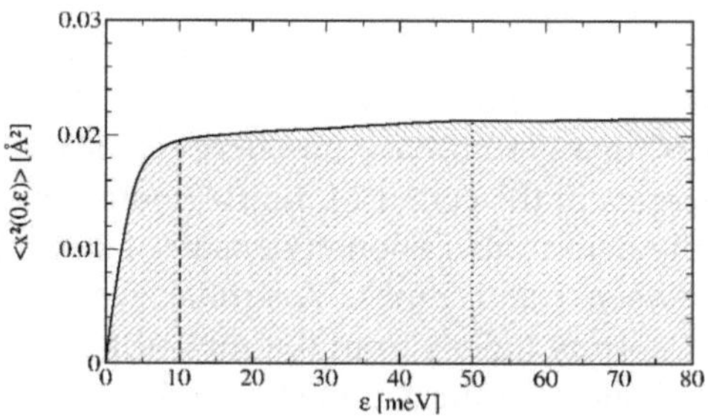

Figure 3. Mean square displacements of the iron in MbCO at 300 K as function of the energy of the contributing phonons. The modes with the energy between $\varepsilon_1 = 0$ meV and the value $\varepsilon_2 = \varepsilon$ are taken into account according to Equation (1). The dashed line marks the $\langle x^2 \rangle$ value with contributions of the modes between 0 and 10 meV, the dotted line the $\langle x^2 \rangle$ value, where all modes up to 50 meV contribute. Note that phonons with a higher energy practically do not contribute to the mean square displacement of the iron atom.

impossible at 40 K. The position of vibrational lines is temperature independent (compare line at $\varepsilon = 40$ meV) indicating a harmonic behavior. This fact was more clearly seen in Mbdeoxy and Mbmet since more temperatures have been measured in these samples [33]. The phonon density of states spectrum obtained from the experiments is shown in Figure 2c. Details of the evaluation can be found in [33]. From the phonon density of states spectrum, $D[\varepsilon]$, one can extract parameters like the mean square displacement, $\langle x^2 \rangle$, of the iron atom:

$$\left\langle x^2[\varepsilon_1, \varepsilon_2] \right\rangle = \frac{\hbar^2}{m_{\mathrm{Fe}}} \int_{\varepsilon_1}^{\varepsilon_2} \frac{D[\tilde{\varepsilon}]}{\tilde{\varepsilon}} \left(\frac{1}{\exp[\tilde{\varepsilon}/k_B T] - 1} + \frac{1}{2} \right) \mathrm{d}\tilde{\varepsilon}. \tag{1}$$

Here $\hbar$ is the Planck constant, k_B the Boltzmann constant, m_{Fe} the mass of the iron and T the temperature. Equation (1) allows the determination of the contributions of the phonons in an energy interval from ε_1 to ε_2 to the mean square displacement. The result for $\varepsilon_1 = 0$ and variable $\varepsilon_2 = \varepsilon$ is presented in Figure 3. There is a strong increase of the mean square displacement adding up the contributions of low energy phonons up to 10 meV (vertical dashed line in Figure 3). The phonons from 10 to about 50 meV increase the $\langle x^2 \rangle$ value only slightly (vertical dotted line), the phonons of higher energies give practically no additional contribution. As an example we give the $\langle x^2[0 \text{ to } 80 \text{ meV}] \rangle$ value at 40 K which is 0.0029 ± 0.0005 Å^2 and at 235 K which is 0.0169 ± 0.0026 Å^2. Note, that the phonon density of states spectrum shown in Figure 2 together with Equation (1) yield a linear increase of the $\langle x^2 \rangle$ values up to room temperature. These values will be compared later with those obtained from the Lamb–Mössbauer factor. The energy of phonon modes at the border of the first Brillouin zone of myoglobin crystals can be estimated within the Debye model. With a sound velocity of 2000 m/s [34] this energy is 1 meV. This is in good agreement with a normal mode analysis where the intramolecular modes also start at about 1 meV [33]. Together with an increase of the density with ε^2 it is clear that the corresponding phonons are acoustic phonons in the Debye limit where the wavelength is so long that the protein can be regarded as a continuous medium. This is endorsed by a normal mode analysis [35], which shows that the lowest normal modes of myoglobin have an energy of about 0.8 meV. These are the optical phonon energies in the center of the Brillouin zone ($K = 0$). Astonishingly, the experimentally determined phonon density of states keeps up a quadratic energy dependence between 1 and 3 meV. A similar average behaviour is seen in the density of states of normal modes from 0.8 meV till 3 meV. The large number of optical phonons dominate over the 3 acoustic modes and obscure any van Hove singularity. In a naive physical picture one may assume that the wavelength of phonons below an energy of 3 meV is too long (it is longer than 27.5 Å) to be sensitive to the details of the molecule structure. Those phonons sense the protein sample as a continuum like the long wavelength acoustic phonons in the Debye model of solid state.

Mean square displacements calculated from the phonon densities explain excellently the low temperature regime (see Figure 5). The slope is temperature independent up to room temperature (compare [33]) and is in reasonable good agreement with mean square displacements calculated from a normal mode analysis and also with those obtained from conventional Mössbauer spectroscopy. However, the strong increase of $\langle x^2 \rangle$ above T_c measured by Mössbauer absorption spectroscopy is not explained by the phonon-assisted Mössbauer effect. Normal mode coordinates are a special way to arrange all degrees of freedom of the atoms of a molecule. In order to get degrees of freedom causing the more than linear increase of the $\langle x^2 \rangle$ values above T_c, a part of the normal mode coordinates has to be rearranged in a way to account for the protein specific dynamics. It can be only a small subset since, in the limit of the experimental error, the phonon density of states spectrum is nearly temperature independent up to room temperature. This is in accordance with theoretical investigations of Go and coworkers (see, e.g., [36] and references therein).

4. Back to Mössbauer absorption spectroscopy

As shown in Section 2 many basic experiments concerning the investigation of protein dynamics by Mössbauer spectroscopy lay decades ago. However, our understanding of the underlying physical principles and the connection with the biological function is still rather incomplete. A comparison with results obtained by other techniques became more and more important. New experiments with better statistics on different proteins had to be performed. The comparison of the results of phonon-assisted Mössbauer effect with conventional Mössbauer spectroscopy has clarified the physics of the low temperature regime. It has also been shown that most of the phonon interactions are still present at physiological temperatures [33]. One now can concentrate on the analysis of the motions, which are important for the function. In the following we discuss some new Mössbauer experiments. The protein specific motions are explained by a two-state model. In the “rigid” state there are only harmonic vibrations, which reveal themselves in the phonon density of states spectrum. The "flexible" state can be thermally populated. In this state the protein performs Brownian oscillations. The backdriving force of the Brownian oscillator is modeled by only one frequency. The theoretical description was already developed in [18, 19, 22].

The Mössbauer spectra are analyzed by the transmission integral given by Equation (2):

$$T(v) = 1 - Rf_s\left[1 - \int_{-\infty}^{\infty} Q(v, E_s, E)\mathrm{e}^{-\sigma_a(E_a,E)n_{\mathrm{Fe}}}\,\mathrm{d}E\right], \tag{2}$$

where $Q(v, E_s, E)$ gives the energy distribution of the source with the Lamb–Mössbauer factor f_s moved with the Doppler velocity v, and n_{Fe} is the number of ^{57}Fe nuclei per area. R corrects for the background. The absorption cross section, σ_a, is given by:

$$\begin{aligned}\sigma_a(E_a, E) &= \sigma_0 \frac{\Gamma_{\mathrm{nat}}}{2}\mathrm{e}^{-k^2\langle x_v^2\rangle}\mathrm{e}^{-k^2\langle x_t^2\rangle} \\ &\quad\times \sum_{N=0}^{\infty} \frac{[k^2\langle x_t^2\rangle]^N}{N!}\frac{(\Gamma_a/2 + N\hbar\alpha_t)}{(E - E_a)^2 + (\Gamma_a/2 + N\hbar\alpha_t)^2}.\end{aligned} \tag{3}$$

Note that Γ_a is not necessarily the natural line width, Γ_{nat} of the Mössbauer transition but may be modified by convolution with a Gaussian in order to take into account inhomogeneities due to conformational substates. This cross section allows the calculation of Mössbauer spectra of the protein in the rigid as well as in the flexible state. The low-temperature regime is described by $N = 0$ and $\langle x_t^2\rangle = 0$. We are back to the standard expression. The dynamics is taken into account by $\langle x_v^2\rangle$. At physiological temperatures, the molecules may be in the flexible state (often called transition state) with a certain probability $p_t(T)$. Quasi-diffusive motions described by a Brownian oscillator start to contribute. This has two consequences: the diffusion in limited space yields a mean square displacement $\langle x_t^2\rangle$ which reduces the intensity of the line with $N = 0$ by $\exp -(k^2\langle x_t^2\rangle)$. The missing intensity is found in additional broad lines where the broadening depends on the factor α_t.

In the low-temperature regime we obtain from a least-squares fit the mean square displacement $\langle x_v^2\rangle$ (compare Figure 4a) which agree with those determined from the phonon density of states spectrum. Since we know from the analysis of the phonon density of states spectrum that $\langle x_v^2\rangle$ increases linearly with temperature up to room temperature we used extrapolated $\langle x_v^2\rangle$ values for the analysis of the physiological regime. The small absorption at temperatures above 200 K enables us to apply the thin absorber approximation. A least-squares fit allows the determination of the mean square displacement $\langle x_t^2\rangle$ and the parameter α_t. Essential for this analysis is the spectral shape, which reflects the additional broad lines (compare Figure 4b). It has to be mentioned that the sum over N converges rapidly and can be terminated at $N = 5$ without making an error of more than 0.1% at $T = 243$ K.

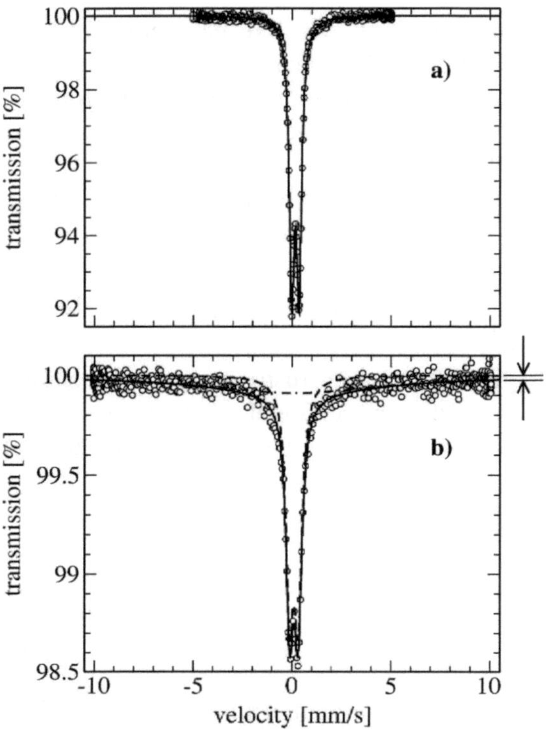

Figure 4. Mössbauer absorption spectrum of MbCO crystals. (a) $T = 80$ K; the solid line is a fit with the transmission integral of a quadrupole doublet. (b) $T = 243$ K; the solid line is a fit of a quadrupole doublet within the thin absorber approximation using the Brownian oscillator model. The dashed line indicates the elastic absorption. The quasi-elastic absorption due to overdamped motion within the flexible state is shown as a dashed dotted line.

The results for the mean square displacement of iron in MbCO crystals and of MbCO in a glycerol/water solution (3/1 glycerol/water volume/volume) are shown in Figures 5 and 6. We have now to extract the probability, $p_t(T)$, of the molecule to be in the flexible state. The temperature dependence of $\langle x_t^2 \rangle$ is determined by two factors, the probability to be in the flexible state and the mean square displacement, $\langle x_{t0}^2 \rangle$ which is performed if the molecule is permanently in the flexible state. In our simple model $\langle x_{t0}^2 \rangle$ increases linearly with T. According to [19] one obtains:

$$\langle x_t^2 \rangle(T) = \langle x_{t0}^2 \rangle \cdot p_t(T) = c \cdot T \cdot p_t(T) \tag{4}$$

and

$$\alpha_t(T) = \alpha_{t0} \cdot p_t(T). \tag{5}$$

The most simple assumption for $p_t(T)$, used in the following data analysis, is a population of the flexible state according to the Boltzmann statistics:

$$p_t(T) = \frac{1}{1 + \exp((\Delta H - \Delta S \cdot T)/k_B T)}. \tag{6}$$

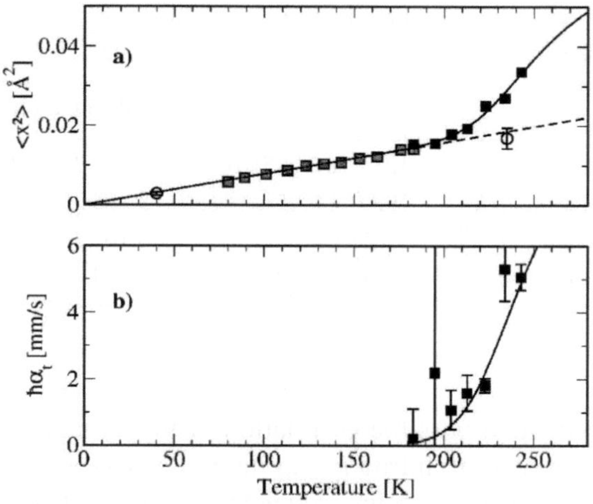

Figure 5. (a) Mean square displacement, $\langle x^2\rangle^{\gamma}$, of the iron in MbCO crystals. Grey squares: $\langle x_v^2\rangle$ due to harmonic vibrations as obtained by the transmission integral. Black squares: $\langle x^2\rangle^{\gamma} = \langle x_v^2\rangle + \langle x_t^2\rangle$ obtained with the thin absorber approximation. $\langle x_v^2\rangle[T]$ has been taken from an extrapolation of the low temperature values. All error bars σ less than 1/3 of the size of the symbols. Open circles: $\langle x_v^2\rangle$ values due to harmonic vibrations as obtained by phonon-assisted Mössbauer effect. (b) Black squares: α_t values characterizing the half width of the broad Lorentzians according to Equation (3).

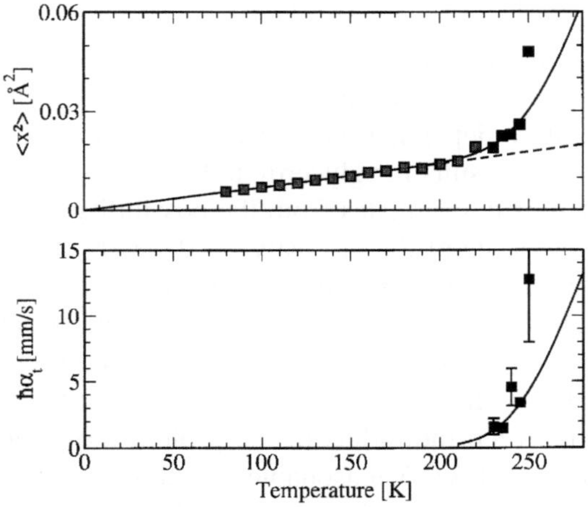

Figure 6. As Figure 5. However, the sample has been MbCO in a glycerol/water solution (3/1 vol/vol).

ΔH and ΔS are the enthalpy and the entropy differences between the rigid and the flexible state, respectively. For the determination of $p_t(T)$ we used a least-squares fit taking into account Equations (4), (5) and (6) simultaneously. The values of $\langle x_t\rangle(T)$ and $\alpha_t(T)$ were weighted according to $1/\sigma^2$ where σ is the standard deviation obtained from the fit of the Mössbauer spectrum. In this way we obtained the values for c, α_{t0}, ΔH and ΔS. They are given in Table I.

Table I. Parameter obtained from the least-squares fit of the Mössbauer spectra in the physiological temperature regime

	MbCO crystals	MbCO in glycerol
c [$\text{Å}^2\,\text{K}^{-1}$]	$(1.078 \pm 0.085)10^{-4}$	$(3.3 \pm 2.5)10^{-4}$
ΔH [kJ mol^{-1}]	29.40 ± 0.45	30.83 ± 1.63
ΔS [$\text{kJ mol}^{-1}\,\text{K}^{-1}$]	0.122 ± 0.002	0.109 ± 0.001
α_{t0} [mm/s]	9.09 ± 0.80	28.4 ± 0.5

In order to investigate a possible correlation of the dynamical transition with a glass transition one may try to introduce an internal viscosity $\eta_t(T)$ for the protein molecule and analyze its temperature dependence. For liquids of the viscosity η, the diffusion constant D of a globular probe of radius r is $D = k_B T/(6\pi\eta_t r)$. With the Einstein equation $D = k_B T/(2m\beta_t)$ and the experimentally results for $\langle x_t^2 \rangle$ and α_t one obtains the temperature dependence of the viscosity valid for our model [22] as

$$\eta_t(T) = \frac{k_B}{6\pi r \cdot c \cdot \alpha_{t0} \cdot p_t(T)} = \frac{\eta_{t0}}{p_t(T)}. \tag{7}$$

The viscosity η_{t0} within the transition state depends on c and α_{t0}. The parameter r, which characterizes the size of the moving segment, was chosen to be 5 Å. The results of the temperature dependent viscosity of MbCO at the iron site are given in Figure 7 together with results for glycerol and pure water. The sole temperature dependence of the viscosity originates from the possibility $p_t(T)$ to be in the flexible state. The absolute scale of the viscosity η_{t0} depends on the radius r, a value, which is somewhat arbitrary. Pastor and Karplus parameterized the Stokes' law of friction for small molecules in a solvent by introducing an effective hydrodynamic radius for each atom [37]. In this way, the degree of solvent contacts of each atom was considered separately. In the case of myoglobin, the effective radius r must be located somewhere between the radius of the protein molecule (~15 Å) and the atomic radius of iron (~1.4 Å). A reasonable choice for the effective radius could be $r = 5$ Å, the mean radius of the heme group. With the data for c and α_{t0} given in Table I, the viscosity of iron within myoglobin in Equation (7) is then scaled by $\eta_{t0} = 21$ P and 2.2 P for crystalline MbCO and MbCO in glycerol, respectively.

5. Discussion

First we can compare the slope of the mean square displacements, $\langle x_v^2 \rangle$ versus temperature as obtained from the phonon density of states spectrum and the Mössbauer absorption experiments, respectively. As seen from Table II there is an excellent agreement. However, the Mössbauer absorption experiments yield only the total

Table II. Comparison of parameters obtained from Mössbauer absorption experiments in the energy space (Mössb) and results from phonon-assisted Mössbauer effect (phon). The lower subscript "v" indicates that the values are obtained from the low temperature regime (rigid state). f is the Lamb–Mössbauer factor

	f_V (phon) $T = 80$ K	f_V (Mössb) $T = 80$ K	$\langle x_v^2 \rangle / T$ (phon) Å^2/K	$\langle x_v^2 \rangle / T$ (Mössb) Å^2/K
MbCO cryst.	0.74	0.71	$(7.2 \pm 1.1)10^{-5}$	$(7.90 \pm 0.14)10^{-5}$
MbCO glyc.		0.74		$(7.12 \pm 0.15)10^{-5}$

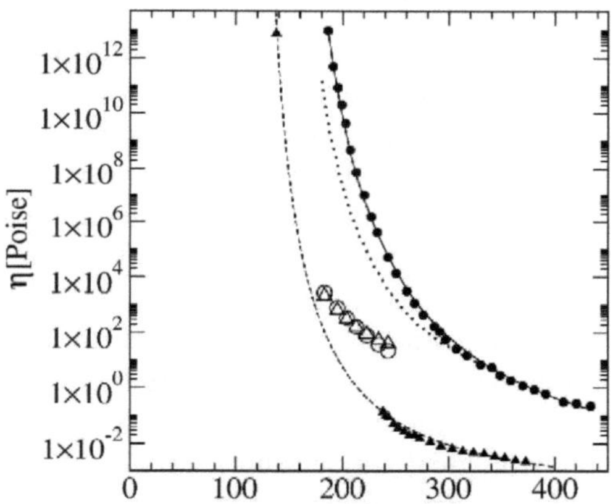

Figure 7. Temperature-dependent internal viscosity of the protein labeled at the iron in MbCO together with the viscosity of solvents. Closed circles: pure glycerol [42], thin line: Vogel–Fulcher law as a subsidiary line. Dotted line: glycerol/water mixture (3/1 vol/vol) [43]. Closed triangles: pure water [44] and pure water at the glass temperature of 138 K, dashed line: modified Vogel–Fulcher law as a subsidiary line. Open circles and open squares: internal viscosity in MbCO in glycerol/water and in MbCO crystals, respectively.

mean square displacement of the iron atom. The contribution of the different mode of motions can not be separated. In contrast, the contribution of each phonon mode i to the mean square displacement $\langle x_i^2(T) \rangle$ can be separately calculated from the density of phonon states. The total $\langle x_v^2(T) \rangle$ is dominated by contributions from modes below 3 meV, which have a wavelength longer than 27.5 Å. As stated in Chapter 3 these are acoustic and quasi acoustic and should depend also on the solvent of the sample. As a consequence the slope $\langle x_v^2 \rangle$ versus T differs by 10% for the crystal sample and the sample with glycerol. Another consequence of the different environments shows up in the T_c values of the samples. While the dynamical transition is at about 180 K in the crystal sample it becomes 210 K in the glycerol/water sample. Note that the glass transition of the glycerol/water mixture is about 170 K (compare Figure 7).

The protein specific motions have been explained by a two-state model where the molecule can be in a rigid and in a flexible state. A similar model has been used to explain the protein dynamics as measured by incoherent neutron scattering [38–40]. Also in this approach the more flexible state is populated according to Boltzmann statistics. A difference to our model is the fact that undamped motions occur in the flexible state while we use strongly overdamped motions in order to explain the quasi-elastic absorption. It should be mentioned in passing that one may use other assumptions for $p_t(T)$. Such investigations are in progress. New experiments supply a great evidence that for many processes within a protein the enthalpy needed to go from the rigid into a flexible state is connected with the fluctuation of the water dipoles. The fluctuations in the flexible state may be entropy driven [41].

The slope c in the flexible state is given in Table I for MbCO in both solvents. If one assumes the Boltzmann model for $p_t(T)$, an occupation of the transition state of 99% can be reached at about 350 K and 435 K for the crystal and glycerol samples, respectively. A further investigation in the upper temperature regime is necessary. However, there are clear limitations. Let us compare the mean square displacements at 245 K in the crystal and the glycerol sample. While in the crystals the $\langle x_t^2 \rangle$ value follows the theoretical curve there is a significant discrepancy between the experimental value and the theory in glycerol. Here, the free diffusion of the whole myoglobin molecule is no longer negligible at this temperature while free diffusion is still absent in the crystals. Crystals are the better choice to perform investigation at physiological temperatures. Finally, the available temperature window is limited by the heat denaturation of the protein.

In the data analysis we have used the fact that the phonon density of states spectrum is practically independent of temperature. Therefore, only few degrees of motional freedom can be responsible for the protein specific motions. According to [36] only a small subspace of about 5% of the normal modes is necessary to represent motions which connect different conformational substates.

Figure 7 shows striking differences between the internal viscosity of myoglobin compared to the viscosity of pure glycerol [42], a 3/1 (vol/vol) glycerol/water mixture [43] and pure water [44, 45]. The internal viscosity of myoglobin labeled at the position of the iron increase much slower with decreasing temperatures than glass forming liquids. At the dynamical transition T_c the viscosity is many orders of magnitude below 10^{13} poise, which characterizes a glass transition. However, the estimation of internal viscosity used here is rather crude and strongly dependent on the model. There are a number of other attempts to discuss protein viscosity. Hence the hydrodynamic radius introduced by Pastor and Karplus was one approach to parameterize the problem [37]. Go and coworkers [46] found by comparative normal mode analysis and molecular dynamics calculations of melittin that solvent water produces many local minima with low-energy barriers within a potential energy surface of the simulation obtained in vacuum. In addition, the low frequency normal modes become overdamped. In our case, the correspondent modes, which couple to the iron in myoglobin, produce the broad lines visible at higher temperatures in Mössbauer absorption (see Figure 4b). The friction of each iron mode depends on the degree of coupling of this mode to atoms, which are in contact with the solvent. In our model we collect these modes into one single mode with frequency ω_t yielding a mean square displacement $\langle x_t^2 \rangle$ in the flexible state. Further work is necessary for a better understanding of the correlation of the dynamical transition temperature and a glass transition.

6. Conclusions

Protein dynamics have been studied by Mössbauer spectroscopy for decades. Nevertheless, there are still many open questions. A substantial progress came from the Mössbauer spectroscopy with synchrotron radiation. Especially the phonon-assisted Mössbauer effect is of great importance to separate protein specific motions from solid state vibrations in proteins. As the result, the functional important motions in the physiological temperature regime can now be studied in more detail. In future, nuclear forward scattering may lead to significant progresses if it becomes possible to extend the time window to times as short as 1 ns without intolerable loss in intensity.

References

1. Antonini, E. and Brunori, M., In: *Hemoglobin and Myoglobin in Their Interactions with Ligands*, Horth-Holland Publishing Company, Amsterdam and London, 1971.
2. Brunori, M., *Trends Biochem. Sci.* **26** (2001), 21.
3. Frauenfelder, H., McMahon, B. H., Austin, R. H., Chu, K. and Groves, J. T., *Proc. Natl. Acad. Sci. USA* **98** (2001), 2370.
4. Kendrew, J. C. and Parrish, R. G., *Proc. R. Soc. A* **238** (1956), 305.
5. Kendrew, J. C., Dickerson, R. E., Strandberg, B. E., Hart, R. G., Davies, D. R., Phillips, D. C. and Shore, V. C., *Nature* **185** (1960), 422.
6. Vojtechovsky, J., Chu, K., Berendzen, J., Sweet, R. M. and Schlichting, I., *Biophys. J.* **77** (1999), 2153.
7. Kachalova, G. S., Popov, A. N. and Bartunik, H. D., *Science* **284** (1999), 473.
8. Srajer, V., Ren, Z., Teng, T., Schmidt, M., Ursby, T., Bourgeois, D., Pradervand, C., Schildkamp, W., Wulff, M. and Moffat, K., *Biochemistry* **40** (2001), 13802.
9. Ostermann, A., Tanaka, I., Engler, N., Niimura, N. and Parak, F. G., *Biophys. Chem.* **95** (2002), 183.
10. Parak, F. G. and Nienhaus, G. U., *Eur. J. Chem. Phys. Phys. Chem.* **3** (2002), 249.
11. Ostermann, A., Waschipky, R., Parak, F. G. and Nienhaus, G. U., *Nature* **404** (2000), 205.
12. Parak, F. and Formanek, H., *Acta Cryst. A* **27** (1971), 573.
13. Berg, A. I., Knox, P. P., Kononenko, A. A., Frolov, E. N., Chrymova, I. N., Rubin, A. B., Lichtenstein, G. I., Goldanskii, V. I., Parak, F., Buckel, M. and Mössbauer, R. L., *Molecularnaja Biologija* **13** (1979), 81.
14. Keller, H. and Debrunner, P. G., *Phys. Rev. Lett.* **45** (1980), 68.
15. Parak, F., *J. de Physique* **41** (1980), C1.
16. Parak, F., Frolov, E. N., Kononenko, A. A., Mössbauer, R. L., Goldanskii, V. I. and Rubin, A. B., *FEBS Lett.* **117** (1980), 368.

17. Parak, F., Finck, P., Kucheida, D. and Mössbauer, R. L., *Hyp. Interact.* **10** (1981), 1075.
18. Parak, F., Knapp, E. W. and Kucheida, D., *J. Mol. Biol.* **161** (1982), 177.
19. Knapp, E. W., Fischer, S. F. and Parak, F., *J. Phys. Chem.* **86** (1982), 5042.
20. Bauminger, E. R., Cohen, S. G., Nowik, I., Ofer, S. and Yariv, J., *Proc. Natl. Acad. Sci. USA* **80** (1983), 736.
21. Nowik, I., Bauminger, E. R., Cohen, S. G. and Ofer, S., *Phys. Rev. Lett.* **50** (1983), 1528.
22. Knapp, E. W. and Parak, F., *J. Chem. Phys.* **78** (1983), 4701.
23. Mayo, K. H., Parak, F. and Mössbauer, R. L., *Phys. Lett. A* **82** (1981), 468.
24. Cohen, S. G., Bauminger, E. R., Nowik, I., Ofer, S. and Yariv, J., *Phys. Rev. Lett.* **46** (1981), 1244.
25. Cohen, S. G., Bauminger, E. R., Nowik, I., Ofer, S. and Yariv, J., In: R. H. Sarma (ed), *Proceedings of the Second SUNYA Conversation in the Discipline Biomolecular Stereodynamics*, Adenine Press, New York, 1982, p. 299.
26. Gerdau, E., Rüffer, R., Winkler, H., Tolksdorf, W., Klages, C. P. and Hannon, J. P., *Phys. Rev. Lett.* **54** (1985), 835.
27. van Bürck, U., Siddons, D. P., Hastings, J. B., Bergmann, U. and Hollatz, R., *Phys. Rev. B* **46** (1992), 6207.
28. Smirnov, G. V., *Hyp. Interact.* **97/98** (1996), 551.
29. Parak, F. and Achterhold, K., *Hyp. Interact.* **123/124** (1999), 825.
30. Keppler, C., Achterhold, K., Ostermann, A., van Bürck, U., Chumakov, A. I., Rüffer, R., Sturhahn, W., Alp, E. E. and Parak, F. G., *Eur. Biophys. J.* **29** (2000), 146.
31. Chumakov, A. I., Rüffer, R., Baron, A. Q. R., Grünsteudel, H. and Grünsteudel, H. F., *Phys. Rev. B.* **54** (1996), R9596.
32. Achterhold, K., Keppler, C., van Bürck, U., Potzel, W., Schindelmann, P., Knapp, E. W., Melchers, B., Chumakov, A. I., Baron, A. Q. R., Rüffer, R. and Parak, F. G., *Eur. Biophys. J.* **25** (1996), 43.
33. Achterhold, K., Keppler, C., Ostermann, A., van Bürck, U., Sturhahn, W., Alp, E. E. and Parak, F. G., *Phys. Rev. E* **65** (2002), 051916.
34. Edwards, C., Palmer, S. B., Emsley, P., Helliwell, J. R., Glover, I. D., Harris, G. W. and Moss, D. S., *Acta Cryst. A* **46** (1990), 315.
35. Melchers, B., Knapp, E. W., Parak, F., Cordone, L., Cupane, A. and Leone, M., *Biophys. J.* **70** (1996), 2092.
36. Kitao, A. and Go, N., *Current Opinion in Structural Biology* **9** (1999), 164.
37. Pastor, R. W. and Karplus, M., *J. Phys. Chem.* **92** (1988), 2636.
38. Bicout, D. J. and Zaccai, G., *Biophys. J.* **80** (2001), 115.
39. Zaccai, G., *Journal of Physics and Chemistry of Solids* **60** (1999), 1291.
40. Zaccai, G., *Science* **288** (2000), 1604.
41. Fenimore, P. W., Frauenfelder, H., McMahon, B. H. and Parak, F. G., *Proc. Natl. Acad. Sci. USA* **99** (2002), 16047.
42. Huck, J., Bondeau, A., Noyel, G. and Jorat, L., *IEEE Trans. Electr. Insul.* **23** (1988), 615.
43. Leyser, H., Physik-Department E13, 1994.
44. Hallet, J., *Proc. Phys. Soc.* **82** (1963), 1046.
45. Bondarenko, N. F., Zheleznyi, B. V., Osipov, Y. A. and Ostapenko, N. S., *Colloid J. USSR* **39** (1977), 207.
46. Kitao, A., Hirata, F. and Go, N., *Chemical Phys.* **158** (1991), 447.

Hyperfine Interactions **144/145:** 223–230, 2002.

Exchange-Driven Magnetic Anomalies in Fe–Zr–B-Based Nanocomposites

W. A. STEER[1], G. D. FORSTER[1], I. P. PARKIN[2], M. A. ROBERTS[3] and Q. A. PANKHURST[1,*]
[1]*Department of Physics and Astronomy, University College London, Gower Street, London WC1E 6BT, England*
[2]*Department of Chemistry, University College London, 20 Gordon Street, London WC1H 0AJ, England*
[3]*CLRC Daresbury Laboratory, Warrington, Cheshire WA4 4AD, England*

Abstract. A series of amorphous Fe–Zr–B-based nanocomposites have been prepared by borohydride chemical reduction. The structural, magnetic and crystallisation properties of these materials have been investigated by Mössbauer spectroscopy, differential scanning calorimetry, X-ray diffraction, EXAFS and SQUID magnetometry. Nanocrystals of iron and iron boride form at ca. 470 °C, while the zirconium remains in an amorphous matrix until the samples are heated above 600 °C, after which crystalline zirconia is formed. No intermetallic iron-zirconium phases are observed, implying that in the as made state the materials comprise nanocomposites of amorphous iron-boron and amorphous zirconium oxide. Unusually large magnetic coercivities are observed, due to interfacial exchange pinning between the magnetically soft Fe–B nanoparticles and a surface passivation layer. Further magnetic anomalies observed in the heat treated samples indicate the influence of exchange interactions between nanocrystalline grains in an amorphous and insulating matrix.

Key words: magneto-nanocomposites, magnetic interactions, nanomagnetism.

1. Introduction

Following important discoveries in the late 1980's, there has been much intensive research into the effects of nanocrystalline structure on the magnetic properties of iron-based materials. Both the magnetically hardest ($Fe_{79}Nd_7B_9$) and softest ($Fe_{79}Zr_7B_9$) materials currently known have a nanocrystalline structure, and their respective coercivities span seven orders of magnitude [1]. In both classes of materials exchange interactions are at the heart of their extreme magnetic properties: in the hard magnets, exchange between the Nd–Fe–B and soft Fe grains, and in the soft magnets, exchange between neighbouring ferromagnetic grains, mediated by an amorphous matrix.

By far the most common technique for creating these specialised materials is the partial crystallisation of amorphous melt-spun ribbon precursors. Although highly successful, this approach has some significant drawbacks for processing: ribbons

* Corresponding author. E-mail: q.pankhurst@ucl.ac.uk.

are typically only 25 μm thick, limited to a maximum of around 0.1 m wide, and are often very brittle. Melt-spinning has limitations with regard to elemental compositions: the requirement of having a single-phase melt limits the available compositions to near-eutectic mixes. Furthermore, melt-spinning deals exclusively with metallic alloys, and as such cannot be used to form the more general class of magneto-nanocomposite materials, such as those comprising both metallic and insulating nanoparticles within a common matrix.

In this context a promising alternative method for creating amorphous precursor alloys, with or without an incorporated insulating matrix, is that of chemical reduction and precipitation [2, 3]. For example, the addition of sodium borohydride solution to a vigorously-stirred solution of iron and zirconium sulphates can result in the precipitation of a structurally amorphous black powder containing iron, zirconium and boron [4, 5]. In this paper we describe investigations of the thermally-induced crystallisation behaviour of a selection of chemically-precipitated Fe–Zr–B samples, and present some preliminary results regarding their exchange-driven magnetic properties.

2. Experimental

A series of Fe–Zr–B based amorphous and nanocrystalline materials was prepared by chemical reduction synthesis, using standard methods described in detail elsewhere [4, 5]. Both aerobic and anaerobic methods were used, the latter involving Schlenk lines and inert atmospheres to minimise exposure to air during synthesis. The products were filtered and partially dried, and then were stored under argon in a glove box. This procedure limits the oxidation of the product to the slow growth of a surface passivation layer [6].

The Fe, Zr and B elemental compositions of each sample were determined by atomic-emission spectroscopy. ^{57}Fe Mössbauer spectra were recorded at room temperature on a conventional transmission spectrometer. Spectra were folded to remove baseline curvature, and calibrated with respect to α-Fe. Differential scanning calorimetry (DSC) measurements were made under argon in stainless steel high-pressure crucibles using a Shimadzu DSC-50, at a heating rate of 10 K/min. Magnetisation loops were recorded at 250 K on a Quantum Design MPMS SQUID magnetometer.

Given the practical difficulties of doing so, the samples were not subjected to an X-ray diffraction (XRD) and EXAFS study during heating. Instead, quartz XRD capillary tubes were loaded with a few milligrams of sample inside the glove box, then removed from the box and sealed by melting in such a way as to maintain a vacuum/argon atmosphere inside the tubes. The content of these tubes was then heat-treated in situ in a tube furnace, heating at 10 K/min (matching the DSC) to a series of set temperatures, up to 750 °C. The tubes were then withdrawn and left to cool in air. Synchrotron radiation XRD measurements were performed using a curved image plate camera [7] on Daresbury Laboratory's station 9.1.

Zr-edge fluorescence-EXAFS experiments were performed on Daresbury's station 9.3, using the same capillary-mounted as made and heat treated samples.

One useful extension to the XRD technique that was used with good effect was 'resonant scattering XRD'. Pairs of diffraction exposures were made at two wavelengths, one below the Zr K-edge at 0.68877 Å and the other at slightly lower energy. Owing to electron resonance effects (variation in f′, because of the energy dependence of f′) using a wavelength very close to and further away from the absorption edge elucidated differential changes in diffraction intensity arising from phases containing a significant fraction of zirconium in their structure relative to the iron peaks, despite increased absorption and fluorescence effects. This approach reveals useful information about the behaviour of the zirconium, even without any prior information, let alone models, of the phases it might be in.

3. Results and discussion

Room temperature Mössbauer spectra are shown in Figure 1 for two representative samples: aerobically prepared $Fe_{59}Zr_7B_{34}$, and anaerobically prepared $Fe_{57}Zr_9B_{34}$. These spectra were fitted using Voigtian profiles to model the distributions in hyperfine parameters associated with the main magnetically split sextet and quadrupole split doublet components, and Lorentzian profiles for a narrow-line sextet due to crystalline α-Fe (see Table I).

As has been discussed previously [4], the main difference between the aerobic and anaerobic preparation is the presence of an air-sensitive Fe(II) complex in the anaerobic samples, a phase which appears not to be incorporated into the bulk nanocomposite and is a bystander with regard to the magnetic properties of the

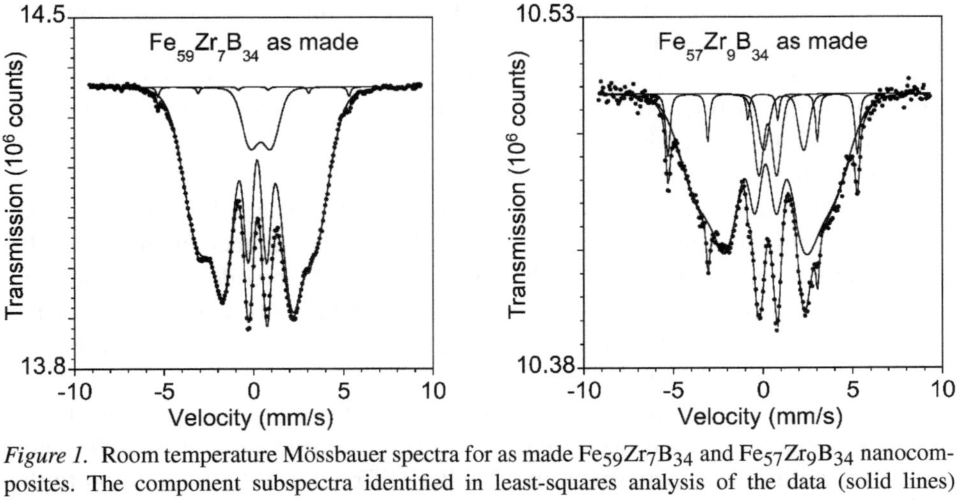

Figure 1. Room temperature Mössbauer spectra for as made $Fe_{59}Zr_7B_{34}$ and $Fe_{57}Zr_9B_{34}$ nanocomposites. The component subspectra identified in least-squares analysis of the data (solid lines) indicate the presence of several iron-bearing phases: amorphous Fe–B or Fe–Zr–B alloy; amorphous Fe oxyhydroxide; crystalline α-Fe; and, for the $Fe_{57}Zr_9B_{34}$ sample, a ferrous iron complex.

Table I. Mössbauer hyperfine parameters obtained from least-squares fitting of the room temperature spectra of the as made $Fe_{59}Zr_7B_{34}$ and $Fe_{57}Zr_9B_{34}$ nanocomposites in Figure 1. The parameters listed include the hyperfine field B_{hf}, isomer shift δ, quadrupole splitting or quadrupole shift Δ, and the Gaussian standard deviation σ associated with distributions in these parameters. The number in parentheses is the error on the last digit

B_{hf} (kOe)	σ_B (kOe)	δ (mm/s)	σ_δ (mm/s)	Δ (mm/s)	σ_Δ	Area	Phase
			Aerobic $Fe_{59}Zr_7B_{34}$				
–	–	0.40(1)	–	1.14(2)	0.8(1)	9(1)	Amorphous Fe oxyhydroxide
331(2)	–	–	–	–	–	1(1)	α-Fe
196(2)	50(2)	0.18(1)	0.16(1)	−0.03(1)	–	90(1)	Amorphous Fe–B or Fe–Zr–B alloy
			Anaerobic $Fe_{57}Zr_9B_{34}$				
–	–	0.31(5)	–	0.92(9)	0.3(1)	6(2)	Amorphous Fe oxyhydroxide
–	–	1.17(5)	0.10(5)	2.31(11)	0.5(1)	7(1)	Fe(II) complex
327(1)	3(2)	–	–	–	–	7(1)	α-Fe
220(10)	69(20)	0.14(3)	0.22(5)	−0.01(2)	–	80(2)	Amorphous Fe–B or Fe–Zr–B alloy

material. A more significant phase regarding the magnetic properties (see later) is the disordered Fe(III) oxyhydroxide, which appears as a ferric doublet. In recent work we have established that this is a surface coating phase arising from inherent passivation of the nanoparticles [6].

Figure 2 shows the crystallisation behaviour in the $Fe_{59}Zr_7B_{34}$ sample as observed by X-ray diffraction. The DSC curve shown for reference highlights the main crystallisation exotherm at about 470 °C. Working from the front of the plot (as-made) to the back (post-750 °C heat-treatment) the sample goes from an amorphous to crystalline state. By 500 °C the iron pattern is substantially developed, and various other minority phases are also evident. In the 750 °C trace the nature of the minority phases has changed somewhat, with some of the broader peaks resolving into discrete sharp peaks. All of these patterns give the lattice parameter for the iron crystals as 2.866(1) Å, which is indistinguishable from the accepted value for pure α-iron within experimental error.

Knowing that the diffractograms of annealed samples show multi-phase materials, considerable attempts were made to identify components by search-match procedures with library patterns in the Powder Diffraction File. In addition to the iron, some good or reasonable matches were found to FeB phases, particularly Fe_2B. Even so, this procedure was not entirely successful – many peaks remained

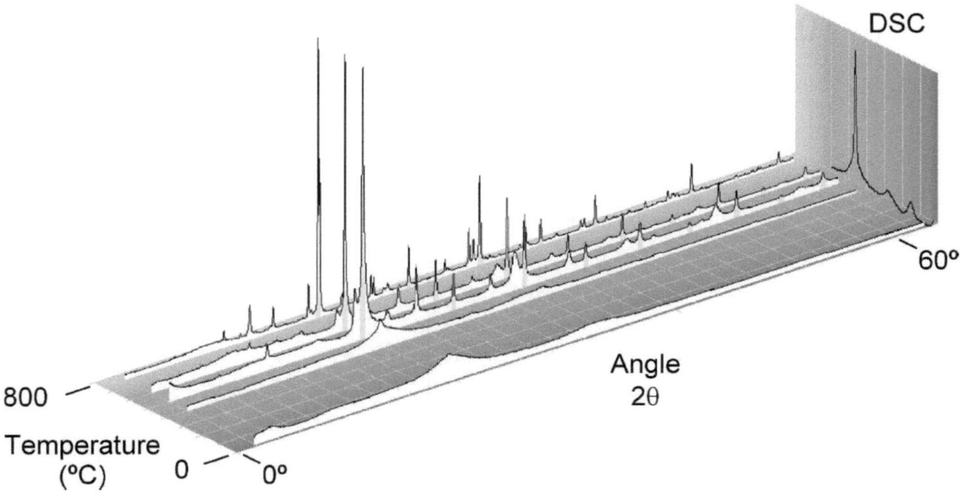

Figure 2. Crystallisation of chemically-precipitated $Fe_{59}Zr_7B_{34}$ nanocomposite, as observed by X-ray diffraction ($\lambda = 0.6920$ Å) and differential scanning calorimetry (DSC).

unidentified, and some of the matches were dubious. This might be explained if we are observing strained or highly substituted phases.

For the resonant XRD studies, pairs of diffraction patterns were normalised on iron-peak intensity, then the other peaks inspected. It proved computationally very difficult to subtract one plot from another, owing partly to different sampling intervals, but mainly due to the problem of defining the background sufficiently accurately. Instead, the pairs of traces were turned through 90° so that the 2θ axis was vertical and then a stereoscopic viewing technique was adopted (see Figure 3 for data on $Fe_{57}Zr_9B_{34}$). In this way the human brain picks out differential peak heights quite naturally by making them appear to lean up from, or sink behind, the plane of the page.

It was found that diffractograms from samples heated to 750 °C showed some peaks with differential heights, while those from 600 °C and below did not – though these had a differential change in low-angle background. This indicates that the zirconium only exists in any crystalline form in the 750 °C samples. We note however that minor substitutions of zirconium in other crystals would not be detected with this technique.

Even in the raw data, the EXAFS spectra of both $Fe_{59}Zr_7B_{34}$ and $Fe_{57}Zr_9B_{34}$ (Figure 4) show that a major change in the local zirconium environment occurs between samples heated to 600 °C and below, and 660 °C and above. Preliminary analysis of the $Fe_{57}Zr_9B_{34}$ data shows that in the low-temperature state, the zirconium is surrounded by a nearest-neighbour shell of 4 oxygen atoms at 2.14 Å in a disordered state, whereas at 660 °C and above it has crystallised to a bulk phase with 4 oxygens at 2.11 Å, 4 oxygens at 2.36 Å and 12 zirconium atoms at 3.64 Å.

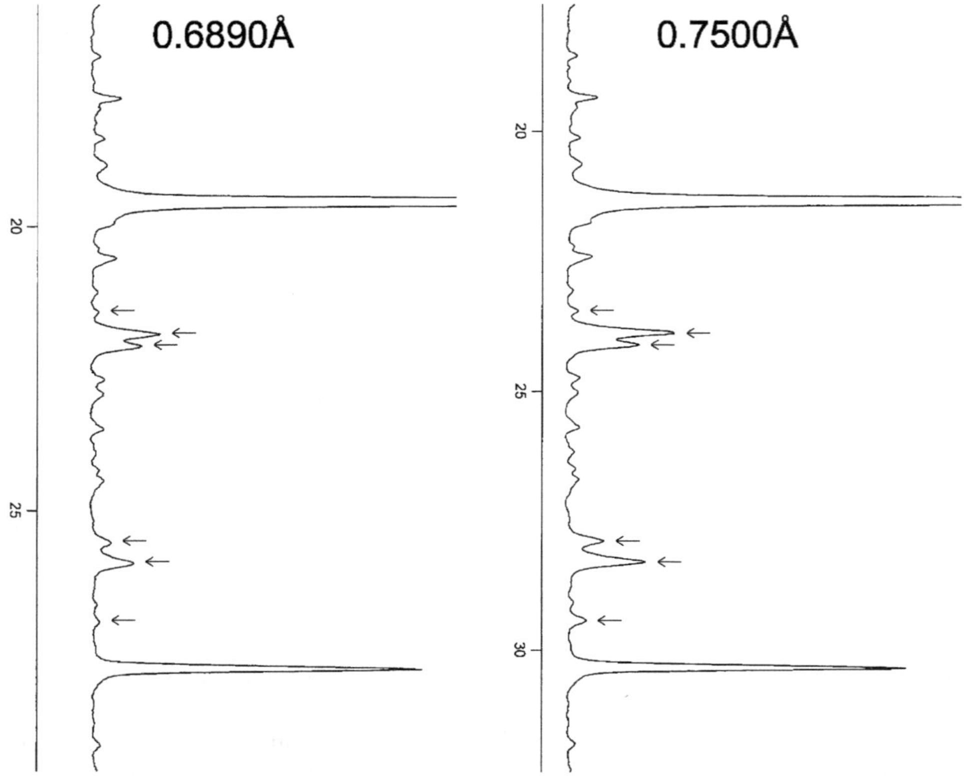

Figure 3. Comparison of two X-ray diffraction patterns for nanocomposite $Fe_{57}Zr_9B_{34}$ following heating to 750 °C, showing the effect of resonant scattering. The pattern on the left was exposed at 0.6890 Å, on the right, at 0.7500 Å. Note the marked change in intensity of the indicated peaks. [To view stereoscopically: hold finger approximately half way between the diagram and face, look at the finger. Slowly move the finger back and forth until the pair of graph traces in the background fuse into one (or three). Remove the finger and allow the eyes to refocus. While the iron peaks will still appear to be flat, the arrowed peaks will appear to slant up from the plane of the page, owing to their differential height.]

The latter is characteristic of a tetragonal ZrO_2 phase (ICSD ref. 31983), which was subsequently simulated in XRD and found to match the XRD data.

Magnetic hysteresis data on the aerobic $Fe_{59}Zr_7B_{34}$ samples measured at 250 K in fields of up to 10 kOe are shown in Table II. A striking and unusual feature, apparent for the as made sample, is the coercivity, which is much higher than the few Oe that would be expected for an amorphous soft ferromagnetic material. This high value cannot be attributed to the very small 1% fraction of α-Fe present in the sample (if it were the hysteresis loop would have a characteristic 'wasp-waisted' or 'pot-bellied' shape, which it does not), and has recently been proved to be due to exchange pinning of the soft ferromagnetic nanoparticles by the surface passivation layer [6].

Table II. Magnetic hysteresis parameters of as made and annealed samples of $Fe_{59}Zr_7B_{34}$ nanocomposite, recorded at 250 K. The parameters listed are the coercivity H_c, remanence σ_r and maximum magnetisation σ_{max} measured in an applied field of 10 kOe. The number in parentheses is the error on the last digit

Sample	H_c (Oe)	σ_r (emu/g)	σ_{max} (emu/g)
As made	141 (5)	5.3 (5)	77.6 (5)
400 °C annealed	131 (5)	8.7 (5)	100.0 (5)
500 °C annealed	335 (5)	13.8 (5)	94.7 (5)

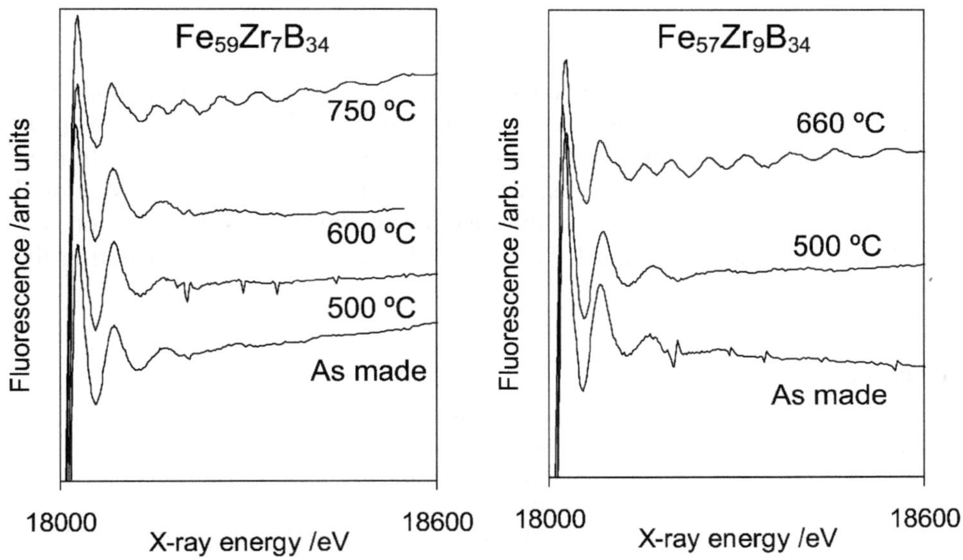

Figure 4. Zr-edge EXAFS data for $Fe_{59}Zr_7B_{34}$ and $Fe_{57}Zr_9B_{34}$ nanocomposites as a function of annealing temperature. The increased oscillation frequencies evident in the data for the 660 °C and 750 °C samples indicates a change in the zirconium environment from disordered to crystalline.

Interesting features are also apparent in the annealed samples. After 400 °C annealing, the coercivity falls while both the remanence and the saturation magnetisation increase. After 500 °C annealing, the coercivity rises again, while the remanence rises further and the saturation magnetisation falls slightly. These effects are reminiscent of those seen in the FINEMET class of ternary alloys [1], where at intermediate annealing temperatures the exchange coupling between nanoscale grains of α-Fe is mediated through an amorphous matrix, leading to ultra-soft magnetic properties.

4. Conclusions

A combination of Mössbauer effect, DSC, XRD, EXAFS and SQUID magnetometry experiments were performed on both as made and heat treated Fe–Zr–B nanocomposites prepared by borohydride reduction. These showed that the main crystallisation exotherm is at 470 °C, and that below this temperature, the samples remain mainly amorphous. At the exotherm, iron and some iron borides crystallise. A crystalline ZrO_2 phase is only seen in samples heated to 660 °C and above; while at the same temperatures there is no evidence of any FeZr intermetallic phases. An unusually large magnetic coercivity in the as made samples indicates exchange pinning behaviour, while the evolution of coercivity, remanence and saturation magnetisation with annealing temperature indicate that other exchange-driven magnetic properties are activated in the heat-treated materials.

Acknowledgements

We thank Neil Cohen and Marianne Odlyha of the ULIRS Thermal Methods Service for the DSC analyses; Ian Harvey and Fred Mosselmans for their help on station 9.3 at Daresbury Laboratory; and David Gleeson of the Royal Institution for assistance with EXAFS analysis. W.A.S. also thanks the Wohlfarth Memorial Fund, the C.R. Barber Trust Fund, the Institute of Physics CMMP division, the UCL Graduate School, and the UCL Physics Dept, for their support during the course of this work.

References

1. Hernando, A., *J. Phys. Condens. Matter* **11** (1999), 9455.
2. Linderoth, S. and Mørup, S., *J. Appl. Phys.* **69** (1991), 5256.
3. Wells, S., Charles, S. W., Mørup, S., Linderoth, S., van Wonterghem, J., Larsen, J. and Madsen, M. B., *J. Phys. Condens. Matter* **1** (1989), 8199.
4. Forster, G. D., Fernández Barquín, L., Pankhurst, Q. A. and Parkin, I. P., *J. Non-Cryst. Solids* **244** (1999), 44.
5. Forster, G. D., Fernández Barquín, L., Bilsborrow, R. L., Pankhurst, Q. A., Parkin, I. P. and Steer, W. A., *J. Mat. Chem.* **9** (1999), 2537.
6. Pankhurst, Q. A., Fernández Barquín, L. and Yédra Martínez, A., unpublished.
7. Roberts, M. A., Finney J. L. and Bushnell-Wye, G., *Mater. Sci. Forum* **278–281** (1998), 318.

Hyperfine Interactions **144/145:** 231–238, 2002.

Investigations on Structure and Transformation of Some Organometallic Compounds in Solution

SZILVIA KÁRPÁTI[1], GYÖRGY VANKÓ[1,*], ANDREA DEÁK[2], ALAJOS KÁLMÁN[2] and ATTILA VÉRTES[1]
[1]*Research Group of Nuclear Techniques in Structural Chemistry, Hungarian Academy of Sciences at Eötvös Loránd University, P.O. Box 32, H-1518 Budapest, Hungary*
[2]*Chemical Research Center of the Hungarian Academy of Sciences, Institute of Chemistry, P.O. Box 17, H-1525 Budapest, Hungary*

Abstract. Earlier NMR investigations revealed that the tetrameric $[Me_3SnL]_4$ compound [Me = methyl group, CH_3; L = cupferronato anion, $PhN_2(O)O^-$] falls apart to monomeric units in chloroform solution. Comparison of Mössbauer spectra of the crystalline and the dissolved form indicated a minor change in structure, but not clearly. From the four possible structures in solution, two could be excluded by the help of partial quadrupole splitting (PQS) model calculations. The final decision between the remaining two possible monomeric structures was made by IR spectroscopy. The authors cite and, using PQS calculations, explain the case of another compound, $[Me_2N(CH_2)_3]_2SnF_2$, the two structural isomers of which had been found to have very similar spectra in spite of a seemingly great difference between the positions of the ligands.

Key words: partial quadrupole splitting, point charge model, organotin(IV) cation, cupferronato ligand, coordination chemistry.

1. Introduction

Cupferronato anion ($PhN_2(O)O^-$) forms complexes with most of the metal ions; it generally coordinates as a bidentate ligand forming chelating coordination patterns. We studied novel compounds consisting of organotin cations and cupferronato ligands by Mössbauer spectroscopy, which are described by the general formula $[R_nSnL_{4-n}]_m$ (R: methyl, Me or phenyl, Ph group; L: $PhN_2(O)O^-$, cupferronato anion). In these compounds, besides the usual chelating ones, the cupferronato ligand also forms bridging – simple and triconnective (bridging chelating) – bonding patterns. The amazing variety of structures (marked by different coordinating modes of the anion and the variation of the coordination number of tin between 5 and 8) makes these compounds very interesting targets for structural studies. As intriguing structural transformations had been expected, we decided to study the structures of the complexes in solution. Therefore, we carried out Mössbauer spectroscopic measurements on the crystalline compounds and their

* Now at the European Synchrotron Radiation Facility, BP 220, F-38043 Grenoble, France.

quick-frozen solutions. Independent NMR-spectroscopic data, model calculations and FTIR spectroscopic investigations also helped the interpretation of the Mössbauer data. Besides structural changes, many interesting chemical transformations of the dissolved complexes were followed using Mössbauer spectroscopy (e.g., reactions with Lewis bases and steps of complex formations), which demonstrate that Mössbauer spectroscopy can be a very efficient tool for preparative chemists. Although this work was fully covered in the talk, due to the large volume of the work we can only focus on a small part of the work here, namely some general aspects on the use of the PQS approach, and on the determination of the structure of $[Me_3SnL]_4$ in chloroform solution. For the complete work the reader is directed to [1].

2. Discussion

It has been widely recognized that Mössbauer spectroscopy can be a useful method in coordination chemistry because of its sensitivity to the structure in the ligand sphere. Furthermore, it is well known that *cis* and *trans* forms of octahedral complexes can be easily distinguished by Mössbauer spectroscopy, because the QS of the *cis* form is about twice of the *trans* form. With our first example, a *cis* $\Leftrightarrow$ *trans* transition, we would like to draw attention to a particular feature of the EFG of organometallic compounds.

An octahedrally ligated organotin compound, $[Me_2N(CH_2)_3]_2SnF_2$, contains two F–Sn, two N–Sn and two C–Sn bonds as seen in Figure 1. Although this compound has *trans* structure in solid as shown in Figure 1(a), it was observed (with NMR spectroscopy) to undergo a partial transformation to the *cis* form in chloroform solution (cf. Figure 1(b)) [2].

Nevertheless, Mössbauer analysis gave similar results on both the solid and the frozen chloroform solution: the measured quadrupole splitting values were $3.79\ \mathrm{mm\,s^{-1}}$ and $3.67\ \mathrm{mm\,s^{-1}}$, respectively, [3]. Not only that the change of QS is far less than expected, but – due to the relatively large linewidths of tin (0.8–0.9 $\mathrm{mm\,s^{-1}}$ in this case), and the standard deviation of fitted line positions ($\pm 0.04\ \mathrm{mm\,s^{-1}}$) – these values do not reflect any significant difference at all.

A quick model calculation using the partial quadrupole splitting (PQS) approach [4–7], which is often used to help the determination of the structure of tin complexes, reveals why the Mössbauer spectroscopy failed to distinguish between the two forms of the complex.

The PQS model, based on the point charge model, considers the ligands of a hybridisation mode as having additive and independent contributions to the EFG at the site of the Sn nucleus. Accordingly, the overall quadrupole splitting (Δ) is calculated by summing up the PQS contributions of the individual ligands that surround the tin ion. In our example, the folowing PQS values were used: $\{C\} = -1.03\ \mathrm{mm\,s^{-1}}$; $\{N\} = +0.01\ \mathrm{mm\,s^{-1}}$; $\{F\} = 0.00\ \mathrm{mm\,s^{-1}}$ [7, 8]. Now, the overall quadrupole splitting values calculated for the *cis* and *trans* forms (which notation

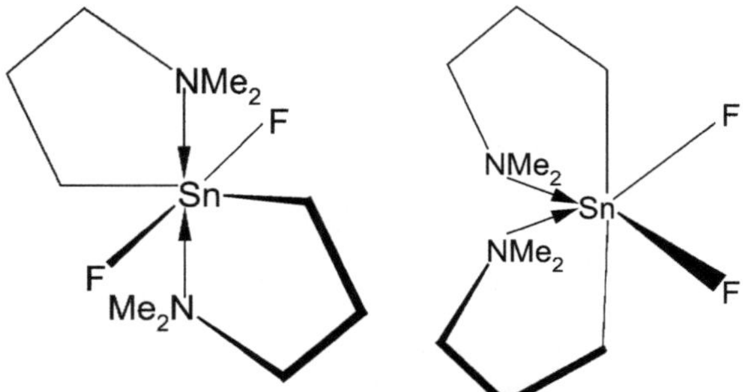

***trans*-F** ***cis*-F**

Figure 1. Structure of the *cis* and *trans* isomers of $[Me_2N(CH_2)_3]_2SnF_2$.

refers here to the relative positions of the halogen atoms) were found nearly the same ($\Delta \approx 4.1$ mm s^{-1}). The V_{zz} is described by same formula for both forms:

$$V_{zz} = 4\{CH_2\} - 2\{F\} - 2\{N\}.$$

Having another look at Figure 1, we can see the reason: there is no difference between the relative carbon positions, since they occupy *trans* positions in both cases. (If they occupied *cis* positions, the quadrupole splitting would be only about $\Delta \approx 2$ mm s^{-1} [9].) So by using the PQS model we can understand why the two forms are practically impossible to distinguish on the sole basis of their Mössbauer-spectra.

What we have learned from the above example is this: the EFG of tin ion in organometallic compounds is basically determined by the Sn–C bonds, when the other coordinated atoms are N, O or halogen. Consequently, in this type of complexes the quadrupole splitting resulted from the interaction of the tin and the ligands depends mainly on the position of the carbon atoms that are bonded directly to the tin atom.

On the other hand, as Mössbauer spectroscopy is quite sensitive to the relative position of the (bonding) carbon atoms in a complex, in our studies presented below, we were able to make good use of this sensitivity of Mössbauer spectroscopy to the relative position of the (bonding) carbon atoms, since the structural changes in solution were always accompanied with the displacement of the carbon atoms [1]. Such changes were reflected as remarkable differences between the spectra of the solid compound and its quick-frozen solution. As an example, we will discuss below the case of the tetrameric compound $[Me_3SnL]_4$ (Figure 2), whose structure showed the most radical change in solution of all the tin complexes studied.

Earlier NMR investigations revealed that the tetrameric complex $[Me_3SnL]_4$ dissociates to monomeric species in $CHCl_3$ solution [10]. For this monomer, the estimated Me–Sn–Me angle was about 110°, a value typical of *cis*-TBP coordination geometry [11].

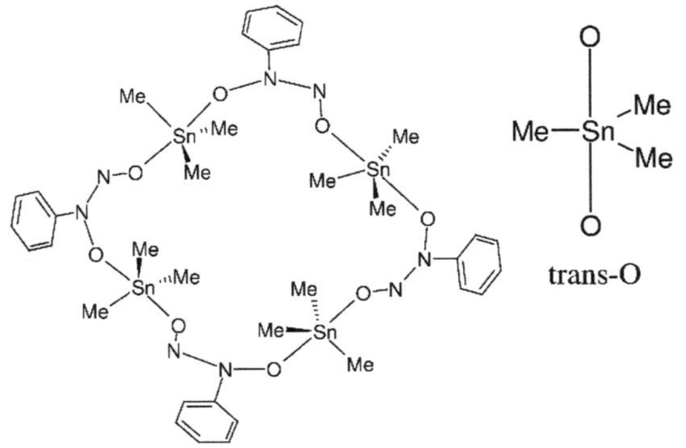

Figure 2. Molecular structure of the compound $[Me_3SnL]_4$. The local geometry of tin is trigonal bipyramidal.

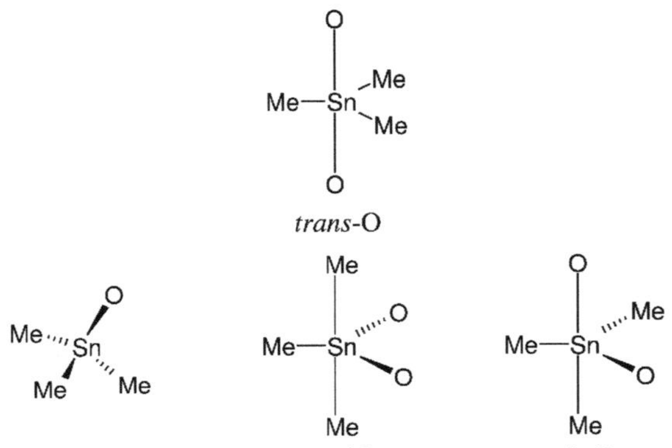

Figure 3. Possible stuctures of $[Me_3SnL]_4$ in solution.

However, the Mössbauer parameters did not vary substantially upon dissolution (the only change was that the quadrupole splitting increased from 3.55 to 3.66 mm s^{-1}), therefore, the change of the coordination sphere could neither be proved nor disaproved by Mössbauer spectroscopy.

So we carefully considered all the possible structures of the solute that could be imagined in solution. If the complex remained intact, then the local geometry around the tin atom should be (*trans*-O)$_2$SnMe$_3$-TBP, just like in the solid state (see Figure 2). (However, this possibility had been excluded by the NMR studies [11].) But if the tetrameric unit falls apart to monomers, then three different structures are conceivable: one with tetrahedral and two with trigonal bipyramidal (TBP) geometry (Figure 3). The fourfold coordination is only possible, if the cupferronato ligand is monodentate, which would be rather unexpected. The rest of the plausible structures in solution contains penta-coordinated tin with TBP geometry, in which the cupferronato anion is a bidentate chelating ligand. The relative arrangement of the ligands can be either (*mer*-Me$_3$)SnO$_2$-TBP (which, with both oxygen atoms of the cupferronato ligand occupying the equatorial position of the TBP ligand sphere, is named after the position of the three Me-groups), or $(cis\text{-O})_2SnMe_3$-TBP (where the oxygen atoms of the cupferronato ligand occupy an apical and an equatorial position in the coordination sphere). The $(trans\text{-O})_2SnMe_3$-TBP, $(mer\text{-Me}_3)SnO_2$-TBP, and $(cis\text{-O})_2SnMe_3$-TBP arrangements will be referred to as "*trans*-", "*mer*-", and "*cis*-TBP structures", respectively (see Figure 3).

To decide which of these structures is actually present in solution, we compared the quadrupole splitting data obtained on the solid and on the solution with typical values from the literature and with the results of PQS calculations.

In the case of TBP coordination geometry, the following calculations are to be made [5]:

$$\begin{aligned}
\left|\Delta[R_3Sn(\mathit{trans}\text{-OX})_2]\right| &= 4\{OX\}^{tba} - 3\{R\}^{tbe}, \\
\left|\Delta[R_3Sn(\mathit{cis}\text{-OX})_2]\right| &= \frac{2\{R\}^{tba} - 8\{R\}^{tbe} + 2\{OX\}^{tba} + 5\{OX\}^{tbe}}{\sqrt{13}}, \\
\left|\Delta[\mathit{mer}\text{-}R_3Sn(OX)_2]\right| &= \frac{-8\{R\}^{tba} - \{R\}^{tbe} + 7\{OX\}^{tbe}}{\sqrt{7}}.
\end{aligned}$$

In the above equations: $\{OX\}^{tba}$, etc. denote the PQS values of the bonding atom (e.g., in the given case the contribution of an oxygen donor atom in a general ligand OX to the quadrupole splitting) with the superscript showing the position of the ligand (*tba*: apical position of trigonal bipyramidal, *tbe*: equatorial position of TBP). By using PQS values most common for ligands with O donor atoms, we found that the estimated quadrupole splitting is unrealistically small. Therefore we used PQS values related to O atoms, that are part of a conjugated system (as is the case with our compounds). Then we obtained reasonable estimates. The results can be seen in Table I.

For the four-coordinated tin, the PQS model gives the following result:

$$\Delta = 2\{OX\}^{tetr} - 2\{R\}^{tetr} = 2.00\ \mathrm{mm\,s^{-1}}.$$

The PQS values used for these ligands are: $\{R\}^{tetr} = -1.37\ \mathrm{mm\,s^{-1}}$, $\{OX\}^{tetr} = -0.37\ \mathrm{mm\,s^{-1}}$ [12].

The small PQS values obtained for the tetrahedral and for the *cis*-TBP structures rule out these arrangements. Albeit the experimental Δ values follow the tendency predicted by the PQS model for a *trans* to *mer* transformation (cf. Table I), the difference is small and therefore we cannot exclude that only a slight alteration of the complex structure takes place in solution, which preserves the coordination pattern. Moreover, we cannot expect a better accuracy than 10% for this approximation, and the relatively close values calculated for the remaining *mer*-TBP and *trans*-TBP structures do not allow us to make an unambiguous decision.

Since these arrangements have a symmetry difference that IR spectroscopy is rather sensitive to, we compared the FTIR spectra of the solid compound and its solutions made with chloroform and *n*-heptane. The spectra, which are presented in Figure 4, reveal numerous differences. First, the Sn–C stretchings give a clear

Table I. Quadrupole splitting values estimated by the PQS method for molecules of the type $R_3Sn(OX)_2$ having a trigonal bipramidal structure

Structure	Δ [mm s^{-1}] (OX = O[a])	Δ [mm s^{-1}] (OX = bipyo[b])	Δ [mm s^{-1}] (O = acac[c])
trans-$(OX)_2$	2.55	3.23	3.27
cis-$(OX)_2$	1.74	2.17	2.25
mer-R_3	3.03	3.66	3.80

[a]O: typical ligand with O donor atom, [b]bipyo: 2,2′-bipyridil-N,N′-dioxide, [c]acac: acetil-acetonate. The following ligand PQS values were used for the calculations [5, 12, 4]: $\{R\}^{tba} = -0.94$ mm s^{-1}, $\{R\}^{tbe} = -1.13$ mm s^{-1}, $\{OX = O\}^{tba} = -0.21$ mm s^{-1}, $\{OX = O\}^{tbe} = -0.09$ mm s^{-1}, $\{OX = \text{bipyo}\}^{tba} = -0.04$ mm s^{-1}, $\{OX = \text{bipyo}\}^{tbe} = 0.15$ mm s^{-1}, $\{OX = \text{acac}\}^{tba} = -0.03$ mm s^{-1} and $\{OX = \text{acac}\}^{tbe} = 0.20$ mm s^{-1}.

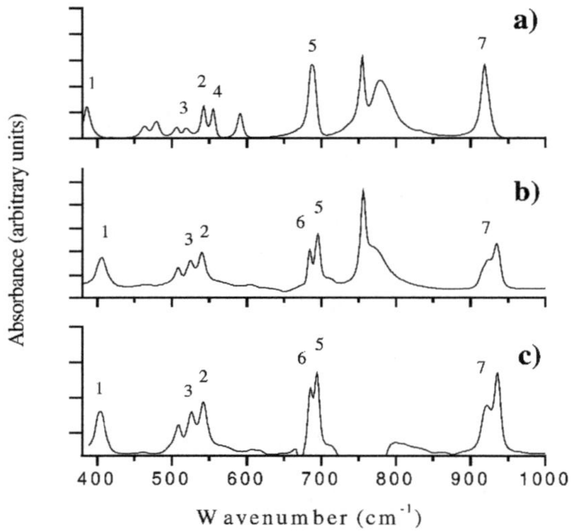

Figure 4. FTIR spectra of the complex $[Me_3SnL]_4$ in (a) solid form, (b) *n*-heptane solution and (c) chloroform solution. Assignments: 1 – ν(Sn–O) stretching, 2 – ν(Sn–C) stretching, 3 – $\nu_S(Sn–C_3)$, 4 – $\nu_{AS}(Sn–C_3)$, 5 – Ph, 6 – O–Sn–O bending and 7 – δ(ONNO). The absorption of the solvent can alter the spectra in the range 646–820 cm^{-1} and 700–800 cm^{-1} in the case of chloroform and *n*-heptane, respectively.

evidence of the altered symmetry. In the tetrameric compound, the three methyl groups are arranged symmetrically in a plane; therefore, due to the selection rules, the symmetric $\nu_S(SnC_3)$ vibration is Raman active and infrared inactive. Indeed, this is a high-intensity feature of the Raman spectra at 520 cm^{-1} [13], but it is also present in the IR spectra with low intensity due to deviations from perfect symmetry. The asymmetric $\nu_{AS}(SnC_3)$ mode, for which the selection rules are just the opposite, is expected to be about 30 cm^{-1} higher [14]. This mode is reflected by an intense band at 555 cm^{-1} in the IR spectra, but it is absent from the Raman spectra [13]. Both modes can be observed in the IR spectrum of the complex $[Me_3SnL]_4$ displayed in Figure 4, just below and above the band of the $\nu(SnC)$ stretching ($\approx$540 cm^{-1}). The latter serves as a good reference when we study the behavior of the $\nu(SnC_3)$ modes.

It is expected that the IR intensity of the $\nu_S(SnC_3)$ mode increases, and the band of the $\nu_{AS}(SnC_3)$ mode disappears upon breaking the threefold planar symmetry of the arrangement of the methyl groups [14, 15]. This is exactly what we can observe in the IR spectra of the solutions: the band at 520 cm^{-1} increases substantially, while the 555 cm^{-1} peak practically disappears.

Furthermore, the wave number of the Sn–O stretching became larger upon dissolution (solid: 387 cm^{-1} $\rightarrow$ in $CHCl_3$: 404 cm^{-1}, in n-heptane: 406 cm^{-1}), which indicates a variation of the Sn–O bonds. The latter values are close to those obtained with compounds containing a bidentate chelating cupferronato anion (400–401 cm^{-1}) [13, 16, 10]. This can be interpreted as a sign that the tetramer dissociated and the coordination of the cupferronato anion switched from bridging to chelating.

The appearance of the band at 680 cm^{-1} in the spectra of the solutions (typical of the O_1–Sn–O_2 bending mode of a chelating O_1O_2-η^2 coordination) as well as the variation of the δ(ONNO) mode at 930 cm^{-1} suggest that the tetramer dissociates in solution and the monomer formed has a *mer*-TBP structure.

3. Conclusions

Results of Mössbauer spectroscopic investigations, PQS calculations and IR spectroscopy showed that the tetrameric compound $[Me_3SnL]_4$ dissociates to monomeric species in chloroform solution, and this monomer has a *mer*-Me_3SnL structure with trigonal bipyramidal geometry.

In the case of the compound $[Me_2N(CH_2)_3]_2SnF_2$, the similarity between the Mössbauer spectra of the two isomers (*cis*-F and *trans*-F) could be interpreted by considering the results of PQS calculations. In fact, in this compound, the quadrupole splitting is practically determined by the contribution of the carbon atoms, which have the same position in both forms, and therefore they have very similar Mössbauer spectra.

References

1. Karpati, S. *et al.*, to be published.
2. Pieper, N., Klaus-Mrestani, C., Schürmann, M., Jurkschat, K., Biesemans, M., Verbruggen, I., Martins, J. C. and Willem, R., *Organometallics* **16** (1997), 1043.
3. Jurkschat, K. *et al.*, to be published.
4. Barbieri, R., Pellerito, L. and Huber, F., *Inorg. Chim. Acta* **30** (1978), L321.
5. Bancroft, G. M., Kumar Das, V. G., Sham, T. K. and Clark, M. G., *J. Chem. Soc. Dalton Trans.* (1976), 643.
6. Clark, M. G., Maddock, A. G. and Platt, R. H., *J. Chem. Soc. Dalton Trans.* (1972), 281.

7. Maddock, A., *Mössbauer Spectroscopy Principles and Applications*, Horwood, Chichester, 1997.
8. Jurkschat, K., Tzschach, A., Weichmann, H., Rajczy, P., Mostafa, M. A., Korecz, L. and Burger, K., *Inorg. Chim. Acta* **179** (1991), 83.
9. Davies, H. G. and Smith, P. J., In: G. Wilkinson, F. Gordon, A. Stone and E. W. Abel (eds), *Comprehensive Organometallic Chemistry*, Vol. 2, Pergamon Press, 1982, p. 523.
10. Deák, A., Venter, M., Kálmán, A., Párkányi, L., Radics, L. and Haiduc, I., *Eur. J. Inorg. Chem.* (2000), 127.
11. Deák, A., Radics, L., Kálmán, A., Párkányi, L. and Haiduc, I., *Eur. J. Inorg. Chem.* (2001), 2849.
12. Nagy, L., Korecz, L., Kiricsi, I., Zsikla, L. and Burger, K., *Struct. Chem.* **2** (1991), 231.
13. Deák, A., Haiduc, I., Párkányi, L., Venter, M. and Kálmán, A., *Eur. J. Inorg. Chem.* (1999), 1593.
14. Cunningham, D., Kelly, L. A., Molloy, K. C. and Zuckerman, J. J., *Inorg. Chem.* **21** (1982), 1416.
15. Silvestru, C., Haiduc, I., Klima, S., Thewalt, U., Gielen, M. and Zuckerman, J. J., *J. Organomet. Chem.* **327** (1987), 181.
16. Párkányi, L., Kálmán, A., Deák, A., Venter, M. and Haiduc, I., *Inorg. Chem. Commun.* **2** (1999), 265.

Hyperfine Interactions **144/145:** 239–248, 2002.

Quantification of Fe^{2+}/Fe^{3+} by Electron Microprobe Analysis – New Developments

H. E. HÖFER
Institut für Mineralogie, Johann Wolfgang Goethe-Universität, Senckenberganlage 28, D-60054 Frankfurt am Main, Germany; e-mail: hoefer@em.uni-frankfurt.de

Abstract. The measurement of the soft FeLα and FeLβ X-ray emission spectra by electron microprobe (flank method) allows the determination of the iron oxidation state in minerals if our new method for self-absorption correction is applied. It can be shown that the FeLβ/FeLα ratios obtained with the flank method correlate with Mössbauer quadrupole splitting and crystallographic data. This can be explained by the successive distortion of the iron coordination polyhedra with cation substitution. Thus, the electron microprobe data unexpectedly reveal details of the electrostatic field at the crystallographic site. Although the resolution with respect to hyperfine parameters is lower compared to other methods, the Fe^{2+}/Fe^{3+} determination by electron microprobe bears several advantages: (i) the small sample amount needed, (ii) the small sampling volume $\leqslant 1\ \mu m^3$, (iii) the simultaneous in situ Fe^{2+}/Fe^{3+} determination *and* quantitative chemical analysis, (iv) the easy routine sample preparation, (v) the ready availability of an electron microprobe in comparison to a synchrotron.

Key words: iron oxidation state, X-ray spectroscopy, electron microprobe, flank method, Mössbauer, garnet.

1. Introduction

Mössbauer spectroscopy is an important and accepted tool in geoscience to determine the iron oxidation state in minerals. However, there are limitations to Mössbauer spectroscopy with respect to sample size so that alternative methods to determine iron speciation on the micrometer scale are being developed. Amongst these are XANES and XAFS with synchrotron (see [1] and references) or X-ray photoelectron spectroscopy (XPS, e.g., [2]) with a relatively poor spatial resolution of several tens to hundreds of μm, only slightly better than the recent development of the Mössbauer milliprobe [3], and transmission electron microscopy (EELS, e.g., [4]) with a resolution of 1 μm. In geoscience, these methods are all useful, but limitations exist such as the ready availability of a synchrotron, the amount of sample material needed or the destructive preparation for TEM.

On the other hand, geoscientists investigate the chemical composition of natural and synthetic minerals routinely by electron microprobe. This method exploits the characteristic X-ray emission of elements in very small sample volumes for quantitative chemical analysis on the micrometer scale. The most sensitive method of detection is in the wavelength dispersive mode. Since such electron microprobes are widely used in material science and geoscience laboratories, the possibility of determining Fe^{2+}/Fe^{3+} with such an instrument would be of general interest. It even offers the possibility of the simultaneous in situ determination of chemical composition and of Fe^{2+}/Fe^{3+} on the micrometer scale without additional sample preparation. Another advantage is the small amount of sample material needed. With Mössbauer spectroscopy, several milligrams of sample are needed depending on the amount of total iron and an electron microprobe quantitative determination for the analysis of the bulk iron content is still required.

There have been several attempts in the past to determine the iron oxidation state by electron microprobe. Yet, only semiquantitative Fe^{2+}/Fe^{3+} estimations were achieved so far (e.g., [5, 6]). Based on these experiences we developed a new method [7] that allows accurate measurement of the critical X-ray intensities that are related to the iron oxidation state in solid samples. This method has been applied successfully for sodic amphiboles by Enders *et al.* [8] and for wüstites by our group [9]. Our latest results with synthetic garnets [10] show that the raw microprobe data can be empirically corrected for self-absorption and that an unambiguous determination of Fe^{2+}/Fe^{3+} is possible by electron microprobe. In the present paper we compare these results on garnet solid solutions with Mössbauer data of the same samples. It will be shown that, even at the low resolution of the electron microprobe compared to Mössbauer spectroscopy with respect to hyperfine parameters, the electron microprobe data hide more structural details than previously expected.

2. Experimental

The synthetic garnets (general formula: $Me_3^{2+}Me_2^{3+}Si_3O_{12}$) are solid solutions with Mg^{2+}, Ca^{2+}, and Mn^{2+} for Fe^{2+} on the dodecahedral sites, and Al^{3+} and Cr^{3+} for Fe^{3+} on the octahedral sites in variable amounts. The garnets were synthesized in several laboratories [11–14] and generously supplied for our investigations. $Fe^{3+}/\Sigma Fe$ contents were either determined by Mössbauer spectroscopy [11, 12], or from stoichiometric calculation.

The samples were mounted with epoxy resin into cylindrical holes in an epoxy pellet of 25 mm diameter and polished with a final diamond powder grain size of 0.25 μm. The minimum amount of sample material needed for electron microprobe investigations are grains of about 10 μm size. This corresponds to the routine preparation of polished sections for petrographic purposes. For the analysis by electron microprobe, the samples are coated with a carbon film of approximately 20 nm. Up to six of these 25 mm polished sections usually with 16 samples each may be placed in a sample holder so that a considerable number of samples (up to 100) may be analysed in one working session.

The results presented here were obtained with a Cameca SX 50 microprobe (Technische Universität Darmstadt) and a Jeol JX 8900 R microprobe (Universität Frankfurt). The accelerating voltage was 10 kV with a beam current of 60 nA and a beam diameter of 3 μm. Under these conditions, the electron beam penetrates to a depth of roughly 1 μm. In general, the electron beam can be focused to $\ll$1 μm so the analyzed volume in which characteristic X-rays are produced is about 0.5–1 μm^3.

3. Results and discussion

In the routine mode for quantitative chemical analysis by electron microprobe, such X-ray emission lines are exploited which are not affected by chemical bonding, i.e., which result from inner-shell electron transitions. If we aim to use the electron microprobe in an as yet non-routine fashion to determine Fe^{2+}/Fe^{3+}, we need to choose X-ray emission lines in which valence electrons are involved. Soft Fe$L\alpha$ and Fe$L\beta$ emission lines are produced by valence band electron transitions and are therefore affected to a large extent by the valency of the emitting Fe ion. Other factors influencing the valence spectra are the chemical bonding and the crystallographic surrounding (i.e. the nearest neighbours as given by the coordination polyhedron as well as the next-nearest neighbours, given by the crystal structure). Although systematic variations of both the intensity and the wavelength with ferric iron content have been reported, it has not been possible to determine the ferric iron content in oxides and silicates from these spectra with sufficient accuracy.

We have made progress by (a) concentrating on samples with the same crystal structure (thereby keeping the crystallographic surrounding, i.e., coordination polyhedra, bonding atoms, bonding type nearly constant) and (b) by developing a measuring technique ("flank method" [7, 9]) that enhances the sensitivity compared to former methods which relied either on peak intensity ratios or on peak shifts.

The flank method is based on determining the Fe$L\beta$/Fe$L\alpha$ intensity ratios in those two positions of the peaks for which the differences between the spectra of pure ferrous or ferric iron-containing samples are most pronounced [7]. Such spectra and their difference spectrum are shown in Figure 1 for two garnets containing either only Fe^{2+} (almandine) or Fe^{3+} (andradite). The shift in energy and the change of the Fe$L\beta$/Fe$L\alpha$ intensity ratio between the Fe^{2+} bearing garnet and the Fe^{3+} bearing garnet is most obvious. The two measuring positions for the flank method are indicated by the vertical lines and correspond to the minimum and the maximum of the difference spectrum of $L\beta$ and $L\alpha$. This yields a high sensitivity by combining both energy shifts and intensity changes. For the actual measurements, 27 700 counts were accumulated at the derived energy positions for $L\alpha$ and $L\beta$. Thus, counting statistics are identical for samples with high or low iron contents. Since a good spectrometer reproducibility is essential for these measurements, we developed a special method to improve the setting of the spectrometer [9].

The Fe$L\beta$/Fe$L\alpha$ intensity ratios determined with the flank method in synthetic garnets vary systematically with the total iron content of the synthetic garnets because of the effect of self-absorption. This effect is produced by the iron atoms

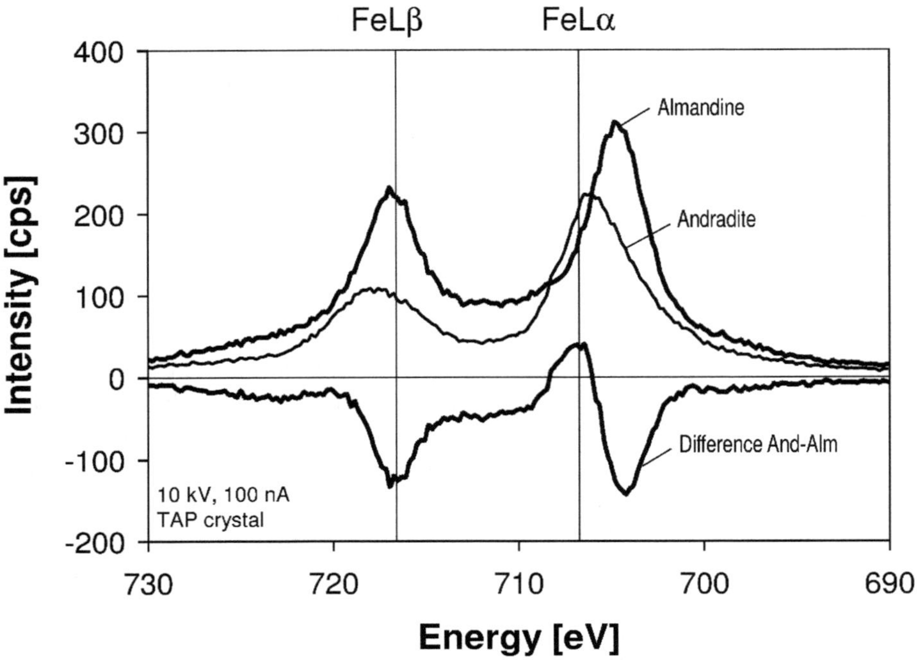

Figure 1. FeLα and FeLβ X-ray emission spectra of almandine $Fe_3^{2+}Al_2^{3+}Si_3O_{12}$ and andradite $Ca_3^{2+}Fe_2^{3+}Si_3O_{12}$. The maxima are shifted to higher energies for the garnet containing ferric iron with a concommitant decrease of the Lβ/Lα intensity ratio. The postions for the flank method measurements are marked by the vertical lines and are situated in the minimum and maximum of the difference spectrum.

in the sample: when measuring X-ray emission next to the FeL absorption edges, an overlap of emission and absorption energy states occurs. The X-ray emission produced in the sample is absorbed by the iron atoms on their path through the sample. Since the FeL absorption edges themselves shift with iron oxidation state, we imply that the changes of X-ray emission in these positions are in fact a result of selective self-absorption by the Fe ion species. By empirically correcting for self-absorption, a linear relationship results between the corrected ratio FeLβ/FeLα and the content of Fe^{3+} for all the analyzed garnet samples, regardless of their chemical composition (Figure 2). Hence Lβ/Lα intensity ratios can be used unambiguously to determine the ferric iron content in garnets. The results also indicate that the self-absorption effect consists of two components which have to be treated separately: one accounting for ferrous iron and the other for ferric iron implying a different mass absorption behaviour for ferrous and ferric iron. Details about the correction procedure will be published elsewhere [10].

To compare the flank method electron microprobe data with the Mössbauer data, we focus on the garnet solid solutions almandine–skiagite ($Fe_3^{2+}Al_2^{3+}Si_3O_{12}$–$Fe_3^{2+}Fe_2^{3+}Si_3O_{12}$) and skiagite–andradite ($Fe_3^{2+}Fe_2^{3+}Si_3O_{12}$–$Ca_3^{2+}Fe_2^{3+}Si_3O_{12}$).

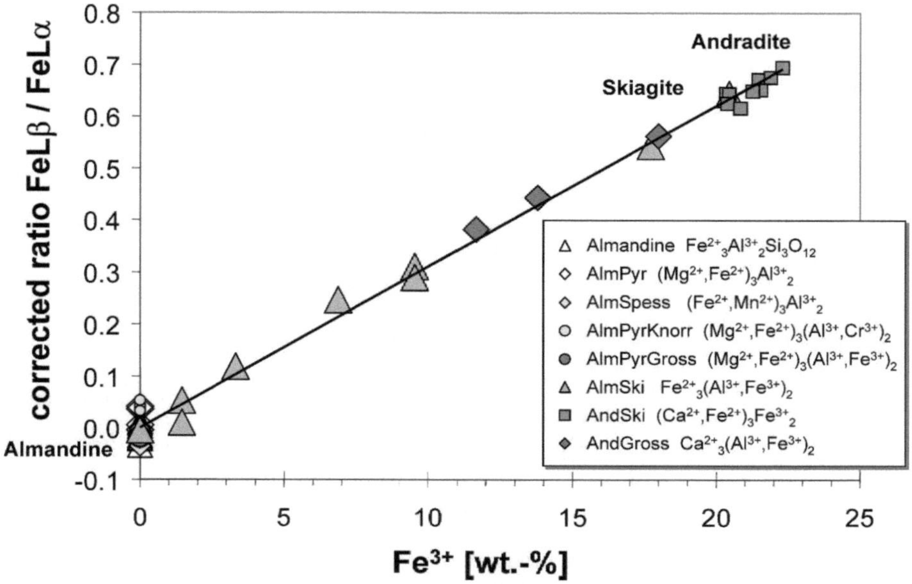

Figure 2. Plot of the FeLβ/FeLα intensity ratios as determined by the flank method and empirically corrected for self-absorption versus the ferric iron content of the garnets. The total iron content in these samples varies from 6 to 50.6 wt.%. The uncertainty in y is about ± 0.03.

They were provided by A. B. Woodland, their Mössbauer paramaters are described in [11, 12].

The ratios FeLβ/FeLα obtained by the flank method (raw data, uncorrected for self-absorption) are plotted in Figure 3 versus $Fe^{3+}/\Sigma Fe$. With increasing substitution of Al^{3+} by Fe^{3+} in the almandine-skiagite solid solution on the octahedral site, the intensity ratio decreases only slightly. In the skiagite–andradite solid solution, Fe^{2+} is substituted by Ca^{2+} on the dodecahedral site which results in a pronounced decrease of Lβ/Lα with $Fe^{3+}/\Sigma Fe$. An almost identical pattern is observed for the quadrupole splitting of Fe^{3+} on the octahedral site obtained from Mössbauer spectra (Figure 4). The Fe^{2+} quadrupole splitting on the dodecahedral site does not vary with $Fe^{3+}/\Sigma Fe$. This behaviour was already observed by Amthauer *et al.* [15] for natural garnets. The Fe^{3+} quadrupole splitting and the electron microprobe data show a comparable dependence on $Fe^{3+}/\Sigma Fe$ and Figure 5 shows a linear correlation between both data sets (only the endmembers skiagite and andradite plot slightly off the correlation line).

In the garnet structure, Fe^{2+} and Fe^{3+} ions occupy different crystallographic sites: the larger divalent cation is coordinated by 8 oxygens in a distorted cubic arrangement conventionally called the dodecahedral site and the smaller trivalent cation is coordinated by 6 oxygens in octahedral geometry. For the present investigation the linking between these polyhedra is of relevance: while the octahedra are isolated from each other, they share 6 of their 12 edges with neighbouring dodec-

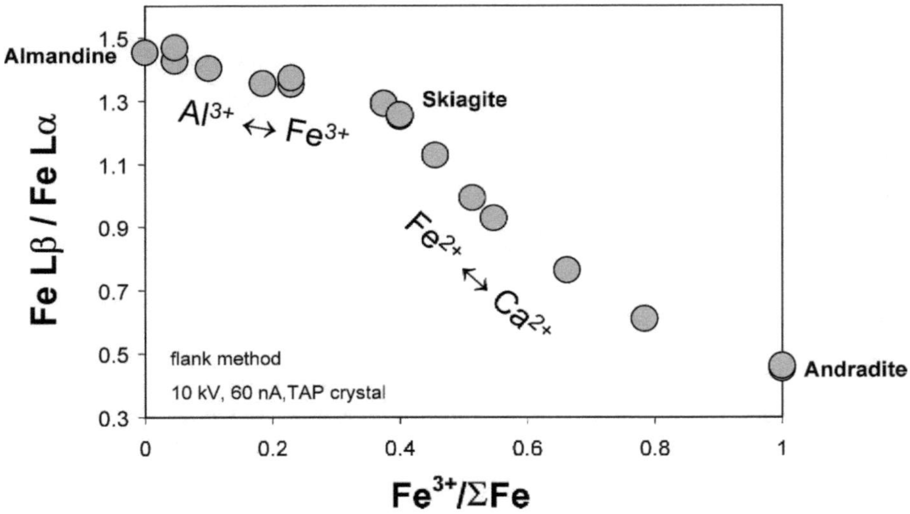

Figure 3. FeLβ/FeLα intensity ratios determined by the electron microprobe flank method versus $Fe^{3+}/\Sigma Fe$. There is only a slight negative correlation for the almandine-skiagite series and a pronounced decrease for the skiagite–andradite series with increasing $Fe^{3+}/\Sigma Fe$.

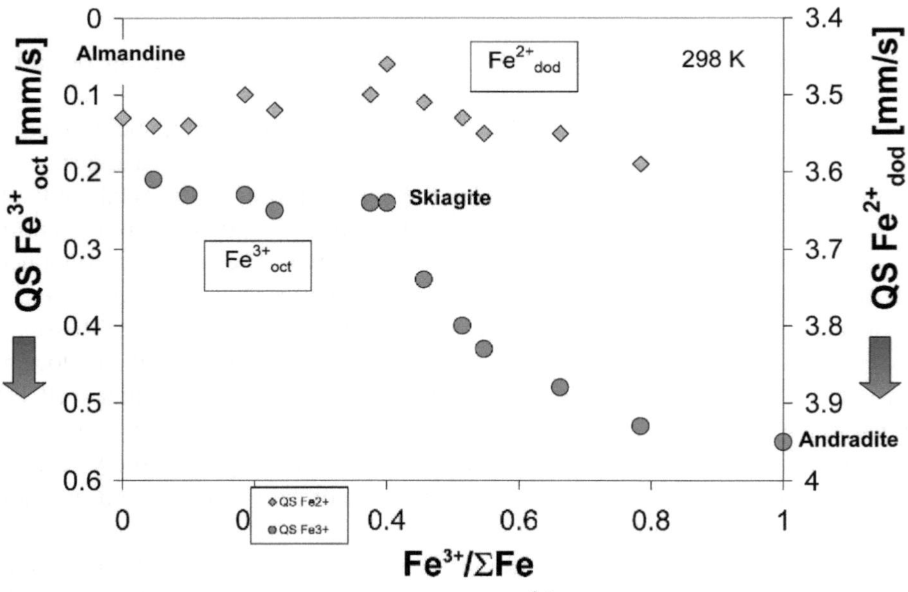

Figure 4. Mössbauer quadrupole splitting (QS) data for Fe^{3+} on the octahedral site (left axis) and for Fe^{2+} on the dodecahedral site (right axis) versus $Fe^{3+}/\Sigma Fe$ for two synthetic garnet solid solution series. The Fe^{2+} QS is independent of the chemical composition while the Fe^{3+} QS shows a similar dependence as the electron microprobe data (Figure 3). Replotted from [12].

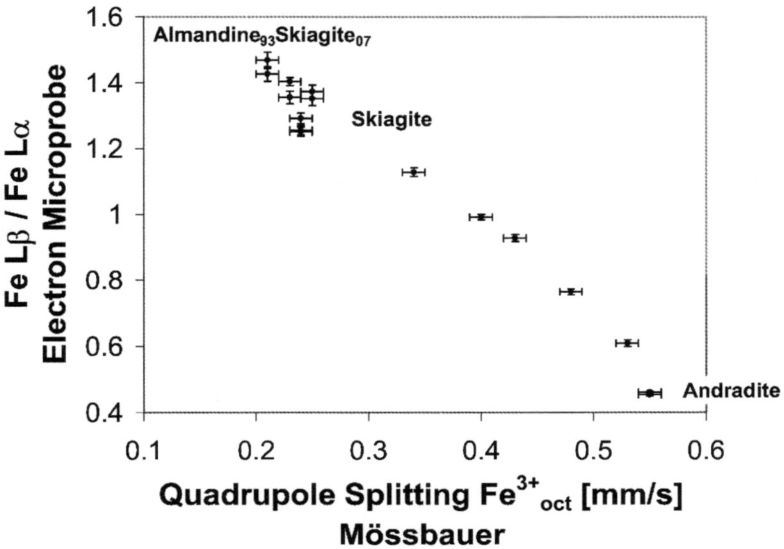

Figure 5. Electron microprobe versus Mössbauer data [12] of synthetic garnets of the almandine-skiagite-andradite solid solution series.

ahedra. Thus, a change of the dodecahedral site geometry due to cation exchange induces a concomitant change of the octahedral site geometry. Conversely, only 4 of 18 dodecahedral edges are shared with neighbouring octahedral sites implying that cation exchange on octahedral sites does not affect the dodecahedral distortion as much as vice versa. Woodland and Ross II [12] showed this relationship in their Figure 5 by plotting the octahedral distortion parameter σ versus the dodecahedral distortion parameter ρ.

This behaviour is reflected in Figure 4, from which we know that the Fe^{2+} quadrupole splitting does not depend on the kind of cations on the dodecahedral and octahedral sites while the Fe^{3+} quadrupole splitting decreases with increasing $Fe^{3+}/\Sigma Fe$. To visualize the effect of cation exchange from an atomic and crystallographic view, we replotted the data from Woodland and Ross II [12] by normalizing the amount of Fe^{2+} and Fe^{3+} to the number of available dodecahedral or octahedral sites per formula unit in the garnet structure. The quadrupole splitting of Fe^{3+} is now plotted versus the filling of the octahedral site by Fe^{3+} (Figure 6(a)) and versus the filling of the dodecahedral site by Fe^{2+} (Figure 6(b)). The replacement of Al^{3+} by Fe^{3+} on the octahedral site (almandine-skiagite) does not change the octahedral site distortion and the Fe^{3+} quadrupole splitting to any extent (Figure 6(a)). The substitution of Fe^{2+} by Ca^{2+} on the dodecahedral site in andradite-skiagite, however, has a large effect on the octahedral site distortion and thus changes the Fe^{3+} quadrupole splitting significantly – it is linear with successive filling of the dodecahedral site by Fe^{2+} (Figure 6(b)). The small deviation from linearity on the andradite-rich side may be caused by the octahedral site distortion which is

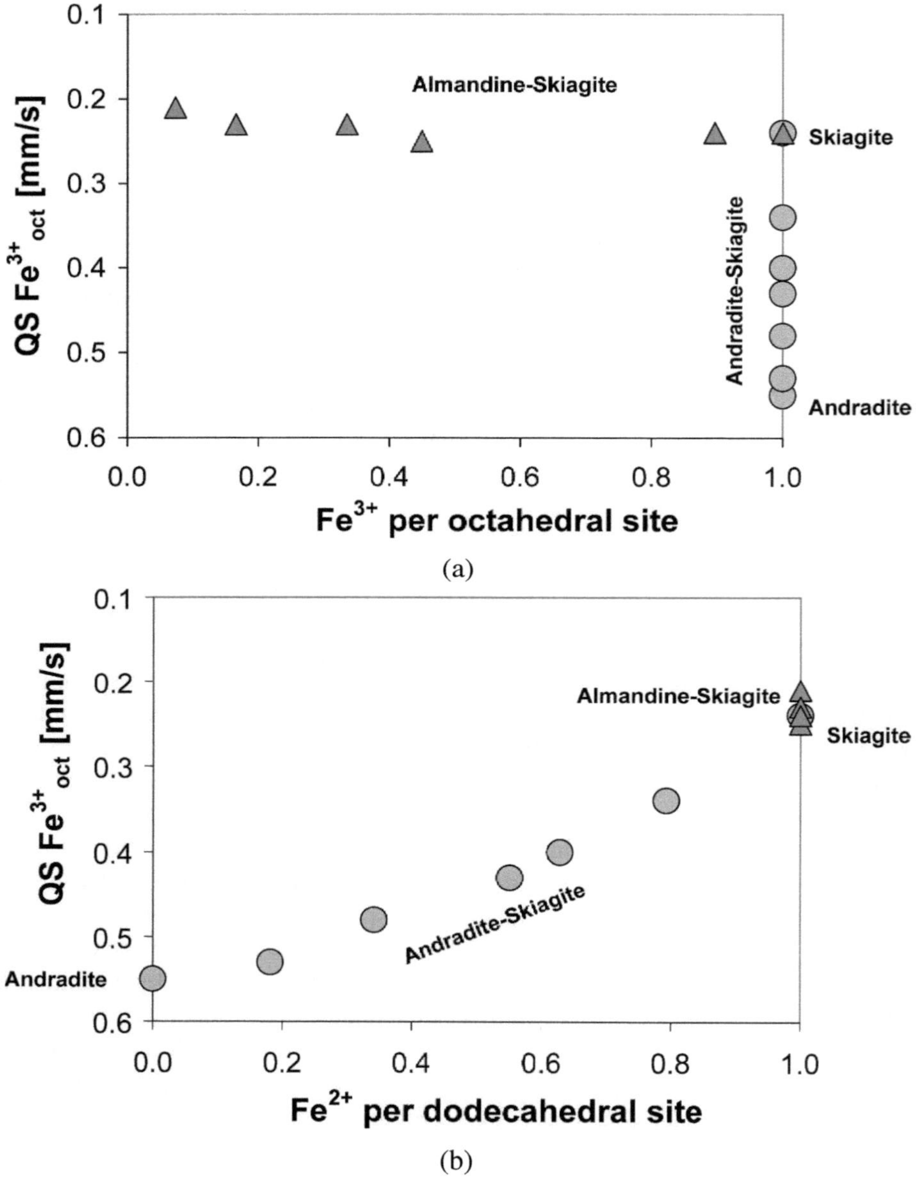

Figure 6. Fe^{3+} Mössbauer quadrupole splitting versus (a) Fe^{3+} normalized to the number of available octahedral sites and (b) Fe^{2+} normalized to the number of available dodecahedral sites in the garnet structure. Replotted from [12].

reported to change from oblate to prolate at the 35 mol% skiagite composition [12]. If this were the case, we should see a pronounced change or a minimum in the relationship. Since it is not the case, we may interpret this in conjunction with [12] that the electric field gradient (EFG) responsible for the Fe^{3+} QS is not only governed by the octahedral site geometry but is also influenced by contributions from further distances.

4. Conclusions

The present work shows that the oxidation state of iron can be determined unambiguously by electron microprobe measurements in garnets if our empirical correction for self-absorption is applied. The comparison with Mössbauer and crystallographic data shows the high sensitivity of the method. We find corresponding relationships between the flank method ratios obtained by the electron microprobe, Mössbauer QS data, and various crystallographic aspects. Since the characteristic X-rays measured in electron microprobe analysis are the result of electronic transitions, the EFG must be related to the valence band spectra exploited for Fe^{2+}/Fe^{3+} determination. Therefore the electron microprobe measurements reveal more details than previously expected. Although these informations have a lower spectral resolution than Mössbauer spectroscopy or XANES, the possibility of determining Fe^{2+}/Fe^{3+} with the microprobe bears many advantages. Amongst them are the high spatial resolution in conjunction with the small analyzed volume (in the order of 1 μm^3), the possibility of the simultaneous chemical analysis of such small volumes, the fastness of the method (~10 min per analysis) and the almost nondestructive preparation of polished mounts.

Acknowledgements

The project was initiated by Gerhard Brey whose continuous interest and discussions are highly appreciated. The paper profited much from his comments. Burkhard Schulz-Dobrick put forward the idea to measure the FeLβ/FeLα ratio not in the peak maxima, but in the flanks of the X-ray emission spectra, an idea which greatly improved sensitivity. The synthetic garnet samples were generously provided by Alan Woodland, Charles Geiger, Thomas Fehr, Anne Feenstra and Alexander Doroshev.

References

1. Wilke, M., Farges, F., Petit, P.-E., Brown Jr., G. E. and Martin, F., *Am. Mineral.* **86** (2001), 714.
2. Raeburn, S. P., Ilton, E. S. and Veblen, D. R., *Geochim. Cosmochim. Acta* **61** (1997), 4531.
3. McCammon, C. A., *Hyp. Interact.* **92** (1994), 1235.
4. van Aken, P.A. and Liebscher, B., *Phys. Chem. Minerals* **29** (2002), 188.
5. Pavićević, M. K., Amthauer, G. and El Goresy, A., *J. Serb. Chem. Soc.* **54** (1989), 359.
6. Armstrong, J. T., *Anal. Chem.* **71** (1999), 2714.

7. Höfer, H. E., Brey, G. P., Schulz-Dobrick, B. and Oberhänsli, R., *Eur. J. Mineral.* **6** (1994), 407.
8. Enders, M., Speer, D., Maresch, W. V. and McCammon, C. A., *Contrib. Mineral. Petrol.* **140** (2000), 135.
9. Höfer, H. E., Weinbruch, S., McCammon, C. A. and Brey, G. P., *Eur. J. Mineral.* **12** (2000), 63.
10. Höfer, H. E. and Brey, G. P., in preparation.
11. Woodland, A. B. and O'Neill, H. St. C., *Am. Mineral.* **78** (1993), 1002.
12. Woodland, A. B. and Ross, C. R. II, *Phys. Chem. Minerals* **21** (1994), 117.
13. Geiger, C. A. and Rossman, G. R., *Phys. Chem. Minerals* **21** (1994), 516.
14. Geiger, C. A. and Feenstra, A., *Am. Mineral.* **82** (1997), 571.
15. Amthauer, G., Annersten, H. and Hafner, S. S., *Z. Kristallogr.* **143** (1976), 14.

Hyperfine Interactions **144/145:** 249–253, 2002.

Mean-Square Amplitudes of Vibration and Anisotropic Motion of Metal Atoms in Iron Organometallics

ROLFE H. HERBER* and ISRAEL NOWIK
Racah Institute of Physics, The Hebrew University, 91904 Jerusalem, Israel

Abstract. Mössbauer recoil-free fraction (f) data for ferrocene [(η^5-$(C_5H_5)_2Fe$] over the temperature range $90 \leqslant T \leqslant 370$ K have been used to calculate the mean-square-amplitude-of-vibration (msav) of the iron atom, and these data have been compared to the thermal factors (U_{ij}) extracted from X-ray diffraction data reported in the literature. The agreement between these two independent data sets is taken as a validation of the interpretation of the Mössbauer data in terms of the mean-square-amplitude-of-vibration of the metal atom in iron organometallic solids. The contribution of three low-frequency IR active modes of ferrocene (ω_{11}, ω_{21}, and ω_{22}) to the $f(T)$ parameter is discussed in detail.

Key words: metal atom motional anisotropy, Gol'danskii–Karyagin effect, recoil-free fraction, ferrocene.

1. Introduction

For solids of diamagnetic, primarily covalent, organometallic compounds of iron, a good deal of useful information can be extracted from the hyperfine interaction parameters – the isomer shift (IS) and quadrupole splitting (QS) – and their temperature dependencies. In addition, recent studies [1, 2] have shown that the dynamical properties of the metal atom, extracted from the temperature dependence of the recoil-free fraction [$f(T)$], and the motional anisotropy related to the Gol'danskii–Karyagin effect (GKE) can provide important characterization of the properties of organometallic solids. However, to validate the connection between these Mössbauer effect derived parameters and those elucidated by other techniques, such as X-ray diffraction, NMR, infrared/Raman, and inelastic neutron scattering, it is worthwhile to effect a comparison between the data derived from several independent techniques. Such a comparison can be provided by reference to the temperature dependence of the recoil-free fraction data on the one hand, and the thermal factors (U_{ij}) extracted from single crystal X-ray diffraction studies, and Raman and inelastic nuclear scattering determined density of vibrational states on the other.

---

* Author for correspondence. E-mail: herber@vms.huji.ac.il.

Because of the centrality of ferrocene [(η^5-$C_5H_5)_2Fe$] in the development of modern organometallic chemistry, as well as the complexity of the molecular-level architecture of this covalent solid, a good deal of X-ray diffraction data at a variety of different temperatures have been reported in the literature and these data are compared to the results obtained from temperature-dependent Mössbauer effect derived data in the present study.

2. Experimental

Ferrocene, [(η^5-($C_5H_5)_2Fe$], was purified by two-fold sublimation at 50°C. Details of sample preparation, temperature control, spectrometer calibration, and data reduction have been described earlier [3, 4]. In addition, the organometallic samples, ground with quartz powder and mixed with BN to ensure random crystallite orientation, were transferred to O-ring closed plastic sample holders, with the powder "sandwiched" between two aluminum foils which served as thermal shields. The closed sample holder was again in turn sandwiched between two aluminum thermal radiation shields and then – in turn – edge-sealed with polymethylmethacrylate glue to ensure a vacuum tight enclosure. As noted earlier [3], this precaution was observed both to prevent sample loss (by vacuum evaporation in the cryostat, especially at high temperatures) which would invalidate the $f(T)$ data, and sublimation from one area of the sample holder to another which would influence the "texture effect" on the sample. In this study, the data for the sublimed and recrystallized samples have been combined to yield a more extensive data set than that reported earlier [3].

3. Results and discussion

The results of a Mössbauer study of ferrocene purified by recrystallization from absolute alcohol have been reported earlier [1]. The present results, pertaining to the temperature dependence of the recoil-free fraction involving the sublimed sample, are in good agreement with those reported earlier for the recrystallized material. Powder X-ray diffraction data at room temperature indicate a complete identity between the recrystallized and sublimed solids. It is also noteworthy that differential scanning calorimetry of ferrocene over the range 250 to 350 K evidence no phase transitions in this region.

The mean-square-amplitude of vibration (msav) of the metal atom in ferrocene can be extracted from the slope of the temperature dependence of the logarithm of the recoil-free fraction, as discussed earlier [1], and then compared to the U_{ij} data reported in single crystal X-ray diffraction studies. This comparison at various temperatures is summarized in Table I. Despite the complexity of the molecular and crystal structure of ferrocene, which evidences two phase transitions at 169 and ~248 K, and the dependence of these structural details on the thermal history of the sample, the correlation between the X-ray diffraction results and those extracted

Table I. Comparison of $k^2\langle x^2\rangle$ values from X-ray diffraction anisotropic thermal factors and temperature-dependent Mössbauer recoil-free fraction data for ferrocene

Temp. K	X-ray	Mössbauer (Sublimed)	Mössbauer (EtOH Recryst.)	Reference
101	0.57	0.69	0.70	[12]
123	0.71	0.84	0.82	[12]
148	0.92	1.00	1.02	[12]
173	1.27	1.17	1.19	[13]
293	2.26	1.96	1.98	[13]
173	1.44			[14]
298	2.13	1.99	2.05	[14]
173	1.49			[15]
298	2.13			[15]
173	1.61			[16]
298	2.20			[16]

from the Mössbauer data is reasonably satisfactory, given the experimental errors in the individual experiments. This agreement infers confidence that the msav extracted from the Mössbauer data at other temperatures can be taken as a reliable measure of the metal atom dynamics in these solids.

It is interesting to note that the temperature dependence of ln f for ferrocene, which is summarized in terms of $k^2\langle x^2\rangle$ (where $k^2 = 4\pi^2/\lambda^2$; $\lambda =$ the wavelength of the Mössbauer gamma ray) in Figure 1, shows significant curvature over the whole temperature range $90 \leqslant T \leqslant 370$ K with the slope increasing with increasing temperature. This curvature arises from the population of several relatively low frequency normal modes of ferrocene which have been observed both in infrared [5, 6] and in inelastic neutron scattering data [7]. The lowest frequency mode involving the relative motion of the Fe atom with respect to the two cyclopentadienyl rings was characterized as an mode at 170 cm^{-1} by Lippincott and Nelson [5], and reported as mode 22 (ω_{22}) in the low temperature (80 to 180 K) IR and Raman data of Bodenheimer and Low [6] at 179 cm^{-1}. Inelastic neutron scattering data at 28 K, and corresponding density functional calculations [7] at 30 K place this mode at 180 and 177 cm^{-1}, respectively, in good agreement with the vibrational spectroscopy results. The next higher density of states maxima occur at 59 and 61.3 meV (corresponding to 476 and 494 cm^{-1}) and have been assigned [5] to modes ω_{11} (A_{2u}) and ω_{21} (E_{1u}) in the infrared spectra at room temperature. It should be noted that these three modes are the lowest frequency modes involving relative motion of the metal atom with respect to the cyclopentadienyl rings, and are thus most significant in accounting for the temperature dependence of $f(T)$.

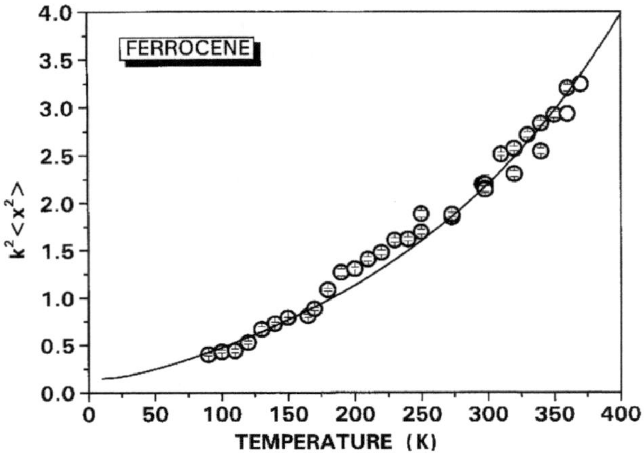

Figure 1. The parameter $k^2\langle x^2\rangle$ for ferrocene as a function of temperature normalized to the 298 K data point extracted from single crystal U_{ij} values. The curvature at higher temperature reflects the increasing participation of the three modes discussed in the text.

The solid line shown in Figure 1 is a single parameter (the rate of reduction with temperature of the energy of the vibrational modes, $\sim -0.1\%\ \mathrm{K}^{-1}$) fit of the Mössbauer data using the relative density of states values of [8] and the energies of the three modes noted above. The relative DOS contributions extracted from a plot of DOS *versus* energy are 36% for the low frequency phonon modes, 13.6% for ω_{22} at 22 meV, and 49.8% for $\omega_{11,21}$ at 59 and 61.3 meV. These contributions were obtained [8] on a microcrystalline powder sample at 30 K, but presumably take on somewhat different values at the higher temperatures of the Mössbauer measurements. The fit was carried out in consonance with Equation 1.43 of reference [9]. In order to account for the $k^2\langle x^2\rangle$ data at higher temperatures, it has been assumed that the lattice softening as T increases can be adequately accounted for by a linear reduction in the vibrational mode energies. It is clear from these results that the vibrational mode softening indicated by the curvature in the $f(T)$ data (Figure 1) for ferrocene – especially at the higher temperatures – arises from the increasing contribution of ω_{11} and $\omega_{21,22}$ to the mean-square-amplitude-of-vibration of the iron atom in this solid as the temperature is increased.

With the validation of the interpretation of the Mössbauer $f(T)$ data in terms of a mean-square-amplitude-of-vibration of the iron atom provided by the results summarized in Table I and the detailed analysis of the $f(T)$ data detailed above, it is clear that temperature-dependent recoil-free fraction data can provide reliable information concerning the metal atom dynamics in ferrocene-related materials. An application to the study of co-facial interactions in di-ferrocenyl complexes [10], and in a *bis*(cyclopentadienyl iron) tetrahydride [11] have been reported in the recent literature.

Acknowledgements

The authors are indebted to Dr. U. Asaf for the powder X-ray diffraction data on ferrocene, and to E. Fartush for the quoted DSC results. We are also grateful to Dr. A. Chumakov for the results of his density of states data at 30 K, and a number of fruitful discussions of his data prior to publication.

References

1. Herber, R. H. and Nowik, I., *Solid State Sci.* **4** (2002), 691.
2. Herber, R. H., In: M. Gielen, R. Willem and B. Wrackmeyer (eds), *Unusual Structures and Physical Properties in Organometallic Chemistry*, J. Wiley and Sons Ltd., Chichester, 2002, p. 207–218.
3. Herber, R. H., *J. Nucl. Radiochem. Sci.* **2** (2001), R5-8 and references therein.
4. Herber, R. H. and Nowik, I., *Hyp. Interact.* **126** (2000), 127.
5. Lippincott, E. R. and Nelson, R. D., *Spectrochim. Acta* **10** (1958), 307; see also Hyams, I. J., *Spectrochim. Acta A* **29** (1973), 1733.
6. Bodenheimer, J. S. and Low, W., *Spectrochim. Acta* **29** (1973), 1733.
7. Kemner, E., de Schepper, I. M., Kearley, G. J. and Jayasooriya, U. A., *J. Chem. Phys.* **112** (2000), 10926.
8. Chumakov, A., private communic.
9. Gol'danskii, V. I. and Makarov, E. F., In: V. I. Gol'danskii and R. H. Herber (eds), *Chemical Applications of Mössbauer Spectroscopy*, Academic Press, New York, 1968, p. 33.
10. Herber, R. H., Nowik, I. and Rosenblum, M., *Organometallics* **21** (2002), 846.
11. Nowik, I. and Herber, R. H., *J. Physics. Chemistry of Solids* **64** (2003), 313.
12. Seiler, P. and Dunitz, J. D., *Acta Cryst. B* **35** (1979), 2020.
13. *Ibid*, p. 1068.
14. Brock, C. P. and Fu, Y., *Acta Cryst. B* **53** (1997), 928, model R1ZH.
15. *Ibid.*, model R2ZHO.
16. *Ibid.*, model R3ZHO.

Hyperfine Interactions **144/145:** 255–259, 2002.

Degradation of the Giant Magnetoresistance in Fe/Cr Multilayers Due to Ar-Ion Beam Mixing

M. KOPCEWICZ[1], F. STOBIECKI[2], J. JAGIELSKI[1,3], B. SZYMAŃSKI[2], M. SCHMIDT[2] and J. KALINOWSKA[1]
[1] *Institute of Electronic Materials Technology, 01-919 Warsaw, Poland; e-mail: kopcew_m@sp.itme.edu.pl*
[2] *Institute of Molecular Physics, Polish Academy of Sciences, 60-179 Poznań, Poland*
[3] *The Andrzej Soltan Institute of Nuclear Studies, 05-400 Świerk/Otwock, Poland*

Abstract. The influence of 200 keV Ar-ion irradiation on the interlayer coupling in the Fe/Cr multilayer system exhibiting the giant magnetoresistance effect (GMR) is studied by conversion electron Mössbauer spectroscopy (CEMS), VSM hysteresis loops, magnetoresistivity and electric resistivity measurements and supplemented by the small-angle X-ray diffraction (SAXRD). The increase of Ar ion dose causes an increase of interface roughness, as evidenced by the increase of the Fe step-sites detected by CEMS as a result of which the GMR gradually decreases and vanishes at doses exceeding 1×10^{14} Ar/cm^2. A degradation of GMR with increasing Ar-ion dose is related to the formation of pinholes between Fe layers and the decrease of the antiferromagnetically coupled fraction.

Key words: giant magnetoresistance, Fe/Cr, ion irradiation, ion-beam mixing, antiferromagnetic interlayer coupling, CEMS.

Despite the fact that since the discovery of the giant magnetoresistance (GMR) in Fe/Cr multilayers [1] many experimental and theoretical studies have been performed on the role of interface structure on GMR, it is still not well understood. It was found that the increase of interface roughness may enhance GMR [2]. However, in contrast to these results, very large GMR has been found in Fe/Cr multilayers with sharp interfaces and low residual resistivity [3]. Interface roughness can be affected by deposition conditions of multilayer structure, substrate temperature and thermal annealing (see, e.g., [4]). It was shown recently that ion irradiation with 500 keV Xe-ions [5] induced initial increase of GMR. However, at high ion doses the GMR was destroyed because the formation of the ferromagnetic coupling in Fe/Cr system. Also the 200 MeV Ag-ion irradiation lead to the decrease of GMR in Fe/Cr multilayers [6].

We studied the modification of the interface structure, induced by Ar-ion irradiation, in Fe/Cr system. The samples have been deposited, using the dc magnetron sputtering, on a SiO_x substrate and consisted of [Fe(3 nm)/Cr(1.1 nm)]$_N$. The total thickness of the Fe/Cr film was about 100 nm. Samples were irradiated at room temperature with 200 keV Ar ions with doses ranging from 1×10^{12} to $5 \times$

10^{14} Ar/cm^2. The range of ions matched well the thickness of the multilayer film. The as-deposited and irradiated samples were characterised by the conversion electron Mössbauer spectroscopy (CEMS), small-angle X-ray diffraction (SAXRD) and VSM hysteresis loops. Magnetoresistance and resistivity were measured using the four probe technique in CIP geometry.

The CEMS spectra were interpreted in terms of the model [7] which correlates four magnetically split spectral components with the Fe sites in the Fe/Cr system: $H_1 \approx 33$ T, corresponds to the bulk Fe sites; $H_2 \approx 30$ T and $H_3 \approx 24$ T, correspond to the "step" sites at the Fe/Cr interfaces, and $H_4 \approx 20$ T, corresponds either to the "perfect" interface sites or to some other "step" positions. The origin of the H_4 component is controversial. As suggested recently [8] this component may correspond to the Fe sites in Cr layer separated by at least two atomic Cr monolayers from the Fe layer. However, it was shown earlier that Fe atoms isolated in Cr are paramagnetic [9].

Ar-ion irradiation induced clear changes in the CEMS spectra of irradiated samples. Already at low ion doses ($(5-25) \times 10^{12}$ Ar/cm^2) the spectral contribution of the H_2 component increases substantially at the expense of the H_1 component what suggests that the interface roughness increases (the number of Fe step-sites increases) (Figures 1a–c, 2). Simultaneously the shape of the hysteresis loop changes: the remanence increases, the hysteresis loop becomes more rectangular and the saturation field decreases (Fig. 3). The decrease of GMR and simultaneous changes in F_{AF} (antiferromagnetically coupled fraction defined as $F_{AF} = 1 - M_R/M_S$,

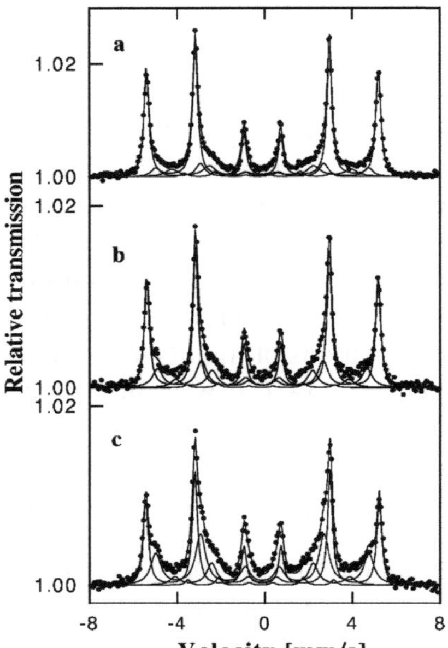

Figure 1. CEMS spectra for as-deposited (a) and irradiated Fe/Cr multilayers at indicated ion doses (b, c).

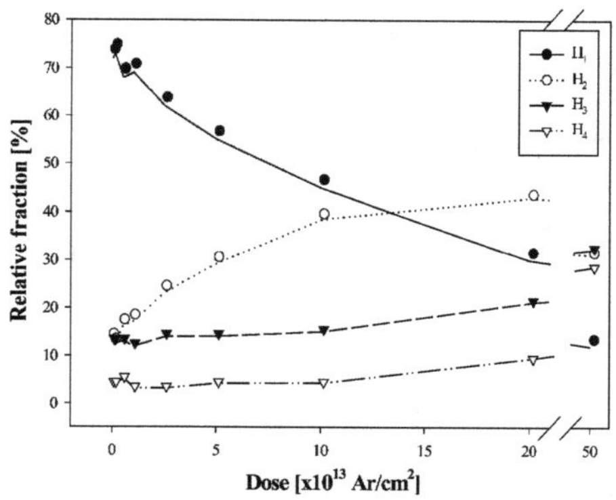

Figure 2. Relative spectral fractions of H_1–H_4 components vs. ion dose.

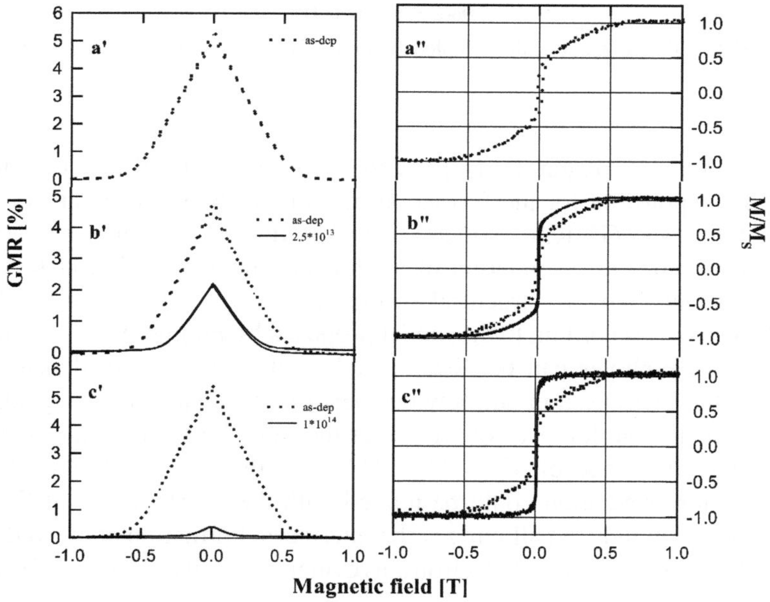

Figure 3. The GMR and hysteresis loops for as-deposited (a′, a″) and irradiated Fe/Cr multilayers at indicated ion doses (b′–c′, b″–c″).

where M_R and M_S are the remanence and the saturation magnetization determined from the hysteresis loops) and saturation fields, H_S, vs. ion dose are shown in Figure 4. As can be seen, the GMR gradually decreases with increasing ion dose and vanishes at doses exceeding 1×10^{14} Ar/cm^2. Increase of ion dose causes a significant increase of the H_2 and H_3 components in the CEMS spectra (Figures 1, 2) which strongly suggests an efficient ion-beam mixing at the Fe/Cr interfaces. At the highest ion dose used (5×10^{14} Ar/cm^2) the mixing is sufficiently strong that almost entire Fe layers are consumed for the formation of FeCr alloy (Figures 1c and 2). A degradation of GMR with increasing Ar-ion dose is related to the significant

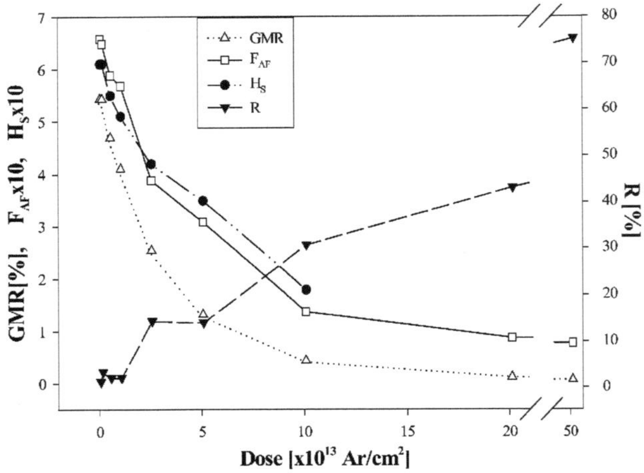

Figure 4. GMR, antiferromagnetically coupled fraction, F_{AF}, saturation field, H_S, and resistivity, R, vs. ion dose.

increase of interface roughness. In particular, the uncorrelated part of roughness seems to be responsible for the formation of pinholes between Fe layers and the decrease of F_{AF} and GMR [6]. Since a very low density of pinholes is sufficient for an almost complete degradation of antiferromagnetic coupling [10] it is very difficult to detect them by structural investigations (e.g., using the small angle X-ray diffraction). According to the model proposed by Fulghum and Camley [10], the presence of pinholes can be documented by the temperature dependence of M_R. This method was used in our previous paper in which a similar degradation of antiferromagnetic coupling was observed for the same Fe/Cr multilayers annealed in the temperature range of 373–523 K [11]. Another cause for degradation of GMR due to ion irradiation could be related with the consumption of Cr and Fe layers for ion-beam induced alloying at the interfaces. Formation of alloyed regions decreases the mean free path of electrons, as evidenced by the increase of resistivity (Figure 4) and reduces GMR.

The most sensitive technique which reflects the change of the interface roughness is the measurement of the magnetoresistance and hysteresis loop. The CEMS is also quite sensitive to the interface structure. However, the SAXRD technique hardly detects the changes induced by ion irradiation, even at high ion doses at which GMR disappears, hysteresis loop shows substantial ferromagnetic coupling and CEMS reveals strong ion-beam mixing and alloying effect at interfaces.

Acknowledgement

Financial support from the NEDO (Japan) grant "Nanopatterned Magnets" is gratefully acknowledged.

References

1. Baibich, M. N., Broto, J. M., Fert, A., Nguyen Van Dau, F., Petroff, F., Eitenne, P., Creuzet, G., Friederych, A. and Chazelas, J., *Phys. Rev. Lett.* **61** (1988), 2462.
2. Fullerton, E., Kelly, D. M., Guimpel, J., Schuller, I. and Bruynseraede, Y., *Phys. Rev. Lett.* **68** (1992), 859.
3. Schad, R., Potter, C. D., Belien, P., Verbanck, G., Mosshchalkov, V. V. and Bruynseraede, Y., *Appl. Phys. Lett.* **64** (1994), 3500.
4. Kopcewicz, M., Lucinski, T., Stobiecki, F. and Reiss, G., *J. Appl. Phys.* **85** (1999), 5039.
5. Kelly, D. M., Schuller, I., Korenivski, V., Rao, K. V., Larsen, K. K., Bottiger, J., Gyorgy, E. M. and van Dover, R. B., *Phys. Rev. B* **50** (1994), 3481.
6. Paul, A., Gupta, A., Chaudhari, S. M. and Phase, D. M., *Vacuum* **60** (2001), 401.
7. Landes, J., Sauer, Ch., Brand, R. A., Zinn, W., Mantl, S. and Kajcsos, Zs., *J. Magn. Magn. Mater.* **86** (1990), 71.
8. Uzdin, V. M., Keune, W., Schörr, H. and Walterfang, M., *Phys. Rev. B* **63** (2001), 104407.
9. Rensing, N. M., Clemens, B. M. and Williamson, D. L., *J. Appl. Phys.* **79** (1996), 7757.
10. Fulghum, D. B. and Camley, R. E., *Phys. Rev. B* **52** (1995), 13436.
11. Nowak, T., Dubowik, J., Stobiecki, F., Szymanski, B., Kopcewicz, M. and Jagielski, J., *Phys. Stat. Sol. (a)* **196** (2003), 45.

Hyperfine Interactions **144/145:** 261–266, 2002.

Magnetic Relaxation and Its Relation to Magnetoresistance in $FeCr_2S_4$ Spinel

Z. KLENCSÁR[1,*], E. KUZMANN[1], Z. HOMONNAY[1], A. VÉRTES[1],
A. SIMOPOULOS[2], E. DEVLIN[2] and G. KALLIAS[2]
[1]*Research Group for Nuclear Methods in Structural Chemistry, Hungarian Academy of Sciences, Department of Nuclear Chemistry, Eötvös Loránd University, Pázmány P. s. 1/a, Budapest 1117, Hungary; e-mail: klencsar@para.chem.elte.hu*
[2]*Institute of Materials Science, NCSR Demokritos, Aghia Paraskevi 153 10, Athens, Greece*

Abstract. The temperature dependence of the local electronic and magnetic state of iron in the chalcogenide spinel $FeCr_2S_4$ has been studied in detail around the Curie temperature with the help of ^{57}Fe Mössbauer spectroscopy. Ferrimagnetic order was found to break down anomalously at $\cong 20$ K below the Curie temperature in $FeCr_2S_4$. Magnetic relaxation of iron magnetic moments was observed with increasing relaxation rate above $T \cong 155$ K. The possible relation of magnetic relaxation to the effect of magnetoresistance is discussed.

Key words: chalcogenides, magnetoresistance, Mössbauer spectroscopy.

1. Introduction

The discovery of colossal magnetoresistance (CMR) in manganese-based perovskites has stimulated intense research on the physical bases of the CMR effect [1–3]. The puzzling existence of a considerable intrinsic magnetoresistance reported [4] in $FeCr_2S_4$ chalcogenide spinels – that do not possess manganese, oxygen, perovskite structure, or even a metal-to-insulator transition – indicates that phenomena other than the double exchange effect [2] should also be considered in the explanation of the effect of magnetoresistance observed in these materials.

$FeCr_2S_4$ has long been known as a ferrimagnet with semiconducting like resistivity characteristics [5–15]. It has a crystal structure of a cubic normal spinel where Fe^{2+} and Cr^{3+} cations occupy tetrahedrally and octahedrally coordinated positions, respectively [16]. The paramagnetic to ferrimagnetic transition occurs in the range $T_C = 170$–180 K [4, 5]. The recent observation of a CMR effect in $FeCr_2S_4$ [4] initiated a reinvestigation of this compound by the means of various methods [17–24]. By the help of ^{57}Fe Mössbauer spectroscopy we investigated the local magnetic and electronic state of Fe^{2+} in $FeCr_2S_4$ as a function of temperature in the range 75–290 K. In this article we focus our attention on the possible relation between the magnetic state of iron in $FeCr_2S_4$ and the magnetoresistance as well as

* Author for correspondence.

the resistivity anomaly observed in this compound at temperatures in the vicinity of the Curie temperature [4]. The detailed analysis of the temperature dependence of the Mössbauer parameters will be published elsewhere [25].

2. Experimental

The $FeCr_2S_4$ sample that we investigated is the same material as that for which macroscopic magnetization, thermopower, resistivity and CMR measurements have been reported by Ramirez *et al.* in [4] which article may be consulted for details of the sample preparation. ^{57}Fe Mössbauer spectroscopy measurements were carried out on powdered samples in transmission geometry. During measurements the temperature of the sample was kept constant with a precision of $\Delta T \cong \pm 0.5$ K. The material was first cooled down to 75 K, then measurements were performed by raising the temperature by $\cong$5 K between subsequent measurements.

3. Results and discussion

Figure 1 shows representative ^{57}Fe Mössbauer spectra of $FeCr_2S_4$ taken at temperatures near the Curie point as well as at $T = 75$ K. Although the Curie temperature,

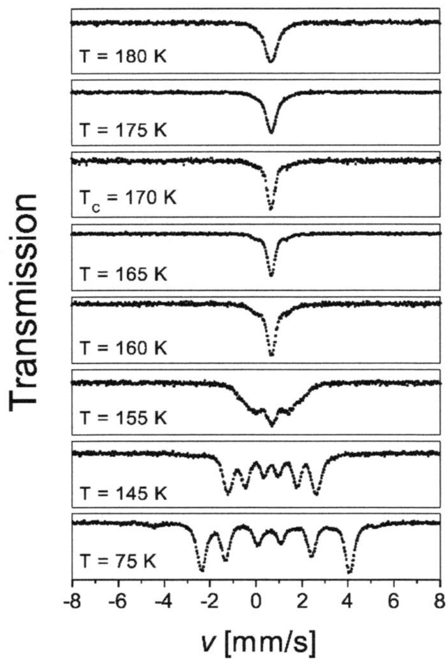

Figure 1. ^{57}Fe Mössbauer spectrum of $FeCr_2S_4$ at selected temperatures near the Curie temperature (T_C) and at $T = 75$ K. The graphs share a common velocity axis. Zero velocity corresponds to the isomer shift of α-iron at 290 K.

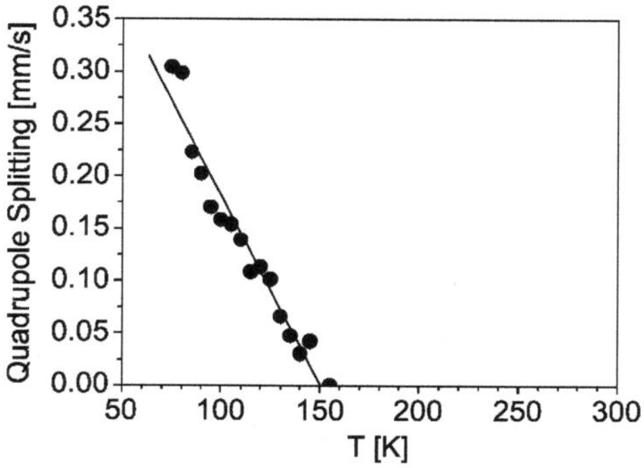

Figure 2. The temperature dependence of the apparent quadrupole splitting of ^{57}Fe in $FeCr_2S_4$. The size of the points is representative for the uncertainty of the displayed values. The line through the points is a guide to the eye.

determined by magnetization measurements, was reported to be $T_C \cong 170$ K for this particular sample [4], the long range magnetic order starts to break down already at $T_R \cong 155$ K, and the Mössbauer spectrum taken at $T = 160$ K consists mainly of a singlet. The existence of a magnetically unsplit singlet component in the Mössbauer spectrum of $FeCr_2S_4$ below the Curie temperature can be explained by the relaxation effect [26], and shows that magnetic order breaks down anomalously in $FeCr_2S_4$.

The onset of the magnetic relaxation observed in the ^{57}Fe Mössbauer spectra of $FeCr_2S_4$ at $T = T_R$ is in correlation with the temperature dependence of the coercive force [22] and anisotropy field [20], both quantities going to zero at $T \cong T_R$. Thus, the magnetic relaxation observed between T_R and T_C arises because the magnetic moment of the iron cations can relax easily between different directions as a consequence of the zero anisotropy field.

The temperature dependence of the apparent ^{57}Fe quadrupole splitting is shown in Figure 2. The quadrupole splitting decreases with temperature and reaches zero at $T \cong T_R$ where the magnetic relaxation sets in. This is in accordance with earlier measurements [10], and shows that in $FeCr_2S_4$ quadrupole splitting of ^{57}Fe is induced by long-range magnetic order. The observation that both the quadrupole splitting and the magnetocrystalline anisotropy [20, 22] goes to zero with increasing temperature at around $T \cong T_R$ indicates that T_R is the temperature where the spin–orbit coupling of Fe^{2+} [7, 10] becomes weak compared to the magnetic exchange interaction.

It was recently pointed out that the persistence of short-range magnetic order above the Curie temperature is indicative of the existence of magnetic polarons [24, 27]. In $FeCr_2S_4$ we tested the existence of a possible short-range magnetic order existing above T_C by a ^{57}Fe Mössbauer measurement (Figure 3) carried out at $T = 186.5$ K, by 16.5 K above the Curie temperature, in 6 T external magnetic field. As shown in Figure 3, $H = 6$ T external magnetic field is enough to establish

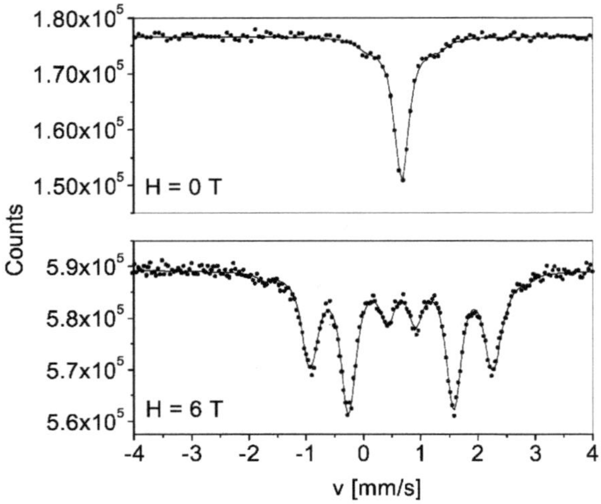

Figure 3. ^{57}Fe Mössbauer spectrum of $FeCr_2S_4$ recorded at $T = 186.5$ K in an external magnetic field of $H = 0$ T (*top*) and $H = 6$ T (*bottom*). The two graphs share a common velocity axis. Zero velocity corresponds to the isomer shift of α-iron at 290 K.

long range magnetic order without the relaxation effect in $FeCr_2S_4$, as a result of which even above the Curie temperature a well-developed magnetic sextet is observed.

The magnetic field displayed by the sextet equals to $H = 9.89(1)$ T, higher than the applied field by around 3.89 T, which latter is the induced hyperfine field. This shows that at $T = 186.5$ K $FeCr_2S_4$ is not in a pure paramagnetic state but it still displays a considerable antiferromagnetic exchange interaction between Fe^{2+} and Cr^{3+}. This observation is in agreement with the existence of magnetic polarons above the Curie temperature in $FeCr_2S_4$ [24].

The electrical resistivity of $FeCr_2S_4$ [4, 21] displays semiconducting characteristics at all temperatures except in the range 150 K $\leqslant T \leqslant$ 180 K where it steeply increases with temperature. Although the resistivity characteristics in this range are often interpreted as being metallic [17, 21], we propose that this range is better seen as a transition range between two different semiconducting states (S_{FM} for $T \leqslant 150$ K and S_{PM} for $T \geqslant 180$ K), in which $FeCr_2S_4$ is semiconducting at all temperatures.

While the state S_{FM} displays a well-developed long range magnetic order, the state S_{PM} corresponds to the paramagnetic state. The transition region between these two states is represented by a magnetic order that breaks down gradually with increasing temperature. It is this temperature range where the negative magnetoresistance observed in $FeCr_2S_4$ is the most pronounced [4].

In conclusion, the resistivity anomaly as well as the observed negative magnetoresistance effect in $FeCr_2S_4$ can be understood in the following way.

Below $T \cong T_R$ nonzero spin–orbit coupling of Fe^{2+} [7, 10] results in considerable magnetic anisotropy [20] and consequent coercive force [22]. The apparent

Fe^{2+} quadrupole splitting displayed in the ^{57}Fe Mössbauer spectra (Figure 2) is also proportional to the strength of the spin–orbit interaction. With increasing temperature spin–orbit coupling becomes weaker and approaches zero at around $T \cong T_R$. As a result, magnetic anisotropy and Fe^{2+} quadrupole splitting becomes zero at the same temperature. Due to the low level of magnetic anisotropy energy, in comparison with the magnetic exchange energy and the thermal energy, at $T \cong T_R$ relaxation of the Fe^{2+} magnetic moments sets in and gets visualized in the ^{57}Fe Mössbauer spectra for $T \geqslant T_R$ (Figure 1).

As revealed by the ^{57}Fe Mössbauer measurement in external magnetic field (Figure 3), short range magnetic correlations exist in $FeCr_2S_4$ even above the Curie temperature. This refers to the existence of magnetic polarons as charge carriers [27]. Thus, it is plausible to assume that in $FeCr_2S_4$ electrical current proceeds through the activated hopping of these quasiparticles. The activated hopping type conduction process results in the semiconducting resistivity characteristics. At the same time, the effective activation energy of hopping of a magnetic polaron is expected to be influenced by the degree of magnetic order. Namely, with increasing magnetic disorder the probability of hopping is reduced, and the effective activation energy of the hopping of magnetic polarons is expected to increase. This explains why electrical resistivity starts to increase with increasing temperature at $T \cong T_R$ where magnetic order starts to break down. The increase of electrical resistivity with increasing temperature continues until the level of magnetic disorder reaches a critical (near maximum) value around $T \cong T_C$. A further increase in temperature results in decreasing electrical resistivity due to the increase in thermal energy. In the transition region 150 K $\leqslant T \leqslant$ 180 K the application of an external magnetic field can recover long range magnetic order, thereby reducing the activation energy of hopping of magnetic polarons. As a consequence, electrical resistivity is reduced and negative magnetoresistance is realized as observed [4].

Acknowledgements

This work was supported by the Hungarian Science Foundation OTKA (F 034837 and T 034839), by the Hungarian Academy of Sciences (AKP 2000-143 2, 4). Z. Klencsár acknowledges the support from the Hungarian Bolyai János research grant.

References

1. Helmolt, R., Wecker, J., Holzapfel, B., Schultz, L. and Samwer, K. *Phys. Rev. Lett.* **71** (1993), 2331.
2. Zener, C., *Phys. Rev.* **82** (1951), 403.
3. Millis, A. J., *Nature* **392** (1998), 147.
4. Ramirez, A. P., Cava, R. J. and Krajewski, J., *Nature* **386** (1997), 156.
5. Shirane, G., Cox, D. E. and Pickart, S. J., *J. Appl. Phys.* **35** (1964), 954.
6. Haacke, G. and Beegle, L. C., *Phys. Rev. Lett.* **17** (1966), 427.

7. Eibschutz, M., Shtrikman, S. and Tenenbaum, Y., *Phys. Lett. A* **24** (1967), 563.
8. Haacke, G. and Beegle, L. C., *J. Phys. Chem. Solids* **28** (1967), 1699.
9. Yagnik, C. M. and Mathur, H. B., *Solid State Comm.* **5** (1967), 841.
10. Hoy, G. R. and Singh, K. P., *Phys. Rev.* **172** (1968), 514.
11. Lotgering, F. K., Van Stapele, R. P., Van der Steen, G. H. A. M. and Van Wieringen, J. S., *J. Phys. Chem. Solids* **30** (1969), 799.
12. Gibart, P., Dormann, J.-L. and Pellerin, Y., *Phys. Stat. Sol.* **36** (1969), 187.
13. Gibart, P., Goldstein, L. and Brossard, L., *J. Magn. Magnet. Mat.* **3** (1976), 109.
14. Van Diepen, A. M., Lotgering, F. K. and Olijhoek, J. F., *J. Magn. Magnet. Mat.* **3** (1976), 117.
15. Brossard, L., Dormann, J. L., Goldstein, L., Gibart, P. and Renaudin, P., *Phys. Rev. B* **20** (1979), 2933.
16. West, A. R., *Basic Solid State Chemistry*, 2nd edn, John Wiley & Sons, New York, 1999, p. 60.
17. Chen, Z., Tan, S., Yang, Z. and Zhang, Y., *Phys. Rev. B* **59** (1999), 11172.
18. Yang, Z., Tan, S. and Zhang, Y., *Solid State Comm.* **115** (2000), 679.
19. Yang, Z., Tan, S., Chen, Z. and Zhang, Y., *Phys. Rev. B* **62** (2000), 13872.
20. Tsurkan, V., Lohmann, M., Krug von Nidda, H.-A., Loidl, A., Horn, S. and Tidecks, R., *Phys. Rev. B* **63** (2001), 125209-1.
21. Park, M. S., Kwon, S. K., Youn, S. J. and Min, B. I., *Phys. Rev. B* **59** (1999), 10018.
22. Yang, Z. R., Tan, S. and Zhang, Y. H., *Appl. Phys. Lett.* **79** (2001), 3645.
23. Tsurkan, V., Hemberger, J., Klemm, M., Klimm, S., Loidl, A., Horn, S. and Tidecks, R., *J. Appl. Phys.* **90** (2001), 4639.
24. Wang, S., Li, K., Chen, Z. and Zhang, Y., *Phys. Rev. B* **61** (2000), 575.
25. Klencsár, Z., Kuzmann, E., Homonnay, Z., Vértes, A., Simopoulos, A., Devlin, E. and Kallias, G., *J. Phys. Chem. Solids* (2003), in press.
26. Vértes, A. and Nagy, D. L. (eds), *Mössbauer Spectroscopy of Frozen Solutions*, Akadémiai Kiadó, Budapest, 1990, p. 15.
27. De Teresa, J. M., Ibarra, M. R., Algarabel, P. A., Ritter, C., Marqulna, C., Blasco, J., Garcia, J., del Moral, A. and Arnold, Z., *Nature* **386** (1997), 256.

Hyperfine Interactions **144/145:** 267–272, 2002.

^{57}Fe Mössbauer Studies of $Sr_{2-x}Ca_xFeReO_6$ Double Perovskite Compounds

A. P. DOUVALIS[1], M. VENKATESAN[1], J. M. D. COEY[1], T. ALAMELU[2] and U. V. VARADARAJU[2]
[1]*Physics Department, Trinity College, Dublin 2, Ireland; e-mail: douvalia@kcd.ie*
[2]*Materials Science Research Centre, Indian Institute of Technology, Madras, India*

Abstract. We present a ^{57}Fe Mössbauer study of the $Sr_{2-x}Ca_xFeReO_6$ double perovskite series, with $x = 0$, 0.1, 0.2, 0.5, 1.0, 1.5 and 2.0. The analysis at 300 and 16 K shows iron in three different metal neighbour environments, due to the misplacement of Fe and Re ions (antisite defects). Judging from the isomer shift and hyperfine magnetic field values, iron ions on the perfect ordered structure show intermediate valence state (between Fe^{3+} and Fe^{2+}), with an effective electronic configuration $3d^{(5+y)}$, where y decreases from 0.3 to 0.2 as x increases from 0 to 2. Changes in quadrupole shift values for $x \geqslant 1.0$ confirm the cubic to monoclinic structural transition observed in these compounds.

Key words: Mössbauer spectroscopy, double perovskites.

1. Introduction

The half metallic ferrimagnetic double perovskites Sr_2FeReO_6 and Sr_2FeMoO_6 show quite good low field magnetoresistive properties at room temperature (RT), and their Curie temperatures (T_C) exceed significantly those of the magnetoresistive manganite perovskites [1, 2]. These materials have potential for novel spin electronic and magnetic sensor device applications [3]. The $T_C = 405$ K of Sr_2FeReO_6 can increase by substitution of Sr^{2+} by Ca^{2+} to the maximum value of $T_C = 540$ K for Ca_2FeReO_6 [4]. However the crystal structure of $Sr_{2-x}Ca_xFeReO_6$ reduces from cubic to monoclinic, and the conductivity from metallic to non-metallic as x increases from 0 to 2 [4]. Fe and Re ions occupy the octahedral O-coordinated B and B′-sites of the $A_2BB'O_6$ double perovskite structure. Hybridisation of only the $\downarrow t_{2g}$ Fe $3d$ and Re $5d$ with the O $2p$ electronic states at the Fermi level is responsible for the half metallic properties of Sr_2FeReO_6 [1]. Investigation of the characteristics of Fe ions is of great importance to the understanding of the mechanisms that are responsible for the magnetic, structural, electronic and magneto-transport properties of these compounds. Here the properties of iron ions in the series $Sr_{2-x}Ca_xFeReO_6$ are investigated using ^{57}Fe Mössbauer spectroscopy.

2. Experimental

Bulk polycrystalline $Sr_{2-x}Ca_xFeReO_6$ (x = 0, 0.1, 0.2, 0.5, 1.0, 1.5, 2) samples were prepared by solid-state reaction following the method used in [4, 5]. Mössbauer spectra (MS) were collected in transmission geometry at temperatures of 300 and 16 K using a constant acceleration mode spectrometer, equipped with a $^{57}Co(Rh)$ source kept at RT, and an APD (DMX-20) closed-loop He cryostat. Calibration of the spectrometer was done using α-Fe, and all isomer shift (IS) values are given relative to α-Fe at RT.

3. Results and discussion

3.1. MÖSSBAUER SPECTROSCOPY

Representative MS of the $Sr_{2-x}Ca_xFeReO_6$ samples for x = 0, 0.5, 1.0, 1.5 and 2 collected at 300 and 16 K are shown in Figures 1(a) and (b), respectively. All MS show well-resolved magnetic split spectral lines at both temperatures. However, a broadening of the outer spectral lines is evident for all spectra both at 300 and 16 K. A set of three components was used to fit the spectra for each compound at both temperatures. This set of components for all spectra is composed of a "main" component, which has the largest absorption area (A), a "secondary" component with lower absorption area relative to the main, and a third "minor" component, which has the lowest absorption area of all. The values of the hyperfine parameters resulting from the fits at 300 and 16 K vary with x. In Figure 2 the x variation of IS, quadrupole shift (2ε), hyperfine magnetic field (B_{hf}), and A found at 16 K is shown. As can be seen from this figure the main component has the highest and the minor component has the lowest IS value for each compound, while the highest B_{hf} value corresponds to the minor component and the lowest to the main component for each spectrum. The secondary component has intermediate IS and B_{hf} values, closer however to the corresponding values of the main component for each compound. Similar behaviour is observed for the corresponding values of the 300 K spectra.

It is known that in the ideal $A_2BB'O_6$ double perovskite crystal structure Fe and Re ions alternate in the non-equivalent B and B′-sites respectively, creating a NaCl-type superstructure [1, 6]. However, misplacements where a Fe ion takes the place of a Re ion on a B′-site and vice-versa, where a Re ion takes the place of a Fe ion on a B-site, can occur in the structure. These defects are known as antisite defects [7, 8]. Consequently not all iron ions experience the same first cation neighbour environment. Iron in the perfect NaCl-type superstructure has a total of six rhenium first cation neighbours at the B′-sites. However, the misplaced iron ions at the B′-sites have a total of six iron ions at the neighbouring B-sites, while these last six iron ions experience an environment of five rhenium and one iron ions at their neighbouring B′-sites. Thus if the concentration of antisite defects (p) is relatively low ($<6\%$), only the iron ions with the three different cationic

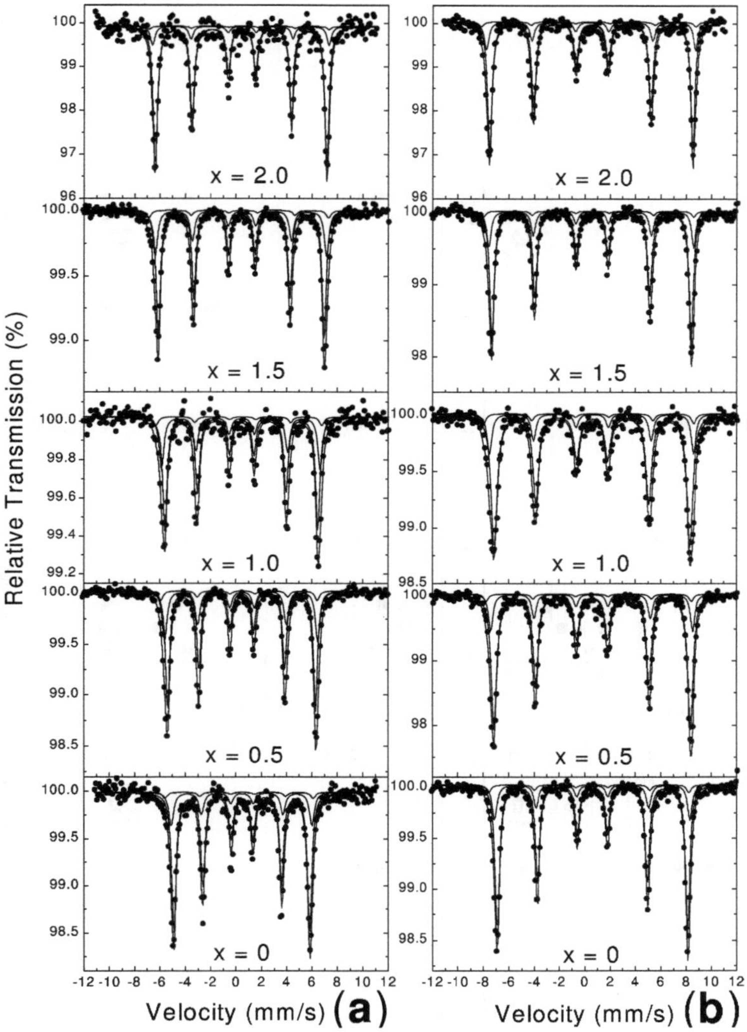

Figure 1. Mössbauer spectra of the $Sr_{2-x}Ca_xFeReO_6$ compounds collected at 300 K (a) and 16 K (b). The continuous lines are best fits to the experimental data.

environments mentioned above are expected to be present in appreciable numbers on the structure of these compounds. These three different kinds of iron ions appear as different components of the Mössbauer spectra due to the influence of their different environment on their hyperfine parameters.

The IS value of the main component is always slightly higher than the value of the secondary component for all compounds. Both are intermediate between the characteristic values found for octahedraly coordinated Fe^{3+} and Fe^{2+} ions in oxides [9, 10]. However, the IS values of the minor component for all compounds correspond to a pure ferric state (see Figure 2(a)). As in the case of other iron-based

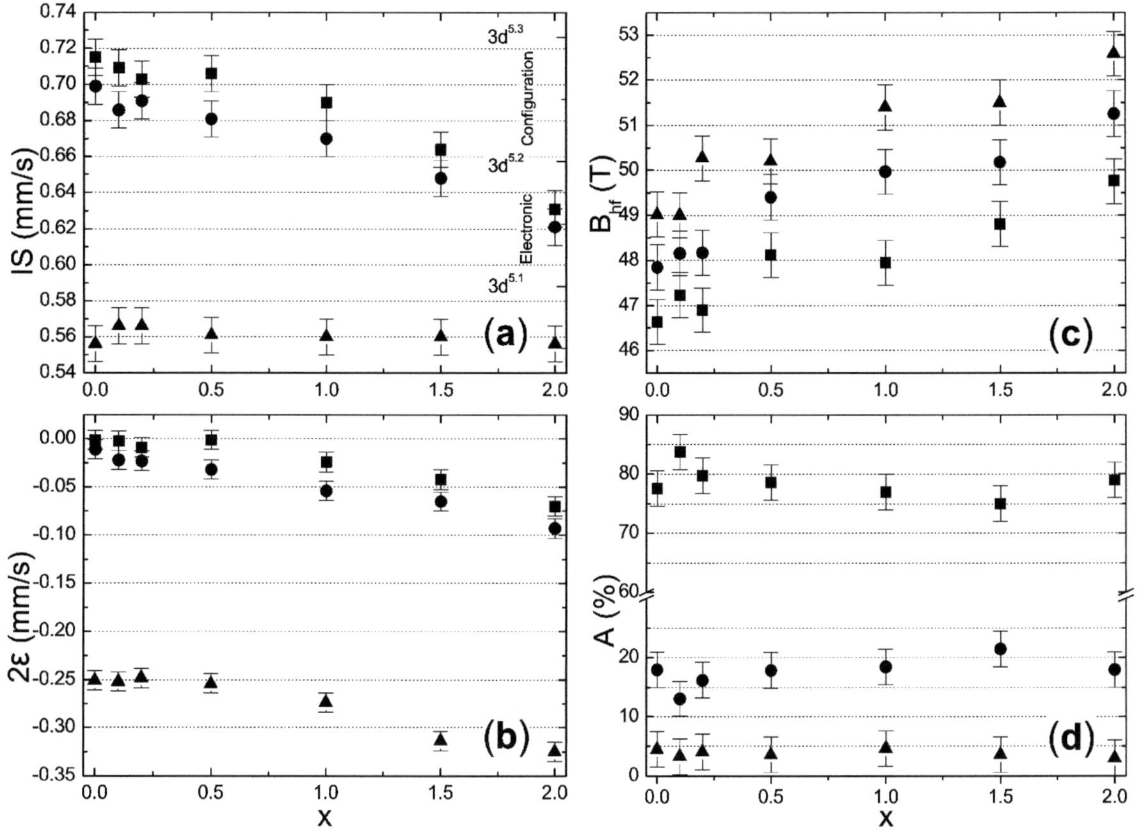

Figure 2. Variation with x of the Mössbauer parameters for the $Sr_{2-x}Ca_xFeReO_6$ compounds as resulted from the fits of the spectra collected at 16 K, (a) isomer shift, (b) quadrupole shift, (c) hyperfine magnetic field, and (d) relative absorption area. The points denote the three components used to fit each spectrum (see text); squares: main, circles: secondary, triangles: minor. Error bars for the Mössbauer parameters are included for each value. The right y axis in Figure 2(a) scales the electronic configuration of iron ions.

double perovskite compounds [7, 8, 11, 12], we attribute the main components off all spectra to iron ions in the perfect NaCl-type superstructure. The secondary components are attributed to iron ions with a total of five rhenium and one iron first cation neighbours, and the minor components are attributed to iron ions with a total of six iron first cation neighbours. The intermediate valence state observed for the main and secondary components results from the additional electronic charge density for these ions, introduced by the hybridisation between the ↓Fe $3d$ (t_{2g}) and Re $5d$ (t_{2g}) electronic states [1]. The slight increase in the valence state of the secondary components comes from the reduction of this hybridisation due to the presence of one iron as first cation neighbour. The minor components show ferric character due to the absence of any hybridisation between the electronic states of iron and rhenium ions.

Figure 2 shows the IS of the main and secondary components to decrease only slightly for $0 < x < 1.0$ and more profoundly for $x \geqslant 1.0$. However, the IS values of the minor components show no change with x variation. The values of 2ε of the main and secondary components are close to zero and show no appreciable variation for $0 < x < 1.0$ but a further increase on their absolute values is shown for $x \geqslant 1.0$. The minor components show the highest absolute 2ε values for each compound because the misplaced iron ions experience a more distorted octahedral oxygen environment, due to the presence of six iron first cation neighbours. However these values are also increased further for $x \geqslant 1.0$. The B_{hf} values of all components increase with x, but the increase is continuous for this parameter. Finally the values of A are nearly independent of x for all components.

The reduction of the IS values for the main and secondary components for $x \geqslant 1.0$ suggests a corresponding relative decrease in the hybridisation of the ↓Fe $3d$ (t_{2g}) and Re $5d$ (t_{2g}) electronic states. Since this hybridisation relates to the conduction mechanism in these compounds [1], its reduction should have an effect on the conducting properties, and indeed an increase in the resistivity has been observed when the x value increases, as the materials $Sr_{2-x}Ca_xFeReO_6$ from metallic ($x = 0$) become non-metallic ($x = 2$) [4]. However the main and secondary components for $x = 2$ do not show the same pure ionic Fe^{3+} character as the third component in this compound. This could be due to the presence of additional electronic charge density on these iron ions, which comes from a different kind of electronic state admixture between the ↓Fe $3d$ (e_g) and Re $5d$ (e_g) states as recently proposed by Wu [13]. Using the IS values of the main components and the limiting IS values of 0.52 mm/s and 1.20 mm/s for $3d^{5.0}$ and $3d^{6.0}$ configurations, respectively [9–12], we can estimate the effective electronic configuration, which for the iron ions on the perfect ordered structure reduces from $3d^{5.29}$ to $3d^{5.16}$ as x increases from 0 to 2. This is shown in Figure 2(a).

A crystal structure transition from cubic to monoclinic occurs for $x > 1.0$ [4]. The enhanced absolute 2ε values of all components for $x \geqslant 1.0$ are indicative of this structural transition. The parallel change on the IS and 2ε values for $x \geqslant 1.0$ suggests that the structural transition should be related to the change of the electronic configuration and to the conductivity reduction. Ca^{2+} ions have much smaller ionic radius than Sr^{2+} ions, and on increasing the Ca^{2+} concentration the structural transition occurs, which effects the electronic and transport properties of these compounds [14, 15].

The B_{hf} values increase with increasing Ca^{2+} concentration but they do not show an abrupt change for $x \geqslant 1.0$. This increase reflects an enhancement on the strength of the magnetic interactions between the magnetic Fe and Re ions as Ca^{2+} concentration increases, which is further confirmed by the linear increase of the T_C values with x [4].

Acknowledgement

This work was supported by the EU growth program through the AMORE project (Advanced Magnetic Oxides for Responsive Engineering).

References

1. Kobayashi, K.-I., Kimura, T., Tomioka, Y., Sawada, H. and Terakura, T., *Phys. Rev. B* **59** (1999), 11159.
2. Kobayashi, K.-I., Kimura, T., Sawada, H., Terakura, K. and Tokura, Y., *Nature* **395** (1998), 677.
3. Coey, J. M. D., In: M. Ziese and M. J. Thornton (eds), *Spin Electronics*, Springer-Verlag, Berlin, Heidelberg, 2001, p. 277.
4. Alamelu, T., Varadaraju, U. V., Venkatesan, M., Douvalis, A. P. and Coey, J. M. D., *J. Appl. Phys.* **91** (2002), 8909.
5. Westerburg, W., Lang, O., Ritter, C., Felser, C., Tremel, W. and Jakob, G., *Sol. State Comm.* **122** (2002), 201.
6. Sleight, A. W. and Weiher, J. F., *J. Phys. Chem. Solids* **33** (1972), 679.
7. Greneche, J. M., Venkatesan, M., Suryanarayanan, R. and Coey, J. M. D., *Phys. Rev. B* **63** (2001), 174403.
8. Venkatesan, M., Grafoute, M., Douvalis, A. P., Greneche, J.-M., Suryanarayanan, R. and Coey, J. M. D., *J. Mag. Magn. Mater.* **242–245** (2002), 744.
9. Greenwood, N. N. and Gibb, T. C., In: *Mössbauer Spectroscopy*, Chapman and Hall, London, 1971.
10. Kawanaka, H., Hase, I., Toyama, S. and Nishihara, Y., *Physica B* **281–282** (2000), 518.
11. Douvalis, A. P., Venkatesan, M. and Coey, J. M. D., *Hyp. Interact.* (2003) (in press).
12. Douvalis, A. P., Venkatesan, M., Coey, J. M. D., Grafoute, M., Greneche, J.-M. and Suryanarayanan, R., *J. Phys.: Condens. Matter* **14** (2002), 12611.
13. Wu, H., *Phys. Rev. B* **64** (2001), 125126.
14. Gopalakrishnan, J., Chattopadhyay, A., Ogale, S. B., Venkatesan, T., Greene, R. L., Millis, A. J., Ramesha, K., Hannoyer, B. and Marest, G., *Phys. Rev. B* **62** (2000), 9538.
15. Kato, H., Okuda, T., Okimoto, Y., Tomioka, Y., Oikawa, K., Kamiyama, T. and Tokura, Y., *Phys. Rev B* **65** (2002), 144404.

Hyperfine Interactions **144/145:** 273–278, 2002.

Surface and Bulk Crystallization of $Fe_{61}Co_{21}Nb_3B_{15}$ Alloy

S. N. KANE[1,*], D. RODIONOV[1,**], B. BERNHARDT[1], L. KRAUS[2] and G. KLINGELHÖFER[1,***]
[1]*Institut für Anorganische und Analytische Chemie, Johannes Gutenburg-Univertät Mainz, StaudinderWeg 9, D-55099 Mainz, Germany; e-mail: klingel@mail.uni-mainz.de*
[2]*Institute of Physics ASCR, Na Slovance 2, CZ-180 40, Praha 8, Czech Republic*

Abstract. Surface and bulk crystallization of $Fe_{61}Co_{21}Nb_3B_{15}$ alloy has been studied. Crystallization (surface and bulk) of specimens starts after annealing at 400°C. Differences in crystalline fraction, order in crystalline phase and environment around Fe-atom are observed at the surface and in the bulk. Crystalline component appearance (both at the surface and in the bulk) after annealing at 480°C is attributed to Fe_3Co.

Key words: miniaturized Mössbauer spectrometer, XRD, DSC, nano-crystalline alloys.

1. Introduction

Nanocrystalline soft magnetic materials produced by devitrification of amorphous ribbons consisting of extremely fine crystalline grains (~10 nm) embedded in a residual amorphous matrix [1], have received much attention in last few years. It has been shown that the random orientation of grains and the tiny grain size results in substantial reduction of magnetocrystalline anisotropy, leading to extremely soft magnetic material [2]. In these materials, contrary to conventional granular materials, the matrix is magnetic and its properties can be changed by increasing the annealing temperature of the parent alloy. In transition metal metalloid metallic glasses it is shown that small amount of Nb helps in inhibiting the grain growth, resulting in nano-crystallization. Crystallization is generally found to start from the surface. Lower stability of the surface against crystallization may be attributed to several factors [3], e.g. a change in the surface composition due to oxidation or surface segregation; a faster atomic transport at the surface [4]. In addition, the oxygen content may stabilize a number of crystalline phases, thus increasing the driving force for crystallization [4]. In magnetic specimens, internal stresses generated by surface crystallization may drastically modify their bulk magnetic prop-

* On leave from: School of Physics, D.A. University, Khandwa road Campus, Indore-452017, India.
** On leave from: Space Research Institute, Profsouznaya Str. 84/32, Moscow, Russia.
*** Corresponding author.

erties [5]. Thus, from the point of view of application, understanding the surface crystallization is of great importance. Influence of the composition on magnetic, magnetoelastic properties as well as crystallization behaviour of $Co_xFe_{84-x}Nb_7B_9$ has been studied previously [6–9]. Up to now only few works have been devoted to the comparison of surface and bulk crystallization behaviour [10–13]. Mössbauer spectroscopy (in transmission and backscattering geometry) is very suitable technique to study nanocrystalline alloys yielding information about structural and magnetic arrangement [14]. In the present work, for the first time we have used Miniaturized Mössbauer spectrometer '*MIMOS II*' [15] to study change occurring on the surface due to surface crystallization. The spectrometer '*MIMOS II*' [15] operates in backscattering geometry measuring the scattered 14.41 keV Mössbauer radiation and the 6.4 keV Fe X-rays. The backscattered 6.4 keV Fe X-rays penetrate (~ few tens of nm) at the very surface of the sample, providing information about change occurring at the surface and hence can be used effectively to monitor crystallization occurring at the surface. Thus, combining transmission and backscattering Mössbauer measurements, one can obtain information about bulk and surface crystallization.

Objective of the present work is to study the surface and bulk crystallization of $Fe_{61}Co_{21}Nb_3B_{15}$ alloys as a function of annealing temperature.

2. Experimental details

Amorphous ribbons of nominal composition $Fe_{61}Co_{21}Nb_3B_{15}$, 10 mm wide and about 20 μm thick, prepared by planar flow casting method, were obtained from the Slovak Academy of Sciences, Bratislava. To get various stages of crystallization, the samples were annealed isochronally in Ar-atmosphere for 1 hour between 350–480°C. Differential Scanning Calorimetery (DSC) measurements were done at a heating rate of 20 K/minute, to get the crystallization temperature. X-ray diffraction (XRD) measurements were done using Cu-K_α radiation. Mössbauer spectra (transmission and backscattered) were recorded at room temperature in constant acceleration mode using ^{57}Co:Rh source. Transmission spectra were recorded using conventional spectrometer, while backscattered spectra from shiny side of ribbons were recorded using miniaturized Mössbauer spectrometer '*MIMOS II*' [15]. Backscattered and transmission Mössbauer spectra consisting of amorphous and crystalline components were fitted assuming a distribution of hyperfine fields.

3. Results and discussions

First crystallization temperature obtained from DSC measurement is 421°C. The as-quenched specimen was X-ray amorphous. The diffractograms taken from wheel side showed a small sharp peak near the (200) reflections of α-Fe, which disappeared after a short etching in diluted HNO_3. This indicates the presence of some crystalline phase in a very thin surface layer on the wheel side of the rib-

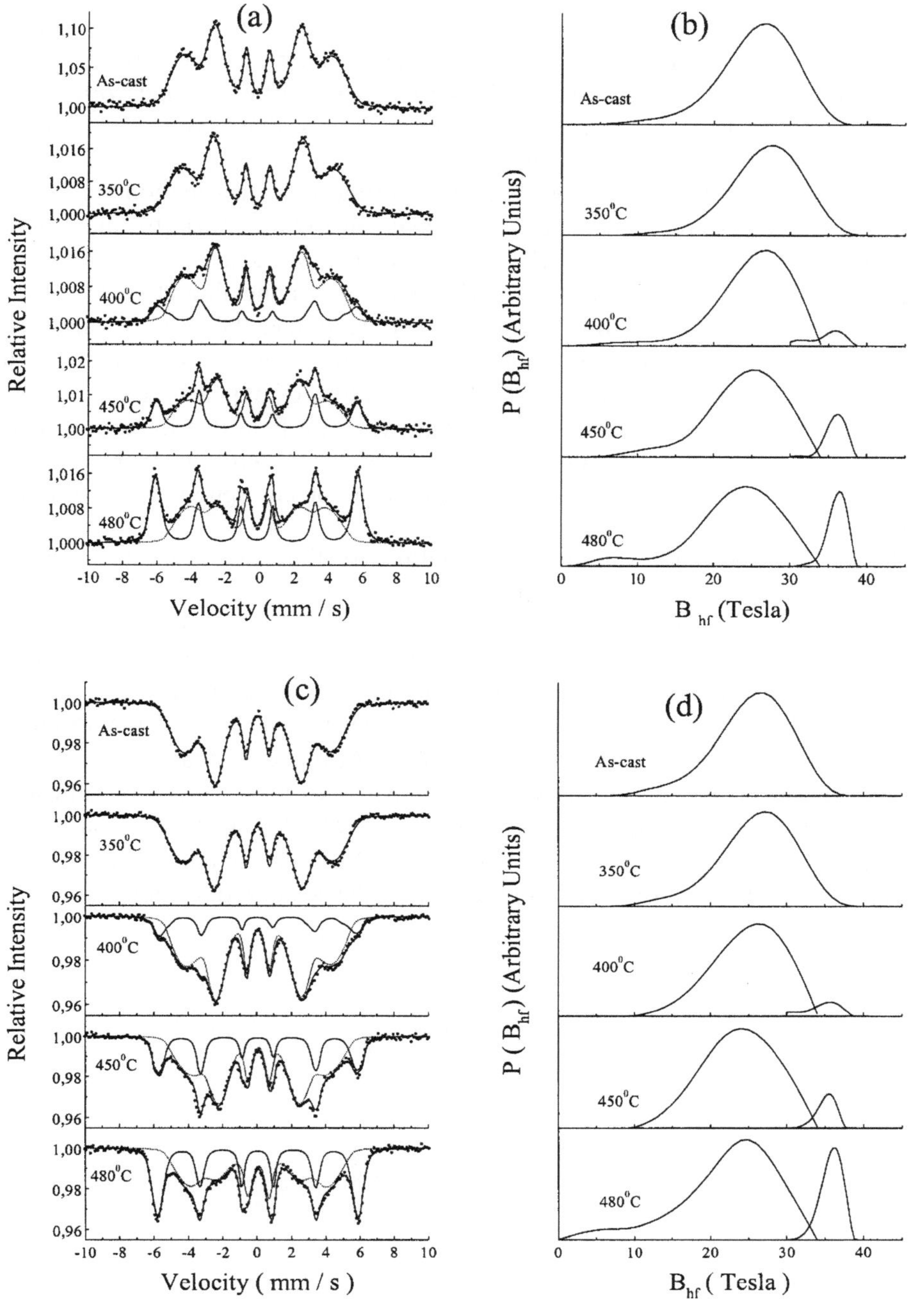

Figure 1. Room temperature transmission (a)–(b) and backscattering (corresponding to 6 keV Fe X-ray) (c)–(d) Mössbauer spectra and corresponding hyperfine field distributions of the as-prepared and annealed (for 1 hour at indicated temperatures) $Fe_{61}Co_{21}Nb_3B_{15}$ specimens.

bon. Crystallization of the specimens starts at 400°C and the XRD patterns of the samples annealed between 400–480°C for one hour show the coexistence of a bcc crystalline phase and a residual amorphous phase. The mean grain size of the bcc crystalline phase was evaluated by Scherrer's equation from the half width of the (110) reflection peak. Mean grain size for specimens annealed at 400, 450 and 450°C is respectively 26.9, 29.5 and 33.3 nm.

Mössbauer spectra and corresponding hyperfine field distribution for surface and bulk for the studied specimens are shown in Figures 1(a)–(b) and (c)–(d), respectively. In accordance with XRD measurements, Mössbauer results also show that crystallization (both surface and bulk) starts after annealing at 400°C. Mössbauer spectra of these specimens consist of a crystalline (bcc-Fe–Co alloy) and an amorphous components, and are fitted well with a single crystalline and a single amorphous component, assuming the distribution of hyperfine fields for both components. Perusal of Figure 1 shows that hyperfine field distribution of amorphous component (both for surface and bulk) is broad, while that of crystalline component, it is sharp. This behaviour is typical for amorphous and crystalline components.

Figure 2(a) shows the variation of crystalline fraction (expressed in %) for surface and bulk (obtained from the fitting of Mössbauer spectra) of the crystallized specimens with annealing temperature. Perusal of Figure 2(a) shows variation in crystalline fraction at the surface and in the bulk of the specimen. Surface initially shows higher crystalline fraction compared to the bulk of the specimen, suggesting that crystallization starts from surface and proceeds to the bulk of the specimen. That fact is in agreement with earlier work [16]. At higher annealing temperature, crystallization in the bulk of the specimen supercedes the surface crystallization. Variation of fractional width of distribution $\Delta H/\langle H\rangle$ (for crystalline component) with annealing temperature is shown in Figure 2(b). The fractional width of the hyperfine field distribution $\Delta H/\langle H\rangle$ (ΔH being the standart deviation and $\langle H\rangle$ the

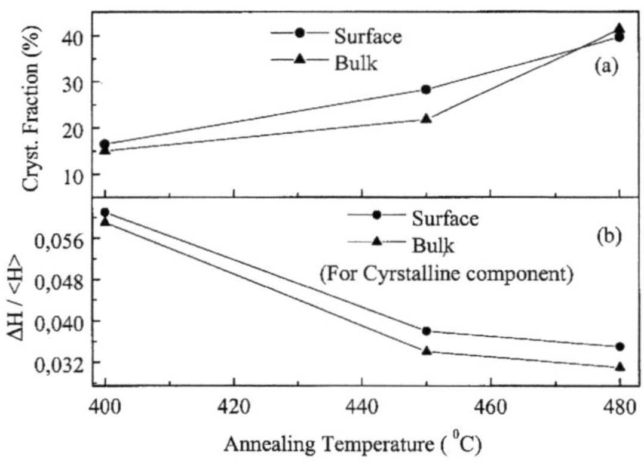

Figure 2. Variation of crystalline fraction (a) and fractional width of distribution (b) of $Fe_{61}Co_{21}Nb_3B_{15}$ for surface and bulk of the specimen with annealing temperature.

average value of the hyperfine field) is a measure of the degree of disorder in the specimen. Decrease of $\Delta H/\langle H\rangle$ with increase in annealing temperature suggests the ordering of crystalline phase. Here also one can observe (from the values of $\Delta H/\langle H\rangle$) that surface shows lower ordering than of surface then bulk. In case of as-cast specimen, $\Delta H/\langle H\rangle$ for surface and bulk is respectively 0.198 and 0.194, suggesting differences in disorder even in as-cast state. Intensity of second and fifth lines relative to innermost lines of the Mössbauer spectrum 'b' is a measure of spin texture in the specimen. In absence of stresses in the specimen, the atomic spins are expected to remain in the ribbon plane due to shape anisotropy, thus making 'b' to have value 4. On the other hand, for a completely random spin alignment, 'b' is expected to be equal to 2. In case of as-cast specimen, value of 'b' for surface and bulk is respectively 2.78 and 2.89, suggesting that atomic spins are preferentially aligned in the ribbon plane, which is in accordance with usual situation [17].

Figures 3(a) and (b) respectively shows the variation of average hyperfine field $\langle H\rangle$ of crystalline and amorphous phase (surface and bulk) with annealing temperature. Perusal of Figure 3(a) shows that $\langle H\rangle$ (for crystalline component) slightly increases with increase in annealing temperature. This variation in the hyperfine field with annealing temperature may be attributed to a variation in the composition of the alloy [18]. After annealing at 480°C the $\langle H\rangle$ of the crystalline component is $\approx$ 35.8 T, which is close to that of Fe_3Co, suggesting that crystalline component formed (both at the surface and in the bulk of the specimen) after this annealing treatment is Fe_3Co. Perusal of Figure 3(b) shows that $\langle H\rangle$ (amorphous component) decreases with increase in annealing temperature. This observed decrease in $\langle H\rangle$ could be attributed to the Nb near neighbours of Fe, causing iron magnetic moment to decrease, thus resulting in a lower hyperfine field. One can also see that $\langle H\rangle$ has slightly lower values for bulk than surface. Variation of $\langle H\rangle$ gives indirect evidence of some changes in environment around Fe-atom at surface and the bulk.

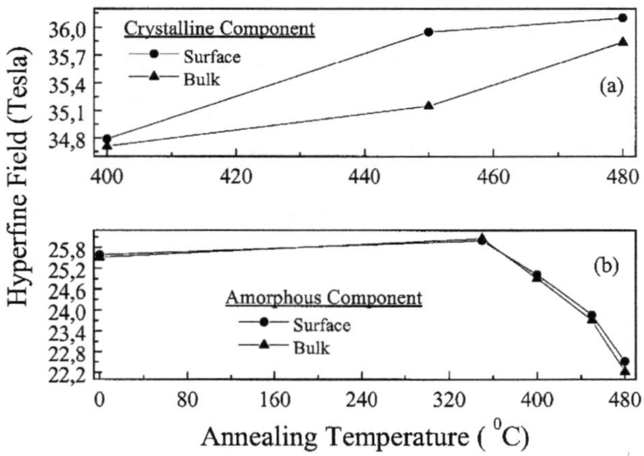

Figure 3. Variation of average hyperfine field of crystalline (a) and amorphous (b) phases (for surface and bulk) of $Fe_{61}Co_{21}Nb_3B_{15}$ with annealing temperature.

4. Conclusions

XRD and Mössbauer spectroscopy has been used to study surface and bulk crystallization of $Fe_{61}Co_{21}Nb_3B_{15}$ alloy. Both XRD and Mössbauer measurements show that crystallization (surface and bulk) of specimen starts after annealing of specimens at 400°C. Crystallization starts from surface and proceeds to the bulk of the specimen. Present studies also give indirect evidence of some changes in environment around Fe-atom at surface and the bulk. Annealing of specimen at 480°C leads to the formation of Fe_3Co phase at surface and bulk of the specimen.

Acknowledgement

Financial support to authors (SNK and DR) from German Space Agency (DLR), Germany is gratefully acknowledged.

References

1. Yoshozawa, Y., Oguma, S. and Yamaguchi, K., *J. Appl. Phys.* **64** (1988), 6044.
2. Herzer, G., *IEEE Trans. Mag.* **26** (1990), 1741.
3. Koster, U., *Mater Sci. Engg.* **97** (1988), 233.
4. Gibson, M. A. and Delamore, G. W., *J. Mater. Sci.* **23** (1988), 1164.
5. Herzer, G. and Hilzinger, H. R., *J. Magn. Magn. Mater.* **62** (1986), 1164.
6. Kraus, L., Soyka, V., Duhaj, P. and Matko, I., *J. de Physique IV* **8** (1998), 95.
7. Kaczkowski, Z., Kraus, L., Soyka V. and Duhaj, P., *Acta Phys. Slovaca* **48** (1998), 687.
8. Kraus, L., Haslar, V., Duhaj, P., Svec, P. and Studnicka, V., *Mater. Sci. Engg. A* **226–228** (1997), 626.
9. Kane, S. N., Gupta, A., Kraus, L. and Duhaj, P., *J. Magn. Magn. Mater.* **215–216** (2000), 375.
10. Kopcewicz, M., Grabias, A. and Nowicki, P., In: P. Duhaj, P. Mrafko and P. Svec (eds), *Proc. Nineth Inter. Conf. on Rapidly Quenched and Metastable Materials*, Supplement, Bratislava, Elsevier, Amsterdam 1997, p. 95.
11. Grabias, A. and Kopcewicz, M., *Materials Sci. Forum* **269–272** (1998), 725.
12. Grabias, A., Kopcewicz, M. and Idzikowski, B., *Mater. Res. Symp. Proc.* **577** (1999), 543.
13. Miglirini, M., Seberini, M. and Greneche, J.-M., *Czechlslovak J. Phys.* **51** (2001), 677.
14. Greneche, J.-M., Miglirini, M. and Slawska-Winiewaska, A., *Hyp. Interact.* **126** (2000), 27.
15. Klingelhöfer, G., In: M. Miglierini and D. Petridis (eds), *Mössbauer Spectroscopy in Materials Science*, Academic Publishers, Amsterdam, 1999, p. 413.
16. Oak, H. N. and Morrish, A. H., *Phys. Rev. B* **23** (1981), 2257.
17. Kane, S. N., Bhagat, N., Gupta, A. and Varga, L. K., *J. Mgn. Magn. Mater.* **167** (1997), 241.
18. Gupta, A., Ruwali, K., Paul, N. and Duhaj, P., *Mater. Sci. Engg. A* **304–306** (2001), 371.

Hyperfine Interactions **144/145:** 279–288, 2002.

Effect of Precipitation Temperature and Number of Iron Atoms per Molecule on the Structure of Hydrated Iron(III) Oxyhydroxide Ferritin Cores Synthesised *in vitro*

WANIDA CHUA-ANUSORN[1], HYANG-RAN MUN[2], JOHN WEBB[2], NICOLE T. GORHAM[1] and TIMOTHY G. ST. PIERRE[1,*]
[1] *School of Physics, The University of Western Australia, Perth, WA 6009, Australia; e-mail: stpierre@physics.uwa.edu.au*
[2] *Chemistry, Division of Science and Engineering, Murdoch University, Perth, WA 6150, Australia*

Abstract. Horse spleen apoferritin was reconstituted with varying numbers of iron atoms per protein shell at 25 and 50°C. Samples with mean particle sizes ranging from approximately 20 to 1100 iron atoms per particle were synthesised. Transmission electron microscopy combined with elemental analysis and protein assays indicated a characteristic iron density for the iron(III) oxyhydroxide of 6.4 ± 0.6 iron atoms per nm^3 thus suggesting a maximum possible loading per ferritin shell of 1700 ± 200 iron atoms. Room temperature Mössbauer spectra of the iron(III) oxyhydroxide particles formed within the ferritin protein shells were doublets with slightly larger quadrupole splittings when precipitated at 25°C compared with those precipitated at 50°C. The quadrupole splittings were not significantly correlated with the number of iron atoms per protein shell although the general trend was for smaller particles to have larger quadrupole splittings. A very small but statistically significant increase in the centre shift of the Mössbauer spectra with increasing particle size was observed. These results, taken together with results from other studies, indicate that reconstituted ferritin cores are highly disordered and low density when compared with those found in native horse spleen ferritin. The results also indicate that raising the temperature of reconstitution increases the structural order within the particles slightly but that increasing the number of iron atoms per ferritin molecule above 300 has little influence on the apparent overall structural order.

Key words: ferritin, iron, Mössbauer spectroscopy, ferrihydrite.

1. Introduction

The ferritins are a family of proteins with a common ultrastructure [1]. They consist of 24 polypeptide chains assembled into 12 dimers which in turn assemble to form an approximately spherical shell-like structure. The outer diameter of the shell is approximately 12 nm while the inner cavity has a diameter of approximately 8 nm. In many biological systems ferritin is used as an iron storage/homeostasis protein. Iron is stored in the central cavity of the protein in the form of a hydrated

* Author for correspondence.

iron(III) oxyhydroxide associated with varying concentrations of inorganic phosphate depending on the biological source of the protein [2]. Most commonly, the mineral core of ferritin has a structure based on that of the mineral ferrihydrite ($5Fe_2O_3 \cdot 9H_2O$). However, the degree of crystallinity and size of the mineral cores vary depending on the source of the protein [2–4]. The size and structure of the nanoscale mineral cores of ferritin are likely to determine their reactivity and hence biological role. For example, larger particles will have a lower surface area per iron atom thus reducing the number of reaction sites per iron atom, while a lower degree of crystallinity probably results in less tightly bound iron ions hence increasing reactivity. The presence of phosphate in the mineral core can lead to a stronger binding of iron [5, 6].

In order to enable more systematic studies of the chemical properties of ferritin mineral cores, it is advantageous to prepare synthetic mineral cores by precipitating iron(III) oxyhydroxide within the protein shells of apoferritin. In this way the number of iron atoms per protein shell and the number of phosphate ions per protein shell can be controlled allowing studies of the effect of particle size and phosphate content on the properties of ferritin [7–9]. It has also been shown that the temperature of the solution during the precipitation process has an effect on the structure of the resulting mineral cores with higher temperatures yielding more crystalline particles [8]. Here we further investigate the effect of reaction temperature on the structure of the mineral cores of ferritin reconstituted *in vitro*. Mössbauer spectroscopy is used to gain information on the structural characteristics of mineral cores with varying numbers of iron atoms precipitated at two different temperatures.

2. Materials and methods

Horse spleen apoferritin in 0.1 M sodium acetate buffer (pH 5.3; Sigma®) was utilized throughout this study. Before use, apoferritin was treated by dialysis at 4°C against 10 mM EDTA to remove any excess metal atoms in the solution followed by dialysis against appropriate buffer at pH7.5 (0.1 M MOPS: 3-[N-morpholino]propane-sulfonic acid; Sigma®). After dialysis, the sample was filtered through a 0.22 μm Millex-GS sterile filter. All chemicals were analytical grade. All glassware was acid-washed and sterilized.

The concentration of apoferritin was firstly determined by using UV/visible spectrophotometry at 280 nm following methods described elsewhere [10]. A Hewlett Packard 8450A UV/VIS spectrophotometer was used for these measurements.

2.1. RECONSTITUTION OF FERRITIN WITH IRON

Apoferritin (1 mg/ml) in MOPS buffer (0.1 M, pH 7.5) was reconstituted by additions of 10 mM $FeSO_4$ solution to give theoretical loadings of 50, 100, 500,

1000, 1500 and 2000 iron atoms per molecule at 25 and 50°C. Both the apoferritin solutions and the Fe(II) solutions were placed in a water-bath for preheating before the addition of Fe(II) solution. The reaction was carried out in a stepwise fashion by loading 50 iron atoms per molecule of protein every hour. The Fe(II) solution was added dropwise to apoferritin gently and slowly. After each addition of 500 iron atoms per molecule of protein, the ferritin solution was kept at the same temperature for 12 hours before further addition of iron.

After reconstitution with iron, each ferritin solution was filtered with a 0.45 μm membrane and was then concentrated with the buffer being replaced with distilled water using an ultrafiltration cell (Amicon Inc.). Finally, the reconstituted ferritin in water was filtered again with a 0.22 μm Millex-GS sterile syringe filter. Drops of the solution were then prepared for transmission electron microscopy (TEM) study with the remainder being freeze-dried for iron content analysis using atomic absorption spectrometry (AA Spectrometer, Spectra AA-40, Varian) and Mössbauer spectroscopic studies.

In addition, reference samples of "2-line" and "6-line" ferrihydrite were prepared following the procedures of Schwertmann and Cornell [11]. These synthetic minerals are so named because of the number of powder diffraction "lines" they exhibit.

2.2. TRANSMISSION ELECTRON MICROSCOPY

The morphology and size of the mineral cores of the reconstituted ferritins were studied using a transmission electron microscope operated at 80 keV. Drops of each of the ferritin samples diluted to about 0.5 mg/ml were allowed to dry onto formvar coated electron microscope copper grids. The grids were air-dried.

The particle sizes of the reconstituted ferritins were measured by using vernier callipers on enlargements of transmission electron micrographs. The largest dimension of each electron dense core was recorded. At least 146 particles were measured for each sample. For some samples it was possible to measure approximately two or three hundred particles. Catalase crystals (Alltech®) which have lattice spacings of 8.75 nm and 6.85 nm, were used to calibrate the magnification of the TEM. Particle volume distributions were calculated from the particle size distributions by assuming spherical shapes for the particles and taking the particle size measurement to represent the diameter of each particle.

2.3. MÖSSBAUER SPECTROSCOPY

Each freeze-dried reconstituted ferritin sample and synthetic ferrihydrite was mounted in 10 mm diameter perspex sample holders. Synthetic ferrihydrite samples were diluted with a small amount of sucrose. The amount of sample was adjusted according to the iron content in each sample, so that the intensity of the 14.4 keV γ-rays was attenuated by a factor of approximately $1/e$ by the sample wherever possible. Mössbauer spectra were recorded at room temperature using a ^{57}Co in Rh foil source mounted on a constant acceleration drive operated with a double-ramp wave-form. The velocity range of ±4 mm/s, was calibrated with respect to the spectrum of an iron metal foil, the centre of the pattern being taken as zero velocity. Spectra were folded to eliminate the parabolic background formed by the variation of solid angle between source and detector. The folded spectra consist of 250 channels with approximately 1×10^7 counts per channel. Spectra were fitted with distributions of doublets of Lorentzian peaks using the method of Rancourt and Ping [12]. The underlying Lorentzian peaks were constrained to have linewidths of 0.184 mm/s. All other parameters were allowed to vary freely.

3. Results

On addition of the Fe(II) solutions to the apoferritin solutions, the resulting solution was found to be clear with a reddish brown colour. No precipitation was observed to occur for loadings of 1500 iron atoms per ferritin shell or less at 25°C and for loadings of 1000 Fe atoms per shell or less at 50°C. At 2000 iron atoms per shell at 25°C and 1500 iron atoms per shell at 50°C, there was evidence for a small amount of precipitation in the resulting solutions. These precipitates were removed during the subsequent filtration phase of preparation. At higher iron loadings and especially at 50°C, precipitation was very evident indicating precipitation of iron external to the protein shells. Figure 1 shows the relationship between the nominal

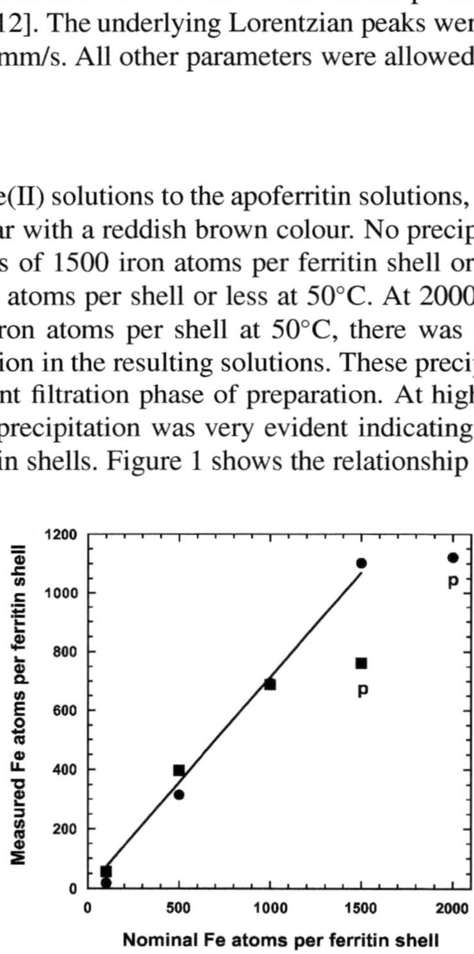

Figure 1. Graph showing relationship between the nominal number of iron atoms per ferritin shell as determined by the preparation method and the number of iron atoms per ferritin shell as measured by atomic absorption spectroscopy of acid digested freeze dried reconstituted ferritin. Squares indicate samples reconstituted at 50°C and circles indicate samples reconstituted at 25°C. The two data points marked with a "p" correspond to preparations where a small amount of precipitation was seen during the reconstitution procedure. The solid line is a straight line fit through the origin to the data points for which no precipitation was seen (gradient = 0.71 ± 0.02). Note that there is both a 50 and 25°C datum point for nominal loadings of 1000 iron atoms per ferritin shell.

number of iron atoms added per protein shell and the number of iron atoms per protein shell deduced from elemental analysis of the iron content of the freeze dried reconstituted ferritins. There is a good linear correlation between the two measures for all samples for which no precipitation was observed. The gradient of the straight line through the origin is 0.71 ± 0.02 which is less than the ideal value of 1. For the two samples that showed a small amount of precipitation during preparation, the ratios of numbers of iron atoms per shell measured for the freeze dried material to the nominal number of iron atoms per shell fall below 0.71.

Examples of transmission electron micrographs of the ferritins reconstituted at 25 and 50°C are shown in Figure 2. The morphology of the particles appears to be slightly irregular with the particles precipitated at 50°C perhaps having a slightly greater degree of irregularity in shape.

The relationship between the number of iron atoms per shell measured for the freeze dried material and the mean particle volume measured by TEM is shown in Figure 3. There appears to be a linear correlation between the two with the linear gradient being 6.4 ± 0.6 iron atoms per nm^3.

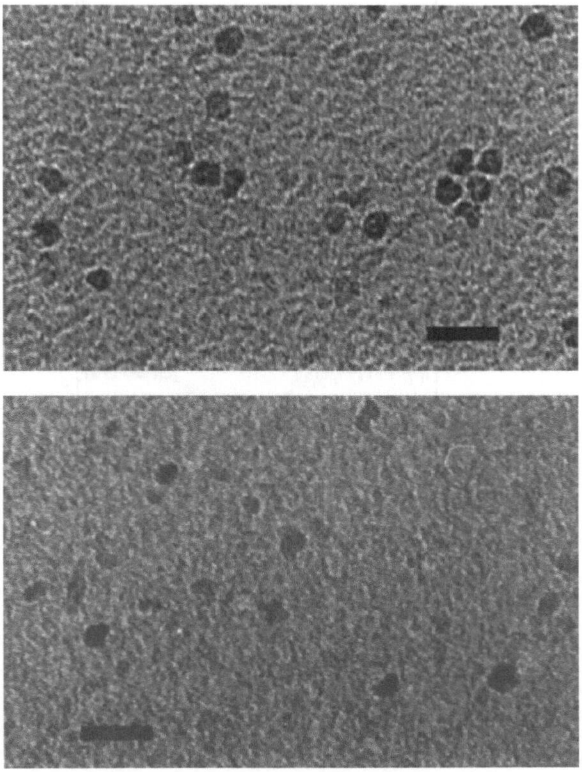

Figure 2. Transmission electron micrographs of ferritin mineral cores reconstituted with a nominal 1500 iron atoms per ferritin shell at (*top*) 25°C and (*bottom*) 50°C. Scale bars are 20 nm.

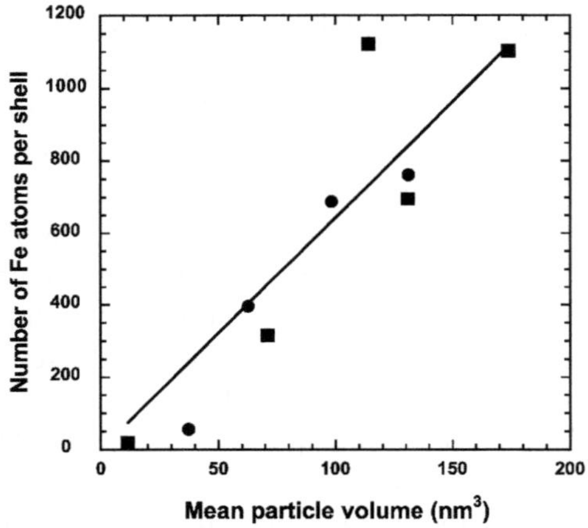

Figure 3. Graph showing the relationship between number of iron atoms per ferritin shell measured by atomic absorption spectrometry of freeze dried samples and mean mineral particle volume measured from transmission electron microscopy images. Square data points correspond to 50°C preparations and circles to 25°C preparations. The solid line is a straight line through the origin fitted to the data (gradient $= 6.4 \pm 0.6$ iron atoms per nm^3).

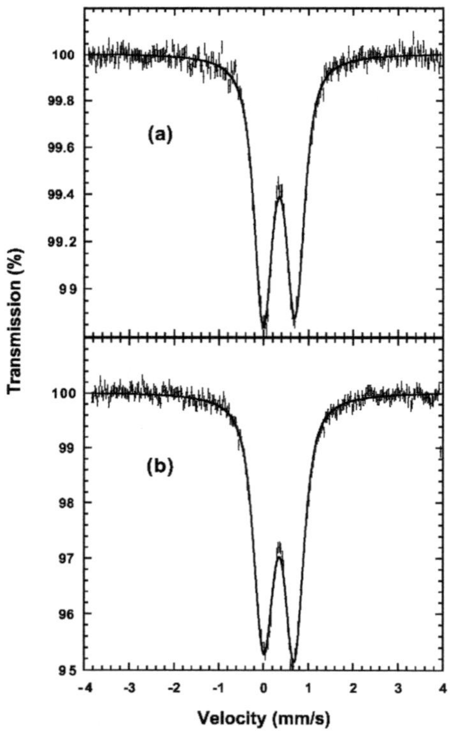

Figure 4. Example Mössbauer spectra of freeze-dried reconstituted ferritins at room temperature. (a) Ferritin reconstituted at 25°C with a measured number of iron atoms per shell of 20. (b) Ferritin reconstituted at 25°C with a measured number of iron atoms per shell of 1100. The solid lines are fits of Voigt profiles to the data.

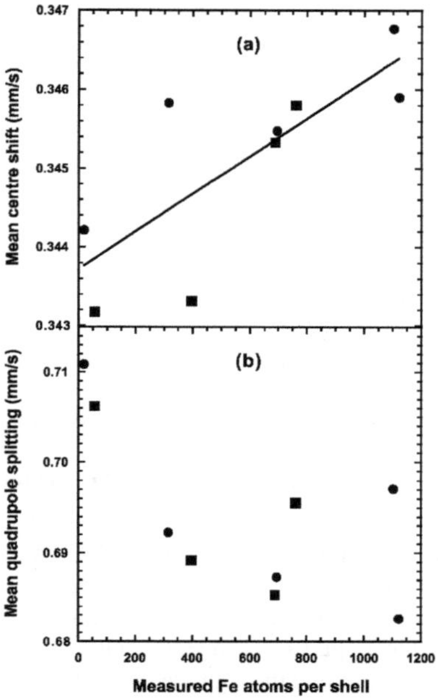

Figure 5. (a) Mean centre shift of Mössbauer spectra vs measured number of iron atoms per ferritin shell. (b) Mean quadrupole splitting vs measured number of iron atoms per ferritin shell. Squares indicate data from 50°C preparations, circles indicate 25°C preparations.

Examples of the Mössbauer spectra obtained from the freeze dried reconstituted ferritin samples at room temperature are shown in Figure 4. The spectra consist of a doublet of broad absorption peaks. The spectra were not well fitted with Lorentzian peaks. Hence the spectra were fitted with a Gaussian distribution of Lorentzian peaks using the method of Rancourt and Ping [12]. A significant ($p = 0.03$) positive Spearman rank order correlation was found between the mean centre shift and measured number of iron atoms per protein shell (Figure 5(a)). However, the magnitude of the changes in centre shift with protein iron loading are extremely small. Variations in quadrupole splitting for the reconstituted ferritins were somewhat larger. Although no statistically significant correlation was found between quadrupole splitting and protein iron loading, the trend showed a generally decreasing quadrupole splitting with increasing iron loading (Figure 5(b)). It was found that the samples prepared at 25°C have consistently larger mean quadrupole splittings than the corresponding samples prepared at 50°C. Furthermore, a high degree of correlation was found between the mean quadrupole splitting values measured for samples resulting from identical preparation conditions but at 25 and 50°C (Figure 6). No significant correlations were found between the width of the quadrupole splitting distribution and protein iron loading.

Mössbauer spectra of the samples of "2-line" and "6-line" ferrihydrite gave mean centre shifts of 0.339 ± 0.002 and 0.346 ± 0.002 mm/s, respectively, and mean quadrupole splittings of 0.721 ± 0.005 and 0.664 ± 0.005 mm/s, respectively.

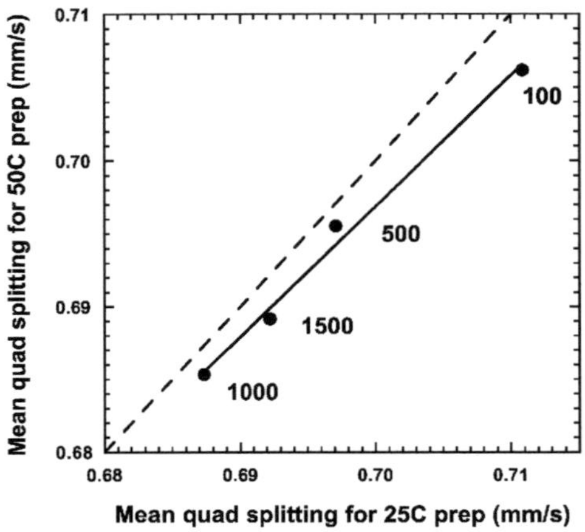

Figure 6. Graph showing relationship between the mean quadrupole splittings measured for ferritins reconstituted at the two different temperatures of 50 and 25°C. Each datum point represents a pair of samples (one prepared at 50°C and the other at 25°C). The labels by each datum point indicate the nominal number of iron atoms per ferritin shell used in each preparation. For each pair of samples, preparation conditions were identical apart from the temperature. The solid line is a straight line fit to the data. The dashed line is a straight line through the origin with a gradient of 1.

4. Discussion

The reddish brown colour that appears in the resulting solutions on addition of Fe(II) solution to apoferritin solution indicates the oxidation of the Fe(II) to Fe(III). The lack of bulk precipitation is an indication that the iron has been oxidised within the aqueous soluble ferritin protein shells. The appearance of discrete particles rather than agglomerations of particles in the TEM images of the dried solutions is consistent with the controlled precipitation of iron(III) oxyhydroxide within the protein shells. The discrepancy between the iron to protein ratios obtained from measurement of the protein concentration with spectrophotometry and those obtained from elemental analysis of the iron content of the freeze dried samples is systematic (Figure 1). The major contribution to the systematic error most likely arises from the calibration of the spectrophotometer since the specific absorption coefficients for apoferritin quoted in the literature vary greatly. As such, the iron to protein ratio measured for the freeze dried protein using atomic absorption spectrometry is probably the more reliable measure.

The particle size distributions measured by transmission electron microscopy indicate maximum particle sizes of approximately 8 nm. This observed maximum limit for the particle size is consistent with the known dimensions of horse spleen apoferritin. The relationship between mean particle volume and measured iron to protein ratio suggests a density of 6.4 ± 0.6 iron atoms per nm^3 for the mineral particles. Taking the internal diameter of horse spleen apoferritin to be 8 nm, these data imply a maximum possible iron loading per protein shell of 1700 ± 200 iron atoms. Native horse spleen ferritin is known to have average iron loadings of up to

3000 Fe atoms per protein shell [4] implying that the density of the iron oxyhydroxide mineral in the native protein is somewhat greater than that for the reconstituted mineral particles in this study. The higher density in the native protein may reflect the slower rate of iron uptake under physiological conditions.

The non-Lorentzian lineshape of the Mössbauer spectra for the reconstituted ferritins is most probably due to (i) the semi-disordered nature of ferrihydrite and (ii) the large fraction of iron atoms at the surfaces of the particles. The fact that the spectra were recorded at room temperature means that the samples are at temperatures much greater than the maximum superparamagnetic blocking temperature of ferritin [8] so that magnetic relaxation effects are negligible. Surface roughness will tend to produce a distribution of iron site environments thus broadening the Mössbauer spectral lines and distorting them from the Lorentzian shape expected for a sample with a unique iron site environment. As an example of the extent of influence of the surface on the Mössbauer spectrum, consider a particle with 1000 iron atoms. Approximately 50% of the iron atoms will be at the surface. Interestingly, there was no significant correlation between the width of the quadrupole splitting distribution (the prime determinant of doublet linewidth) and the protein iron loading possibly indicating that the range of iron site environments within the cores of the particles is just as great as the range encountered at the surfaces. This situation would be the case if the internal structure of the cores was reasonably disordered. The very small but significant increase in the centre shift of the Mössbauer spectra as the iron to protein ratio is increased (Figure 5(a)) suggests that iron site environments within the cores of the particles have slightly greater centre shifts than those at the surfaces. The mean centre shifts for the reconstituted ferritins were closer to the mean centre shift measured for the "2-line" ferrihydrite than that measured for the "6-line" ferrihydrite. The trend of decreasing mean quadrupole splitting with increasing iron load was not statistically significant. The mean quadrupole splittings for most of the reconstituted ferritins fall approximately midway between those found for the "2-line" and "6-line" ferrihydrites. The mean quadrupole splittings for the two reconstituted ferritins with the smallest nominal loadings had mean quadrupole splittings closer to that for the "2-line" ferrihydrite. The lack of significant correlation between the mean quadrupole splitting and ferritin iron loading may again reflect a high degree of disorder within the core of the particles resulting in core site quadrupole splittings being similar to those at surface sites.

In previous studies [8] it was shown that more crystalline particles were produced when ferritin was reconstituted at higher temperatures. Electron diffraction was used to show that ferritin mineral particles reconstituted at 55°C were more crystalline (up to 5 electron powder diffraction rings) than those reconstituted at 4°C (up to 2 electron powder diffraction rings). Since more disordered materials are expected generally to yield larger quadrupole splittings, the quadrupole splitting data for the ferritins reconstituted under identical conditions but at two different temperatures in this study (Figure 6) are consistent with the previous observations of the effect of temperature of reconstitution on the resulting crystallinity of the mineral particles. The samples produced at 50°C consistently have lower quadrupole splittings than their counterparts produced at 25°C.

In conclusion, the iron(III) oxyhydroxide cores of horse spleen ferritin reconstituted in a stepwise fashion at 25 and 50°C appear to have a high degree of structural disorder, with the degree of disorder within the cores of the particles being similar to that at the surface of the particles. The resulting mineral particles are somewhat less dense than those found in native horse spleen ferritin, probably because of the higher degree of crystallographic disorder compared with the native protein (native horse spleen ferritin typically contains mineral cores that exhibit at least 5 or 6 powder diffraction rings [4]). The Mössbauer spectral data are consistent with the observation that reconstitution at higher temperatures produces mineral particles with a slightly higher degree of crystallographic order. For particles with 300 or more iron atoms, the number of iron atoms per protein shell does not appear to have a major influence on the apparent degree of crystallographic disorder.

Acknowledgements

This work was partly funded by the Australian Research Council. We would also like to thank Mr Adam Fleming for assistance with use of the atomic absorption spectrometer.

References

1. Harrison, P. M. and Arosio, P., *Biochim. Biophys. Acta* **1275** (1996), 161.
2. St. Pierre, T. G., Webb, J. and Mann, S., In: S. Mann, J. Webb and R. J. P. Williams (eds), *Biomineralization: Chemical and Biochemical Perspectives*, VCH, Weinheim, 1989, p. 295.
3. St. Pierre, T. G., Tran, K. C., Webb, J., Macey, D. J., Heywood, B. R., Sparks, N. H., Wade, V. J., Mann, S. and Pootrakul, P., *Biol. Metals* **4** (1991), 162.
4. Mann, S., Bannister, J. V. and Williams, R. J. P., *J. Mol. Biol.* **188** (1986), 225.
5. Watt, G. D., *Anal. Biochem.* **99** (1979), 399.
6. Watt, G., Frankel, R. B., Papaefthymiou, G. C., Spartalian, K. and Stiefel, E. I., *Biochemistry* **25** (1986), 4330.
7. Mann, S., Williams, J. M., Treffry, A. and Harrison, P. M., *J. Mol. Biol.* **198** (1987), 405.
8. St. Pierre, T. G., Chan, P., Bauchspiess, K. R., Webb, J., Betteridge, S., Walton, S. and Dickson, D. P. E., *Coord. Chem. Rev.* **151** (1996), 125.
9. Williams, J. M., Danson, D. P. and Janot, C., *Phys. Med. Biol.* **23** (1978), 835.
10. Hofmann, T. and Harrison, P. M., *J. Mol. Biol.* **6** (1963), 256.
11. Schwertmann, U. and Cornell, R. M., *Iron Oxides in the Laboratory: Preparation and Characterization*, VCH, Weinheim, 1991.
12. Rancourt, D. G. and Ping, J. Y., *Nucl. Inst. Methods Phys. Res. B* **58** (1991), 85.

Hyperfine Interactions **144/145:** 289–296, 2002.

Mössbauer Spectroscopy in the Geosciences: Highlights and Perspectives

CATHERINE A. McCAMMON
Bayerisches Geoinstitut, Universität Bayreuth, D-95440 Bayreuth, Germany; e-mail: catherine.mccammon@uni-bayreuth.de

Abstract. Mössbauer spectroscopy in the geosciences is evaluated in the context of past achievements, current applications, and future prospects. The pattern of past publications shows the maturing of Mössbauer spectroscopy as an established technique, and an evolution from the simple description of minerals towards the study of dynamic processes. Current applications continue to exploit the many unique advantages of Mössbauer spectroscopy, and a survey of recent publications showcases some of the highlights. Future possibilities are discussed within the context of existing capabilities that remain to be exploited, ongoing developments, and innovations on the horizon.

Key words: mineralogy, review, high pressure.

1. Introduction

The start of the new millennium has motivated numerous appraisals of various research fields, particularly in the context of past successes, and the opportunities, directions, and problems of the future. Several workshops have focused on the field of mineralogy, highlighting its increasing interdisciplinary nature and encroachment into the fields of condensed-matter physics, biology and materials science. The nature of the techniques used to study minerals has evolved, and now embraces a large range of modern physics methods [1]. Mössbauer spectroscopy is one of these methods, and since its discovery in 1958, has been applied to a wide range of problems in the geosciences. Applications cover a large number of sub-disciplines, including descriptive mineralogy, mineral physics, environmental studies, volcanology and other mantle processes, applied mineralogy, oceanography and extraterrestrial studies. Geoscience publications account for roughly 10% of all published Mössbauer papers, a proportion that has remained more or less constant over the years. But what of the future? Can such an activity level be maintained? Are there aspects of Mössbauer spectroscopy still to be exploited by geoscientists? Are there new innovations on the horizon? This article explores some answers to these questions.

2. Historical perspective

The historical record can be evaluated with the help of data maintained by the Mössbauer Effect Data Center, Asheville (USA). A search under the subject heading "terrestrial and extraterrestrial minerals" (which includes most geoscience papers) shows an interesting pattern in publication rate that reflects research activity for the most part, but is also subject to other factors as discussed below (Figure 1).

Mössbauer spectroscopy became relevant to the geosciences in 1959 with the discovery of the effect in ^{57}Fe [2, 3]. Iron is an important element in the Earth, not only because is it present in large quantities, but also because it is the most abundant element by a large margin that has a variable valence state. Measurement of oxidation state provides an indication of oxygen activity, which is important to numerous physical and chemical properties and processes, among others the evolution of the Earth's atmosphere.

The number of mineralogy Mössbauer publications increased dramatically in 1966 and 1967, partly due to the discovery that Mössbauer data could be used to elucidate structural details of minerals such as cation distribution [4]. A pair of subsequent papers identified the empirical correlation between hyperfine parameters and the coordination and geometry of crystallographic sites [5, 6], and thus one of the most common applications of Mössbauer spectroscopy in mineralogy was born. The number of Mössbauer publications increased again in 1970 due to the return of the lunar samples, providing a bonanza for Mössbauer spectroscopists, e.g., [7]. Another increase in the number of Mössbauer publications occurred in 1975 due to two large proceedings volumes that arose from Mössbauer conferences, where such papers accounted for more than 30% of mineralogy Mösssbauer publications in that year. The peaks and troughs in subsequent years reflected the biennial frequency of conferences, with a reduction in published papers occurring after the recent decision to restrain proceedings volumes.

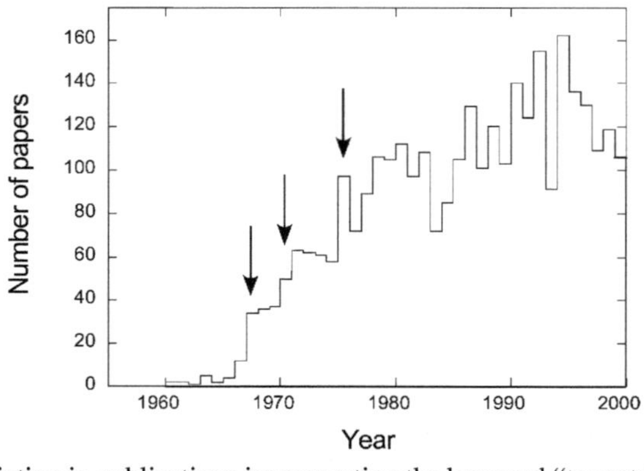

Figure 1. Annual variation in publications incorporating the keyword "terrestrial and extraterrestrial minerals" according to data maintained by the Mössbauer Effect Data Center, Asheville (USA). Arrows indicate years that saw a significant increase in publication rate.

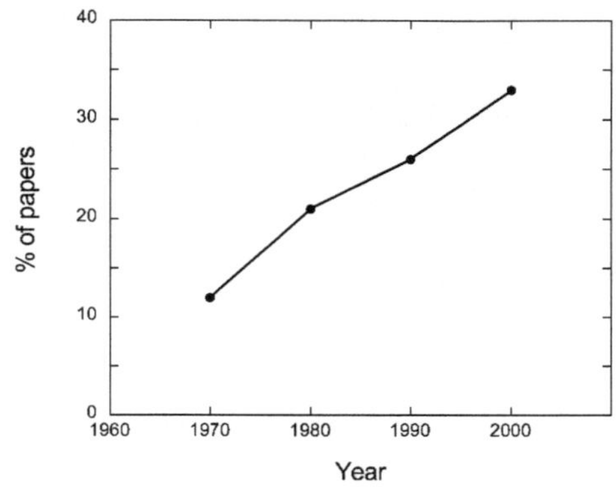

Figure 2. Annual variation in the percentage of mineralogy Mössbauer papers that describe studies of dynamic processes.

There has been an evolution in the nature of Mössbauer studies within the geosciences. A survey of papers from past decades reveals a trend away from the simple description of minerals to studies that investigate dynamic phenomena (Figure 2). Indeed within the geosciences in general there has been a trend away from descriptions of places and things to studies of dynamic systems and processes, e.g., [8].

3. Current applications

Mössbauer spectroscopy has many advantages that set it apart from most other techniques used in the geosciences. It is specific to the nucleus being examined, and hence is able to isolate those phases containing the Mössbauer isotope, which is ^{57}Fe in more than 95% of mineralogical studies. The energy resolution of ^{57}Fe is extremely high, better than 1 part in 10^{13}, which provides a sensitive probe of the atomic environment. It is able to distinguish between the different valence states of iron, which sets it apart from most other analytical techniques. It also provides structural information, such as the coordination and geometry of iron crystallographic sites. It is a short range probe, which allows the study of poorly crystallised materials such as glasses, as well as providing information complementary to that from long-range probes such as X-ray diffraction. It is highly sensitive, which enables data to be collected for minerals with low iron concentrations. Finally, Mössbauer spectroscopy can be applied at a large range of pressures and temperatures, enabling the study of samples at conditions approximating those both on the Earth's surface and within its interior.

These advantages are reflected in the nature of Mössbauer studies that have been made in the geosciences. Investigations can be classified roughly into one of three areas: structure (related to crystallographic, electronic or magnetic structure), phase analysis and oxidation state. While structural studies constitute mainly investigations of the static nature of minerals, phase analysis and valence state measure-

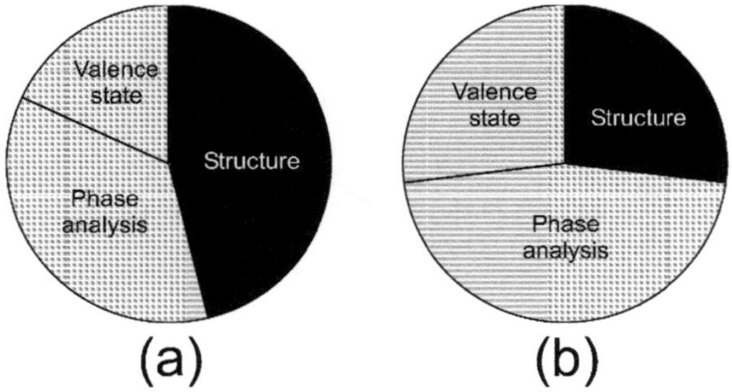

Figure 3. Distribution according to their nature for mineralogical investigations in the years 2000–2002 for (a) studies involving the static description of a material; and (b) studies focusing on dynamic processes.

ments include many studies of dynamic processes, such as phase transformations, oxidation and diffusion (Figure 3).

Mössbauer spectroscopy has taken its place in some laboratories alongside techniques such as the electron microprobe and X-ray diffraction in the routine analysis of iron-bearing samples. The description of new iron-bearing minerals often requires information on the structural and electronic state of iron, and Mössbauer spectroscopy is generally the tool of choice [9]. Measurements of physical properties often involve sample characterisation, particularly of oxidation state, since it can affect electrostatic charge balance and equilibrium defect concentration. Properties highly sensitive to such effects include rheology, transport behaviour, solubility of volatiles and partitioning of major and trace elements (to name just a few). Recent measurements on (Mg,Fe)O, for example, found a dependence of Fe^{3+} concentration on elastic properties [10]. Mössbauer spectroscopy is a crucial component in some geosensors, for example, the oxybarometer which is used to infer oxygen fugacity in equilibrium mineral assemblages containing garnet [11, 12]. Mössbauer measurements play an important role in some potential geothermometers [13, 14] and geospeedometers [15–17]. Indeed the only viable geothermometer for eclogite assemblages, garnet-clinopyroxene, is strongly influenced by Fe^{3+} concentration [18, 19], which is most reliably measured using Mössbauer spectroscopy.

There are many highlights in the past three years of Mössbauer applications in the geosciences, but in the limited space available here it is impossible to mention more than just a few. In the spirit of providing a snapshot of the field, however, the following list is given to showcase the range of studies that has been reported:

(1) investigation of low-temperature iron- and nickel-catalysed reactions showed that production of metal is sufficient to catalyse the generation of appreciable amounts of methane over geologic time [20];

(2) second-order Doppler shift measurements have been used to determine the isotopic fractionation of iron [21];

(3) zoning of ferric iron was used to infer the metasomatic history of kimberlites, with information on the preservation of diamond [22];
(4) analysis of Antarctic lake sediments revealed a possible biomarker for use on Mars [23];
(5) structural studies of volcanic glasses were used to constrain the thermal and temporal history of magma chambers [24];
(6) measurements in a diamond anvil cell provided evidence for chemical reactions at the mantle–core boundary [25];
(7) analysis using backscattered techniques showed the non-uniform distribution of phases in lateritic ore which is relevant to mineral concentration processes [26];
(8) analysis of natural tourmalines from Brazil were used to constrain their geological history [27]; and
(9) analysis of material from the Cretaceous–Tertiary boundary layer showed evidence for a rapid change in redox conditions after the impact [28].

4. Future perspectives

New developments in Mössbauer spectroscopy can arise from advances in instrumentation, theory, data analysis, or from new supplies of absorbers. The following section examines such possibilities within the following framework: (1) existing capabilities that have not been fully exploited by geoscientists; (2) advances that are currently under development; and (3) innovations on the distant horizon.

One aspect of Mössbauer spectroscopy that has improved in recent years is spatial resolution. The development of the Mössbauer milliprobe, for example, has allowed improvements in spatial resolution by more than two orders of magnitude [29, 30], while the imaging multiple Mössbauer spectrometer enables the study of inhomogeneities on a similar scale [31]. These techniques are seriously underutilised, however, despite the many potential applications. The history of many earth systems, for example, is written in the compositional variation of zoning patterns that can be observed in minerals. Coarse zoning (on a length scale of at least 50 μm) can already be studied using the above methods, but zoning on a finer scale will have to wait for new technique developments. Further improvement of spatial resolution may be anticipated with developments in synchrotron studies of nuclear forward scattering [32], while other possibilities include development of a Mössbauer electron microscope which would focus conversion electrons using conventional electron optics [33].

Another aspect of Mössbauer spectroscopy that could be more fully utilised by geoscientists is time resolution. Studies with existing techniques are possible over a wide range of time scales, including *in situ* studies of phase transitions on time scales of tens to hundreds to seconds, studies of transport properties such as diffusion on time scales of microseconds, and studies at the intrinsic time scale of the Mössbauer effect in ^{57}Fe to investigate processes such as relaxation and electron transfer. Existing facilities for forward nuclear scattering at synchrotron installations reduce this timescale even further. A time dependent multiple Mössbauer spectrometer has recently been described that provides time resolution in the range from microseconds to seconds by tracking changes in the chemical or magnetic state of iron atoms that are subject to cyclic perturbation [34].

Mössbauer spectroscopy can be performed under a wide range of pressure and temperature conditions. Such *in situ* studies are crucial to studies of dynamic effects such as diffusion, or phases that cannot be quenched to ambient conditions. Existing equipment allows experiments to be performed to approximately 120 GPa, and addition of a cryostat extends temperature capability from 4.2 to 300 K. Studies at room pressure are possible to at least 1200°C, depending on the Debye temperature of the material, but are ultimately limited by lattice vibrations, which reduce the recoil-free fraction with increasing temperature. Measurements at high pressures and temperatures have been made using externally heated diamond anvil cells, and extensions to current temperature and pressure capabilities are underway. Developments in nuclear forward scattering provide accelerated data collection that could allow pressure and temperature limits to be further expanded, raising the tantalising possibility of collecting Mössbauer spectra at temperatures and pressures of the Earth's mantle.

The number of research groups that report results of Mössbauer experiments applied to the geosciences is relatively small. This is partly due to the challenge of data analysis, despite the widespread availability of computer-based fitting methods. It is not trivial in many cases to relate the extracted hyperfine parameters to the underlying structure and properties of the material without a significant amount of experience, and many pitfalls have been documented, e.g., [35]. Nevertheless, there are many relatively routine applications of Mössbauer spectroscopy that could benefit from an integrated approach to spectral analysis, and the current development of automated data analysis through the use of genetic algorithms, fuzzy logic and artificial neural networks [36] may provide the breakthrough that is needed to broaden the use of Mössbauer spectroscopy in the geosciences.

Absorbers for Mössbauer studies in the geosciences can be obtained from a wide variety of sources. Many come from the Earth, from depths anywhere from the surface down to approximately 200 km, although there are a few reports of rocks from greater depths. Within the past decade a source of material from the deep mantle (i.e. below 670 km) has been discovered – within diamonds. When diamonds crystallise in the Earth's mantle, they can enclose minerals from their surroundings and protect the minerals from alteration during their rapid ascent to the surface. A number of recently discovered deep diamonds (reviewed in [37]) contain inclusions that are large enough (i.e. greater than 100 μm) for study using Mössbauer spectroscopy. In addition, a completely new suite of samples is poised to become available through missions planned to Mars, Venus, the asteroids, and comets. In addition to natural samples, a number of absorbers used in geoscience measurements are synthesised in the laboratory. Large volume presses can simulate conditions within the Earth to depths reaching the lower mantle, while the diamond anvil cell can reach conditions approximating those in the centre of the core. Advances in environmental control (for example, oxygen fugacity) within these devices will provide a continuing supply of absorbers for Mössbauer experiments.

5. Summary

This paper has examined applications of Mössbauer spectroscopy in the context of the past, present and future. A survey of current uses for Mössbauer spectroscopy indicates that research activity will likely be maintained at current levels or higher for many years to come, with the potential for future developments to increase the use of Mössbauer spectroscopy in the geosciences.

Acknowledgements

I am grateful to T. Greaves and J. Cashion for preprints of their papers prior to publication.

References

1. Hemley, R. J., *Science* **285** (1999), 1026.
2. Pound, R. V. and Rebka, Jr., G. A., *Phys. Rev. Lett.* **3** (1959), 554.
3. Schiffer, J. P. and Marshall, W., *Phys. Rev. Lett.* **3** (1959), 556.
4. Bancroft, G. M., Maddock, A. G. and Burns, R. G., *Nature* **212** (1966), 913.
5. Bancroft, G. M., Maddock, A. G. and Burns, R. G., *Geochim. Cosmochim. Acta* **31** (1967), 2219.
6. Bancroft, G. M., Burns, R. G. and Stone, A. J., *Geochim. Cosmochim. Acta* **32** (1968), 547.
7. Abelson, P. (ed), *The Moon Issue*, *Science* Vol. 167, Amer. Assoc. Adv. Sci., Washington DC, 1970.
8. Lawton, J., *Science* **292** (2001), 1965.
9. Hawthorne, F. C., *Am. Mineral.* **68** (1983), 287.
10. Jacobsen, S. D., Reichmann, H.-J., Spetzler, H. A., Mackwell, S. J., Smyth, J. R., Angel, R. J. and McCammon, C., *J. Geophys, Res.* **107** (2002), 10.1029/2001JB000490.
11. Luth, R. W., Virgo, D., Boyd, F. R. and Wood, B. J., *Contrib. Mineral. Petrol.* **104** (1990), 56.
12. Gudmundsson, G. and Wood, B. J., *Contrib. Mineral. Petrol.* **119** (1995), 56.
13. Nell, J., Pollak, H. and Lodya, J. A., *Hyp. Interact.* **91** (1994), 601.
14. Fuchs, Y., Lagache, M. and Linares, J., *Am. Mineral.* **83** (1998), 525.
15. Redfern, S. A. T., Henderson, C. M. B., Wood, B. J., Harrison, R. J. and Knight, K. S., *Nature* **381** (1996), 407.
16. Kroll, H., Lueder, T., Schlenz, H., Kirfel, A. and Vad, T., *Eur. J. Mineral.* **9** (1997), 705.
17. Brizi, E., Molin, G., Zinazzi, P. F. and Merli, M., *Am. Mineral.* **86** (2001), 271.
18. Galazka-Friedman, J., Bauminger, E. R. and Bakun-Czubarow, N., *Hyp. Interact.* **112** (1998), 223.
19. Sobolev, V. N., McCammon, C. A., Taylor, L. A., Snyder, G. A. and Sobolev, N. V., *Am. Mineral.* **84** (1999), 78.
20. Medina, J. C., Butala, S. J., Bartholomew, C. H. and Lee, M. L., *Geochim. Cosmochim. Acta* **64** (2000), 643.

21. Polyakov, V. B. and Mineev, S. D., *Geochim. Cosmochim. Acta* **64** (2000), 849.
22. McCammon, C. A., Griffin, W. L., Shee, S. H. and O'Neill, H. S. C., *Contrib. Mineral. Petrol.* **141** (2001), 287.
23. Bishop, J. L., Lougear, A., Newton, J., Doran, P. T., Froeschl, H., Trautwein, A. X., Korner, W. and Koeberl, C., *Geochim. Cosmochim. Acta* **65** (2001), 2875.
24. Petrini, R., Forte, C., Orsi, G., Piochi, M., Pinzino, C. and Pedrazzi, G., *Contrib. Mineral. Petrol.* **140** (2001), 532.
25. Dubrovinsky, L., Annersten, H., Dubrovinskaia, N., Westman, F., Harryson, H., Fabrichnaya, O. and Carlson, S., *Nature* **412** (2001), 527.
26. Sen, R., Sarkar, A., Bannerjee, S. and Spottiswood, D. J., *Neues Jahrb. Mineral.-Monatsh.* (2002), 319.
27. De Oliveira, E. F., Castaneda, C., Eeckhout, S. G., Gilmar, M. M., Kwitko, R. R., De Grave, E. and Botelho, N. F., *Am. Mineral.* **87** (2002), 1154.
28. Verma, H. C., Upadhyay, C., Tripathi, A., Tripathi, R. P. and Bhandari, N., *Meteor. Planet. Sci.* **37** (2002), 901.
29. McCammon, C. A., Chaskar, V. and Richards, G. G., *Meas. Sci. Technol.* **2** (1991), 657.
30. McCammon, C. A., *Hyp. Interact.* **92** (1994), 1235.
31. Smith, P. R., Cashion, J. D. and Brown, L. J., *Hyp. Interact.* **71** (1992), 1503.
32. Gerdau, E. and DeWaard, H. (eds), *Nuclear Resonant Scattering of Synchrotron Radiation, Hyp. Interact.* Vols. 123–124, Kluwer Academic Publishers, Dordrecht, 1999.
33. Rancourt, D. and Klingelhöfer, G., Fourth Seeheim Workshop on Mössbauer Spectroscopy Abstracts, 1994, p. 129.
34. Greaves, T. L., Cashion, J. D., Benci, A. L., Jaggi, N. K., Hogan, C. and Bond, A. M., *Hyp. Interact.*, in press.
35. Rancourt, D. G., *Hyp. Interact.* **117** (1998), 3.
36. de Souza, P. A., *Lab. Robot. Automation* **11** (1999), 3.
37. McCammon, C., *Science* **293** (2001), 813.

Hyperfine Interactions **144/145:** 297–306, 2002.

Interplay of Antiferromagnetic Coupling and Spin Crossover in Dinuclear Iron(II) Complexes

ANA B. GASPAR[1], VADIM KSENOFONTOV[1], HARTMUT SPIERING[1], SERGEY REIMAN[1], JOSÉ A. REAL[2] and PHILIPP GÜTLICH[1]
[1] *Institut für Anorganische und Analytische Chemie, Johannes Gutenberg Universität, Staudinger Weg 9, D-55099 Mainz, Germany*
[2] *Departament de Química Inorgánica/Institut de Ciéncia Molecular, Universitat de València, Doctor Moliner 50, E-46100 Burjassot, València, Spain*

Abstract. This article reports on the study of the interplay between magnetic coupling and spin transition in 2,2′-bipyrimidine (bpym)-bridged iron(II) dinuclear compounds. Coexistence of both phenomena has been observed in $[Fe(bpym)(NCS)_2]_2$bpym, $[Fe(bpym)(NCSe)_2]_2$bpym and $[Fe(bt)(NCS)_2]_2$bpym (bpym = 2,2′-bipyrimidine, bt = 2,2′-bithiazoline) by the action of external physical factors namely pressure or electromagnetic radiation. Competition between magnetic exchange and spin crossover has been studied in $[Fe(bpym)(NCS)_2]_2$bpym at 6.3 kbar. LIESST experiments carried out in $[Fe(bpym)(NCSe)_2]_2$bpym and $[Fe(bt)(NCS)_2]_2$bpym at 4.2 K have shown that is possible to achieve dinuclear molecules with different spin states in this class of compounds.

Key words: spin transition, Fe(II) dinuclear complexes, LIESST effect, magnetic field Mössbauer spectroscopy.

1. Introduction

Molecular materials have emerged in many areas of material science for designing new magnets [1], molecular assemblies for data storage [2], conductors and superconductors [3], or nonlinear optical (NLO) materials [4]. An equally important, but more immediately realisable objective is the production of materials which exhibit useful bulk electronic/photonic properties that can be switched by stimulating changes at the molecular and/or macroscopic levels. For instance, to gather magnetic exchange and spin transition phenomena in the same compound or polymeric network could afford new switching materials with considerable amplification of the signal. In this respect, 2,2′-bipyrimidine (bpym)-bridged iron(II) dinuclear compounds represent a first step in this line. Six coordinate iron(II) spin crossover compounds represent a class of switchable molecular systems. The electronic spin-crossover phenomenon is a type of molecular magnetism [5] that occurs in some six-coordinate first-row transition metal complexes as the result of an electronic instability that is driven by external physical factors (temperature, pressure, or electromagnetic radiation) [6] that induce structural changes at both the molecular and lattice levels [7]. Six coordinate iron(II) ($3d^6$) spin-crossover compounds

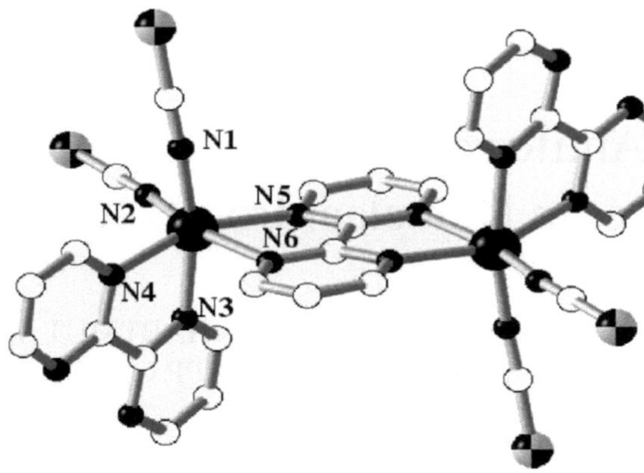

Figure 1. Crystal structure of the molecule of $[Fe(bpym)(NCS)_2]_2bpym$.

are among the most investigated systems; they reversibly change from diamagnetic ($S = 0$) to paramagnetic ($S = 2$) spin states. This so-called spin crossover phenomenon (SC) is observed both in solution and in the solid state. In solution, the molecular process occurs without interactions between spin-changing molecules, whereas in solid state, cooperative intermolecular interactions can lead to phase transitions. Cooperativity stems from the large difference in the metal-to-ligand bond lengths (ca. 0.2 Å) and changes of shapes of the spin changing molecules. Cooperativity is a measure of how efficiently the intramolecular changes are transmitted far into the crystal. Hence, it is the reason of the observed discontinuity in the magnetic and optical properties and eventually of thermal hysteresis, which confers potential memory effect to the SC systems.

Dinuclear iron(II) complexes of the type $[Fe(L)(NCX)_2]_2bpym$ (L: 2,2′-bipyrimidine, 2,2′-bithiazoline, X: S, Se) (see Figure 1) were first published by Real *et al.* [8], but has recently found revival due to observations we have made during our pressure studies on these compounds. The synthetic work of Real *et al.* has shown that the spin state of the iron centers and the magnetic coupling can be controlled by chemical means. By appropriate choice of the ligands the original antiferromagnetic coupling can be suppressed due to the onset of thermal spin crossover in one or even both iron centers as can be seen in Figure 2. This, of course, is a consequence of the change of ligand field strength brought about by electronic and/or steric influence. Pressure studies which we have recently performed on these compounds have nicely demonstrated that the ligand field strength can be altered in an equivalent but much more controlled manner [9]. A dinuclear compound with both iron(II) centers in the high spin state ($S = 2$), which are antiferromagnetically coupled over the whole temperature range, has been observed to undergo thermal spin transition in only one of the two centers under hydrostatic pressure. This is unusual in view of the fact that both centers have, according to X-ray structure determination, entirely equivalent surroundings. The appearance of spin transition in only one center of the bridged pair, finally ending

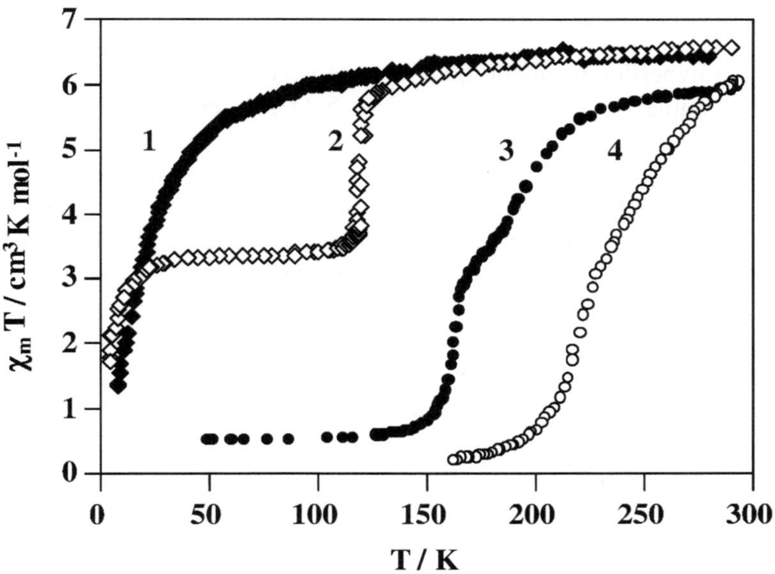

Figure 2. Temperature dependence of $\chi_M T$ for $[Fe(bpym)(NCS)_2]_2bpym$ (**1**), $[Fe(bpym)(NCSe)_2]_2bpym$ (**2**), $[Fe(bt)(NCS)_2]_2bpym$ (**3**) and $[Fe(bt)(NCSe)_2]_2bpym$ (**4**).

at a plateau with all pairs present in LS–HS pair formation, supports the conclusions drawn from earlier work [8] that LS–HS pair formation is a preferred process arising from a synergetic effect between intramolecular and cooperative intermolecular interactions. Recently a method of direct monitoring of spin state and magnetic coupling in dinuclear iron(II) compounds was developed [10]. The molecular fractions of high spin (HS) and low spin (LS) components in such compounds have been deduced mainly from magnetic susceptibility data. Zero field Mössbauer spectroscopy of dinuclear compounds allows the amount of HS and LS components to be determined, but does not allow the distinction between HS iron complexes in LS–HS and HS–HS pairs in the paramagnetic state to be made. However, this distinction becomes possible if Mössbauer measurements are carried out in an external magnetic field. Applying this method the corresponding amounts of HS–HS, LS–HS and LS–LS pairs in $[Fe(bpym)(NCS)_2]_2bpym$, $[Fe(bpym)(NCSe)_2]_2bpym$ and $[Fe(bt)(NCS)_2]_2bpym$ (bpym = 2,2′-bipyrimidine, bt = 2,2′-bithiazoline) at 4.2 K were identified. Further, the combination of applied field MS and LIESST effect on dinuclear compounds demonstrates the effectiveness of this approach and pointed out like high pressure experiments the synergism between antiferromagnetic coupling and spin transition present in the title iron(II) dinuclear compounds.

In the present article we will discuss the interplay between magnetic coupling and spin transition initiated by the influence of physical factors like pressure or light in some derivatives of the bpym-bridged family, concretely in $[Fe(bpym)(NCS)_2]_2bpym$ (**1**), $[Fe(bpym)(NCSe)_2]_2bpym$ (**2**) and $[Fe(bt)(NCS)_2]_2bpym$ (**3**).

2. Background of the method

The effective hyperfine magnetic field, H_{eff}, at iron nuclei of a paramagnetic non-conducting sample in an external magnetic field H_{ext} consists of several contributions: $H_{\text{eff}} = H_{\text{ext}} + H_{\text{c}} + H_{\text{L}} + H_{\text{D}}$. The Fermi contact term H_{c} may be estimated through the expectation value of the atomic spin moment $\langle S\rangle$ by the relation $H_{\text{c}} = -220\langle S\rangle$ kOe [11]. The orbital term may be expressed as $H_{\text{L}} = 560(g - 2)\langle S\rangle$ where g is the Landé splitting factor; the term H_{D}, arising from the dipolar interaction of the nucleus with the spin moment of the atom, may be estimated as $H_{\text{D}} \cong H_{\text{L}}/14$ [11]. Thus, the effective magnetic field on iron nuclei may be estimated as: $H_{\text{eff}} \cong H_{\text{ext}} - [220 - 600(g - 2)]\langle S\rangle$. The obvious difference between the expectation values $\langle S\rangle$ for Fe(II) in the LS and in the HS state in HS–LS or HS–HS pairs enables one to distinguish unambiguously between the dinuclear units consisting of two possible spin states in an external magnetic field. Considering the experimental conditions, it is obvious that to observe a pronounced difference for paramagnetic samples, the measurements should be carried out at low temperatures in order to slow down the electronic spin relaxation. In addition, the strength of the external magnetic field should be sufficient to allow a differentiation of spin states of pairs to be made during the characteristic time window of a Mössbauer experiment ($\sim 10^{-7}$ s). Relevant experimental and theoretical studies of the behavior of Fe(II) complexes in external magnetic fields are given in [12, 13].

3. Experimental results and discussion

Coexistence of antiferromagnetic coupling and spin crossover in $[Fe(bpym)(NCS)_2]_2$bpym(**1**) clearly follows from magnetic susceptibility measurements at $P = 6.3$ kbar. Figure 3 shows the magnetic properties of **1** at different pressures. At ambient pressure (1 bar), **1** does not reveal the spin crossover behavior and the two iron(II) ions being in the HS–HS state antiferromagnetically coupled, with the spin singlet ground state completely populated at low temperatures. This behavior remains as pressure is increased up to 3.8 kbar. When pressure attains 6.3 kbar, an incomplete thermal spin crossover appears around $T_{\text{c}} = 100$ K. At $P = 8.9$ kbar, the spin conversion curve transforms into a typical one-step transition without hysteresis with 50% of molecules remaining in the HS state. The curve $\chi_{\text{M}}T$ vs. T is interesting in that a wide plateau is reached between ca. 25 and 120 K. The fact that $\chi_{\text{M}}T$ is temperature independent between 120 and 25 K indicates that only one iron center of the dinuclear unit has been converted to the LS state. The decrease of $\chi_{\text{M}}T$ at temperatures below 25 K is due to the zero field splitting (ZFS) of the iron(II) ions remaining in the HS state. Further increase of pressure to 11 kbar shifts the transition curve until $T_{\text{c}} \approx 200$ K, with the main features of the transition at $P = 8.9$ kbar being retained. The experimental curves of **1** at 1 bar and 8.9 kbar have been simulated with a full-matrix diagonalization routine using the Hamiltonian (1) for the HS–HS coupled species a and b ($P = 1$ bar) and the

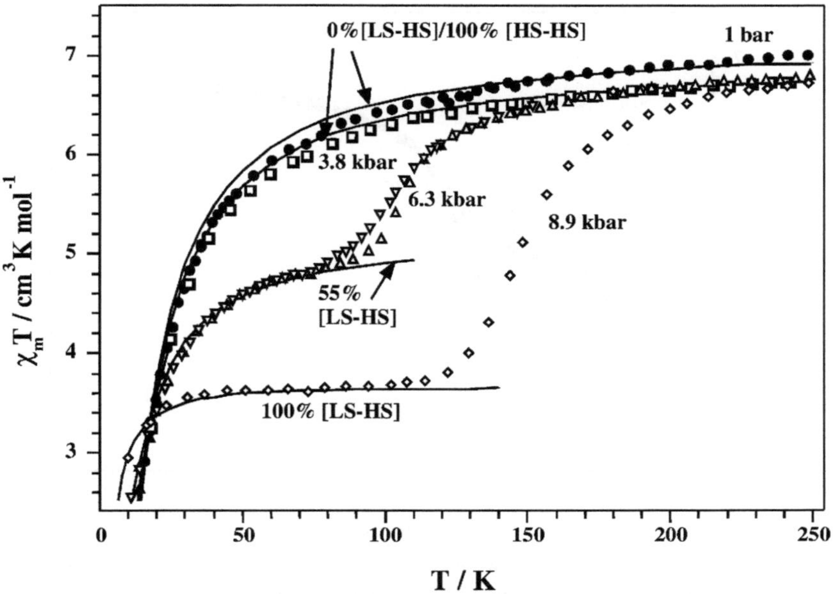

Figure 3. Temperature dependence of $\chi_M T$ $[Fe(bpym)(NCS)_2]_2bpym$ (**1**) at different pressures. The solid lines, together with estimated concentrations of LS–HS and HS–HS species, correspond to the calculations using the Hamiltonians (1) and (2).

Hamiltonian (2) for the LS–HS uncoupled species in the axial ZFS approximation with $S = 2$.

$$\hat{H} = -J\hat{S}_a\hat{S}_b + \hat{S}_a D_a \hat{S}_a + \hat{S}_b D_b \hat{S}_b + \beta\big(\hat{S}_a g_a + \hat{S}_b g_b\big)H, \quad (1)$$

$$\hat{H} = D\big[\hat{S}_z^2 - \tfrac{1}{3}S(S+1)\big] + g\beta H\hat{S}. \quad (2)$$

J is the coupling constant characteristic of the HS–HS species. S, S_a and S_b and $D_a = D_b = D$ are the local spin operators and anisotropy tensors, respectively. The best set of parameters which match with the experimental data at 1 bar are $J = -4.5$ cm^{-1} and $D = 8$ cm^{-1}. The best fit to the experimental data at 8.9 kbar corresponds to $D = 10$ cm^{-1} and $g = 2.19$. A linear combination of two extreme behaviors at 1 bar and 8.9 kbar has been used to simulate the magnetic behavior at 6.3 kbar. The estimated concentrations of LS–HS species were around 55%. At this pressure the incompleteness of spin conversion is directly related with the low temperatures at which it takes place ($T < 100$ K). As is well known, at low temperature the dynamics of spin conversion becomes considerably blocked and due to this fact the complete spin transition is suppressed. It explains the coexistence of HS–HS and LS–HS pairs in **1** at 6.3 kbar which is reflected in the thermal dependence of $\chi_M T$. In order to investigate the competition between magnetic interaction and spin transition phenomena in **1** quenching experiments have been performed. Figure 4 displays the magnetic behavior of the quenched sample at increasing temperatures.

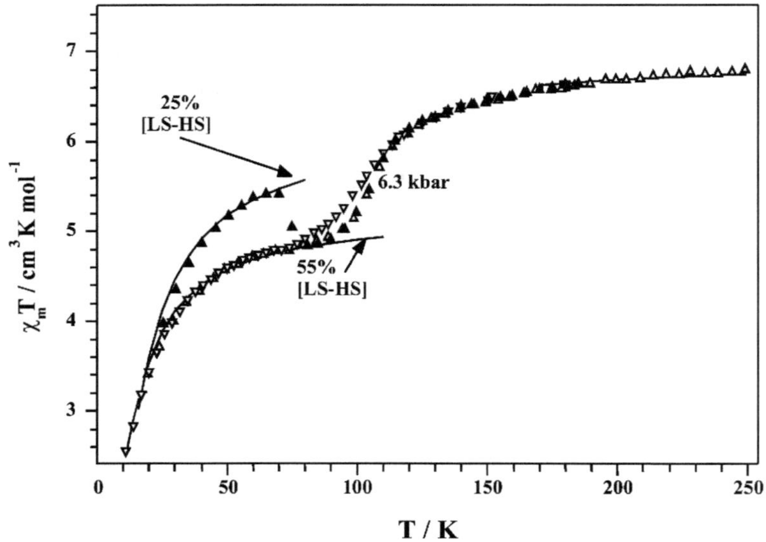

Figure 4. $\chi_M T$ versus T plots for [Fe(bpym)(NCS)$_2$]$_2$bpym (**1**) at $P = 6.3$ kbar. Sample was cooled from 300 to 2 K at 2 K min^{-1} (open triangles). Thermal quenching: sample was cooled from 300 to 5 K at 100 K min^{-1} and then warmed slowly (black triangles).

From the thermal dependence of $\chi_M T$ it can be inferred that HS–HS entities can be "frozen" in the metastable state at low temperatures. Heating the sample above ca. 60 K leads to reentrance to the stable state, which consists mostly of LS–HS dinuclear species. Two main factors, namely, antiferromagnetic intramolecular interactions and elastic interactions, should be taken into account to clarify the reason that stabilizes the metastable state. Considering the low value of $J \approx -4.5$ cm^{-1} of the former one in comparison with the reentrance temperature $T \approx 60$ K and the unusually slow kinetics of the reentrant process with relatively fast kinetics of an equilibrium spin transition taking place at higher temperature, one can conclude that the reentrance is an essentially thermally activated process and that the crystal lattice is substantially involved. The kinetics of the spin transition in dinuclear compounds confirms, that this is the structural rearrangement, associated to the spin changing process, which is responsible for the trapped HS–HS metastable species and not the magnetism. If it was the magnetic interaction which traps the metastable state, the dynamics of the reentrant process would be much faster. In other words, elastic interactions rather than magnetic coupling drives the transformation of HS–HS $\leftrightarrow$ LS–HS under pressure. Further, this experimental result evidence, as we noticed above, that the coexistence of both phenomena should be explained considering the kinetics effects associated with the spin transition below 100 K.

The synergism between antiferromagnetic coupling and spin transition has been observed in [Fe(bpym)(NCSe)$_2$]$_2$bpym (**2**) and [Fe(bt)(NCS)$_2$]$_2$bpym (**3**) under light irradiation, i.e., making use of the LIESST effect. According to preliminary

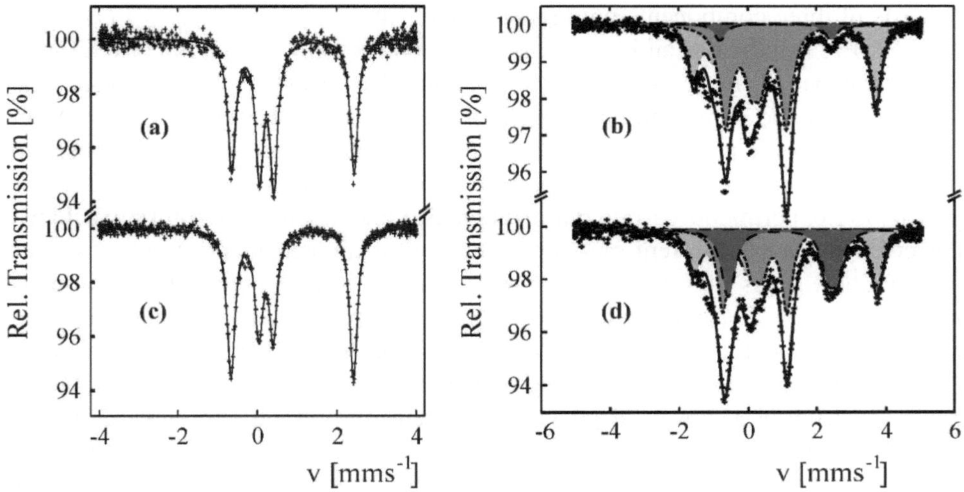

Figure 5. ^{57}Fe Mössbauer spectra of $[Fe(bpym)(NCSe)_2]_2$bpym (**2**) at recorded at 4.2 K in zero-field (a), at 4.2 K in a magnetic field of 50 kOe (b), and after light irradiation in zero-field (c) and after light irradiation in a magnetic field of 50 kOe (d). LS in LS–HS and LS–LS pairs (grey), HS in HS–LS pairs (light grey), HS in HS–HS pairs (dark grey).

studies the ground state of **2** should consist of LS–HS species. The zero-field spectrum at 4.2 K (Figure 5(a)) reflects the nearly "one-half" spin transition according to the area fractions of the HS (48.0%) and LS (52.0%) components with parameters $\delta_{HS}(2) = 0.86(1)$ mm/s, $\Delta_{HS}(2) = 3.11(2)$ mm/s, and $\delta_{LS}(2) = 0.22(1)$ mm/s, $\Delta_{LS}(2) = 0.36(1)$ mm/s, respectively. Measurements in a magnetic field of 50 kOe at 4.2 K reveal features which are not seen in paramagnetic spectra and may be interpreted as follows. The total spectrum consists of three components as can be seen in Figure 5(b). One of them (relative intensity $x = 52.0\%$) with isomer shift and quadrupole splitting being equal to $\delta_{LS}(2)$ and $\Delta_{LS}(2)$ is identified as the "fingerprint" of the LS state which has $H_{eff} \approx H_{ext}$. The second low-intensity ($y = 4.0\%$) broadened doublet with parameters $\delta_{HS}(2)$ and $\Delta_{HS}(2)$ and $H_{eff} = 14$ kOe, corresponds to iron(II) ions in antiferromagnetically coupled HS–HS pairs. The third component (relative intensity $z = 44.0\%$) with parameters $\delta_{HS}(2)$ and $\Delta_{HS}(2)$ should be unambiguously assigned to the HS state in LS–HS pairs, because the measured effective magnetic field at the iron nuclei of 81 kOe clearly originates from a spin quintet ground state of iron(II) ($S = 2$). As a result, the complete distinction of dinuclear units becomes possible. It follows from the area fractions of the subspectra intensities that at 4.2 K sample **2** contains $2z = 88.0\%$ LS–HS, $y = 4.0\%$ HS–HS and $(x-z) = 8.0\%$ LS–LS pairs. After irradiation of the sample during one hour with $\lambda = 514$ nm at 4.2 K, the Mössbauer spectrum of the paramagnetic compound **2** shows a decrease in the intensity of the LS species (41.0%) in favour of an increase of the HS species (59.0%) as is depicted in Figure 5(c). The Mössbauer spectrum recorded subsequently in a magnetic field (Figure 5(d)) consists of three components with relative intensities $x = 41.0\%$, $y = 17.0\%$ and $z = 42.0\%$. It indicates an increase of the amount of HS–HS pairs (17.0%) with a simultaneous decrease of LS–HS (83.0%) and disappearance of LS–LS species. The expected and registered appearance of the metastable HS species in the form of HS–HS pairs is a consequence of the light irradiation and proves the possible coexistence of magnetic coupling and spin transition in the same compound.

LIESST experiments carried out in **3** revealed like in **2** the possibility to achieve the coexistence of different electronic states in a dinuclear entity composed of iron(II) ions. As is illustrated in Figure 6(a) at 4.2 K in **3** practically all iron(II) ions are in the LS configuration. The Mössbauer parameters obtained from the fitting of the spectrum are: $\delta_{LS}(3) = 0.357(1)$ mm/s, $\Delta_{LS}(3) = 0.452(2)$ mm/s. After irradiation of the sample during one hour ($\lambda = 514$ nm) at 4.2 K, the Mössbauer spectrum of **3** shows a decrease in the intensity of the LS species (62.0%) in favour of an increase of the HS species (32.0%) (Figure 6(b)). Measurements in a magnetic field of 50 kOe at 4.2 K allows to identify the nature of metastable states and may be interpreted as follows. As is depicted in Figure 6(c) the total spectrum consists of three components. One of them with isomer shift and quadrupole splitting being equal to $\delta_{LS}(3)$ and $\Delta_{LS}(3)$ is identified as the "fingerprint" of the LS state which has $H_{eff} \approx H_{ext}$. The second low-intensity broadened doublet with parameters $\delta_{HS}(3)$ and $\Delta_{HS}(3)$ and $H_{eff} = 14$ kOe, corresponds to iron(II) ions in antiferromagnetically coupled HS–HS pairs. The third component with parameters $\delta_{HS}(3)$ and $\Delta_{HS}(3)$ should be unambiguously assigned to the HS state in LS–HS pairs, because the measured effective magnetic field at the iron nuclei of 85 kOe clearly originates from a spin quintet ground state of iron(II) ($S = 2$). It follows from the area fractions of the subspectra intensities that at 4.2 K sample **3** contains $2z = 12.0\%$ LS–HS, $y = 26.0\%$ HS–HS and $(x - z) = 62.0\%$ LS–LS pairs. It is worth noting that in **3** the photoinduced species correspond to HS–HS and LS–HS pairs. The appearance of LS–HS species should be interpreted in terms of a synergetic effect between intramolecular and intermolecular cooperative interactions which stabilizes energetically the mixed pair [14].

4. Conclusion

The present dinuclear compounds have shown to be very fortunate systems to study the interplay between magnetic coupling and spin transition. Our experiments demostrated the possibility to achieve different electronic states in dinuclear complexes by influence of external physical factors namely pressure or ligth. In **1** the coexistence of both phenomena is a consequence of the kinetics effects associated with the spin transition below 100 K. On the other hand, LIESST effect in **2** and **3** provokes the photoswitching between spin pairs, LS–HS → HS–HS, and LS–LS → HS–HS and LS–LS → LS–HS, respectively, where HS–HS metastable species are antiferromagnetically coupled. However, the photoconversion does not

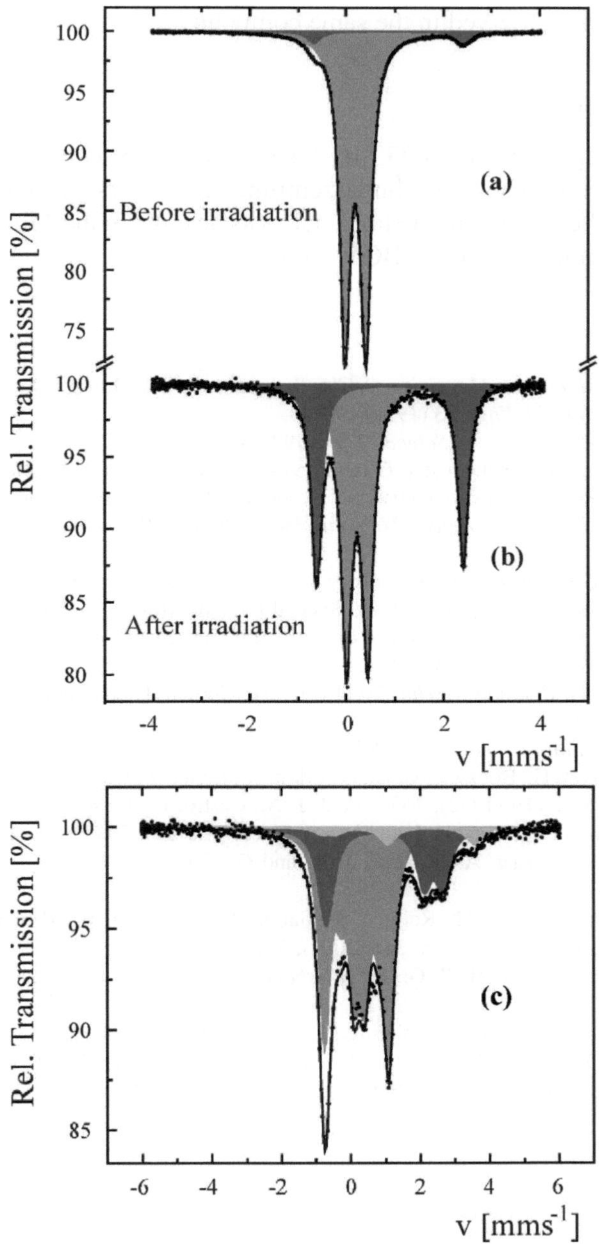

Figure 6. ^{57}Fe Mössbauer spectra of [Fe(bt)(NCS)$_2$]$_2$bpym (**3**) at recorded at 4.2 K in zero-field before (a) and after light irradiation (b). ^{57}Fe Mössbauer spectra of (**3**) recorded at 4.2 K in a magnetic field of 50 kOe after light irradiation (c). LS in LS–HS and LS–LS pairs (grey), HS in HS–LS pairs (light grey), HS in HS–HS pairs (dark grey).

involve all molecules which is the reason why dinuclear molecules with different electronic states are observed in the same compound.

Acknowledgements

The financial help from the DFG, the Fonds der Chemischen Industrie and the Materialwissenschaftliches Forschungszentrum der Universität Mainz is gratefully acknowledged. We also thank the financial assistance from the Ministerio Español de Ciencia y Tecnologiá (project BQV 2001-2928).

References

1. Kahn, O., In: *Molecular Magnetism*, VCH, Weinheim, 1993; Miller, J. S. and Epstein, A. J., *Angew. Chem. Int. Ed. Engl.* **33** (1994), 385.
2. Kahn, O. and Martinez, C. J., *Science* **279** (1998), 44.
3. Optical nonlinearities in chemistry, *Chem. Rev.* **1** (1994), 94; Zyss, J., In: *Molecular Nonlinear Optics: Materials, Physics and Devices*, Academic Press, Boston, 1994; Nalwa, H. S. and Miyata, S., In: *Nonlinear Optics of Organic Molecules and Polymers*, CRC Press, New York, 1997.
4. Cassoux, P. and Miller, J. S., In: *Chemistry of Advanced Materials*: *An Overview*, Wiley–VCH, New York, 1998, p. 19; *J. Mater. Chem.*, Special Issue on Molecular Conductors **5** (1995), 10.
5. Gütlich, P., *Struct. Bonding (Berlin)* **44** (1981), 83; Gütlich, P., Hauser, A. and Spiering, H., *Angew. Chem. Int. Ed. Engl.* **33** (1994), 2024.
6. König, E., *Struct. Bonding (Berlin)* **76** (1991), 51; Gütlich, P. and Hauser, A., *Coord. Chem. Rev.* **97** (1990), 1.
7. König, E., *Prog. Inorg. Chem.* **35** (1987), 527.
8. Real, J. A., Bolvin, H., Bousseksou, A., Dworkin, A., Kahn, O., Varret, F. and Zarembowitch, J., *J. Am. Chem. Soc.* **115** (1992), 4650; Real, J. A., Castro, I., Bousseksou, A., Verdaguer, M., Burriel, R., Castro, M., Linares, J. and Varret, F., *Inorg. Chem.* **36** (1997), 455.
9. Ksenofontov, V., Gaspar, A. B., Real, J. A. and Gütlich, P., *J. Phys. Chem. B.* **105** (2001), 12266.
10. Ksenofontov, V., Spiering, H., Reiman, S., Garcia, Y., Gaspar, A. B., Moliner, N., Real, J. A. and Gütlich, P., *Chem. Phys. Lett.* **348** (2001), 381.
11. Greenwood, N. N. and Gibb, T. G., In: *Mössbauer Spectroscopy*, Chapman and Hall, London, 1971.
12. Zimmermann, R., Ritter G. and Spiering, H., *Chem. Phys.* **4** (1974), 133.
13. Zimmermann, R., Ritter, G., Spiering, H. and Nagy, D. L., *J. Phys.* **35** (1974), C6.
14. Gaspar, A. B., Ksenofontov, V., Spiering, H., Reiman, S., Real, J. A. and Gütlich, P., in preparation.

Hyperfine Interactions **144/145:** 307–323, 2002.

Photoswitching of Nitroprussides

V. RUSANOV[1,2], SV. STANKOV[1,2] and A. X. TRAUTWEIN[2]
[1] *University of Sofia, Faculty of Physics, Department of Atomic Physics, 5 James Bourchier Blvd., 1126 Sofia, Bulgaria*
[2] *Universität zu Lübeck, Institut für Physik, Ratzeburger Allee 160, 23538 Lübeck, Germany*

Abstract. For 25 years the two metastable electronic states (SI and SII), accessible in the nitroprusside anion $[Fe(CN)_5NO]^{2-}$ (NP) by irradiation with blue polarized light, have been the subject of intense investigations from several research groups. This review article summarizes the most important experimental results in the field of photochromatic effects investigated by nuclear resonant methods. The experimentally determined data have been compared with the results derived from density functional theory calculations. The majority of results convincingly indicates: (i) the SI state is represented by the isonitrosyl conformation (Fe–O–N) of the (NO^+)-group and (ii) the SII state is represented by a dynamic conformation, in which a bending vibration of Fe–O–N as well as a rotation of the (NO^+)-group around the main molecular axis simultaneously occur. A possible formation of a new third metastable state observed in rare-earth nitroprussides, which is stable at room temperature, will be discussed.

Key words: photoswitching, optical bistability, structural changes, density-functional theory, nuclear resonant methods.

1. Introduction

By irradiation of a single crystal of sodium nitroprusside, $Na_2[Fe(CN)_5NO]\cdot 2H_2O$ (SNP), and subsequent investigation by Mössbauer spectroscopy Hauser and co-workers discovered in 1977 the existence of a metastable state in the $[Fe(CN)_5 NO]^{2-}$-anion (NP) [1]. Surprisingly this state has practically infinite lifetime below 200 K. Heating the crystal above this temperature it disappears and only the ground state of the NP anion stays. SNP is one of the velocity calibration standards in Mössbauer spectroscopy [2] and its crystal structure features are very well known [3]. The nitroprusside anion in SNP, as a relatively simple structural unit, is used as a model complex in many theoretical studies [4]. The Mössbauer parameters such as isomer shift (IS), quadrupole splitting (ΔE_Q), and the recoilless absorption probability, the so called Lamb–Mössbauer factor (f_{LM}), have been measured with high accuracy for all three principle crystal directions [5]. SNP can easily be grown as a large single crystal with high optical quality and size up to $5 \times 5 \times 10$ cm^3. Therefore this crystal has been used for systematic studies of the nature and accessibility of light-induced metastable states.

2. Experimental results and discussions

2.1. POPULATION CONDITIONS OF METASTABLE STATES

By irradiation with light in the spectral region 350–580 nm two metastable molecular conformations (SI and SII) can be populated, i.e. at temperatures below 200 K SI and below 150 K SII. About 45% of the NP anions can be transferred into SI by irradiation with blue light (457.9 nm) from an Ar^+-laser and only a few percent into SII. The latter, however, can be populated up to about 30% by irradiation with infrared light from a Nd-YAG-laser (1064 nm) subsequent to the population of SI. Both states can be completely and reversibly transferred to the ground state (GS) by irradiation with red light (632.8 nm) from a He–Ne-laser or by heating the crystal to the decay temperature of the corresponding state. To reach saturation of SI the integrated light intensity must be about 2500 J/cm^2. Only about 250 J/cm^2 are necessary to transfer SI into SII and partially back to GS. Maximum populations can only be reached if the electric field vector **E** of the irradiating light is perpendicular to the main molecular axis N–C–Fe–N–O. This feature implies that suitable crystal cuts are necessary if the illumination is performed by polarized laser light. The efficiency in generation of SI and SII is independent on the angle between the wave vector of the irradiating light and the N–C–Fe–N–O axis as long as **E** is perpendicular to this axis [6]. The NP anion exhibits reddish colour in the ground state, populated by SI with 45% it posses light brown-green colour and populated by SII with 25% it exhibits black colour. This photochromatic and photorefractive behaviour of SI and SII are of great interest for optical information storage with extremely high density [7] or holographic data storage [8].

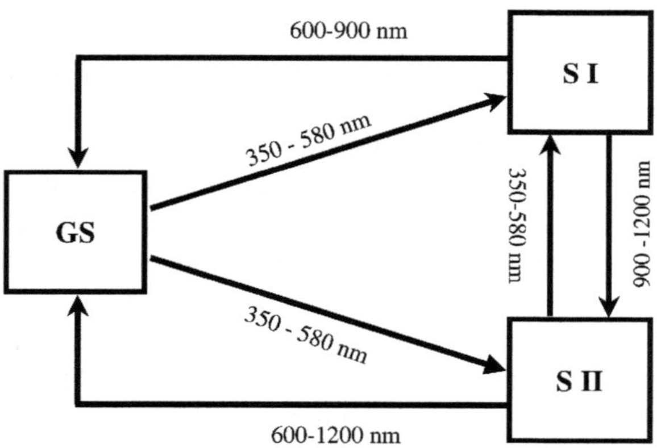

Figure 1. Population and transfer conditions of GS, SI and SII. All wavelengths are given in nm. Using blue light SI can be produced directly from GS and via SII. Irradiation with infrared light transforms SI back into GS and partially into SII, which by further irradiation goes to GS as well. Both SI and SII can be transferred back into GS by red light. These processes are reversible and can be repeated without damaging the crystal.

The first detailed investigation of population conditions has been made applying the differential scanning calorimetry method [6, 9]. Figure 1 summarizes the wavelength-population and wavelength-transfer conditions between GS, SI and SII.

2.2. MÖSSBAUER PARAMETERS OF METASTABLE STATES

The Mössbauer parameters of the GS of the NP anion at 77 K are: isomer shift IS $= -0.191(3)$ mm/s (relative to the α-Fe standard at room temperature) and quadrupole splitting $\Delta E_Q = +1.716(3)$ mm/s. The corresponding parameters for SI (Figure 2(a)) are IS $= +0.178(3)$ mm/s and $\Delta E_Q = +2.755(3)$ mm/s and for SII (Figure 2(b)) IS $= +0.194(3)$ mm/s and $\Delta E_Q = +2.862(3)$ mm/s. Infrared exposure with a Nd-YAG-laser transfers SII into GS (Figure 2(c)) [10, 11]. The insets in Figure 2 schematically represent the possible structural features of GS, SI and SII in the NP anion, which are discussed in details below.

The Mössbauer parameters of SI and SII have been measured in lithium, potassium [12] and barium nitroprusside single crystals, as well as in solid (frozen) solutions containing NP anions [13]. The metastable states have been produced in Ru-complexes [14] like $K_2[Ru(NO_2)_4(OH)(NO)]$, Os-complexes [15] and in other complexes with different ligands, different cations and varying numbers of solvent molecules [16]. Recently guanidinium nitroprusside $(CN_3H_6)_2[Fe(CN)_5NO]$ (GNP) was announced to be especially suitable for investigations of the photochromatic effects [17] because it does not contain crystal water, all nitroprusside anions have almost the same orientation within the single crystal, and its structure is well known [18]. The structural features in GNP allow high population up to 27% by very simple irradiation conditions, i.e. with incoherent, nonpolarized blue light (450 ± 35 nm) produced by light-emitting diodes (LED) [17].

The requirement to keep the **E** vector of the laser light and the molecular axis N–C–Fe–N–O perpendicular to each other, which is necessary for achieving maximum population of SI and SII, is fulfilled for all crystallographic directions in GNP. Using polarized Ar^+-laser light we have performed two population experiments with wavelengths 476.5 nm and 457.9 nm [19]. The Mössbauer spectra of an irradiated **c**-cut GNP single crystal with thickness 480 μm are shown in Figure 3. Both spectra are taken at the orientation angle of $\Omega \cong 45°$ between $\mathbf{k}_\gamma$ and the crystallographic **c**-axis. Calculated population coefficients are $p_{457.9} = 36.2(1)\%$ and $p_{476.5} = 28.2(1)\%$. A detailed investigation of the dependence of the population coefficient on the wavelength of the exciting light was performed for SNP only [6]. Maximum population of SI of about 47% is achieved with 457.9 nm [10]. With this wavelength almost all nitroprussides have been irradiated, but populations similar to those in SNP have not been observed [12, 13]. It seems that in GNP the maximum population of SI can be achieved with wavelengths of about 410 nm, which is shorter than the optimal one for SNP. Unexpectedly, the line intensity ratio for SI, calculated for both spectra (Figures 3(a) and (b)), differs remarkably from those for

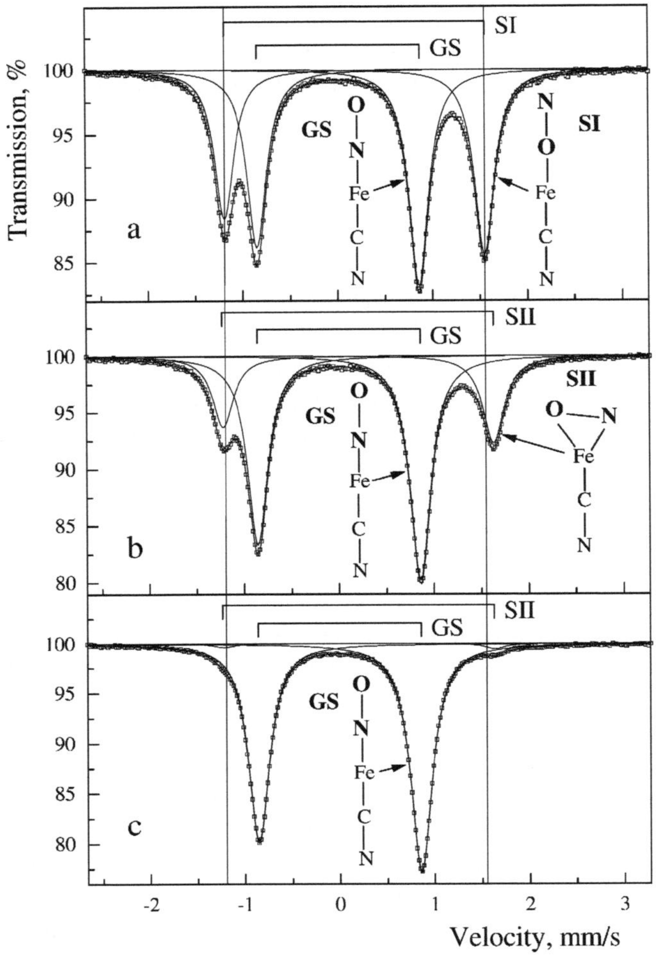

Figure 2. Mössbauer spectra obtained from a 350 μm thick **b**-cut SNP single crystal. The wave vector of the γ-quanta $\mathbf{k}_\gamma$ lies in the mirror **ab**-plane at an angle of 45° relative to the **a**- and **b**-axis: (a) populated with SI to about 42% by irradiation with an Ar^+-laser, $\lambda = 457.9$ nm, wave vector of the laser light $\mathbf{k} \| \mathbf{b}$, electric field vector of the laser light $\mathbf{E} \| \mathbf{c}$-axis, and light intensity $Q \cong 3000$ J/cm^2; (b) population of SII after transfer of SI to SII (and partially to GS) by irradiation with Nd-YAG-laser, $\lambda = 1064$ nm, $\mathbf{k} \| \mathbf{b}$, $\mathbf{E} \| \mathbf{c}$, and $Q \cong 250$ J/cm^2; (c) depopulation of SII after illumination with Nd-YAG laser with $Q \cong 4000$ J/cm^2. The vertical lines mark the positions of SI lines. The velocity scale is related to SNP in GS at 77 K.

the ground state GS. The observed differences of the line intensity ratios for GS and SI indicate differences in the orientation of the electric field gradient (EFG) tensor in GS and SI for GNP, which has not been observed in SNP [10]. Therefore measurement of the EFG parameters by angle-dependent Mössbauer spectroscopy is strongly desirable.

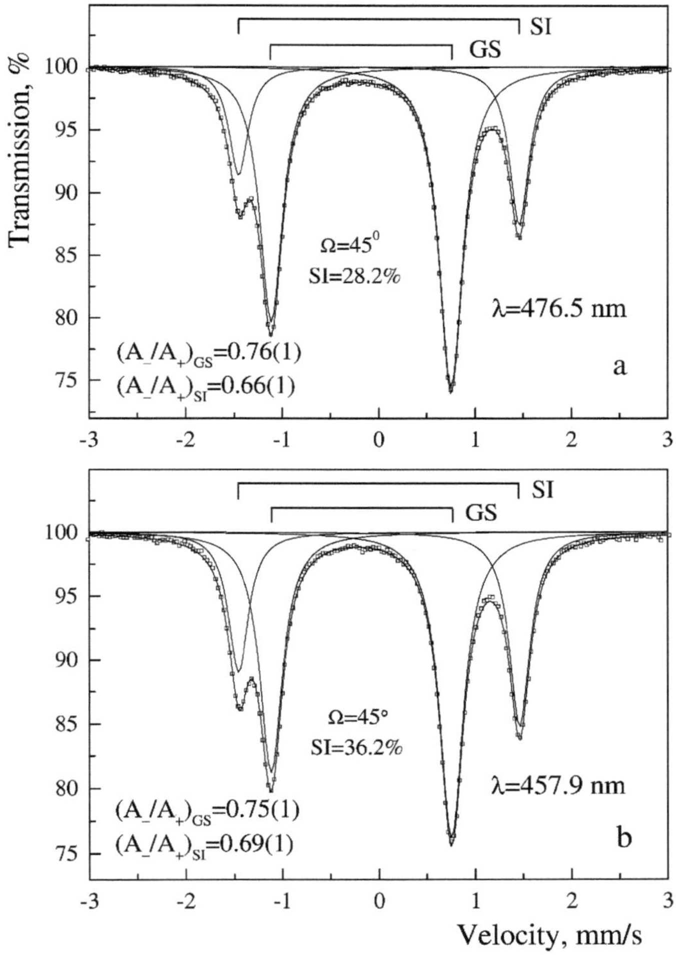

Figure 3. Mössbauer spectra of a **c**-cut GNP single crystal after irradiation with (a) 476.5 nm and (b) 457.9 nm polarized laser light from Ar^{+}-laser, **k**‖**c**, **E**‖**b**-axis, and $Q \cong 3000$ J/cm^2. The wave vector of the γ-quanta $\mathbf{k}_\gamma$ lies in the mirror **ac**-plane at an angle $\Omega = 45°$ relative to the **a**- and **c**-axis. The velocity scale is related to α-Fe at room temperature. The population coefficients for SI as well as the line intensity ratios for GS and SI are shown.

2.3. STRUCTURAL CHANGES OF THE NP ANION IN THE METASTABLE STATES

Up to now two contradictory explanations of the nature of metastable states have been proposed by several research groups. Güdel has suggested that either a large structural change or a multielectron promotion is required to explain the remarkable stability of metastable states of the NP anion [20]. The results from temperature dependent measurements of the quadrupole splitting between 4.2 and 293 K prove with high accuracy that the light induced metastable states are, like the normal NP molecular GS, characterized by an electronic closed shell configuration

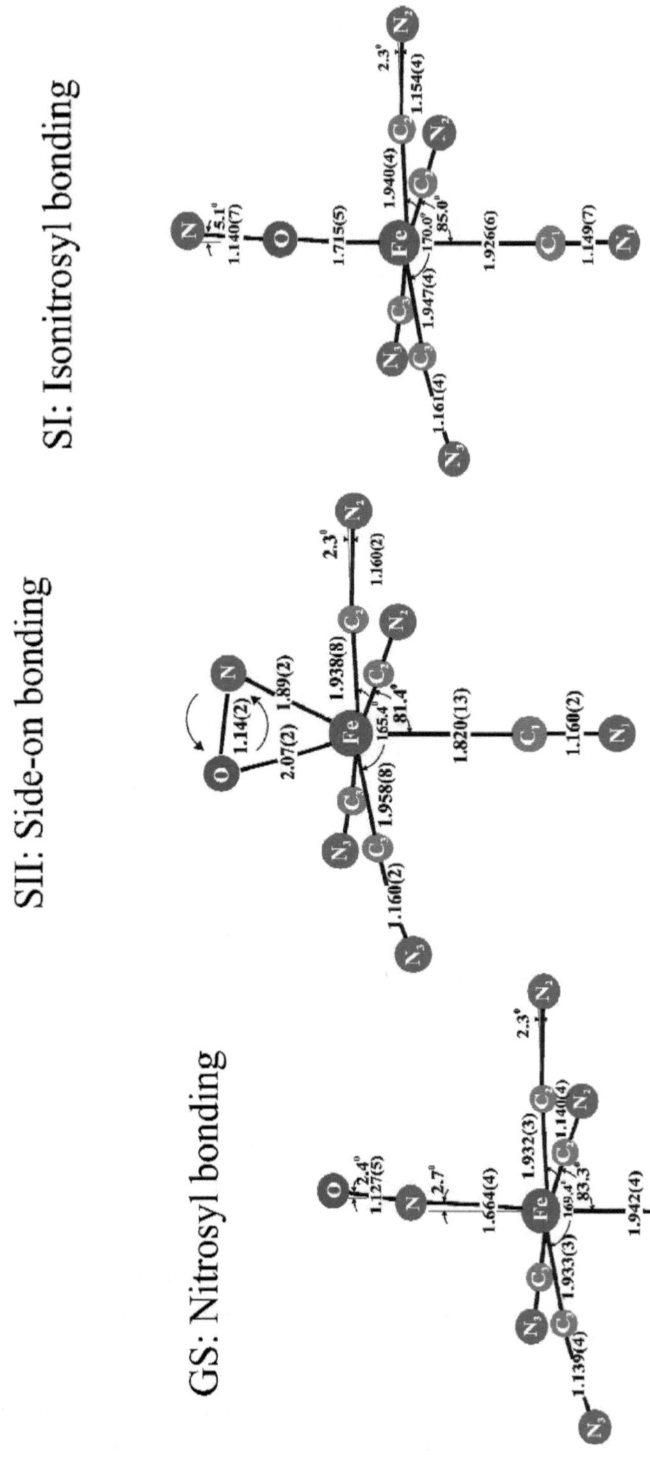

Figure 4. Bond lengths (in Å) and angles for GS, SI and SII of the NP anion obtained at 50 K [23].

and a Fe(II) state with spin zero [13]. By neutron diffraction studies [21, 22] no significant structural changes between SI, SII and GS have been found. The main molecular axis N–C–Fe–N–O remained unchanged. Based on various X-ray diffraction studies and extended Hückel calculations Carducci *et al.* presented totally different results and provided a physical explanation of the nature of the observed

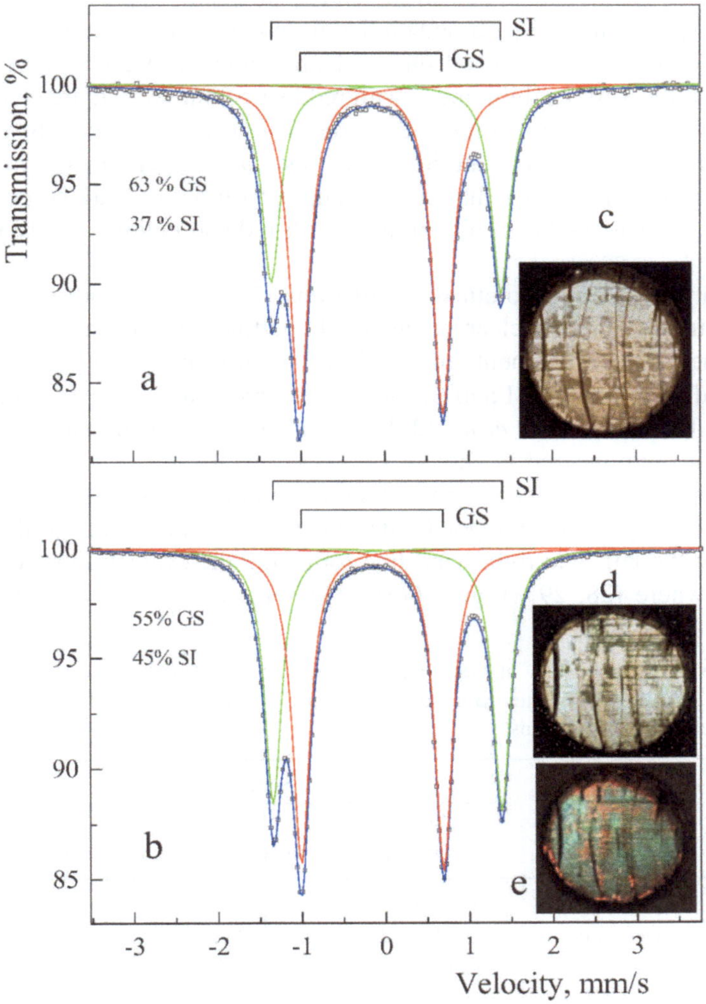

Figure 5. Mössbauer spectra of a **b**-cut SNP single crystal after irradiation with 457.9 nm polarized laser light from Ar^+-laser, $\mathbf{k}\|\mathbf{b}$, $\mathbf{E}\|\mathbf{c}$-axis; (a) $Q \cong 3000$ J/cm^2; (b) after additional irradiation from the opposite side of the crystal, $Q \cong 7000$ J/cm^2. The velocity scale is related to α-Fe at room temperature. The population coefficients for GS and SI are shown. Photographs of the crystal (c) in GS at 77 K, and (d, e) populated up to 45% of SI taken in (d) transmitted light and (e) scattered light. Many cracks and areas with reddish color (i.e. with weak SI population) are visible. The diameter of the crystal is 5 mm, crystal thickness 490 μm.

metastable states [23]. In the SI state the $(NO)^+$-group is inverted with the isonitrosyl structure N–C–Fe–O–N, and in the SII state a side-on molecular conformation of the $(NO)^+$-group has been proposed (Figure 4). The re-investigation of the existing neutron diffraction data confirms again that neither an exchange of the NO-atoms in the nitrosyl group from Fe–N–O to Fe–O–N for SI could be observed nor a 90° bending of the NO-bond for SII [24]. Possible explanation of the discrepancy between the results obtained from neutron-diffraction [21, 22, 24] and X-ray diffraction experiments [23] is based on the morphology of the single crystals after temperature contraction. Single crystals with an average SI population of even 45% within the whole crystal exhibit areas with reddish colour (which are characteristic for weak SI population) as visualized by the inset (e) in Figure 5. Structural information on the metastable state SI on the basis of diffraction experiments with a beam cross-section smaller than a few square millimetres may be obscured by the fact that the narrow beam hits an area with reddish colour (nearly GS of the crystal).

Several independent theoretical calculations [25–27] of molecular dynamics, electronic structure and nuclear resonance hyperfine parameters, using the proposed geometric rearrangements of the $(NO)^+$-group for SI and SII, indicate that the proposed structures of SI and SII are in reasonable agreement with the experimental results of Carducci *et al.* [23]. All calculated and measured Mössbauer parameters are summarized in Table I.

An attempt to support the isonitrosyl structure of SI by angle-dependent Mössbauer spectroscopy has been made by investigating the ratio t_-/t_+ of the effective thickness for the left- and right-hand line following the evaluation procedures described elsewhere [28, 29] (Figure 6(a)). Measurements of the orientation angle

Table I. Mössbauer parameters of GS, SI and SII as reported from different research groups [25–27]. The IS values are related to α-Fe standard at room temperature. The theoretical values are derived from DFT calculations

NP comp.	Parameter	State			Reference
		GS	SI	SII	
		Theory/exp.	Theory/exp.	Theory/exp.	
SNP	IS, mm/s	0.00/0.00	0.12/0.18	0.17/0.20	[25]
	ΔE_Q, mm/s	+1.83/+1.72	+2.55/+2.75	+2.81/+2.85	
	η	0.03/< 0.02	0.04/< 0.03	0.44/< 0.1	
	EFG $\times 10^{21}$, V/m^2	+7.60/+9.69	+14.6/+16.5	−15.2/+17.1	[26]
	η	0.03/0.02	0.13/0.03	0.51/0.2	
GNP	IS, mm/s	–/−0.25	–/−0.01	–	[27]
	ΔE_Q, mm/s	+1.81/+1.85	+2.78/+2.93	−2.99/–	
	η	0.00/0.01	0.00/–	0.68/–	

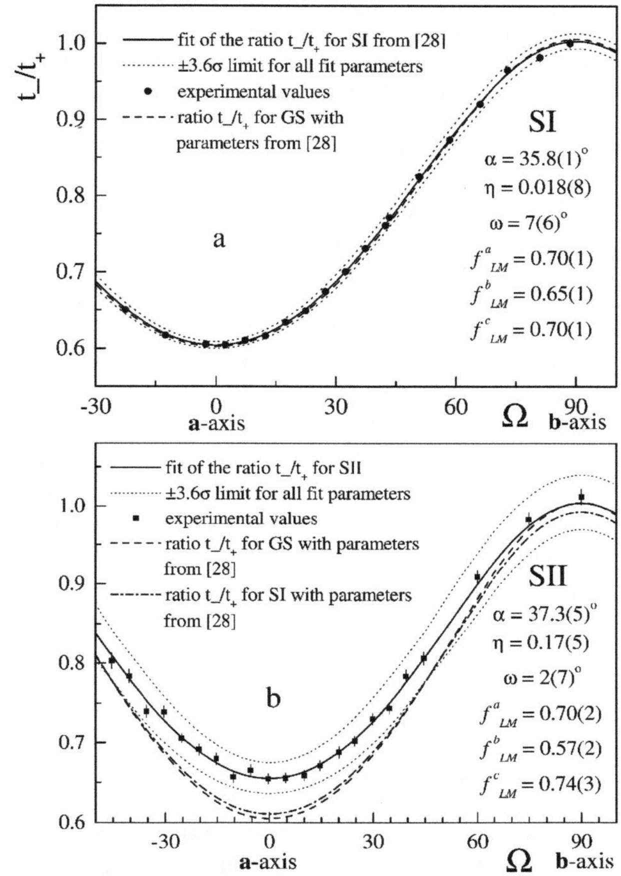

Figure 6. Angle-dependent ratio of the effective thickness of the left-hand low-energy line ($\pm 1/2 \rightarrow \pm 1/2$ transition) t_- relative to the effective thickness of the right-hand high-energy line ($\pm 3/2 \rightarrow \pm 1/2$ transition) t_+ (a) for SI (45%) and GS (55%) and (b) for SII (25%) and GS (75%) (and for comparison for SI). The fit parameters for SI and SII are given in the inset.

α of the EFG tensor, the orientation angle ω of the mean-square displacement (MSD) tensor and the Lamb–Mössbauer factors f_{LM}^{a}, f_{LM}^{b}, and f_{LM}^{c} for all principle crystallographic directions have been performed. The obtained results for GS and SI are very similar because the orientation of the mean molecular axes Fe–O–N and Fe–N–O in the two cases, SI and GS, with respect to the principle crystallographic directions is practically the same, only changes of the quadrupole splitting and isomer shift are observed.

Similar angle-dependent measurements have been performed for SII [11]. The results together with the fitted parameters for SI and SII are presented in Figure 6(b). All available data about the Mössbauer parameters of GS, SI, and SII are summarized in Table II.

Table II. Summarized data for Mössbauer parameters of the NP anion: isomer shift IS (relative to SNP in GS at 77 K), quadrupole splitting ΔE_Q, sign of the main component V_{zz} of the electric field gradient EFG, orientation angle α of the EFG, asymmetry parameter η of EFG, orientation angle ω of the mean-square displacement tensor MSD, and Lamb–Mössbauer factors f^a_{LM}, f^b_{LM}, and f^c_{LM} for the principal crystallographic directions of GS, SI, and SII

Parameters	States				
	GS (only) from [28]	GS (+SI) from [28]	SI (+GS) from [28]	GS (+SII) from [11]	SII (+GS) from [11]
Sign of EFG	+	+	+	+	+
ΔE_Q [mm/s]	1.716(3)	1.714(3)	2.755(3)	1.720(3)	2.862(3)
IS [mm/s]	0.000	+0.001(3)	+0.178(3)	−0.001(3)	+0.194(3)
α [°]	35.3(1)	35.4(1)	35.8(1)	35.6(2)	37.3(5)
η	0.021(8)	0.024(8)	0.018(8)	0.026(8)	0.17(5)
ω [°]	14(6)	14(6)	7(6)	17(10)	2(7)
f^a_{LM}	0.69(1)	0.68(1)	0.70(1)	0.675(7)	0.70(2)
f^b_{LM}	0.66(1)	0.65(1)	0.65(1)	0.648(7)	0.57(2)
f^c_{LM}	0.69(1)	0.69(1)	0.70(1)	0.696(7)	0.74(3)
Population, **a**, **b**, **c**-cut [%]	100, 100, 100	58, 53, 84	42, 47, 16	73, 74, 87	27, 26, 13

2.4. MÖSSBAUER PARAMETERS DISCUSSION

We first discuss the values of the isomer shift and of the quadrupole splitting. Theoretical calculations [25–27] provide values of these parameters for GS, SI, and SII, which are close to the measured values. The calculated sign of the EFG for GS is positive, which has been confirmed experimentally (with the applied-magnetic-field method) for sodium [30], guanidinium [31] and barium [32] nitroprussides. The overall agreement between calculated and measured values for SI and SII is a convincing argument for the conformational properties of the metastable states in SNP, i.e. the isonitrosyl structure of SI and the side-on bonded conformation for SII.

The isonitrosyl structure of SI is supported [33, 34] by infrared and Raman studies using isotope (^{15}N, ^{18}O, ^{54}Fe) substituted nitroprussides. The observed downshift of some vibrational frequencies upon isotope substitution is in good agreement with the normal-co-ordinate analysis and with the results from density functional theory (DFT) calculations.

The GNP single crystal holds a special place in the investigation of the metastable states in NP complexes. The opportunity to produce SI by non-polarized light, using blue light-emitting diodes allowed us to perform an experiment applying Nuclear Inelastic Scattering (NIS) using synchrotron radiation. The experimental set-up is shown on Figure 7(d). The **c**-cut GNP single crystal (enriched to 25%

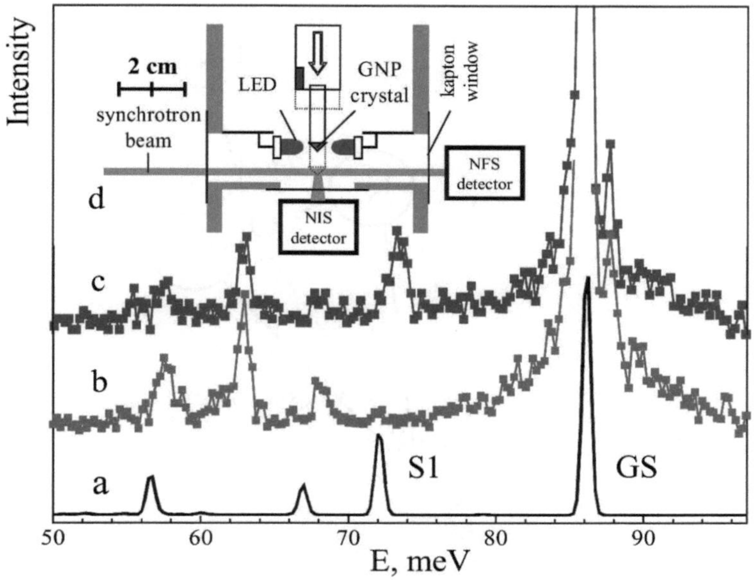

Figure 7. (a) NIS spectrum of the NP anion derived from DFT calculations assuming 25% population of SI; (b) experimental NIS spectra of non-irradiated GNP single crystal in GS; (c) experimental NIS spectra of GNP after irradiation with unpolarized, non-monochromatic light (450 nm) from light-emitting diodes; (d) schematic presentation of the experimental set-up.

in the ^{57}Fe isotope) was mounted on a sample holder surrounded by two LEDs of 450 nm wavelength and 200 mcd intensity [35]. The simulated NIS spectrum using DFT calculations, Figure 7(a), has been compared with the experimental one recorded after irradiation of the crystal by the LEDs (reaching a population of the SI of about 10%), Figure 7(c), and the obtained spectrum after heating the crystal up to 250 K for 30 min (GS only), Figure 7(b). The predicted Fe–O stretching vibrational mode of SI at 72 meV, assuming the isonitrosyl structure has been experimentally confirmed with high precision. Thus, IR, Raman and NIS results [33–35] confirm the isonitrosyl conformation of SI proposed by Carducci *et al.* [23].

For SII two out of three DFT calculations have yielded a negative sign for the main component V_{zz} of the EFG [26, 27], but in the third calculation [25] the sign is positive. In the latter case the Fe–C–N axial direction defines the orientation of V_{zz}. After irradiation of crystals (**a**- and **c**-cut), the line intensity ratio for SII is approximately the same as for GS. If the sign of V_{zz} for SII were negative, the two lines should change their position and the line-intensity ratio for SII should become the reciprocal value of the ratio for GS, which has not been observed experimentally. The angle-dependent Mössbauer measurements in the mirror **ab**-plane (Figure 6(b)) unambiguously show that the sign of the EFG for SII is positive [11].

The orientation angle α of the EFG for SII was found to be only about 2° larger than the corresponding values measured for GS and SI. The geometry of the side-on bonded $(NO)^{+}$-group in SII is presented in Figure 8 according to the X-ray

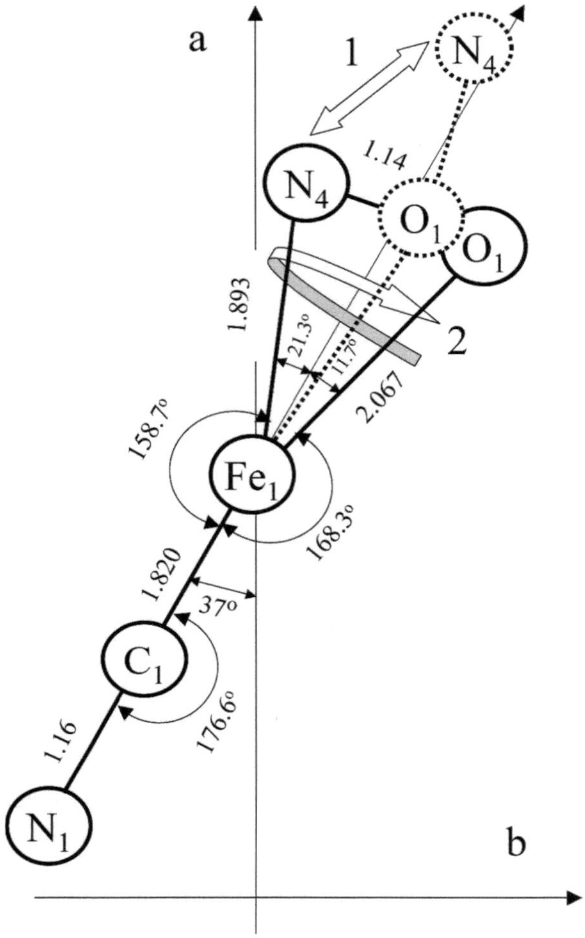

Figure 8. Bond lengths (in Å) and angles for SII "staggered" conformation [23]. Arrow 1 represents the bending mode between static side-on and nearly ideal isonitrosyl bonding of the (NO^+)-group and arrow 2 the simultaneous rotation of the (NO^+)-group around the N_1–C_1–Fe_1 axis.

results [23]. The axial C_1N_1-group lies in the mirror **ab**-plane and the N_1–C_1–Fe_1 direction makes an angle of about 37° with the **a**-crystallographic axis. If the $(NO)^+$-group for SII were linear as for GS (N–C–Fe–N–O) or for SI (N–C–Fe–O–N) this angle would determine the orientation of the EFG. The other four equatorial CN-ligands are inclined by 45° with respect to the crystallographic mirror **ab**-plane. In case of a side-on bonded $(NO)^+$-group the triangle of bonds $Fe_1N_4O_1$ lies in the **ab**-plane, and this is one of the possible "staggered" conformations in which the projection of the $(NO)^+$-group lies between both N–C–Fe–C–N equatorial chains. The Fe_1N_4- and the Fe_1O_1-bonds make angles of 21.3° and 11.7° with the linear nitrosyl axis, respectively, and these are the maximum possible and expected changes of the orientation angle α of the EFG. If the triangle of bonds $Fe_1N_4O_1$ is perpendicular to the mirror **ab**-plane (another possible "staggered" conformation) the main EFG component would probably no longer lie in the **ab**-plane, but its projection on the mirror **ab**-plane makes the angle $\alpha = 37°$ with the **a**-crystallographic direction. The $(NO)^+$-group in the other possible side-on bonded "eclipsed" states is nearly parallel to one of the equatorial N–C–Fe–C–N chains and the triangle of bonds $Fe_1N_4O_1$ is inclined by 45° to the crystallographic mirror **ab**-plane. It is not known which of the possible discussed conformations are actually present in nature. Probably, in different NP anions different conformations of SII could exist.

DFT calculations for SII yield relatively high values for the asymmetry parameter η, i.e. 0.44 [25], 0.51 [26] and 0.68 [27], as compared to the measured value 0.17(5) for SNP. The reason for this discrepancy is maybe the static side-on "eclipsed" or "staggered" conformation of the $(NO)^+$-group assumed for the calculations. The calculated energy barrier between these conformations is small, i.e. about 0.01 eV or even smaller [36] than $k_BT = 0.0067$ eV at $T = 78$ K. Thus, the $(NO)^+$-group in the side-on bonded "eclipsed" conformation could be thermally activated to the "staggered" conformation and vice versa. This suspected dynamic behaviour of the $(NO)^+$-group would reduce both the asymmetry parameter and the quadrupole splitting [37].

An alternative interpretation of experimental data suggest that the $(NO)^+$-group in SII tilts between the nearly ideal octahedral isonitrosyl conformation of SI and a second state with up to 90° bending of the Fe–O–N bond [11]. Within this picture the nature of SII is represented by a continuous oscillation between the bent and the isonitrosyl structure (arrow 1 in Figure 8). The remarkable Mössbauer line broadening of about 12% reported for SII [10] could be considered as an evidence for a dynamic process of the $(NO)^+$-group around the Fe atom. The frequency of this bending vibration is expected to be in the range from 10 to 100 MHz. Within the half-life time of the 14.4 keV Mössbauer level, which is about 0.1 μs, several oscillations could occur with the net result that the Fe nucleus registers a mean value for asymmetry parameter and quadrupole splitting. Taking the values for the static isonitrosyl conformation of SI, $\eta \cong 0$ and $\Delta E_Q = +2.75$ mm/s [10], and for the static side-on bonded SII, $\eta = 0.68$ [27] and $\Delta E_Q = +2.86$ mm/s [10], the dynamic molecular conformation (switching at high frequencies between the two static states) yields: $\eta \cong 0.34$, $\Delta E_Q \cong 2.85$ mm/s, a positive sign for EFG, a change in the orientation angle α by about 10°, and a slight reduction of the Lamb–Mössbauer factors.

Other more complicated models for the dynamic behaviour of the $(NO)^+$-group for SII cannot be excluded [11]. One of these cases, which represents a combined oscillation (arrow 1) and rotation (arrow 2) of the $(NO)^+$-group, is shown in Figure 8. In this case the parameters take values ($\eta \cong 0.17$, $\Delta E_Q \cong 2.85$ mm/s, positive sign of the EFG, the orientation angle α only a few degrees larger than that measured for GS and SI) which are comparable with the experimental findings for SII. If the bending vibrational mode of the $(NO)^+$-group in this case is predominantly active in the **b**-direction of SNP, only for this direction lower values

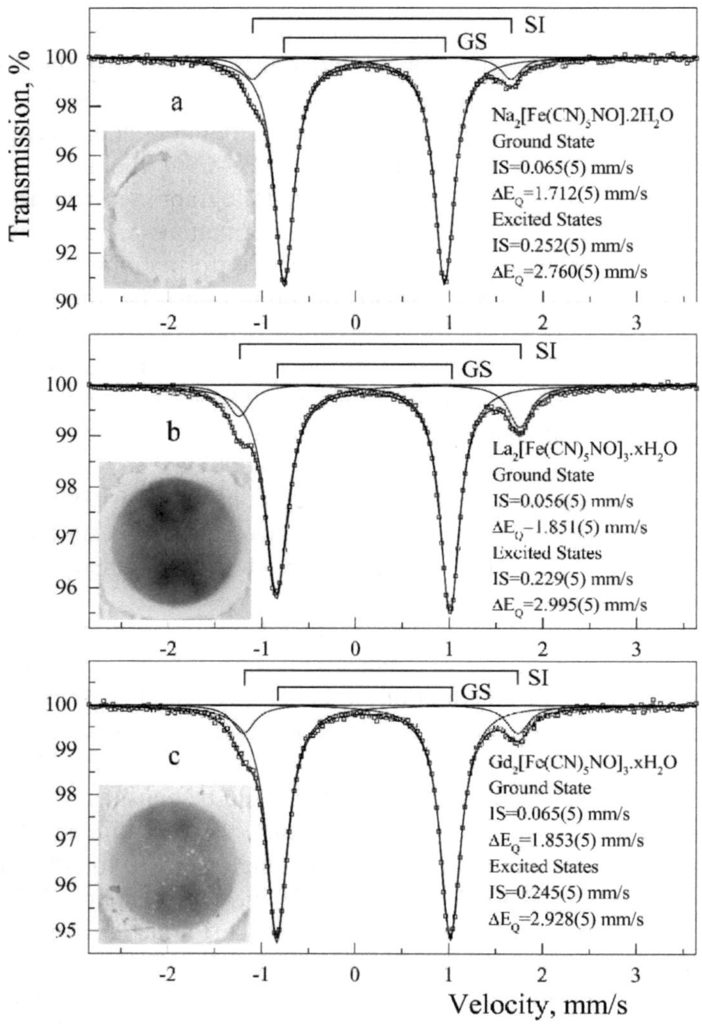

Figure 9. Mössbauer spectra taken at 77 K of powder samples of (a) sodium, (b) lanthanum and (c) gadolinium nitroprussides irradiated with blue light from LEDs. The velocity scale is related to SNP in GS at room temperature. The insets show photographs of the corresponding rare-earth nitroprussides taken at room temperature after irradiation.

of the Lamb–Mössbauer factor are expected, which is also in agreement with the experimental results (Table II).

2.5. PHOTOSWITCHING OF RARE-EARTH NITROPRUSSIDES

The photoswitching experiments have been extended to rare-earth nitroprussides (RNP). Mössbauer spectra of irradiated (with LEDs) powder samples at 77 K of sodium, lanthanum and gadolinium nitroprussides are shown in Figure 9 together with images of the samples taken at room temperature after the irradiation. In all nitroprussides the metastable states are populated up to 10–15%, which is typical for powder samples. The Mössbauer spectra taken at room temperature show only a quadrupole doublet characteristic for GS. The color change occurs only in a thin surface layer of some micrometer thickness and therefore can not be detected by transmission Mössbauer spectroscopy. It is well known that all nitroprussides change their colour by irradiation at room temperature, forming Prussian blue analogues after photochemical reaction. After irradiation at 77 K the SNP does not change its colour (Figure 9(a)) while, depending on the time of irradiation, the RNP change from light grey to black colour (Figures 9(b) and (c)). However, it does not change to blue colour, which would be typical for Prussian blue analogues. It seems that in these compounds, besides the common SI and SII states, a third state is produced by irradiation.

A possible explanation for the appearance of the third state is based on an internal photochemical redox reaction [38, 39]. In the Prussian blue analogue cobalt-iron cyanide the metal centres Co(III) and Fe(II) are in the low-spin diamagnetic $S = 0$ state, which by irradiation with red light changes into a paramagnetic state. The irradiation induces an electron transfer from Fe to Co through the CN-bridge (with approximately 5 Å length), changing the oxidation and therefore the spin state of both metal centres, i.e. to Fe(III) ($S = 1/2$) and Co(II) ($S = 3/2$), respectively. As a consequence magnetization below 16 K is remarkably increased. Thermal treatment or irradiation with blue light can restore the original situation almost completely.

A similar light-induced mixed valence state accompanied by a change of the colour has been observed in iron nitroprusside, in which the iron atoms are present as cations as well [40, 41]. This result, however, should be considered with caution because the irradiation has been performed at room temperature and Prussian blue analogues could have been formed.

The iron and the rare-earth cations in RNP are also connected via cyano bridges (distance of about 5 Å) and therefore a possible photoinduced molecular magnetization can not be excluded. This effect can be observed only in nitroprussides the cation of which has a vacant electronic state and therefore could accept the electron transmitted via the cyano bridge. This is not the case for SNP in which this effect has not been observed (Figure 9(a)).

3. Conclusions

All nitroprussides can be switched by irradiation with blue light into two metastable states SI and SII, which are stable below 200 and 150 K, respectively. The ground state GS as well as the metastable states SI and SII are diamagnetic, and therefore light-induced spin-crossover of Fe(II) can not explain the photoswitching effects observed in nitroprussides [42]. Instead, the photoinduced metastable states in nitroprussides (and also in other NO-containing materials such as NO-heme proteins [36]) have been identified as linkage isomerism (isonitrosyl and side-on conformations of the $(NO)^+$-group).

Mössbauer parameters of GS and SI derived from DFT calculations are in quantitative agreement with the measured values. For SII such agreement is lacking. The reason for this discrepancy is most likely due to the assumed static side-on conformation of the $(NO)^+$-group in the DFT calculations. The experimental Mössbauer parameters of SII are successfully reproduced by calculations when taking into account dynamic properties of the $(NO)^+$-group. In our opinion, in SII bending vibrations [11] as well as rotations of the $(NO)^+$-group occur [25, 26], the latter around the main molecular axis. A crucial experimental test of SII by nuclear inelastic scattering of synchrotron radiation with SNP single crystals (similar to that performed for SI [35]) is in progress and might contribute additional experimental information about the molecular dynamics of the $(NO)^+$-group in SII. Further, modified tests with *ab initio* molecular dynamic simulations are desirable in order to elucidate more details and to confirm the validity of the liberation-rotation model of SII. Experiments at low temperature (4.2 K) could "freeze" SII in a static side-on conformation and warrant further investigation.

A colour change, which is stable at room temperature, has been observed in rare-earth nitroprussides. An explanation of this phenomenon by light-induced molecular magnetization has been proposed. Measurements of the magnetic susceptibility confirming the molecular magnetism in RNP are in progress. A possible reversion of the colour change by thermal treatment and irradiation with different wavelengths is under investigation.

Acknowledgements

We would like to thank the Alexander von Humboldt Foundation, AvH (V. R.), the German Academic Exchange Serves, DAAD (Sv. S.) and the German Research Foundation, DFG (A. X. T.) for supporting this research. We specially thank the Organizing Committee of the 5th Seeheim Workshop on Mössbauer Spectroscopy and personally Prof. Dr. P. Gütlich for the invitation to present this contribution at the Seeheim Workshop and for financial support.

References

1. Hauser, U., Oestreich, V. and Rohrweck, H. D., *Z. Phys. A* **280** (1977), 17, 125; **284** (1978), 9.
2. Spijkerman, J. J., Ruegg, F. C. and DeVoe, J. R., *Tech. Rep. Ser. I. A. E. A.* **50** (1966), 254.
3. Bottomly, F. and White, P. S., *Acta Crystallogr. B* **35** (1979), 2193.
4. Gómez, J. A. and Guenzburger, D., *Chem. Phys.* **235** (2000), 73.
5. Grant, R. W., Housley, R. M. and Gonser, U., *Phys. Rev.* **178** (1969), 523.
6. Woike, Th., Krasser, W., Zöllner, H., Kirchner, W. and Haussühl, S., *Z. Phys. D* **25** (1993), 351.
7. Woike, Th., Kirchner, W., Schetter, G., Barthel, Th., Kim, H. and Haussühl, S., *Opt. Commun.* **106** (1994), 6.
8. Haussühl, S., Schetter, G. and Woike, Th., *Opt. Commun.* **114** (1995), 219.
9. Zöllner, H., Woike, Th., Krasser, W. and Haussühl, S., *Z. Kristallogr.* **188** (1989), 139.

10. Woike, Th., Kirchner, W., Kim, H., Haussühl, S., Rusanov, V., Angelov, V., Ormandjiev, S., Bonchev, Ts. and Schroeder, A. N. F., *Hyp. Interact.* **77** (1993), 265.
11. Rusanov, V., Woike, Th., Kim, H., Angelov, V., Bonchev, Ts. and Haussühl, S., *Solid State Commun.* **126** (2003), 457.
12. Rusanov, V., Angelov, V., Angelova, J., Bonchev, Ts., Woike, Th., Kim, H. and Haussühl, S., *J. Sol. State Chem.* **123** (1996), 39.
13. Hauser, U., Klimm, W., Reder, L., Schmitz, T., Wessel, M. and Zellmer, H., *Phys. Lett. A* **144** (1990), 39.
14. Woike, Th., Zöllner, H., Krasser, W. and Haussühl, S., *Solid State Commun.* **73** (1990), 149.
15. Guida, J. A., Piro, O. E. and Aymonino, P. J., *Inorg. Chem.* **34** (1995), 4113.
16. Coppens, P., Fomitchev, D., Carducci, M. and Culp, K., *J. Chem. Soc., Dalton Trans.* (1998), 865.
17. Rusanov, V., Winkler, H., Ober, C. and Trautwein, A. X., *Eur. Phys. J. B* **12** (1999), 191.
18. Retzlaff, C., Krumbe, W., Dörffer, M. and Haussühl, S., *Z. Kristallog.* **189** (1989), 141.
19. Rusanov, V., Stankov, Sv., Angelov, V., Koop, N., Winkler, H. and Trautwein, A. X., *Nucl. Instr. Meth. B* **207** (2003), 205.
20. Güdel, H. U., *Chem. Phys. Lett.* **175** (1990), 262.
21. Rüdlinger, M., Schefer, J., Chevrier, G., Furer, N., Güdel, H. U., Haussühl, S., Heger, G., Schweiss, P., Vogt, Th., Woike, Th. and Zöllner, H., *Z. Phys. B* **83** (1991), 125.
22. Schefer, J., Woike, Th., Haussühl, S. and Fernandes-Dias, M., *Z. Kristallogr.* **212** (1997), 29.
23. Carducci, M. D., Pressprich, M. R. and Coppens, P., *J. Am. Chem. Soc.* **119** (1997), 2669.
24. Schefer, J., Woike, Th., Imlau, M. and Delley, B., *Eur. Phys. J. B* **3** (1998), 349.
25. Delley, B., Schefer, J., Woike, Th., *J. Chem. Phys.* **107** (1997), 10067.
26. Blaha, P., Schwarz, K., Faber, W. and Luitz, J., *Hyp. Interact.* **126** (2000), 389.
27. Rusanov, V., Grünsteudel, H., Paulsen, H., Winkler, H., Meyer-Klaucke, W. and Trautwein, A. X., *Hyp. Interact.* **126** (2000), 159.
28. Woike, Th., Imlau, M., Angelov, V., Schefer, J. and Dellej, B., *Phys. Rev. B* **61** (2000), 12249.
29. Timmermann, A., Thesis, Ph. D., Universität zu Köln, 1981.
30. Oosterhuis, W. T. and Lang, G., *J. Chem. Phys.* **50** (1969), 4381.
31. Rusanov, V., Stankov, Sv., Angelov, V. and Schünemann, V., *Nucl. Instr. Meth. B* **170** (2000), 443.
32. Stankov, Sv., Rusanov, V., Angelov, V., Haussühl, S. and Trautwein, A. X., *Hyp. Interact.* **141/142** (2002), 223.
33. Morioka, Y., Takeda, S., Tomizawa, H., Miki, E., *Chem. Phys. Lett.* **292** (1998), 625.
34. Chaćon Villalba, M. E., Güida, J. A., Varetti, E. L. and Aymonino, P. J., *Spectrochim. Acta A* **57** (2001), 367.
35. Paulsen, H., Rusanov, V., Benda, R., Herta, C., Schünemann, V., Janiak, C., Dorn, T., Chumakov, A., Winkler, H. and Trautwein, A. X., *J. Am. Chem. Soc.* **124** (2002), 3007.
36. Coppens, P., Novozhilova, I. and Kovalevsky, A., *Chem. Rev.* **102** (2002), 861.
37. Paulsen, H., personal communication.
38. Sato, O., Iyoda, T., Fujishima, A. and Hashimoto, K., *Science* **272** (1996), 704.
39. Verdaguer, M., *Science* **272** (1996), 698.
40. Morioka, Y., Hisamitsu, T., Inoue, H., Yoshioka, N., Tomizawa, H. and Miki, E., *Bull. Chem. Soc. Jpn.* **71** (1998), 837.
41. Inoue, H., Yoshioka, N. and Morioka, Y., *Hyp. Interact. (C)* **3** (1998), 113.
42. Gütlich, P., Garsia, Y., Woike, Th., *Coord. Chem. Rev.* **219** (2001), 839.

Hyperfine Interactions **144/145:** 325–346, 2002.

Intracluster and Intercluster Interactions in Nanostructures

I. P. SUZDALEV
N. N. Semenov Institute of Chemical Physics, Russian Academy of Sciences, ul. Kosygina 4, Moscow 119991, Russia

Abstract. The formation and properties of three types of nanostructures: solid-state nanostructures, matrix nanostructures and nanostructures under high pressure with shear were studied. Nanocluster peculiarities: solid–liquid state and first-order (jump-wise) magnetic phase transitions were considered. The influence of cluster–cluster interactions, cluster–matrix interactions and cluster defects on cluster atomic dynamics, cluster melting, cluster critical sizes, Curie or Néel points and the character of magnetic phase transitions were investigated.

Key words: nanocluster, nanostructure, atomic motion, melting point, magnetic phase transitions, Mössbauer spectroscopy.

Introduction

The properties of a material are known to change significantly on passing from macrostructures to microstructures the size of which lies in the nanometer range. Thus the crystal lattice parameter, the heat capacity, the melting point and the electrical conductivity of the nanocluster differ from those of the corresponding macrocrystals. In addition they acquire new optical, magnetic and electronic characteristics [1–3]. However, the properties of nanostructures are determined not only by the cluster size but also by the mode of their organization, cluster matrix and intercluster interactions.

This paper includes the study of iron oxide nanostructures formed: (1) by solid-state chemical reactions, (2) by matrix isolation of nanoclusters, and (3) by action of high pressure with a shear. The cluster atomic dynamics and magnetic phase transitions were studied in these nanostructures and characteristics of intercluster and intracluster interactions were derived.

1. Nanosystems based on solid-state chemical reactions

Nanoclusters are formed by solid-state chemical reactions, for example by thermal decomposition of iron salts. Decomposition of iron salts at a temperature above some critical (or threshold) values starts with the formation of a mobile active reaction medium in which nucleation of nanoclusters occur [4]. The process of cluster formation can be compared with seed formation in confined space in solution or melt (a cage or pore). In this case the role of this confinement can be played by diffusion restriction, which prevents the perturbation of the mother medium caused by the change in the cluster size from migrating over the nucleation time (t_n) by a distance longer than $L \sim (Dt_n)^{1/2}$ (D is diffusion coefficient).

The nanocluster system can be fabricated with two types of nanostucture with weak (free cluster) and strong (sintering clusters) intercluster interactions. For nucleation of a small cluster with weak interactions the dependence of Gibbs energy ΔG_f on cluster radius R for one cluster can be described by [4]

$$\Delta G_f = \alpha S_f - \left(\Delta\mu + \ln\frac{n_0 - n}{N}\right)\rho V_f, \tag{1}$$

where α is the density of the surface energy of the cluster (α = const), ρ is the substance density in the cluster, $\Delta\mu$ is the change in the chemical potential of the system caused by the transfer of a single iron atom from the medium to the cluster, N are atoms of media, n_0 are iron atoms being contained in the cluster, $S_f = 4\pi R^2$, $V_f = 4/3\pi R^3$.

The function ΔG_f passes through a maximum at $R = R_{CR}$ (R_{CR} is the critical radius of a seed in the nucleation) and through a minimum at $R = R_{max}$ (R_{max} is the cluster radius corresponding to a stable equilibrium state of the system) (Figure 1). Upon further growth of clusters they come into contact and the boundaries between them vanish (sintering). The onset of cluster sintering corresponds to a discrete transition from isolated or weekly interacting clusters to strongly interacting, sintering clusters. If the half of the distance R_t between the cluster center at the stage

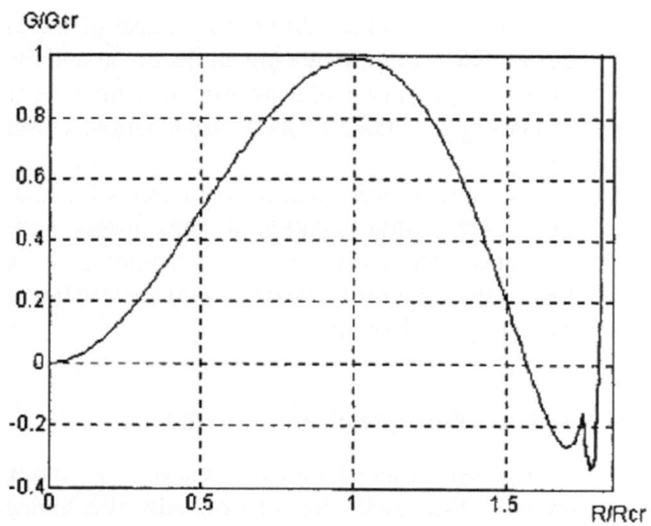

Figure 1. Change in the Gibbs energy G/G_{CR} upon nucleation, cluster growth and cluster sintering according to relations (1) and (3) in R/R_{CR} units.

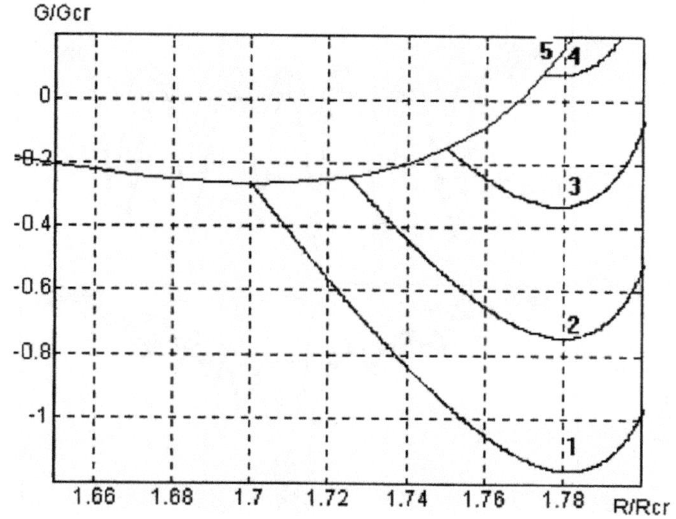

Figure 2. Change in the Gibbs energy G/G_{CR} in the region of the onset of cluster sintering vs. R/R_{CR} at different R_t values (which is a half of of distance between the cluster centers) $R_t = 1.7R_{\mathrm{CR}}$ (1), $1.73R_{\mathrm{CR}}$ (2), $1.75R_{\mathrm{CR}}$ (3), $1.77R_{\mathrm{CR}}$ (4), $1.8R_{\mathrm{CR}}$ (5).

of sintering is equal to $R =$ const, the surface area (S_{S}) and the volume (V_{S}) of the clusters which has formed contacts with k neighbors can be calculated by

$$\begin{aligned} S_{\mathrm{S}} &= 4\pi R^2 + k2\pi R(R_t - R), \\ V_{\mathrm{S}} &= \frac{4}{3}\pi R^3 + k\frac{1}{3}\pi\left(-2R^3 + 3R^2R_t - R_t^3\right). \end{aligned} \tag{2}$$

Corresponding to the change in the Gibbs energy at the stage of sintering, i.e., when $R > R_t$

$$\Delta G_{\mathrm{S}} = \alpha S_{\mathrm{S}} - \left[\Delta\mu + \ln\frac{n_0 - n}{N}\right]\rho V_{\mathrm{S}}. \tag{3}$$

Thus there are two regions in nanostructure formation: for $R < R_t$ expression (1) and for $R > R_t$ expression (3) are available. The calculations give that for $k = 6$ at $R_t \leq 1.7R_{\mathrm{CR}}$ sintering occurs without a barrier; in a system less densely populated by clusters at $R_t = 1.75R_{\mathrm{CR}}$ the transfer to sintering requires overcoming a potential barrier and in more less densely populated system at $R = 1.8R_{\mathrm{CR}}$ sintering does not occur at all (Figure 2).

The differential thermal analysis (DTA) and differential thermal gravimetric (DTG) curves recorded during thermal decomposition of iron oxalate $F_2(C_2O_4)_3 \cdot 5H_2O$ in air exhibit two minima at $T_D = 200$ °C and 260 °C [4]. The first minimum corresponds to dehydration processes of iron complexes, mobile medium starts to form in which nucleation and growth of iron oxide clusters takes place. The second minimum is apparently due to the onset of the sintering of iron oxide clusters. The cluster sizes were determined from specific surface area (BET), using data of X-ray

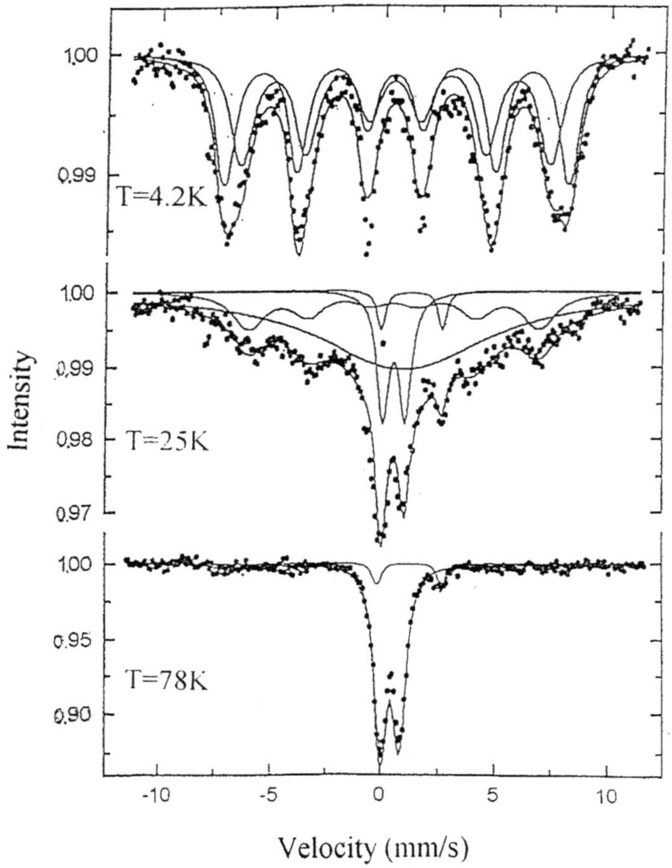

Figure 3. Mössbauer spectra of iron oxide nanoclusters obtained at $T_D = 215$ °C at different temperatures.

diffraction analysis, atomic force microscopy (AFM) and Mössbauer spectroscopy. Nanocluster sizes were estimated from superparamagnetic broadening of hyperfine structure lines for Mössbauer spectra of iron oxide nanoclusters [5] obtained at $T_D = 215$–235 °C. Mössbauer spectra for iron oxide nanoclusters fabricated at $T_D = 215$ °C are shown in Figure 3. One can observe the typical superparamagnetic relaxation with the broadening of HFS lines and their shifting to the middle of spectrum that corresponds to the nanocluster size 2–3 nm. The BET measurements give the same results. These nanoclusters correspond to the region of cluster sizes with $R < R_t$ (Figure 1) and to the week cluster–cluster interactions. The sizes of nanoclusters at the beginning of sintering were estimated by AFM.

AFM image of iron oxide nanocluster obtained at $T_D = 265$ °C is presents in Figure 4. The scale of the image – 500 nm × 500 nm permits to estimate the cluster sizes between 30 and 50 nm. X-ray analyses argue about the presence of αFe_2O_3–γFe_2O_3 with the same sizes [6]. Mössbauer spectra of αFe_2O_3–γFe_2O_3

Figure 4. Atomic force microscopy image for strongly interacting iron oxide nanoclusters obtained at $T_D = 265$ °C with resolution 500 nm.

nanostructure obtained at $T_D = 265$ °C characterized by strong cluster–cluster interactions in region $R > R_t$ are shown in Figure 5. No superparamagnetic behavior is marked for these big nanoclusters up to $T = 300$ K. Mössbauer spectra are composed of magnetic hyperfine structure components for αFe_2O_3–γFe_2O_3 structure corresponding to bulk αFe_2O_3 and γFe_2O_3 and quadrupole doublet. The origin of these phenomena will be interpreted in Section 5.

2. Nanoclusters prepared using matrix isolation

The formation of clusters from solution in closed pore under weak cluster matrix interaction can be treated in the frame of relation (1) as for the nucleation of cluster during the first region of solid-state chemical reaction. The limited concentration

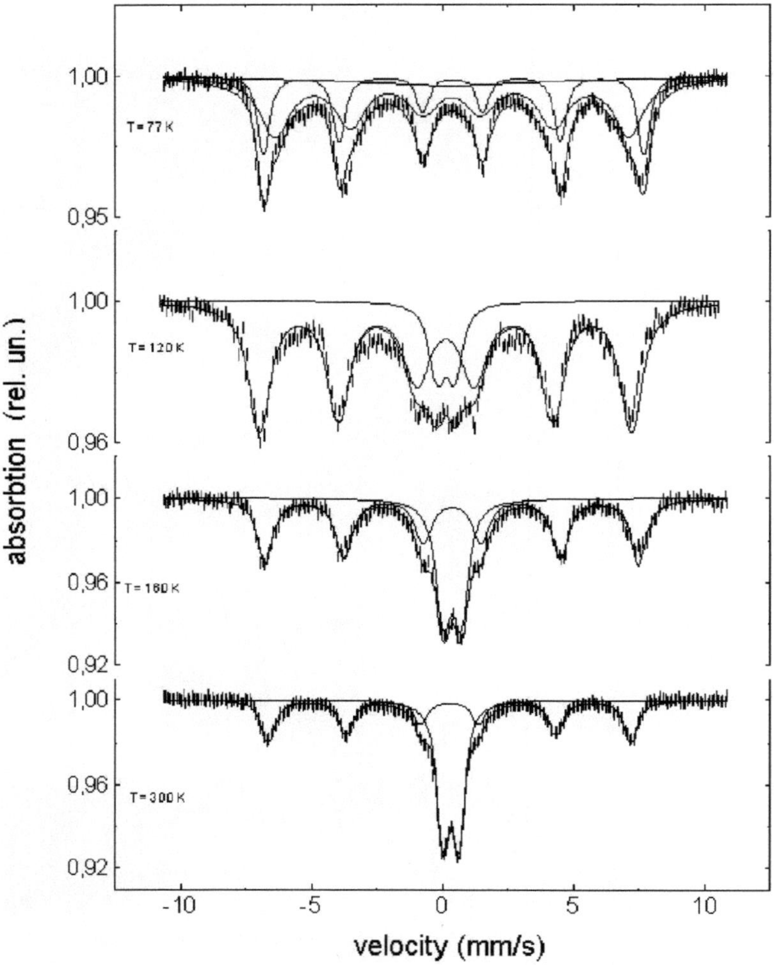

Figure 5. Mössbauer spectra of αFe_2O_3–γFe_2O_3 nanostructure obtained at $T_D = 265$ °C at different temperatures.

of the solute inside the pore entails the existence of limiting volume of the cluster $n_{\max}$, after this volume has been reached, cluster growth becomes energetically unfavorable and ceases. One more important influencing cluster formation and growth in a pore is pore heterogeneity. Heterogeneity decreases the energy barrier to the formation of a stable seed of a cluster having critical volume [7]. For the nucleation in the pore $n_{\min} = 32\alpha^3/(3\rho^2\Delta\mu^3)$, energetic barrier corresponding $n_{\min} - \Delta G_m = 16\pi\alpha^3 K/(3\rho^2\Delta\mu^2)$ and

$$n_{\max} = \frac{n_0 \Delta\mu}{2} + \mu, \tag{4}$$

where n_0 – number of atoms of dissolved substance, e.g., iron, n – number of cluster atoms, $K < 1$ – factor heterogeneity. For the fabrication of isolated nanoclusters with weak cluster matrix interactions one need to use very low concentrations of dissolved substance and absence of chemical sorbent – adsorbate interactions. The nucleation and the formation of iron hydroxide (ferrihydrite) clusters in pore matrix-copolymer styrene and divinilbenzene copolymer (polysorb) with a specific surface area up to 1000 m^2/g and pore size of 7–50 nm have been investigated [8]. The pore surface in polysorbs has no functional groups, therefore the influence of chemisorptions on cluster formation can be neglected. Polysorbs were impregnated with ferric chloride solutions with different iron concentration and flowed by air containing an ammonia admixture (0.008–0.08%). As a result insoluble fer-

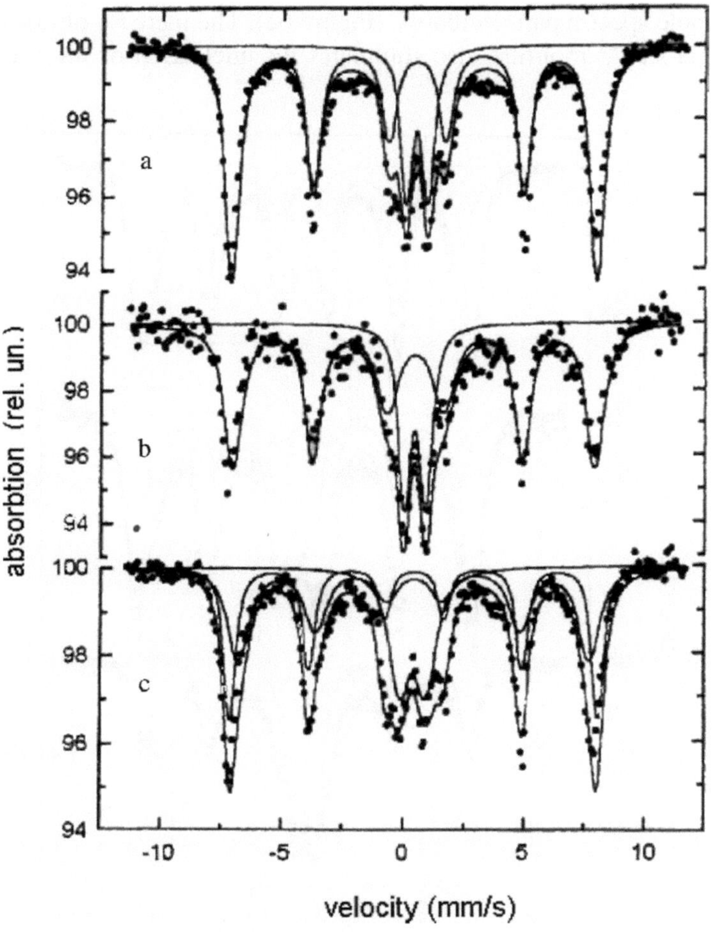

Figure 6. Mössbauer spectra of ferrihydrite clusters ($c = 1.2 \cdot 10^{-3}$ mol/litre) in polysorb pores with sizes 13 nm (a,b) and 20 nm (c) and computer fitting at different temperatures: (a) 4.2, (b) 10, (c) 4.2 K.

rihydrite clusters are nucleated after pH change of medium. This matrix isolation technique allows two cluster systems to be produced, namely, a system with isolated clusters and one with clusters interacting with one another (if several cluster occur in a pore) when iron concentration or pore size increases. Figures 6 and 7 show the Mössbauer spectra of ferrihydrite clusters obtained in 13- and 20-nm polysorb pores. The spectrum recorded at 4.2 K consists of two parts – the central quadruple doublet and magnetic hyperfine structure with relative narrow lines (with allowance for the disorder of cluster structure). As temperature is raised, some of HFS of the spectrum is converted into a paramagnetic doublet without entering the intermediate states corresponding to the superparamagnetic behavior. The increase of pore size up to 20 nm (Figure 6c) at the same concentration $1.2 \cdot 10^{-3}$ mol/litre leads to the nucleation of several clusters in one pore and to the appearance of cluster–cluster interaction. Mössbauer spectra in this case possess more broad line and quadrupole spectra part decreases. (Figure 6c). The increase of iron concentration up to $1.2 \cdot 10^{-2}$ mol/litre also stimulates the nucleation of several clusters in

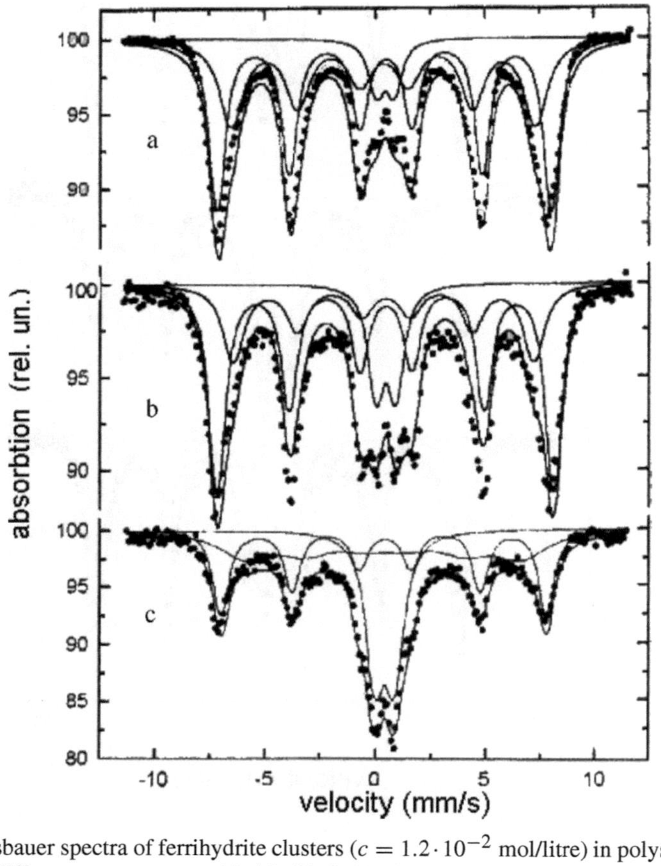

Figure 7. Mössbauer spectra of ferrihydrite clusters ($c = 1.2 \cdot 10^{-2}$ mol/litre) in polysorb pores with sizes 13 nm at different temperatures: (a) 4.2, (b) 6, (c) 10 K.

one pore, the appearance or the increase of cluster–cluster interactions (Figure 7). The appearance of superparamagnetism at 10 K (Figure 7c) made it possible to estimate cluster size from the broadening of the HFS in the spectrum that give the cluster size ~2–3 nm. The evaluation of the size of iron cluster formed in pore from iron concentration ($c = 1.2 \cdot 10^{-3}$ mol/litre) gives cluster size ~1–2 nm for isolated clusters. The comparison of very small (2–3 nm) ferrihydrate clusters (Figures 6, 7) with big (30–50 nm) iron oxide clusters (Figure 5) gives the idea about unusual phenomena that should be explained in term of jump wise transitions between HFS and quadrupole doublet with temperature or cluster size variation without intermediate superparamagnetic state (see Section 5).

3. Nanosystems, formed by action of high pressure with a shear

The two previous methods of nanocluster fabrication deal with the organization of nanostructures from nanoclusters. The action of high pressure with a shear is a technique that leads to a nanostructuration of bulk materials, to a decreasing of nanocluster size in nanostructures and to the appearance of numerous defects and strains. The defect density in nanostructures plays an important role for the definition of mechanical, electronic and magnetic properties in nanostructures [3].

In the difference of atoms and molecules in a bulk crystal, nanoclusters in nanostructure possess by real surface. Therefore for the thermodynamics consideration of a defect nanostructure one need to consider cluster surface as a source of defects, surface energy and chemical potential.

In the equilibrium distribution of defects between the cluster and the interface the chemical potential of defects inside of cluster

$$\mu_V = \mu_{V0}(P) + kT \ln c_V \tag{5}$$

becomes equal to chemical potential of defects in interface

$$\mu_S = \mu_{S0}(P_0) + kT \ln c_S, \tag{6}$$

where P, c_V and P_0, c_S are pressure and defect concentration inside of cluster and in interface, respectively. If the summary number of defects, n_0, and maximum of defect concentration, c_0, are constants such that

$$c_S = c_0 - \frac{1}{q} c_V, \tag{7}$$

where $q(r) = (1 + r/R)^3 - 1$ and r is the thickness of interface layer.

At the calculation of equilibrium concentration one should take into account that surface pressure in cluster is $P - P_0 = 2\alpha/R$, then from $\mu_V = \mu_S$

$$\frac{c_V}{c_0} = \frac{1}{1/[(1 + r/R)^3 - 1] + \exp[(\Delta\mu_0(P_0) + 2\alpha v/aR)/kT]}, \tag{8}$$

where v – volume of defect, a – constant lattice for bulk material.

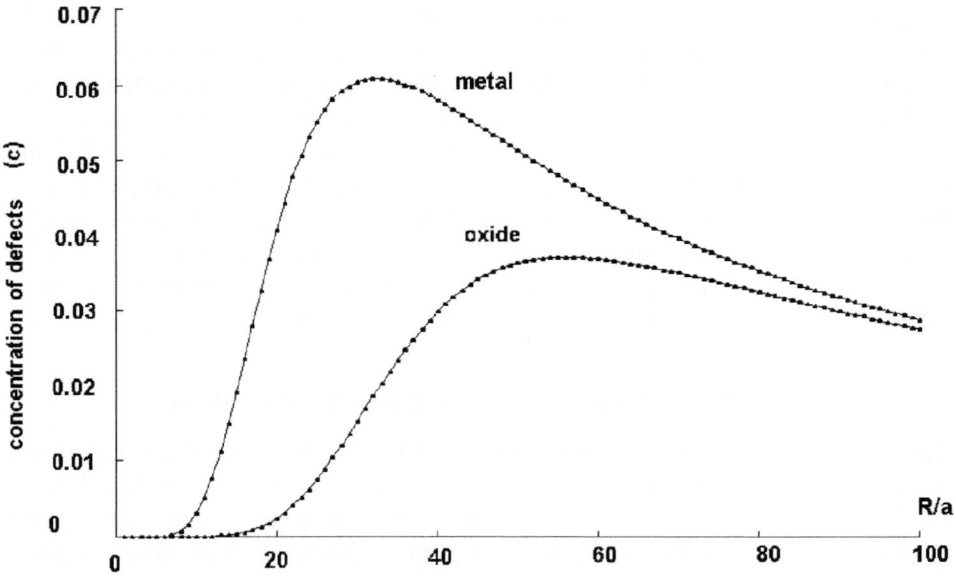

Figure 8. Density of defects c_V in cluster vs. cluster size (in the R/a units).

Two types of nanostructures were studied: (1) on the base of nanoclusters αFe_2O_3–γFe_2O_3 with sizes 30–50 nm, and (2) on the base of metal Eu nanostructures. Both nanosystems (1), (2) were fabricated by pressure 2 GPa and the shear on Bridgman anvil up 240–120 °C correspondingly.

Figure 8 shows the concentration defect dependence on relative cluster sizes for iron oxide and metal europium. One can see the maxima of the concentration of defects in the region 20–40 nm for Eu metal and 40–60 nm for iron oxides. These maxima change drastically the magnetic properties of the nanostructures (see Section 5).

4. Cluster atomic dynamics

Intracluster atomic dynamics defines many properties of nanoclusters; for example, heat capacity and melting. One of the most important characteristics is the mean-square displacement of atoms. The increase of mean-square displacements for surface atoms of cluster leads to an increase of heat capacity and a decrease of melting point. There are numerous microscopic and quantum-statistical methods as well as methods based on the thermodynamical models and computer simulations of motions in atomic mobility studies. Thermodynamic formalism permits to consider the cluster melting and to bind it with the values of mean-square displacement and Mössbauer spectra data [5, 9, 10]. The states of the cluster formed during solid-state chemical reactions or in nucleation from solution in the pores are determined

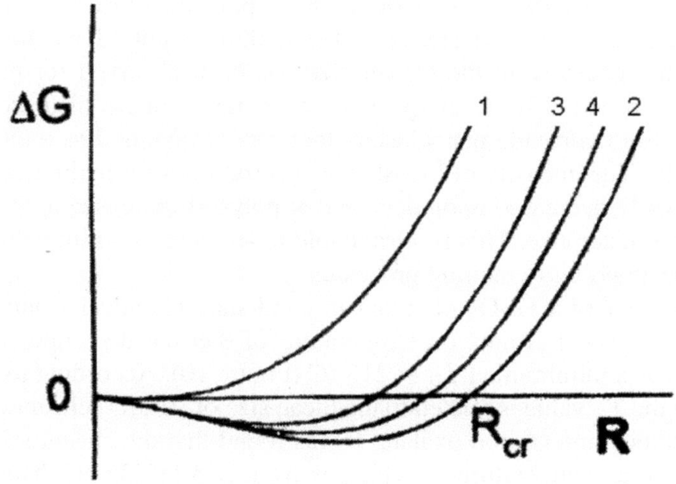

Figure 9. Change in the Gibbs Energy an ensemble of separate atoms in a solid (1), for isolated clusters (2) and for a systems of interacting clusters (3, 4) occurring in the solid and liquid state vs. cluster size. (3) Clusters interacting with one another. (4) Cluster after treatment with surfactants.

by the Gibbs free energy (1)–(3). The difference between the free energies of the cluster in liquid and solid states is

$$\Delta G = 4\pi R^2 \Delta\alpha + \frac{4}{3}\pi R^3 \rho \Delta\mu, \tag{9}$$

where $\Delta\alpha = \alpha_l - \alpha_s$ is the difference between surface energy and $\Delta\mu = \mu_l - \mu_s$ is the difference between the chemical potentials for liquid and solid states of cluster. It is obvious that for melting $\Delta\alpha < 0$ and $\Delta\mu > 0$. The $\Delta G(R)$ curves are shown in Figure 9 and the critical cluster size corresponding to the transition from solid to liquid state ($\Delta G = 0$) is calculated from the formula

$$R_{\mathrm{CR}} = \frac{3\Delta\alpha}{\rho\Delta\mu}. \tag{10}$$

When $R < R_{\mathrm{CR}}$ a cluster occurs in liquid state (Figure 9). As follows from (10), solid–liquid transition state cannot be realized for separate atoms in matrix (for $\Delta\alpha = 0$), when a cluster has not be formed. Meanwhile the $\Delta\alpha$ value should be the greatest for free cluster. Any intercluster interactions would decrease this value and hence, the R_{CR} value, which determines the region of possible fluctuations of solid and liquid states. The decrease of melting point of cluster is equal

$$\frac{\delta T}{T} \approx \frac{3\Delta\alpha}{\rho q R}, \tag{11}$$

where q = latent heat of the phase transition, T_0 is a melting point for bulk material.

Dependence (11) indicates that the melting point of cluster decrease with a decrease in cluster size. A decrease in the melting point (down to room temperature) with a decrease in the cluster size has been observed for gold, tin and CdS nanoclusters [11–13]. A sharp increase of mean square displacements was observed for iron hydroxide nanoclusters that can be qualified as melting of nanocluster [7, 10]. The intensity of Mössbauer spectral lines for hydroxide clusters in polysorbs shapely decreased upon decrease of polysorb pore size and with the decrease of nanocluster size. This is attributable to an increase in intracluster atomic mobility corresponding to melting processes.

For nanocluster of γFe_2O_3 obtained in solid-state chemical reactions the dependence of this spectral area on temperature of thermal decomposition of iron oxalate (T_D) has a minimum at $T_D \approx 215$ °C (Figure 10). According to Mössbauer spectroscopy the T_D value is related to the mean size of γFe_2O_3 cluster prepared by thermal decomposition of iron oxalate. It was found that the cluster size increases from 2 to 4 nm as temperature T_D changes from 215 to 235 °C. The right-hand branch of the curve in Figure 10 is related to decrease in spectral area (following

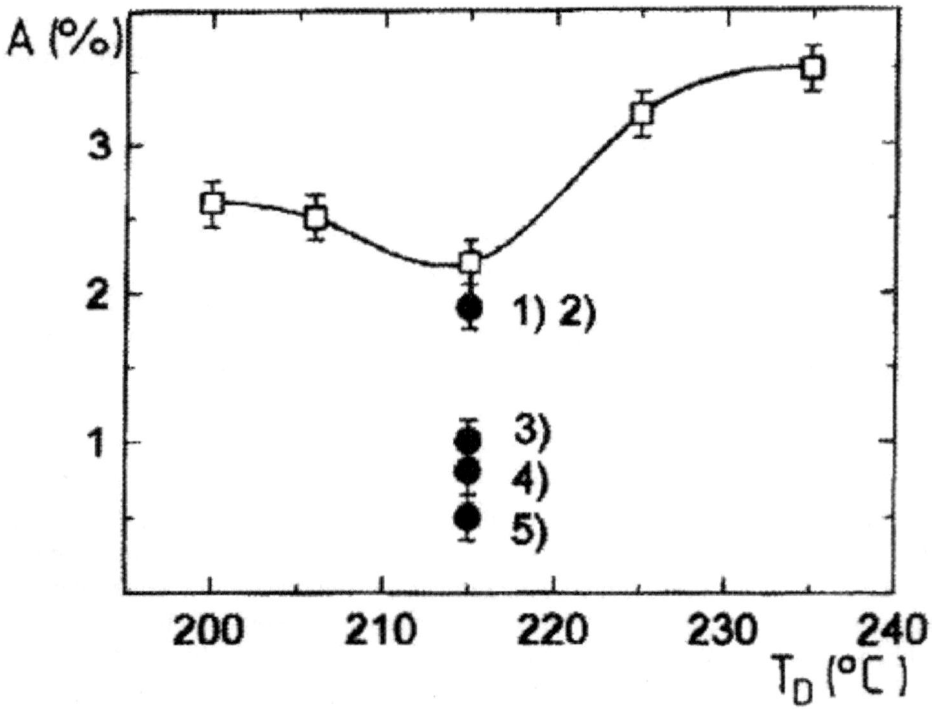

Figure 10. Spectral surface area of Mössbauer spectra of nanosystems formed from iron oxide clusters prepared by thermal decomposition of iron oxalate vs. temperature of thermal decomposition T_D. The light dots are initial experimental data. The dark dots correspond to surfactant action: (1) ethanol, (2) isopropanol, (3) Twin-65, (4) butanol, (5) SDS.

in decrease in T_D) and to increase in atomic mobility in γFe_2O_3 clusters similarly to the observed decrease in line intensity for ferrihydrate. The left-hand ascending branch in the curve $A = f(T_D)$ is related to further decrease in T_D and, correspondingly, to decrease in the size of cluster down to their disappearance to give separate atoms, dimers and trimers of Fe^{3+} (see Figure 9, $\alpha = 0$). Thus the minimum at the $A = f(T_D)$ curve observed at $T_D = 215$ °C appears to be due to the onset of formation of an iron oxide clusters.

As noted above, the atomic mobility is strongly influenced by the intercluster interactions. They decrease the surface energy. However, the intercluster interactions and strains can be reduced by the action of surfactants or mechanochemical treatment. It can be seen from Figure 10 that the spectral area markedly decreases after action of surfactants. Thus, weakening of the intercluster interactions results in an increase in the intercluster atomic mobility and in a decrease in the melting point. Relation (11) can be used to estimate the R_{CR} value corresponding to the onset of melting of γFe_2O_3 clusters and compare it with the R value observed experimentally from Mössbauer spectra. Using the value $\rho = 5.18 \cdot 10^6$ g/m^3, $\Delta\alpha \cong 1$ J/m^2, $q \cong 2 \cdot 10^{-19}$ J [14, 15] it was obtained $R = aT_0/(\delta T)$, where $a = 0.1$–1.0 nm. As a result, for $T_m = 300$ K, $R \sim 0.1$–1.2 nm, which is in good agreement with the value $d \sim 2$ nm (see Section 1) found from Mössbauer spectra.

It also noteworthy that the existence the specific solid–liquid transition state in which atoms possess enhanced mobility can provide an interpretation of the effect of adsorption decrease in the material strength (Rehbinder effect) [16] at the atom

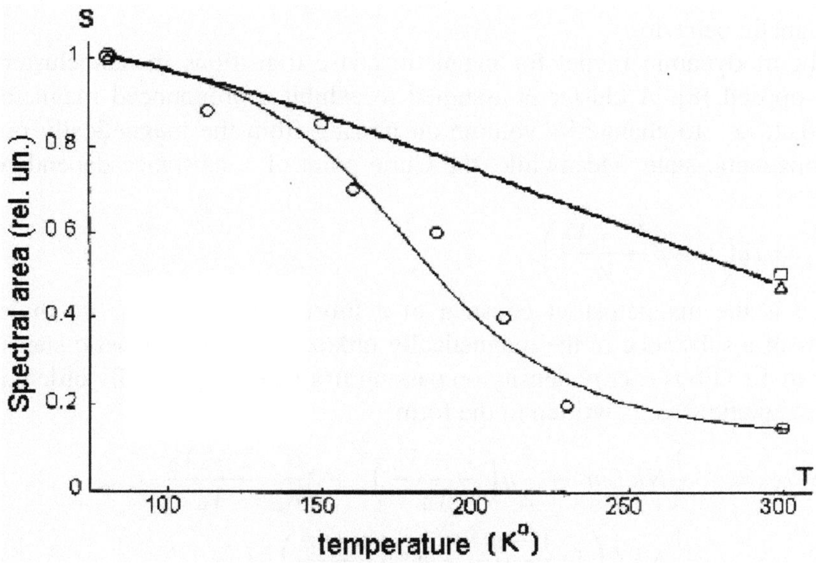

Figure 11. Temperature dependences of relative Mössbauer spectra area for Eu_2O_3 and Fe_2O_3 nanoclusresin pores of carbon. (1) ○ – small Eu_2O_3 nanoclusters, (2) △ – large Eu_2O_3 nanoclusters, (3) □ – Fe_2O_3 nanoclusters. Solid line – Debye model for Eu_2O_3 nanoclusters at $\theta_D = 290$ K.

and molecular level due to wetting and deformation of its surface under the action of a surfactant.

Intracluster atomic mobility was studied for Eu_2O_3 and Fe_2O_3 nanoclusters in carbon matrix. Eu_2O_3 and Fe_2O_3 nanoclusters were nucleated in carbon micropores and calcinated at 600 °C [17]. For heterogeneous nanosystem including europium and iron oxide nanoclusters was obtained the temperature dependence of the area of Mössbauer spectra, $A = f(T)$. A sharp temperature dependence of A value was observed for small europium oxide nanoclusters that was interpreted as melting of nanocluster (Figure 11). In accordance with the thermodynamic approach this phenomenon is followed by the decrease of melting point for Eu_2O_3 up to 200 K in comparison with the bulk (about 2500 K). The relation (11) can be used by the same manner as for nanoclusters Fe_2O_3 for the estimation of critical size Eu_2O_3 nanoclusters ~2 nm [17]. Temperature dependences of A-value for larger Eu_2O_3 and Fe_2O_3 clusters are different (Figure 11). They show solid-state behavior with Debye temperature $\theta = 290$ K (Eu_2O_3) and $\theta = 300$ K (Fe_2O_3).

5. Magnetic phase transitions

As was shown previously isolated ferrihydrate nanoclusters 1–2 nm in porous materials (Figures 6 and 7) and strong interacting αFe_2O_3–γFe_2O_3 nanoclusters 30–50 nm (Figure 5) possess the same unusual magnetic behavior – the jump-like magnetic phase transitions from magnetic (magnetic hyperfine structure) to paramagnetic state (quadrupole doublet). This transformations can be described in terms of the first-order magnetic phase transitions in nanoclusters without superparamagnetic behavior.

A thermodynamic model for magnetic phase transitions in nanoclusters has been proposed [8]. A cluster is assumed to exhibit a pronounced magnetostriction effect, i.e., to change its volume on passing from the magnetically ordered to paramagnetic state. Meanwhile, the Curie point of a substance depends on its volume

$$T_c = T_0\left(1 + \beta\frac{V - V_0}{V_0}\right), \tag{12}$$

where β is the magnetostrict constant of a substance, V and V_0 are molecular volumes of a substance in the magnetically ordered and paramagnetic states. The change in the Gibbs energy density on passing from the magnetically ordered state to paramagnetic state is written in the form

$$\begin{aligned}\Delta G_V = {} & -\frac{1}{2}NkTm^2 + \frac{1}{2}\eta\left(\frac{V - V_0}{V_0}\right)^2 + 2\frac{\alpha}{R}\cdot\frac{V - V_0}{V_0} \\ & + \frac{1}{2}NkT\left(\ln\frac{1 - m^2}{4} + m\ln\frac{1 + m}{1 - m}\right),\end{aligned} \tag{13}$$

where N is the number of atoms with spin 1/2, $m = M/M_S$ is the relative magnetization value, M_S is the saturation magnetization, η is the volume compressibility, and α is the surface tension. The first term in Equation (13) corresponds to the exchange energy density in the molecular field approximation, the second term reflects the deformation energy, the third term corresponds to the surface energy and the fourth one matches the entropy term for the temperature dependence of magnetization. Minimization of the free energy and solution of the set of equations $d\Delta G_V/dV = 0$, $d\Delta G_V/dm = 0$ gives

$$\frac{T}{T_{co}} = m\left(\frac{1}{3}\gamma m^2 + 1\right)\left(\frac{1}{2}\ln\frac{1+m}{1-m}\right)^{-1}, \tag{14}$$

where $T_{co} = T_0(1 - \beta\eta P)$ shows the change in T_0 due to pressure $P = 2\alpha/R$, $\gamma = 3/2NkT_0\eta\beta^2(1 - \beta\eta P)^{-1}$.

Figure 12 presents the temperature dependence $m = f(T/T_{co})$ for different values of the parameter γ, which determines the character of the magnetic phase transition. For $\gamma < 1$, the phase transition corresponds to a second-order magnetic transition and when $\gamma \geq 1$, it is a first-order magnetic transition. The γ value can be represented as $\gamma = T_{cc}/T_{co}$, where $T_{cc} = 3/2NKT_0^2\eta\beta^2$. If the cluster size $R \leq R_{CR}$, then $\gamma \geq 1$, the cluster passes from the magnetically ordered to the paramagnetic state upon a first-order magnetic phase transition.

$$R_{CR} = \frac{2\alpha\beta\eta}{1 - T_{cc}/T_{co}}. \tag{15}$$

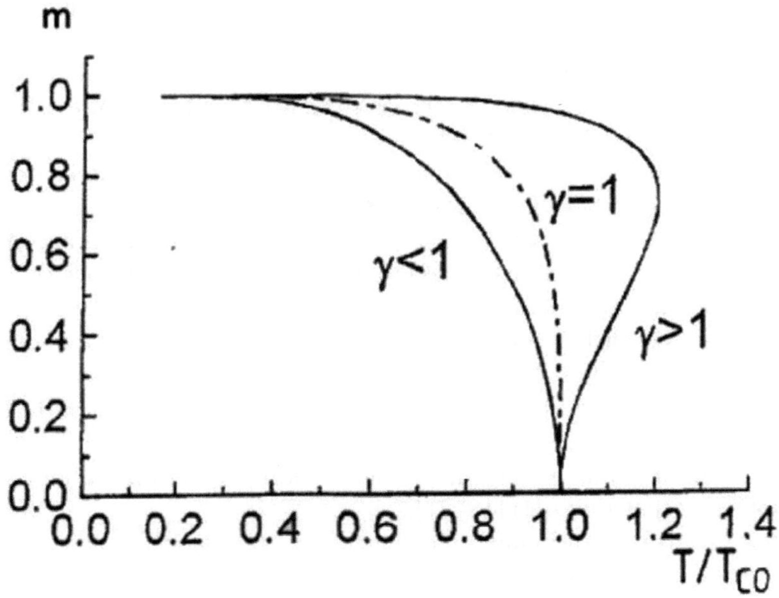

Figure 12. Temperature dependence of cluster magnetization for different γ: (1) $\gamma < 1$, (2) $\gamma = 1$, (3) $\gamma > 1$.

For larger clusters, only the second-order magnetic phase transition takes place (Langeven type). It can also manifest itself as superparamagnetism of temperatures $KT > K_a V$. The first- and second-order magnetic phase transitions were observed, for example, for the ferrihydrite $Fe_5HO_8 \cdot 4H_2O$ clusters in the polysorb pores [8]. The matrix isolation method allows two cluster systems to be produced, namely, a system with isolated clusters and one with clusters interacting with one another (if several cluster occurs in a pore). Figures 6 and 7 show the Mössbauer spectra of the ferrihydrite clusters obtained in 13- and 20-nm polysorb pores. The spectrum at 4.2 K corresponds to quadrupole doublet and consists of two parts – the central component (the doublet) and the magnetic hyperfine structure with relatively narrow lines (with allowance for the disorder of the cluster structure). As the temperature is raised, some of the HFS of the spectrum is converted into a paramagnetic doublet, without entering the intermediate states corresponding to the superparamagnetic behavior. The results presented in Figure 6 suggest that a jump-wise first-order magnetic phase transition occurs in ferrihydrite clusters in the temperature range of 4.2–10 K. An increase in the iron concentration results in an increase in the cluster size; nevertheless, the spectral pattern peculiarities at 4.2 and 6 K does not change (Figure 7).

In the spectra recorded at 10 K (Figure 7) for large clusters ($c = 1.2 \cdot 10^{-2}$ mol litre^{-1}) formed as a 13 nm pore, the absorption line is broadened, which corresponds to the intermediate forms (between the quadrupole doublet and HFS), peculiar to superparamagnetism. Thus, at 10 K, a second-order magnetic transition is observed for large clusters, in addition to the first-order magnetic phase transition. For smaller clusters ($c = 1.2 \cdot 10^{-3}$ mol litre^{-1}), line broadening is absent and second-order magnetic phase transitions (superparamagnetism) do not show themselves (see Figure 6). The appearance of superparamagnetism at 10 K made it possible to estimate the critical cluster size from the broadening of the HFS lines in the spectrum (Figure 7): $R_{CR} \approx 2$ nm.

Estimation of the cluster firmed in a pore from the concentration of iron atoms ($c = 1.2 \cdot 10^{-3}$ mol litre^{-1}) gives the value $R \approx 1$–2 nm (for isolated cluster).

Further growth of cluster and enhancement of their interactions is accompanied by appearance of second-order magnetic phase transitions. The lines of HFS are markedly broadened and the first-order phase transition can no longer be distinguished against the background of the second-order magnetic phase transitions. Estimation of R_{CR} using relation (15) gives $R_{CR} \sim 1$ nm, which coincides in order of magnitude with the results obtained in experiments. The effect similar to the first-order magnetic phase transition were also observed in the αFe_2O_3, $\alpha FeOOH$ nanoclusters and in proteins, ferritin and hemosiderin, wich include the magnetic core $(FeOOH)_8 \cdot (FeOOPO_3H)$ [18–21].

Another parameter determining the character of the magnetic phase transition is the external excess pressure. It has been noted above that the critical temperature of phase transition decreases upon an increase in pressure due to the change in the surface tension. *A* change with external pressure by ΔP entails a change in the magnetic transition temperature for cluster by the value

$$\frac{T_{co}}{T_0} = \frac{\Delta P}{2\alpha/R - 1/(\beta\eta)}. \tag{16}$$

Thus, a decrease in T_{co} increases the probability of the first-order magnetic phase transition.

The study of two nanosystems – an ion exchange sulfro resin (sulfo cation exchange resin) incorporating iron hydroxide clusters with a size of 3 nm and a globular protein (ferritin) incorporating iron hydroxide clusters with a size of 6 nm has been reported [22]. In these systems, the interaction of clusters with the polymeric and biopolymeric networks changed due to hydration. The excess pressure on the cluster was determined from the deformation of the polymeric matrix [from the increase in water volume which accompanied its crystallization (freezing)]. This pressure is calculated from the expression

$$\Delta P = \frac{1}{\eta_m}\frac{\Delta V}{V}. \tag{17}$$

For $\Delta V/V = 0.09$ (for water) and $1/\eta_m = 5$–10 GPa it was found that $\Delta P =$ 0.45–0.9 GPa, which is comparable in magnitude with the surface pressure in a 1.5 nm cluster ($\sim$1 GPa). Using relation (16) can estimate the decrease in T_{co} under the action of pressure. Assuming that $P = 1$ GPa, $\beta = 10^2$, $\mu = 10^{-11}$ m^2/J^{-1} and $1/N \sim 10^{-28}$ m^3 (the data were taken from publications [22]), we have $\Delta T_{co} =$ $(0.1$–$1.0)T_{co}$. Thus, ΔT_{co} can reach a value of the order of T_{co}.

The ΔT_{co} value was estimated from Mössbauer spectra of sulfo cation exchange resin (in the range of $T = 4.2$–30 K) and ferritin (in the range of 4.2–78 K). Hydration followed by freezing was found to be accompanied by a decrease in T_{co}. This value was determined from spectra in which the paramagnetic state and magnetic HFS of spectra were equal. The following values were obtained: $T_{co} \approx 9$–10 K for the hydrated sulfo cation exchange resin, $T_{co} \approx 13$–14 K for the dry cation exchange, $T_{co} = 30$ K for hydrated ferritin and $T_{co} = 34$ K for dry ferritin. Comparison of the T_{co} values for the dry and hydrated samples provides the temperature shift $\Delta T_{co} \approx 3$–4 K, which is in line with thermodynamic estimates. The first-order magnetic phase transition arises also in nanosystems containing larger nanocluster (30–50 nm). These systems are formed in solid-phase chemical reactions (see Figures 1 and 2), for example, in cluster sintering, and are subject to substantial intercluster interactions. The phase transitions observed in these systems are similar to those in the ferrihydrite cluster (1–3 nm) in a polymeric matrix. Figure 5 shows the Mössbauer spectra of the αFe_2O_3–γFe_2O_3 nanosystem. The spectra indicate that jump-wise magnetic transitions from the magnetically ordered state (magnetic HFS) to the paramagnetic state (central doublet) can take place in this system. These magnetic transitions take place at distributed critical temperatures $T_{co} = 120$–300 K, which are lower than the temperatures T_{co} for bulk samples of α- and γ-Fe_2O_3 (856 and 965 K, respectively). The decrease in T_{co} cannot be due to superparamagnetism because of the large cluster size. Superparamagnetic relaxation times are much larger than characteristic Mössbauer time $\sim 10^{-7}$ s. As a consequence, magnetization disappears abruptly, which is related to the transition of the magnetic HFS of αFe_2O_3–γFe_2O_3 into a paramagnetic doublet (Figure 5).

The presence of the first-order magnetic phase transition in the αFe_2O_3–γFe_2O_3 nanosystems is related to the strong intercluster interactions, which bring about substantial stress in the nanosystem. This stress results in the generation of enormous pressures (up to several GPa). The pressure arises during sintering [23] of the α- and γ-Fe_2O_3 cluster, the unit cells of which have different specific volumes and under action of defects and dislocations. As has been shown above (see Figure 8) the density of defects has a maximum at a cluster size of exactly 40–60 nm.

For the explanation of the first-order magnetic phase transition in αFe_2O_3–γFe_2O_3 nanosystem the thermodynamic model taking into account the influence of cluster defect density was developed [24].

In this model

$$T_c = T_{co}(1 + \beta c_V), \tag{18}$$

where $\beta < 0$ is constant, T_c, T_{co} – Curie or Néel points of cluster with defects and without defects, c_v – defect concentration in cluster. The thermodynamic potential of the system may be expressed by

$$\begin{aligned} G &= \Delta\mu n_V + \alpha S(n_V) - \frac{1}{2} N_0 k T_{co}(1 + \beta c_V) m^2 \\ &\quad + NKT\left(\frac{1}{2}\ln\left(1 - \frac{m^2}{4}\right) + m \ln\frac{1+m}{1-m}\right), \end{aligned} \tag{19}$$

where $\Delta\mu = \Delta\mu^0 + kT\ln(N_S n_V/(N_V n_S))$ is the difference of chemical potential for defects situated inside and outside the cluster, N_V, N_S – numbers of atom in cluster volume and surface, respectively, n_V, n_S – concentrations in cluster volume and surface respectively, α – the surface tension on cluster–media interface, S – cluster surface, N – number of spins in cluster, $m = M/M_0$ – relative magnetic moment of cluster, M – magnetic moment of cluster. If the conditions $dG/dn_V = 0$ and $dG/dm = 0$ are simultaneously fulfilled, the equilibrium may be expressed by the equation connecting implicitly the magnetization and temperature

$$\begin{aligned} &1 + \frac{\beta c_0 q(R)}{1 + q(R) K \exp[(V\delta p(R)/kT_{co} - 0.5(n\cdot\beta)m^2)(T_{co}/T)]} \\ &- \frac{0.5}{m}\ln\left(\frac{1+m}{1-m}\right)\frac{T}{T_{co}} = 0, \end{aligned} \tag{20}$$

where $q(R) = [(1 + r/R)^3 - 1] = [(1 + (r/a)(a/R))^3 - 1]$, $\delta p(R) = 2\alpha/R$, $n = N/N_V$ is the concentration of spins in cluster, a is the lattice constant, r is the thickness of intercluster media surrounding cluster the calculation of the magnetization curve $m(T/T_{co})$. The following parameters are used: $\beta c_0 = -50$, $k =$

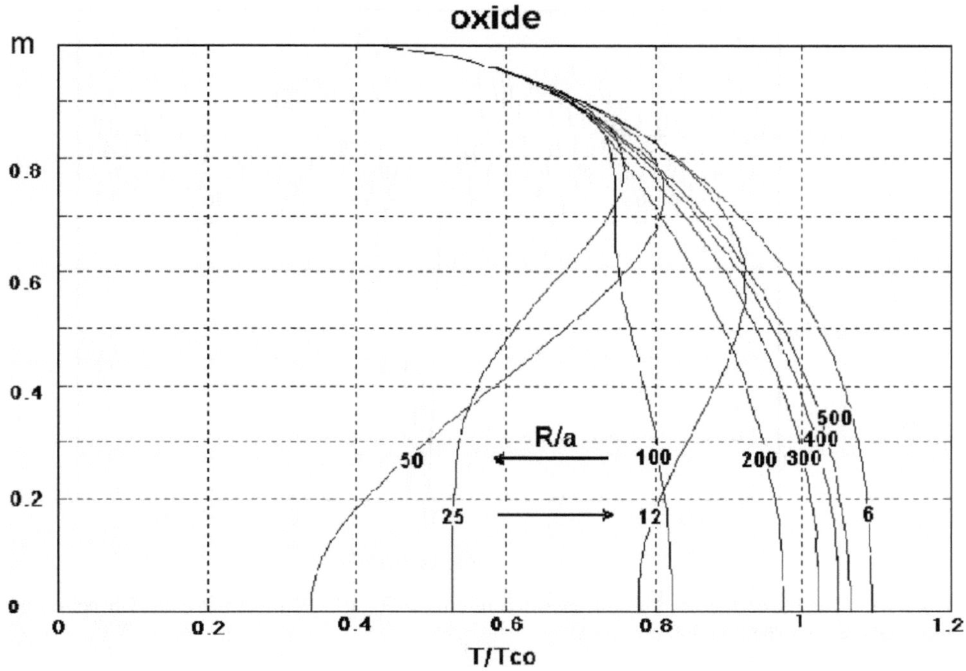

Figure 13. Temperature dependence of calculated cluster magnetization at different R/a.

$\exp(-\Delta\mu^0/kT_{co}) = 0,01$, $n\beta = -3$, $r/a = 1$, $2\alpha V/(akT_{co}) = 50$ (Figure 13). Temperature dependence of magnetization for $R = 500a$–$6a$ shows the character of magnetic phase transition changes notably from the smooth curves typical for of second-order magnetic ($R/a = 500, 400, 300, 200$) to the jump like curves characteristic of the first-order magnetic phase transition ($R/a = 100, 50, 25, 12$) and back to second-order magnetic phase transition for ($R/a = 6$).

The decrease of Néel temperature with decrease of nanocluster size was obtained in nanostructures fabricated by high pressure with shear action. This technique generates nanostructures with nanocrystallite sizes up to 10 nm [25] and high defect density. The same αFe_2O_3–γFe_2O_3 nanocluster structure (about 5 wt.%) with metal Eu (~95 wt.%) composition was treated under pressure 2 GPa and the shear 240°. The results are shown in Figure 14. The absence of HFS supports the idea of the decreasing of cluster size under pressure and shear action and the decrease of Neel temperature for magnetic phase transitions.

The results with the increasing of Néel point was observed for Eu metal and composites. The action of high pressure with shear was studied for metal Eu; αFe_2O_3–γFe_2O_3 nanocluster structure (about 5 wt.%) + Eu metal; adamantine (about 10 wt.%) + Eu metal. The bulk Eu metal possesses the first-order magnetic phase transition at 80 K. Mössbauer spectra of ^{119}Eu are shown in Figure 15. One can see that magnetic state and HFS for metal Eu disappeared but the broadening of

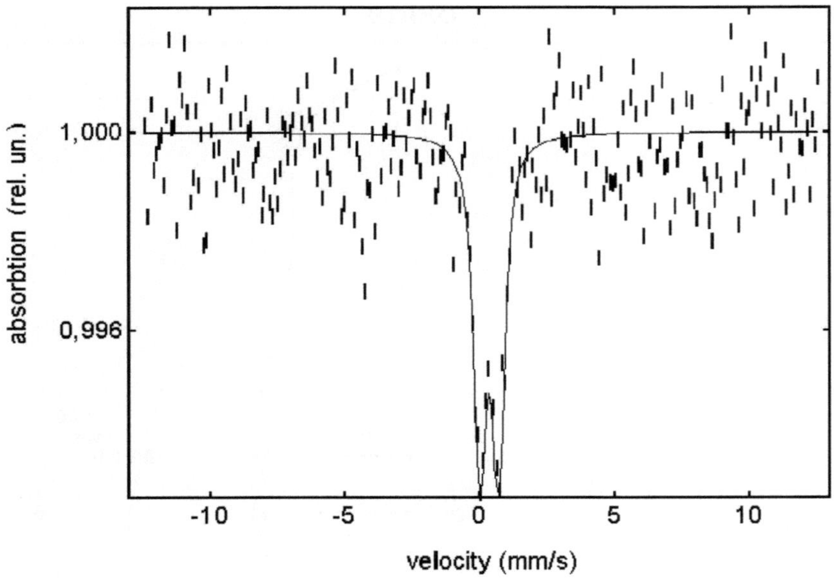

Figure 14. Mössbauer spectra ^{57}Fe of nanostructure αFe_2O_3–γFe_2O_3 + Eu metal after action of high pressure with shear at 300 K.

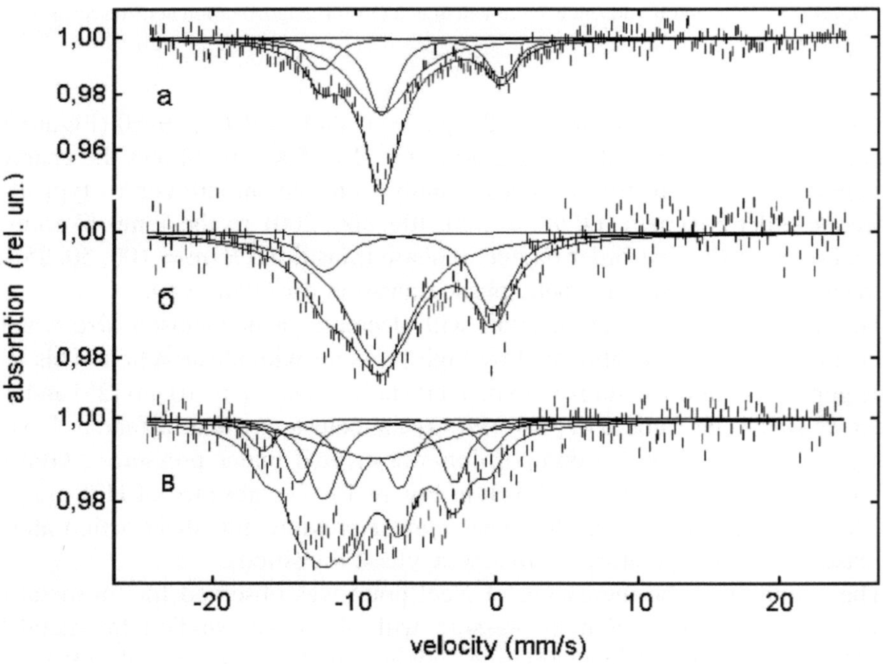

Figure 15. Mössbauer spectra ^{151}Eu nanocomposites after action of high pressure with shear at 90 K: (a) nanostructure on Eu metal, (b) nanostructure on Eu metal + αFe_2O_3–γFe_2O_3, (c) nanostructure on Eu metal + adamantine.

paramagnetic line is still conserved. More pronounced line broadening effect is observed for αFe_2O_3–γFe_2O_3 composition and for the adamantine nanocomposition even resolved HFS is fixed. All these results can be interpreted by the increasing of Néel temperature of Eu nanocomposite and the appearance of the second-order magnetic phase transitions after the action of high pressure with shear. As was shown in Figure 8 the maximum of defect density exists for 20–40 nm cluster size range and influences the changing the first-second order magnetic phase transitions. Thus for the larger and the smaller than 20–40 nm should be observed second-order magnetic phase transitions and Curie or Néel points should be increase. Therefore after action of high pressure with shear the decrease of Eu magnetic domain nanocrystalites (smaller, for example, then 20–40 nm) takes place and former first-order magnetic phase transition converts to the second order with inceasing of Néel points. The imbedding of αFe_2O_3–γFe_2O_3 nanoclusters and adamantine moleculars into Eu matrix permit to distribute a shear over all volume of composite and to conserve generated defects and stresses after the release pressure. Thus one should mark that nanoclusters and nanostructure materials permits to change the character of magnetic phase transitions and to decrease or to increase critical temperature of magnetic phase transitions.

References

1. Prigogine, I. and Rice, S. (eds), *Evolution of Size Effect in Chemical Dynamics*, Part 2, Wiley, New York, 1988.
2. Haberland, H. (ed), *Clusters of Atoms and Moleculas*, Springer, Berlin, 1994.
3. Suzdalev, I. P. and Suzdalev, P. I., *Russian Chem. Rev.* **70** (2001), 177.
4. Suzdalev, I. P., Maksimov, Yu. V., Buravtsev, V. N., Imshennik, V. K., Kazakevitch, A. G. and Novochikhin, S. V., *Kolloid. Zh.* **62** (2000), 257.
5. Suzdalev, I. P., Buravtsev, V. N., Imshennik, V. K., Maksimov, Yu. V., Matveev, V. V., Volynskaya, A. V., Trautwein, A. X. and Winkler, H., *Z. Phys. D* **36** (1996), 163.
6. Suzdalev, I. P., Maksimov, Yu. V., Novichkhin, S. V., Buravtsev, V. N. and Imshennik, V. K., *Khim. Fiz.* **19** (2000), 105.
7. Suzdalev, I. P., Buravtsev, V. N., Imshennik, V. K. and Novichikhin, S. V., *Khim. Fiz.* **12** (1993), 55.
8. Suzdalev, I. P., Buravtsev, V. N., Imshennik, V. K., Maksimov, Yu. V., Matveev, V. V., Volynskaya, A. V., Trautwein, A. X. and Winkler, H., *Z. Phys. D* **37** (1996), 55.
9. Suzdalev, I. P., Buravtsev, V. N., Imshennik, V. K., Maksimov, Yu. V., Trautwein, A. X. and Winkler, H., *Hyp. Interact. C* **1** (1996), 290.
10. Suzdalev, I. P. and Shklovskaya, N. I., *Colloid. Surf. A* **108** (1996), 315.
11. Buffet, Ph. and Borel, J. P., *Phys. Rev. A* **13** (1976), 22.
12. Osima, Y. and Takaynagi, K., *Z. Phys. D* **27** (1993), 287.
13. Goldstein, A. N., Echer, C. M. and Alivisatos, A. P., *Science* **256** (1992), 1425.
14. Grigorov, I. R. and Meilichov, U. S. (eds), *Fizicheskie Velichiny, Spravochnik (Physical Values, Handbook)*, Energoatomizdat, Moscow, 1991.
15. Ormont, B. F., *Dokl. Akad. Nauk SSSR* **124** (1959), 132.
16. Rehbinder, P. A., Surface effects in dispersed systems, In: *Physicochemical Mechanics*, Nauka, Moscow, 1979, p. 142.

17. Suzdalev, I. P., Buravtsev, V. N., Imshennik, V. K., Maksimov, Yu. V. and Novichikhin, S. V., *Zh. Fiz. Chim.* **75** (2001), 1832.
18. Kundig, W., Bommel, H., Constabaris, G. and Lindquist, R. H., *Phys. Rev.* **143** (1966), 327.
19. Shinjo, T., *J. Phys. Soc. Jpn*, **21** (1966), 917.
20. van der Kraan, A. M. and van Loef, L. I., *Phys. Lett.* **20** (1966), 614.
21. Bell, S. H., Meir, H. P., Dickson, D. P. E., Gibson, J., Sharp, G. A. and Peters, T. V., *Biochem. Biophys. Acta* **787** (1984), 227.
22. Suzdalev, I. P., Plachinda, A. S., Buravtsev, V. N., Maksimov, Yu. V., Reiman, S. I., Khromov, V. I. and Dmitriev, D. A., *Khim. Fiz.* **17** (1998), 145.
23. Geguzin, Yu. G., *Fizika Spekanja (The Physics of Caking)*, Nauka, Moscow, 1984.
24. Suzdalev, I. P., Buravtsev, V. N., Imshennik, V. K. and Maksimov, Yu. V., *Scripta Mater.* **44** (2001), 1937.
25. Valiev, R. Z., *Nanostruct. Mater.* **6** (1995), 73.

Hyperfine Interactions **144/145:** 347–357, 2002.

Magnetic Properties of Nanoparticles of Antiferromagnetic Materials

STEEN MØRUP[1], CATHRINE FRANDSEN[1], FRANZ BØDKER[1,*], STINE NYBORG KLAUSEN[2], KIM LEFMANN[2], PER-ANKER LINDGÅRD[2] and MIKKEL FOUGT HANSEN[3]
[1] *Department of Physics, Bldg. 307, Technical University of Denmark, DK-2800 Kgs. Lyngby, Denmark*
[2] *Materials Research Department, Risø National Laboratory, DK-4000 Roskilde, Denmark*
[3] *Mikroelektronik Centret, Bldg. 345E, Technical University of Denmark, DK-2800 Kgs. Lyngby, Denmark*

Abstract. The magnetic properties of antiferromagnetic nanoparticles have been studied by Mössbauer spectroscopy and neutron scattering. Temperature series of Mössbauer spectra of non-interacting, superparamagnetic hematite nanoparticles were fitted by use of the Blume–Tjon relaxation model. It has been found that the magnetic anisotropy energy constant increases significantly with decreasing particle size. Neutron scattering experiments on similar samples give new information on both superparamagnetic relaxation and collective magnetic excitations. There is good agreement between the values of the parameters obtained from Mössbauer spectroscopy and neutron scattering. In samples of interacting hematite nanoparticles, the relaxation was significantly suppressed. The Mössbauer data for these samples are in accordance with a mean field model for an ordered state of strongly interacting particles. Mixing nanoparticles of hematite with CoO nanoparticles resulted in suppression of the superparamagnetic relaxation, whereas NiO nanoparticles had the opposite effect.

Key words: Mössbauer spectroscopy, magnetic nanoparticles, superparamagnetic relaxation, neutron scattering.

1. Introduction

Most experimental studies of the magnetic properties of nanoparticles have been focused on ferromagnetic and ferrimagnetic particles because of their importance for magnetic data storage, in ferrofluids, etc. Furthermore, the theoretical models for superparamagnetic relaxation have been derived for ferromagnetic particles. Antiferromagnetic nanostructured materials are interesting both from a fundamental point of view and because they have important applications in devices in which exchange bias is utilised, such as spin valves. In this paper, we give a brief review of results of studies of antiferromagnetic nanoparticles obtained by use of Mössbauer spectroscopy and neutron scattering.

* Present address: Danfysik, Møllehaven 31, DK-4040 Jyllinge, Denmark.

2. Non-interacting particles

Several studies of both ferrimagnetic and antiferromagnetic nanoparticles have shown that inter-particle interactions can have a strong influence on the magnetic properties of samples containing magnetic nanoparticles [1–5]. Therefore, in order to study the magnetic properties of non-interacting nanoparticles it is necessary to prepare samples with negligible inter-particle interactions. Since long range dipole–dipole interactions are small because of the small magnetisation of antiferromagnetic particles, it may be sufficient to coat the particles with a thin layer of non-magnetic material such that short range exchange interaction can be avoided.

Often the magnetic anisotropy energy of a nanoparticle is assumed to be uniaxial and given by the expression

$$E(\theta) = KV \sin^2 \theta, \tag{1}$$

where θ is the angle between the (sublattice) magnetisation direction and an easy direction of magnetisation, K is the magnetic anisotropy constant, and V is the particle volume. The superparamagnetic relaxation time, i.e. the average time between transitions between the minima of the anisotropy energy, is expected to follow the Néel–Brown expression [6, 7]:

$$\tau = \tau_0 \exp\left(\frac{\Delta E}{k_B T}\right), \tag{2}$$

where k_B is Boltzmann's constant and T is the temperature. ΔE is the height of the energy barrier separating two minima corresponding to easy directions of magnetisation. For a particle with the magnetic energy given by Equation (1), $\Delta E = KV$. In ferromagnetic nanoparticles, τ_0 depends slightly on temperature [7] and is typically in the range 10^{-12}–10^{-9} s. No theoretical expressions for the value of τ_0 for antiferromagnetic particles have been published.

As an example of Mössbauer spectra of non-interacting particles, Figure 1(a) shows spectra of 9 nm hematite nanoparticles coated with oleic acid and suspended in heptane. At 25 K, the spectrum is dominated by a sextet with relatively sharp lines, indicating a relaxation time $\tau \gtrsim 10^{-7}$ s. At 180 K, the spectrum consists of a doublet, indicating relaxation times $\tau \lesssim 10^{-9}$ s. At intermediate temperatures, the spectra can be described as superpositions of sextets and doublets with relatively narrow lines, and there is no significant contribution from particles with relaxation times in the range 10^{-9}–10^{-8} s, which should give contributions with considerable line broadening [1]. This can be explained by a relatively broad particle size distribution in conjunction with the exponential size dependence of the relaxation time. For a particle with magnetic energy given by Equation (1) one finds from Equation (2) that a small volume change ΔV will give rise to a change in relaxation time given by

$$\Delta\tau = \tau(V + \Delta V) - \tau(V) = \tau(V)\left[\exp\left(\frac{KV}{k_B T}\frac{\Delta V}{V}\right) - 1\right]. \tag{3}$$

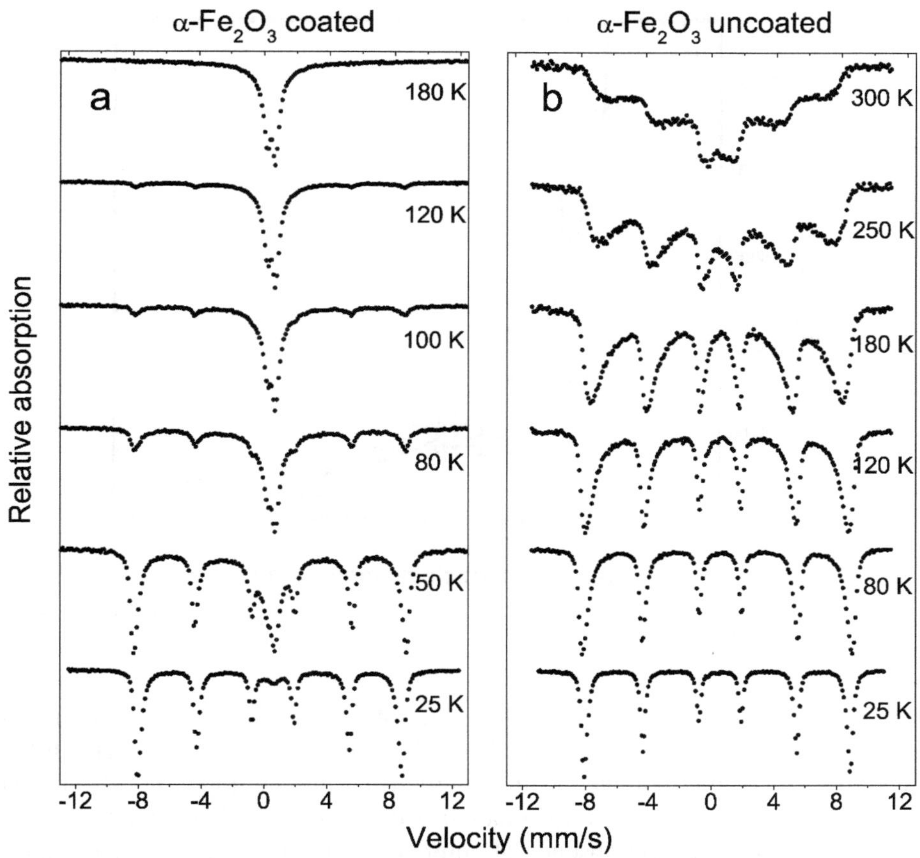

Figure 1. Mössbauer spectra of 9 nm hematite nanoparticles obtained at the indicated temperatures. (a) Particles coated with oleic acid and suspended in heptane. (b) Uncoated particles prepared by drying an aqueous suspension.

For non-interacting hematite nanoparticles, τ_0 is of the order of 10^{-11} s [8]. Therefore, for a particle with a superparamagnetic relaxation time of the order of the time scale of Mössbauer spectroscopy ($\sim 5 \cdot 10^{-9}$ s), we find from Equation (2) that the value of the parameter $KV/k_{\rm B}T$ is approximately 6. According to Equation (3), a change of the particle diameter by only about 10% will then result in a change of the relaxation time, which is sufficient to transform the spectrum to a doublet or a sextet. The samples of hematite nanoparticles studied by us typically have a broad size distribution [9]. Therefore, at a given temperature only a small fraction of the particles in a sample will have relaxation times in the range 10^{-9}–10^{-8} s, for which the spectra are substantially broadened. The spectra therefore appear as superpositions of sextets and doublets with relatively sharp lines.

We have fitted temperature series of spectra of non-interacting hematite nanoparticles simultaneously [4, 8, 10] with the Blume–Tjon model [11] for Mössbauer relaxation spectra. We assumed that the relaxation time is given by the Néel–Brown

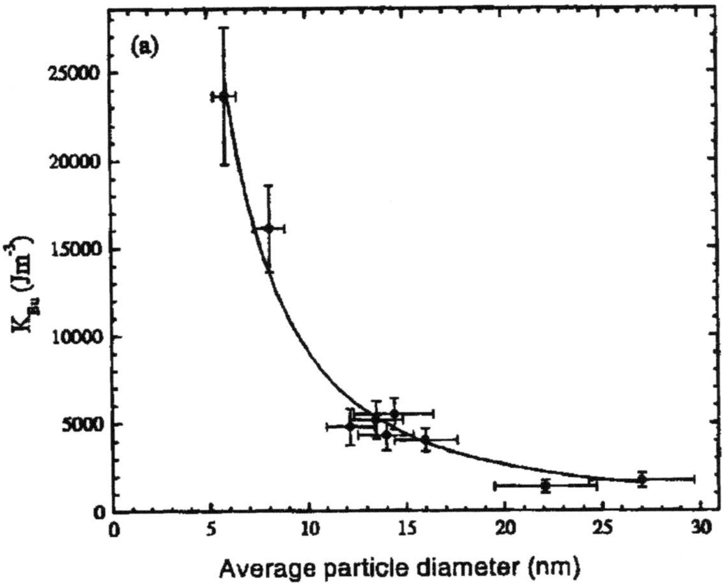

Figure 2. The magnetic anisotropy constant for hematite nanoparticles as a function of the particle size. Adapted from [8].

expression (Equation (2)). In the fits, the values of ΔE and τ_0 were free parameters. The volume distribution was estimated from electron micrographs, and in the fitting procedure, it was approximated with a log-normal distribution. The fits showed unambiguously that the relatively broad size distribution and the small value of τ_0 can explain the apparent absence of components with broad lines. The studies showed, as illustrated in Figure 2, that the magnetic anisotropy energy constant increases by a factor of about 10, when the size decreases from about 25 nm to about 6 nm [8]. It is likely that the major contribution to this increase is the surface anisotropy. The results also showed that the value of τ_0 decreases with decreasing particle size.

At low temperatures, where the superparamagnetic relaxation time is long compared to the time scale of Mössbauer spectroscopy, the Mössbauer spectra of samples of magnetic nanoparticles consist of sextets. However, the magnetic splitting is reduced compared to the bulk value because of the influence of collective magnetic excitations (i.e. fluctuations of the sublattice magnetisation in directions around an easy direction of magnetisation) [12, 13]. These fluctuations can be described as a combination of precession of the (sublattice) magnetisation vector in the anisotropy field and transitions between precession states with different precession angles. If both processes are fast compared to the time scale of Mössbauer spectroscopy, the observed magnetic hyperfine field is at low temperatures given by [13]

$$B_{\mathrm{obs}} \cong B_0\left(1 - \frac{k_{\mathrm{B}}T}{2\kappa V}\right). \tag{4}$$

Table I. The magnetic anisotropy parameters $\kappa V/k_B$ and $\Delta E/k_B$ and the value of τ_0 for 15 nm hematite particles obtained from superparamagnetic relaxation (SPM) and collective magnetic excitations (CME)

	$\kappa V/k_B$ (K)	$\Delta E/k_B$ (K)	τ_0 (s)	Ref.
SPM (Mössbauer)		600 ± 150	$6 \cdot 10^{-11}$	[10]
CME (Mössbauer)	620 ± 50		–	[10]
SPM (neutron scatt.)		500 ± 200	$1.4 \cdot 10^{-11}$	[15, 16]
CME (neutron scatt.)	600 ± 200		–	[15, 16]

Here B_0 is the magnetic hyperfine field in the absence of collective magnetic excitations, and may be close to the bulk magnetic hyperfine field. The parameter κ depends on the magnetic anisotropy constant(s) and may also be influenced by inter-particle interactions. For a particle with magnetic energy given by Equation (1), $\kappa = K$. According to Equation (4), B_{obs}/B_0 should depend linearly on temperature, and this has in fact been found in experimental studies of nanoparticles of, for example, magnetite [12] and hematite [4, 14]. (In hematite nanoparticles, the sublattice magnetisation is to a large extend confined in a plane and the magnetic energy is more complex than that given by Equation (1) [10].)

It should be noticed that the fits of the whole temperature series of Mössbauer spectra are mainly sensitive to the superparamagnetic relaxation, which is governed by the height of the energy barriers, ΔE (although the reduction of the hyperfine fields due to collective magnetic excitations was also included in the fitting model [4, 8, 10]). When analysing only the reduction of the hyperfine field due to collective magnetic excitations at low temperatures, one obtains information on the parameter κV, which is related to the magnetic energy near the bottom of the minima. Therefore, if the magnetic anisotropy is different from that given in Equation (1), the results for the energy barriers, obtained using the two methods, may not be identical. The values, obtained using the two different types of analysis for 15 nm hematite nanoparticles, are given in Table I. In this case there is good agreement between the values.

Another experimental technique, which is useful to study superparamagnetic relaxation and collective magnetic excitations, is neutron scattering [15–17]. Figure 3 shows, as an example, energy scans from a triple-axis neutron spectrometer taken around the pure (1 1 1) antiferromagnetic reflection for 15 nm hematite nanoparticles [15]. The zero field scans (a) show a relatively narrow quasielastic peak centered at zero energy transfer and two broad inelastic peaks at energy transfers around $\varepsilon = \pm 0.2$ meV. The width of the quasielastic peak increases with increasing temperature, because of the finite lifetime of the magnetisation orientations in superparamagnetic particles. Thus the superparamagnetic relaxation time can be estimated from the line width of the quasielastic peak. It is notable that neutron scattering is sensitive to fluctuations at the time scale 10^{-14} s $> \tau > 10^{-7}$ s. It is thus suitable for the study of superparamagnetic relaxation, which is so fast, that

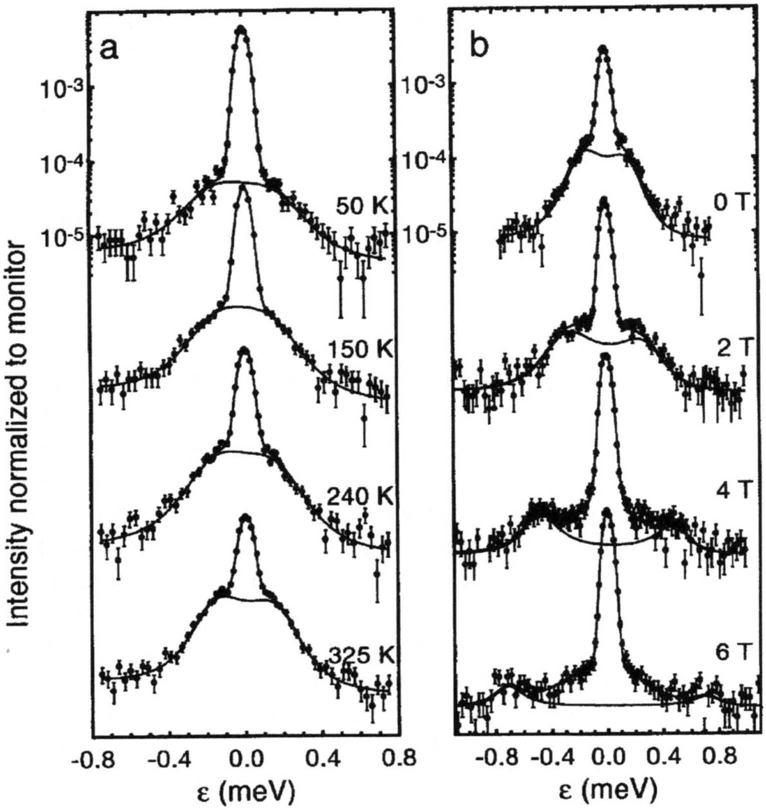

Figure 3. Inelastic neutron scattering data obtained at the indicated temperatures at zero applied magnetic field (a) and at 268 K in various applied magnetic fields (b). Adapted from [15].

Mössbauer spectroscopy becomes insensitive to the relaxation rate. The inelastic peaks are caused by collective magnetic excitations and their positions correspond to the energy change associated with a transition between two neighbouring precession states. Thus the position of the inelastic peaks are at $\varepsilon = \pm\hbar\omega_{AF}$ where ω_{AF} is the antiferromagnetic resonance frequency [15, 16], which is related to the parameter κ. It can be seen that the area of the inelastic peaks increases with increasing temperature. This is due to the temperature dependence of the population of the precession states. When magnetic fields are applied, the inelastic peaks move to higher energies (Figure 3(b)), because the magnetisation vectors precess in an effective field, which has contributions from both the anisotropy field and the applied field. The values of the parameters ΔE, κV and τ_0, obtained from the analysis of the neutron data, are given in Table I. There is a good agreement between the values of ΔE and κV obtained from the analysis of the neutron data and the Mössbauer data. The uncertainty of the estimated values of τ_0 is about one order of magnitude and therefore the values given in the table also agree within the experimental uncertainty.

Neutron scattering has also appeared very useful for studies of the temperature dependence of the sublattice magnetisation in nanoparticles of, for example, NiO. It has been shown that the Néel temperature of NiO nanoparticles is lower than the bulk value and that the temperature dependence of the sublattice magnetisation follows a mean-field model for finite-sized particles [18].

3. Inter-particle interactions

Interactions between antiferromagnetic nanoparticles have in several cases been found to have a surprisingly strong influence on the superparamagnetic relaxation [1, 4, 14, 19–21]. Figure 1(b) shows Mössbauer spectra of interacting 9 nm hematite particles. These particles were taken from the same batch as those yielding the spectra in Figure 1(a), but they were uncoated, suspended in water and then allowed to dry at room temperature such that they may interact. The large differences between the spectra in Figures 1(a) and (b) illustrate that inter-particle interactions result in much slower relaxation than seen in the sample of non-interacting particles. Moreover, the shapes of the spectra in the two figures differ significantly. Instead of the appearance of a doublet at relatively low temperatures as in Figure 1(a), the lines of the sextet in Figure 1(b) become increasingly broadened as the temperature is raised and the average hyperfine field decreases considerably.

The magnetic energy of a particle i, which interacts with its neighbours j, may be written

$$E_i = KV_i \sin^2\theta - \mathbf{M}_i \cdot \sum_j K_{ij}\mathbf{M}_j, \tag{5}$$

where $\mathbf{M}_i$ and $\mathbf{M}_j$ represent the (sublattice) magnetisation of the particles i and j, respectively, and K_{ij} is an effective exchange coupling constant due to exchange coupling between surface atoms belonging to neighbouring particles. If the first term in Equation (5) is predominant, superparamagnetic relaxation may take place between the easy directions at $\theta = 0$ and $\theta = \pi$. However, if the interaction is strong such that the second term is predominant, there may be only one energy minimum for the particle i, and the interactions may result in an ordered (collective) state at low temperatures [1, 4, 13, 21]. At finite temperatures, the (sublattice) magnetisation may then fluctuate around the direction corresponding to this energy minimum. The magnetic properties of such samples have been calculated by use of a mean field model in which the summation in the second term in Equation (5) is replaced by an average value [13, 21]. The properties of the sample can be characterised by an order parameter, which describes the degree of ordering of the (sublattice) magnetisation directions of the particles [4, 21]. The temperature dependence of the order parameter can then be calculated by use of Boltzmann statistics [4, 13, 21]. If the fluctuations of the sublattice magnetisation directions are fast compared to the time scale of Mössbauer spectroscopy, the magnetic splitting in the spectra will be proportional to the order parameter. In a sample of more or less randomly packed nanoparticles one would expect a broad distribution of exchange coupling constants. Therefore, at a given temperature, the order parameter is different in different parts of the sample. This variation can explain the asymmetric line broadening of the lines in the sextet [4, 21]. The model has been used to explain the Mössbauer data for interacting hematite [4] and goethite nanoparticles [21]. If the neighbouring particles are randomly oriented one might expect that the contributions from different neighbouring particles in the summation of Equation (5) partially cancel and therefore the interaction field should be small. However, recent high resolution transmission electron micrographs of TiO_2 nanoparticles have revealed that there can be a large tendency for neighbouring nanoparticles to align with parallel lattice planes [22]. It is possible that a similar tendency is present in samples of hematite nanoparticles and this may explain the strong magnetic coupling.

An alternative explanation of the slower relaxation in samples of interacting particles might be that the effective energy barrier, ΔE, increases with increasing strength of interactions [5, 23]. The shapes of the spectra of interacting particles are, however, as illustrated in Figure 1, very different from those of non-interacting particles. This suggests that the relaxation mechanism is fundamentally different from that of particles with relaxation behaviour governed by Equation (2). Furthermore, the physical model [23], which has been used to explain the apparent increase of the magnetic anisotropy, is based on assumptions, which may not be fulfilled [24].

The influence of interactions between hematite nanoparticles has also been studied by neutron scattering. The studies have shown a large reduction of the relative area of the inelastic peaks when the particles interact. Qualitatively this is in accordance with the presence of an ordered state as discussed above. The analysis of the data is in progress [25].

4. Interactions between nanoparticles of different materials

In recent studies it has been found that interactions between nanoparticles of different antiferromagnetic materials can have unexpected effects on the superparamagnetic relaxation time [20, 26]. The samples of interacting nanoparticles were prepared by suspending mixtures of uncoated particles in water, exposing them to ultrasound, and allowing them to dry at room temperature. Figure 4 shows spectra of hematite nanoparticles at 180 K for a sample consisting solely of hematite particles and of hematite particles from the same batch mixed with similar amounts of NiO or CoO nanoparticles. It is clearly seen that NiO results in faster relaxation whereas CoO results in suppression of the relaxation compared to the pure hematite sample.

In the presence of inter-particle interactions, the expression for the reduction of the magnetic hyperfine field due to collective magnetic excitations may be written [4, 14]:

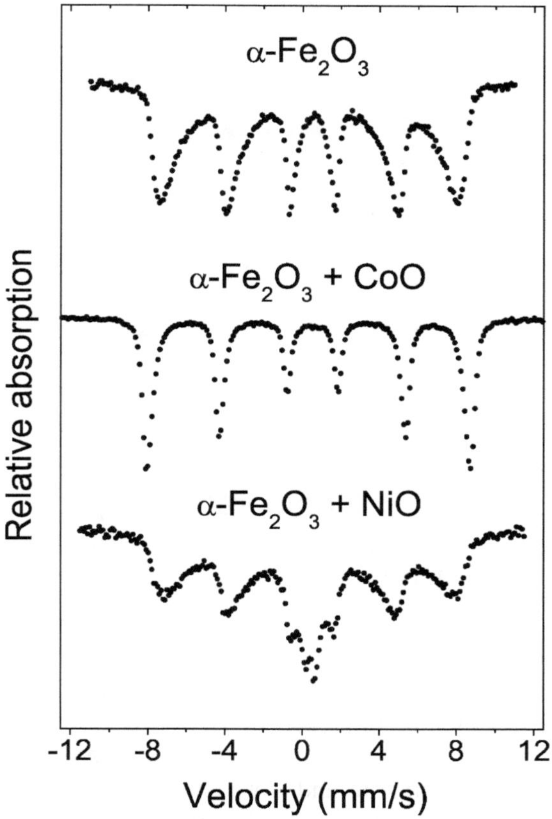

Figure 4. Mössbauer spectra of 9 nm hematite nanoparticles obtained by drying aqueous suspensions of particles of pure hematite and mixtures of hematite particles and CoO or NiO nanoparticles. The spectra were obtained at 180 K.

$$B_{\text{obs}} = B_0\left(1 - \frac{k_B T}{E_a + E_{\text{int}}}\right), \tag{6}$$

where E_a is related to the magnetic anisotropy energy ($= 2KV$ if the anisotropy energy is given by Equation (1)) and E_{int} is proportional to the interaction strength. The temperature dependence of B_{obs} at low temperature for the samples of pure hematite nanoparticles and the mixture of hematite and CoO nanoparticles is shown in Figure 5. Data for the mixture with NiO are not included, because the sample showed a partial Morin transition at low temperatures [20, 26] and the hyperfine fields are therefore not comparable to those of the other two samples. For both samples, a linear temperature dependence of B_{obs} is found, but the slopes differ considerably. In the sample with CoO nanoparticles, the variation of B_{obs} with temperature is small and close to the bulk behaviour. In the sample consisting solely of hematite nanoparticles, the slope is much larger indicating a smaller interaction

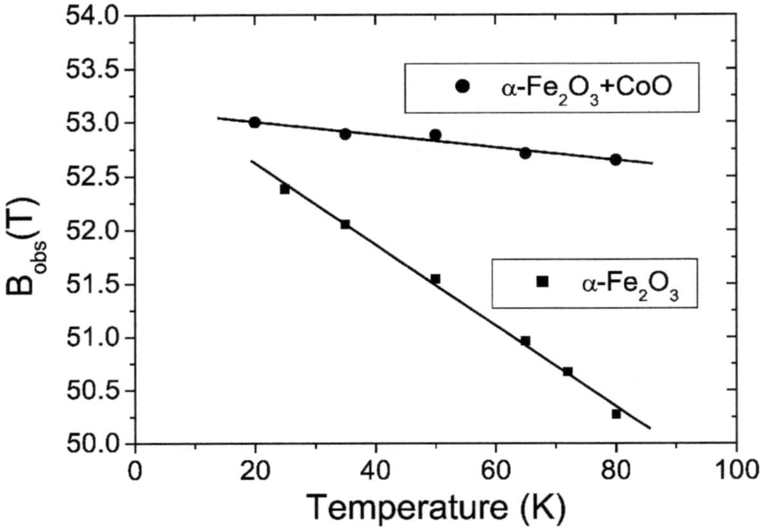

Figure 5. Temperature dependence of the magnetic hyperfine field of 9 nm hematite nanoparticles mixed with CoO nanoparticles and pure hematite.

energy. These results are in accordance with the differences in the strength of interactions, which were found from the analysis of the spectra at 180 K (Figure 4).

The origin of the different influence of NiO and CoO on the relaxation of hematite nanoparticles is not understood in detail. It is possible that a strong exchange coupling between hematite and CoO nanoparticles prevent the relaxation of the hematite particles because their sublattice magnetisation is coupled to the sublattice magnetisation of the CoO particles, which have larger anisotropy energy [20, 26]. The smaller magnetic anisotropy of NiO particles may be the reason for the faster relaxation of hematite particles when they are mixed with NiO [26].

Acknowledgements

The work was supported by the Danish Technical Research Council and the Danish Natural Science Research Council.

References

1. Mørup, S., *Hyp. Interact.* **90** (1994), 171.
2. Mørup, S. and Tronc, E., *Phys. Rev. Lett.* **72** (1994), 3278.
3. Mørup, S., Bødker, F., Hendriksen, P. V. and Linderoth, S., *Phys. Rev. B* **52** (1995), 287.
4. Hansen, M. F., Bender Koch, C. and Mørup, S., *Phys. Rev. B* **62** (2000), 1124.
5. Dormann, J. L., Fiorani, D. and Tronc, E., *Adv. Chem. Phys.* **98** (1997), 283.
6. Néel, L., *Ann. Geophys.* **5** (1949), 99.
7. Brown Jr., W. F., *Phys. Rev.* **130** (1963), 167.
8. Bødker, F. and Mørup, S., *Europhys. Lett.* **52** (2000), 217.

9. Fisker, R., Carstensen, J. M., Hansen, M. F., Bødker, F. and Mørup, S., *J. Nanoparticle Research* **2** (2000), 267.
10. Bødker, F., Hansen, M. F., Bender Koch, C., Lefmann, K. and Mørup, S., *Phys. Rev. B* **61** (2000), 6826.
11. Blume, M. and Tjon, J. A., *Phys. Rev.* **165** (1968), 446.
12. Mørup, S. and Topsøe, H., *Appl. Phys.* **11** (1976), 63.
13. Mørup, S., *J. Magn. Magn. Mater.* **37** (1983), 39.
14. Mørup, S. and Ostenfeld, C. W., *Hyp. Interact.* **136** (2001), 125.
15. Hansen, M. F., Bødker, F., Mørup, S., Lefmann, K., Clausen, K. N. and Lindgård, P.-A., *Phys. Rev. Lett.* **79** (1997), 4910.
16. Hansen, M. F., Bødker, F., Mørup, S., Lefmann, K., Clausen, K. N. and Lindgård, P.-A., *J. Magn. Magn. Mater.* **221** (2000), 10.
17. Lefmann, K., Bødker, F., Klausen, S. N., Hansen, M. F., Clausen, K. N., Lindgård, P.-A. and Mørup, S., *Europhys. Lett.* **54** (2001), 526.
18. Klausen, S. N., Lindgård, P.-A., Lefmann, K., Bødker, F. and Mørup, S., *Phys. Stat. Sol. A* **189** (2002), 1039.
19. Bødker, F., Hansen, M. F., Bender Koch, C. and Mørup, S., *J. Magn. Magn. Mater.* **221** (2000), 32.
20. Ostenfeld, C. W. and Mørup, S., *Hyp. Interact. C* **5** (2002), 83.
21. Mørup, S., Madsen, M. B., Franck, J., Villadsen, J. and Koch, C. J. W., *J. Magn. Magn. Mater.* **40** (1983), 163.
22. Lee Penn, R. and Banfield, J. F., *Geochim. Cosmochim. Acta* **63** (1999), 1549.
23. Dormann, J. L., Bessais, L. and Fiorani, D., *J. Phys. C* **21** (1988), 2015.
24. Hansen, M. F. and Mørup, S., *J. Magn. Magn. Mater.* **184** (1998), 262.
25. Klausen, S. N., Frandsen, C., Lefmann, K. and Mørup, S., *J. Magn. Magn. Mater.* (in press).
26. Frandsen, C. and Mørup, S., *J. Magn. Magn. Mater.* (in press).

Hyperfine Interactions **144/145:** 359–363, 2002.

An Interesting Spin-State Transition for [Fe(PPIX)OH] Induced by High Pressure in a Diamond Anvil Cell

V. J. CORNELIUS[1], P. J. TITLER[1], G. R. FERN[1], J. R. MILLER[1], J. SILVER[1], M. J. SNOWDEN[1] and C. A. MCCAMMON[2]
[1]*School of Chemical and Life Sciences, University of Greenwich, Central Avenue, Chatham Maritime, Kent ME4 4TB, UK*
[2]*Bayerisches Geoinstitut, Universität Bayreuth, Bayreuth, D-95440, Germany*

Abstract. A high-pressure Mössbauer spectroscopic study of [^{57}Fe(PPIX)OH] is reported under conditions where the pressure is increased and then released. Data were recorded over the pressure range from room pressure to 7.1 GPa using a modified Merril–Bassett Diamond Anvil cell. [^{57}Fe(PPIX)OH] exhibits an asymmetric quadrupole doublet at room temperature and pressure, caused by population of higher energy levels known as Kramer's doublets. Under the application of pressure a $S = 5/2$ high spin to $S = 5/2, 3/2$ admixed spin state transition of the Fe(III) site is seen to occur above pressures of 2.2 GPa. This follows a general trend observed with other porphyrin compounds. This is thought to be evidence of movement of the ligand towards the iron atom and movement of the iron atom towards the porphyrin plane. Further evidence for this motion is found in the decrease in the asymmetry of the original site, which is caused by changes in population of the energy levels of the Kramer's doublets. At the highest reported pressures a reversal in asymmetry is observed for the inner $S = 5/2$ quadrupole doublet.

Key words: Mössbauer spectroscopy, high pressure, porphyrin, spin state, Kramer's doublet.

1. Introduction

As a continuation of previous studies [1, 2] a high-pressure Mössbauer spectroscopic study of [^{57}Fe(PPIX)OH] is reported. Data were recorded from room pressure up to 6.8 GPa using a standard diamond anvil cell (DAC). The choice of sample reflects the biological importance of iron(II/III) protoporphyrin centres (haems) to mammalian biology. A study of this compound by Drickamer *et al.* [3] concluded that pressure induced a transition from Fe(III) to Fe(II), but this interpretation is not supported by data now available. The room temperature atmospheric pressure spectrum exhibits a narrow asymmetric signal [4], fitted as an asymmetric quadrupole doublet (see Figure 1). When the temperature is lowered the spectrum becomes symmetric and ultimately reverses into an opposite type of asymmetry [5]. These effects are caused by relaxation of the hyperfine coupling between the nuclear spin and the components of the electronic Kramers' doublets [6]; the mech-

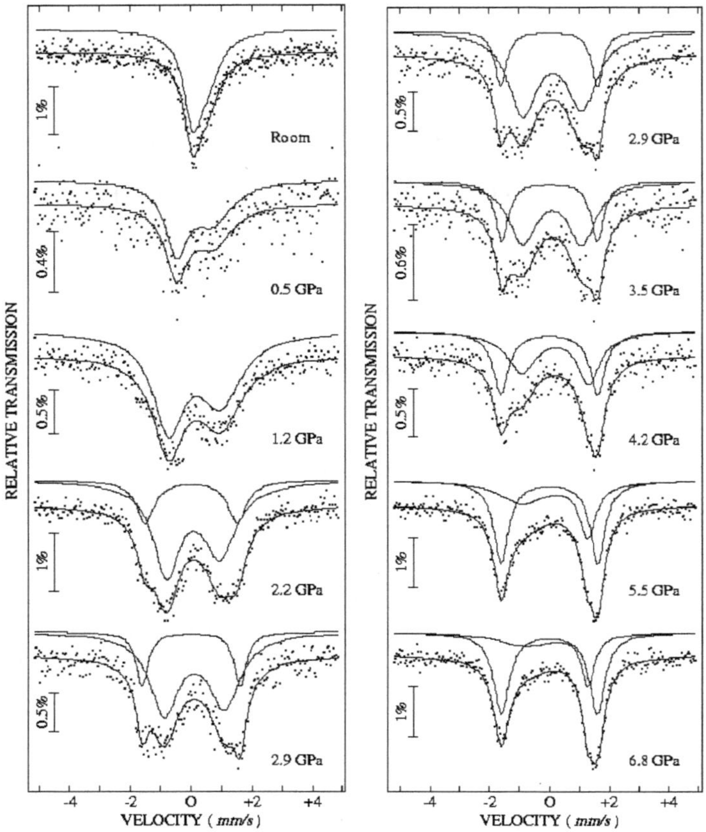

Figure 1. Mössbauer spectra of [Fe(PPIX)OH] at 298 K across the pressure range 1×10^{-4} to 6.8 GPa.

anism has been explained by Clauser and Blume [7] and refined by Dattagupta [8], in so far as temperature induced changes are concerned. However, as reported in this work pressure-induced changes go beyond the limits of Dattagupta's theory.

2. Experimental

High-pressure Mössbauer spectra were recorded using a modified Merrill-Bassett diamond anvil cell (DAC) under hydrostatic pressure using the previously reported system [9]. 4 : 1 MeOH : EtOH was used as the hydrostatic pressure fluid [10], and the ruby fluorescence method was used to measure the pressure in the DAC [11].

Table I. Mössbauer spectroscopic parameters for [Fe(PPIX)OH] at 298 K over the pressure range from 1×10^{-4} to 6.8 GPa

Pressure (GPa)	δ (mm/s)	ΔE_Q (mm/s)	Γ (mm/s)#	L2 : L1*	Area (%)
		$S = 5/2$ site			
1×10^{-4}	0.30(2)	0.44(5)	0.67(4)	1.65(9)	100(2.6)
0.50(5)	0.39(6)	1.18(10)	0.91(13)	1.71(24)	100(9.6)
1.20(12)	0.38(2)	1.69(3)	1.24(5)	1.36(5)	100(2.86)
2.20(22)	0.33(1)	1.73(7)	1.05(6)	1.25(5)	76.0(6.82)
2.90(29)	0.37(2)	1.96(7)	1.18(9)	1.08(6)	77.8(5.46)
3.50(35)	0.36(3)	1.96(13)	1.14(15)	0.98(19)	67.8(12.1)
4.20(42)	0.46(2)	2.23(4)	1.16(13)	0.78(10)	54.4(6.3)
5.50(55)	0.44(6)	2.16(14)	1.68(15)	0.40(10)	43.0(5.6)
6.80(68)	0.49(9)	2.09(19)	2.06(21)	0.26(3)	34.6(4.7)
		$S = 5/2, 3/2$ site			
2.20(22)	0.26(2)	3.01(6)	0.71(10)	s	24.0(6.0)
2.90(29)	0.26(2)	3.22(4)	0.48(7)	s	22.4(4.7)
3.50(35)	0.28(2)	3.18(6)	0.54(11)	s	32.2(9.7)
4.20(42)	0.26(2)	3.22(4)	0.62(7)	s	45.6(5.7)
5.50(55)	0.26(2)	3.22(2)	0.60(4)	s	57.0(5.1)
6.80(68)	0.26(2)	3.19(2)	0.64(3)	s	65.4(4.7)

Full width at half mean height of L1 the low velocity line; * ratio of high velocity peak to low velocity peak line widths; s = fitted as a symmetric doublet.

3. Results and discussion

The data are presented in Table I and Figure 1. The first observation is that the initial application of pressure broadens and splits the spectrum, which may indicate minor modification of the electronic environments of the iron(III) sites as the iron-containing molecules are forced closer together. As pressure is increased further, at 2.2 GPa a second site with a larger quadrupole splitting becomes clearly visible. This site continues to develop in % area with pressure but stabilises above 2.9 GPa with spectral parameters of quadrupole splitting (ΔE_Q) = 3.2 mm/s and isomer shift (δ) = 0.26 mm/s. This site is assigned as a $S = 5/2, 3/2$ admixed spin state of the Fe(III) complex, and mirrors previously reported results for high-pressure spin-state transitions in the monomer [Fe(PPIX)Cl] [1, 2]. The second noticeable effect of pressure is the change in asymmetry of the inner $S = 5/2$ site, which decreases with the application of pressure to 3.5 GPa, where the inner site is symmetric. At higher pressures the asymmetry reverses so that the low velocity line becomes the broader one.

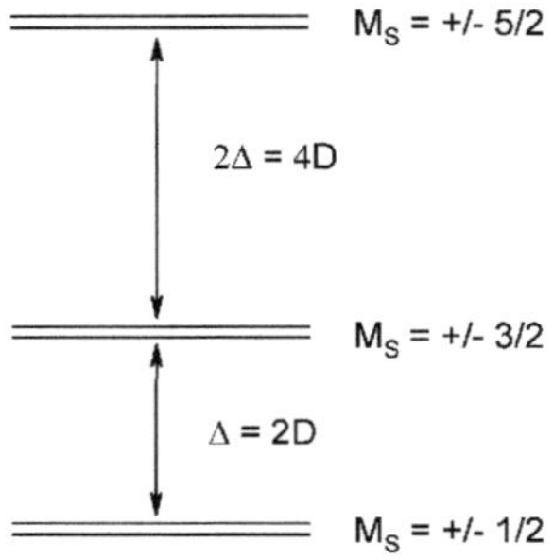

Figure 2. Zero field splitting of ^{6}S Fe^{3+} ground state by an axial crystal field.

If it is assumed that the iron atom moves towards the porphyrin plane under the applied pressure, this will have the effect of increasing the equatorial component of the primary crystal field with two direct consequences. Firstly, increasing the separation between the $d_{x^2-y^2}$ orbital and the d_{zx}, d_{yz} orbitals enables one electron resident in the high-energy $d_{x^2-y^2}$ orbital to drop and pair with an electron in the low energy orbital, creating a $S = 3/2$ state (see Figure 3). Spin–orbit coupling may give rise to a "quantum mechanical spin admixture" [12], a state with properties intermediate between those of $S = 5/2$ and $S = 3/2$. Such a state involves the electron spin and the charge density being effectively shared between the $d_{x^2-y^2}$ orbital and the d_{yz} and d_{zx} orbitals as depicted in Figure 3, and accounts for the appearance of the second site. Secondly, the zero field splitting at the first site increases. This causes the $M_S = \pm 5/2$ and $M_S = \pm 3/2$ energy levels (Figure 2) to depopulate, thus reducing the observed asymmetry in the Mössbauer spectra until only the lowest $M_S = \pm 1/2$ energy state is populated and the spectrum is symmetric at 3.5 GPa (Figure 1). At higher pressures the effect of off-diagonal terms in the hyperfine interaction matrices (mentioned earlier as occurring at low temperature [4–8]) is manifest as a reversed asymmetry. However, in this case where the effect is pressure-induced, the result is more pronounced (Figure 1), with the low-velocity line of the Mössbauer spectrum broadening significantly. Further increases in pressure result in this line almost disappearing by broadening into the base line of the spectrum. Using Dattagupta's equations it has not been possible to fit all the spectra to a consistent set of parameters. It is possible that this is not a failure of the original theory but due to numerical approximations necessary at that time.

4. Conclusions

The room pressure spectrum of [Fe(PPIX)OH] is typical for a Fe(III) $S = 5/2$ high spin signal manifesting the effects of Kramers' populations. The application of pressure generates a second site which is assigned as spin admixed $S = 5/2, 3/2$. The spin admixed site is caused by a relative decrease in the axial component of the crystal field. This is thought to be caused by the Fe atom moving towards the plane

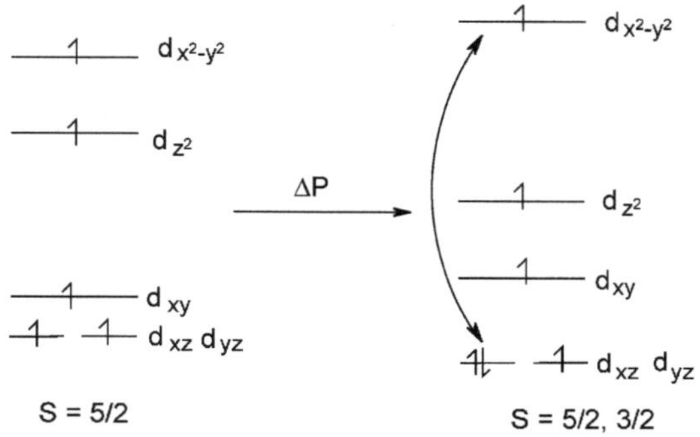

Figure 3. Change of primary crystal field spliting with pressure.

and an increase in crystal field splitting contribution by the 4 equatorial ring *N* atoms. However, some of the original site signal persists even at 6.8 GPa. Reversal of asymmetry in the Mössbauer spectrum for this $S = 5/2$ site is observed, and this is partly explained by current theory. However, this theoretical model cannot fully describe our spectra and further studies are in progress.

Subsequent to the conference presentation further experimental data confirm that the effects described here are reversible with some evidence for hysteresis.

Acknowledgements

High-pressure experiments were performed at the Bayerisches Geoinstitut under the E. C. "Human Capital and Mobility" – Access to large scale facilities programme. (Contract No. ERBCHGET940053 to D. C. RUBIE.)

References

1. Silver, J., Anandan, S., Davies, D. A., Fern, G. R., Marsh, P. J., Miller, J. R., Titler, P. J. and McCammon, C. A., *Hyp. Interact.*, in press, Given as Plenary Lecture, ICAME, 2001.
2. Anandan, S., Silver, J., Fern, G. R., Titler, P. J. and McCammon, C. A., *Hyp. Interact. C*, in press.
3. Grenoble, D. C., Frank, C. W., Bancroft, C. B. and Drickamer, H. G., *J. Chem. Phys.* **55** (1971), 1633.
4. Maracondi, C., Straub, D. K. and Epstein, L. M., *J. Am. Chem. Soc.* **94** (1972), 4162.
5. Fitzsimmons, B. W., Sams, J. R. and Tsin, T. B., *Chem. Phys. Lett.* **38** (1976), 588.
6. Gibb, T. C., *Principles of Mössbauer Spectroscopy*, Chapman and Hall, London and New York, 1976, p. 32.
7. Clauser, M. J. and Blume, M., *Phys. Rev. B* **3** (1971), 583.
8. Dattagupta, S., *Phys. Rev. B* **12** (1975), 3584.
9. McCammon, C. A., In: A. K. Singh (ed), *Recent Trends in High Pressure Research*, Oxfords and IBH, New Dehli, 1982, p. 824.
10. Jayaraman, A., *Rev. Mod. Phys.* **55** (1983), 1.
11. Barnett, J. D., Block, S. and Piermarinni, G. J., *Rev. Sc. Instrum.* **44** (1973), 1.
12. Gupta, G. P., Lang, G., Scheidt, W. R., Geiger, D. K. and Reed, C. A., *J. Chem. Phys.* **85** (1986), 5212 and references therein.

Hyperfine Interactions **144/145:** 365–370, 2002.

Can Mössbauer Spectroscopy Reveal the Origin of Magnetic Soil in Denmark?

H. P. GUNNLAUGSSON[1,*], J. P. MERRISON[1], L. A. MOSSIN[2], P. NØRNBERG[2], J. SANDEN[1], E. UGGERHØJ[3] and G. WEYER[1]
[1]*Department of Physics and Astronomy, Aarhus University, Ny Munkegade, DK-8000 Århus, Denmark; e-mail: hpg@phys.au.dk*
[2]*Department of Earth Sciences, Aarhus University, Ny Munkegade, DK-8000 Århus, Denmark*
[3]*Institute of Storage Ring Facilities, Aarhus University, Ny Munkegade, DK-8000 Århus, Denmark*

Abstract. The rare occurrence of magnetic soil in Denmark is normally explained by the hypothesis that the site has been burned. However, at some sites formation by means of organic processes has been suggested. In this contribution we present results of combined Mössbauer transmission and conversion electron spectroscopies in order to seek an answer to the question of the origin of magnetic soil in Denmark. It turns out that the sample preparation is of extreme importance, and we will document what precautions have to be taken in order to get meaningful results. When these issues are understood, and taken into consideration, Mössbauer spectroscopy seems to be a promising tool to gain insight in soil forming processes.

Key words: superparamagnetism, soil, hematite, maghemite, CEMS.

1. Introduction

At several locations, e.g., a site called Salten Skov in Jutland, Denmark, magnetic soil ($\sigma_S > 1$ Am^2/kg) has been found [1]. The particle size and the magnetic properties [1, 2] are similar to those observed on Mars (see, e.g., [3] and references therein) making the soil an interesting analogue. The soil has been used in wind tunnel experiments simulating the environmental conditions on the surface of Mars [4]. X-ray diffraction and Mössbauer spectroscopy have revealed that the soil contains the magnetic iron oxides hematite (α-Fe_2O_3) and maghemite (γ-Fe_2O_3) ($\sim$30% of the spectral area) in mixtures with the antiferromagnetic goethite (α-FeOOH). The particle size of the soil has been determined to be $\sim$2 μm when dispersed in an ultrasound bath. By means of transmission electron microscopy (TEM) and the line broadening of the X-ray diffraction lines, it has been found that individual particles consist of smaller, nano-sized particles of iron oxides which show superparamagnetic blocking on the timescale of Mössbauer spectroscopy.

---

* Author for correspondence.

The presence of hematite and particularly of maghemite in Danish surface sediments (e.g., at a site called Næsset) is normally explained by a de-hydration of goethite by soil burning (forest fires or metal working). However, it has been speculated whether the presence of hematite and maghemite at some sites (e.g., Salten Skov) might have their origin from micro-biologically or inorganically produced ferrous oxides, known to transform into hematite and/or maghemite over years at room temperature [5].

If the magnetic minerals have formed via soil burning, intuition suggests that the surface of the 2 μm-sized particles might be more affected than the bulk. In this contribution we have tested this suggestion by Conversion Electron Mössbauer Spectroscopy (CEMS). Since the conversion electrons released in the decay of the 14.4 keV state of ^{57}Fe can only penetrate about 0.25 μm of material, CEMS is sensitive to the surface of the individual particles. By comparison of the CEM spectra to bulk spectra obtained by conventional transmission Mössbauer spectroscopy (TMS), information on the thermal history of the sample might thus be obtained.

2. Experimental and samples

Mössbauer spectra were recorded using a ^{57}Co:Rh source moved relative to the sample on a conventional constant acceleration drive system. CEM spectra were obtained by placing the sample on a double adhesive conducting tape in a parallel plate avalanche detector [6]. Isomer shifts are given relative to α-Fe at room temperature.

Samples from two sites have been used in this investigation, Salten Skov, described above, and Næsset. At Næsset, there is strong evidence for site burning. The magnetic soil is found in circular spots, up to $\sim$2 m in diameter, and the mineralogy and distribution of material indicates unstable conditions. Such features are common to prehistoric iron working at other sites. Annealing studies were performed in air on samples from Salten Skov to see which reactions take place.

3. Results

Figure 1 shows selected Mössbauer spectra from Salten Skov samples. The spectra have been analysed in terms of three spectral components indicated in Figure 1: A sextet labelled Gt, assigned to goethite, a sextet labelled Mh/Ht, assigned to maghemite/hematite mixtures and a quadrupole-split component labelled Fe(III) assigned to ferric (super-) paramagnetic iron oxide(s). The sextet components have been analysed with Split-Lorentzian lineshapes [7], which simulate the asymmetric spectral lines quite well and enable the determination of area fractions for the different components. We do not try to separate the hematite and maghemite components in view of their rather similar hyperfine parameters and the rather broad asymmetric lineshapes. The annealing shows no prominent effect for temperatures below 200°C. At 250°C the goethite is spontaneously dehydrated and the spectra

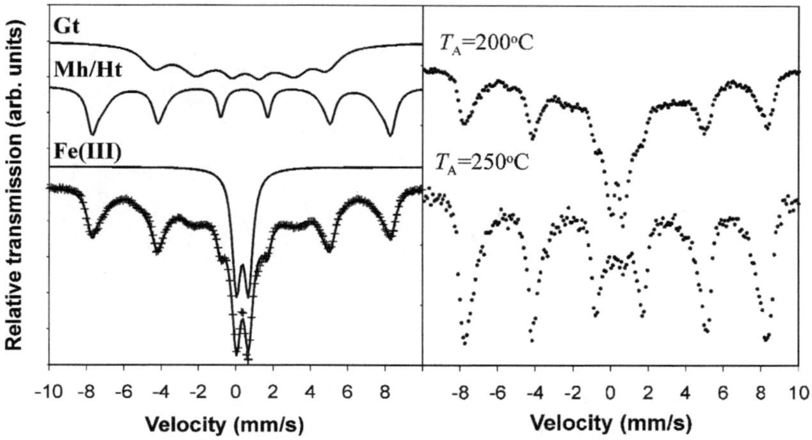

Figure 1. Left: Room temperature TM spectra of bulk samples from Salten Skov. The solid lines show individual fitting components and their sum. *Right*: Room temperature TM spectra of the Salten Skov sample annealed for 1 hour at the temperatures indicated.

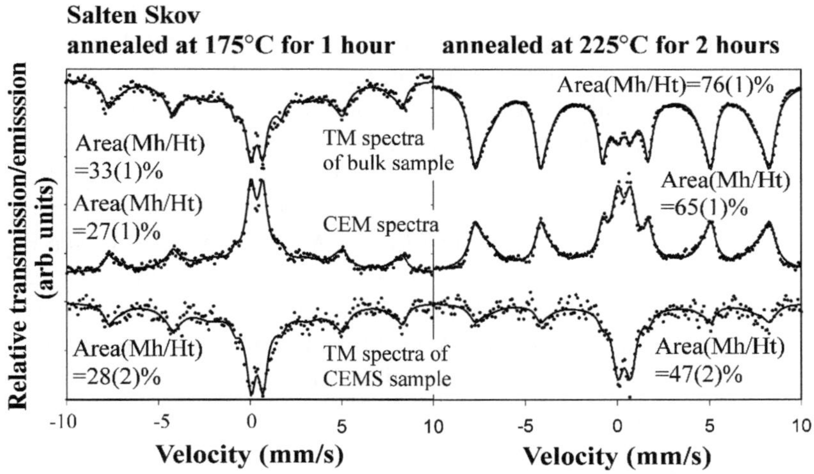

Figure 2. Series of CEM and TM spectra of annealed samples from Salten Skov as indicated.

are dominated by the Mh/Ht spectral component. This indicates what one should expect when comparing the CEM and TM spectra.

Figure 2 shows a comparison of CEM and TM spectra for annealed samples from Salten Skov.

There are clearly different TM results obtained when measuring on CEMS or bulk samples. It is of importance to note that no separation has taken place in the sample preparation; the only difference lies in the way that the CEMS sample is prepared. This is done by gently "smearing" out the material, to form a homogeneous surface suitable for CEMS measurements. Although the preparation is extremely gentle, clearly, changes are made to the sample. An easy way to demon-

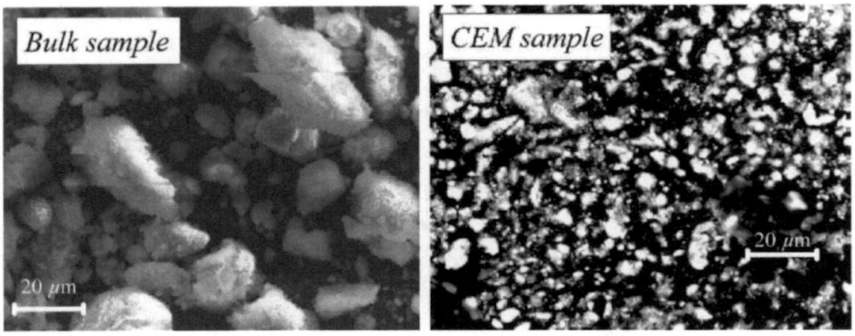

Figure 3. Representative SEM pictures of a sample from Salten Skov. Here the sample annealed at 225°C for 2 hours. *Left*: bulk sample, *right*: CEMS sample.

strate this is the following. The samples can be smeared onto transparency foils (about 40 μg/cm^2), and recollected by using plastic tools. It is then found that the spectrum is different, dominated by a doublet. Mössbauer measurements in high magnetic field (not shown) verify that the doublet is due to superparamagnetic particles. Two plausible hypotheses can explain these results. During annealing, the dehydration of the goethite could lead to enhanced interactions between the maghemite/hematite particles. Secondly, one could imagine that during annealing, crystallites grow together, and start to interact via exchange interactions. These new bonds are then easily broken apart in both cases during the CEMS sample preparation. We favour the latter hypothesis as it explains the irreversibility of the process. If the method is to be applied to study the surface properties of the particles in comparison to the bulk, one needs to document the properties of the samples used, by, e.g., using Scanning Electron Microscopy (SEM), see Figure 3, and perform both the TMS and CEMS measurements on the same sample.

From Figure 3 it is seen that while the bulk sample does not shows the individual particles, but aggregates, these have been broken apart in the preparation of the CEMS sample, where individual 2 μm particles are seen and the pictures show that the CEMS and TMS performed on the CEMS sample are giving us information on the surface and bulk, respectively.

Having pointed out the importance of the sample preparation, one can turn attention to the comparison between the TMS and CEMS results. For the sample annealed at 175°C, no major difference is seen between the measurements, suggesting homogeneous particles. For the sample annealed at 225°C, the sextet component is enhanced by roughly 40% in the surface of the CEMS sample, indicating stronger annealing effects in the surface of the particles. Figure 4 shows series of analogues spectra obtained on a sample from the site Næsset.

4. Discussion and conclusions

The results allow proposing a model for the reactions that take place during and after annealing. During annealing at temperatures above 200°C the goethite is spontaneously dehydrated and hematite (possibly maghemite) is formed. The in-

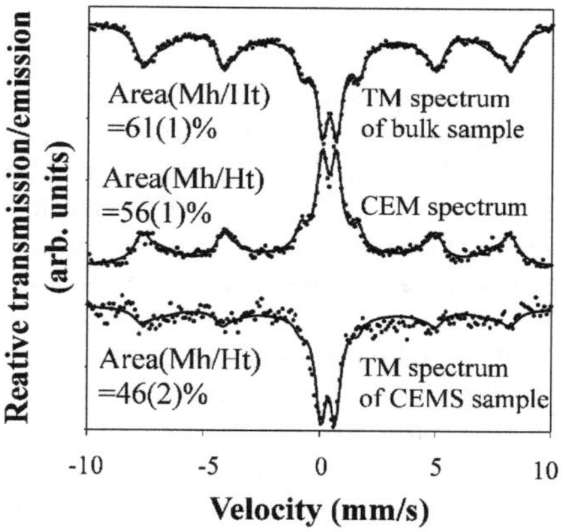

Figure 4. CEM and TM spectra of a sample from Næsset.

dividual nano-sized particles interact; either via exchange interactions or because they have been brought closer together in the dehydration process. The resulting bulk Mössbauer spectra are dominated by a sextet component. During the CEMS sample preparation, most of the newly formed interacting particles are broken apart but those which have been exposed to the largest thermal effect (in the surface layers of the 2 μm particles) have a better chance of surviving this treatment. The resulting Mössbauer spectra are dominated by a superparamagnetic doublet component. The difference in the CEM and TM spectra for the same samples in the spectra of the annealed samples from Salten Skov and the sample from Næsset gives evidence that in both cases the material transformation is more advanced in the surface of the particles.

This difference is not observed for unannealed samples from Salten Skov, and the 2 μm particles appear homogeneous on the scale of CEMS. This allows the conclusion that the formation of the maghemite/hematite mixtures is unlikely to be due to thermal effects during, for example, forest fires. The results here show that the combination of CEMS and TMS to investigate the magnetic soils is a promising method to shed light on the origin of the magnetism in the soil.

There is another important consequence of the work presented here. Mössbauer spectroscopy is widely utilized to characterize iron oxides found in soils [8]. Among the parameters of interest is the particle size, but information on this parameter is, for example, obtained by determination of superparamagnetic blocking temperatures. However, the results here show that this determination may depend on the sample preparation, and in order to obtain reliable results, this has to be tested.

Acknowledgements

This work has been supported by the Danish Natural Research council. Prof. S. Mørup is acknowledged for fruitful discussions and experimental help.

References

1. Nørnberg, P. *et al.*, manuscript in preparation.
2. Bertelsen, P., PhD thesis, University of Copenhagen, 2001.
3. Bell III, J. F., McSween, Jr., H. Y., Crisp, J. A., Morris, R. V., Murchie, S. L., Bridges, N. T., Johnson, J. R., Britt, D. T., Colombek, M. P., Moore, H. J., Ghosh, A., Bishop, J. L., Anderson, R. C., Brükner, J., Economou, T., Greenwood, J. P., Gunnlaugsson, H. P., Hargraves, R. M., Hviid, S., Knudsen, J. M., Madsen, M. B., Reid, R., Rieder, R. and Soderblom, L., *J. Geophys. Res.* **105**(E1) (2000), 1721.
4. Merrison, J. P., Bertelsen, P., Frandsen, C., Gunnlaugsson, H. P., Knudsen, J. M., Lunt, S., Madsen, M. B., Mossin, L. A., Nielsen, J., Nørnberg, P., Rasmussen, K. R. and Uggerhøj, E., accepted for publication in *J. Geophys. Res.* (2002).
5. Murad, E. and Schwertmann, U., *Amer. Miner.* **65** (1980), 557.
6. Weyer, G., In: I. J. Gruverman and C. W. Seidel (eds), *Mössbauer Effect Methodology*, Vol. 10, Plenum Press, New York and London, 1967, p. 301.
7. Morris, R. V., Agresti, D. G., Lauer Jr., H. V., Newcomb, J. A., Shelfer, T. D. and Murali, A. V., *J. Geophys. Res. B* **94** (1989), 2760.
8. Murad., E., *Hyp. Interact.* **111** (1998), 251.

Hyperfine Interactions **144/145:** 371–379, 2002.

The Miniaturized Mössbauer Spectrometer MIMOS II for Extraterrestrial and Outdoor Terrestrial Applications: A Status Report

G. KLINGELHÖFER[1,*], B. BERNHARDT[1], J. FOH[1,2], U. BONNES[2], D. RODIONOV[1], P. A. DE SOUZA[1,3], CH. SCHRÖDER[1], R. GELLERT[1], S. KANE[1], P. GÜTLICH[1] and E. KANKELEIT[2]
[1] *Institute Inorganic and Analytical Chemistry, Johannes Gutenberg-University Mainz, Germany; e-mail: klingel@mail.uni-mainz.de*
[2] *Institute Nuclear Physics, Darmstadt University of Technology, Darmstadt, Germany*
[3] *CVRD Group, Brazil*

Key words: Mössbauer spectroscopy, Mars, industrial applications.

In May and July 2003 both the European space agency ESA and the American space agency NASA will launch space missions to Mars. The ESA lander Beagle 2 and the two NASA Mars-Exploration-Rovers (MER) will explore the Martian surface with a set of sophisticated instruments. Part of the payload will be our miniaturized Mössbauer spectrometer MIMOS II. It operates in backscattering geometry and meets the requirements for space application of low mass ($\leqslant$500 g), small volume (coke can size), and low power consumption ($\leqslant$3 W). Main goals are the determination of the oxidation state of iron and the iron mineralogy on the surface. This information will contribute to a much deeper understanding of the evolution of the planet Mars, its surface and atmosphere, and the history of water. The MIMOS II flight units for MER were delivered in April 2002 to the NASA Jet Propulsion Laboratories (JPL), California, for integration to the Rovers. After some more testing of the complete Rover system the spacecraft will be shipped to the Kennedy Space Center early February 2003. The first launch will be in May 2003 and the second launch in late June on early July 2003. The flight unit for the ESA Mars-Express Beagle lander was delivered to ESA by the end of May 2002 for integration to the lander in late November/early December 2002. The launch is scheduled for June 2003 from Baikonur, Kazakhstan. The instrument MIMOS II is also under consideration for an ESA space mission to Mercury in 2009, and it is part of the ESA exobiology multi-user facility to be launched as part of one of the next lander Mars missions after 2005.

---

* Author for correspondence.

Besides these extraterrestrial projects the instrument MIMOS II has been used already recently for a series of terrestrial projects, including the analysis of archaeological artifacts such as Greek vases, the *in situ* monitoring of Fe oxidation states in the field, and the investigation of Rock paintings in Brazil [7]. Furthermore in cooperation with industry (CVRD group) from Brazil a copy of the flight instrument is being used to monitor air pollution properties in some industrial areas in the state of Espírito Santo (north of Rio de Janeiro).

1. The miniaturized instrument MIMOS II

Developed originally for extraterrestrial applications, the miniaturized Mössbauer spectrometer, named MIMOS II [1], is reduced in size and weight, in comparison to ordinary laboratory setups. The MIMOS II sensor head weighs approximately 400 g, with a volume of $(50 \times 50 \times 90)$ mm^3, and holds two γ-ray sources: the stronger for experiments and the weaker for calibrations (Figure 1). The collimator (in sample direction) also shields the primary radiation off the detectors. Around the drive four detectors are mounted. The detectors are made of Si-PIN-photodiodes in chip form (100 mm^2, thickness of 0.5 mm). The control unit is located in a separate electronics board. This board is responsible for the power

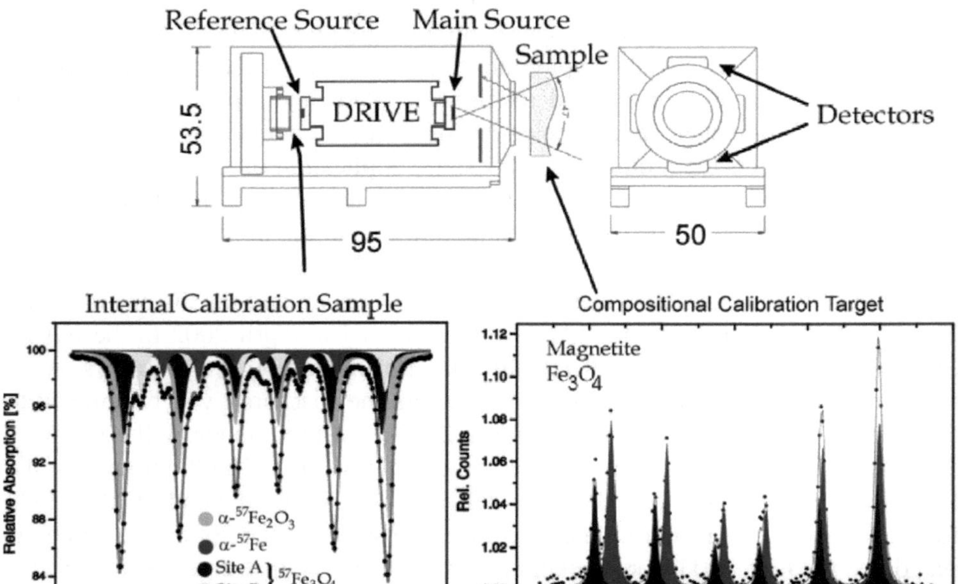

Figure 1. Drawing of the MIMOS II sensor head (dimensions in mm). The drive where both reference and main ^{57}Co/Rh sources are placed is shown in detail. A possible reference sample (internal calibration sample) and a spectrum from the Compositional Calibration Target (CCT, magnetite) for the Mössbauer spectrometer on board of the Mars Exploration Rovers (MER) are also depicted.

supply to the sensor head, generation of the drive's velocity reference signal, the accumulation of the detector pulses to record the spectrum, data storage and communication with the host computer. Mössbauer parameters are temperature dependent. Especially for small particles exhibiting superparamagnetic behaviour, the Mössbauer spectrum may change drastically with temperature. The temperature variation for one spectral accumulation will not be larger than $\pm 10^{\circ}$C. When larger variation occurs, spectra for different temperature ranges are stored separately, resulting in an increase in the total data volume and decrease of the statistical quality for the individual sub-spectra.

2. Implemented data analysis system

After more than four decades after the discovery of the Mössbauer effect, more than 400 minerals were studied at different temperatures. Their Mössbauer parameters were reported in the literature, and have been recently collected in a data bank [2–4]. Previous Mars-missions, namely Viking and Mars Pathfinder revealed Si, Al, Fe, Mg, Ca, K, Ti, S and Cl to be the major constituents in soil and rock elemental composition of the red planet. More than 200 minerals already studied by Mössbauer spectroscopy contain significant amounts of these elements. A considerable number of Mössbauer studies were also carried out on meteorites and on Moon samples. Looking backward in the studies of the whole Mössbauer community, we have collected a specific library containing Mössbauer parameters of those possible Mars minerals. The selected minerals are organized according to their Mössbauer parameters, main site substitutions, behaviour as a function of temperature, and a ranking as expected to be found on Mars. Fe-bearing Mars-analogue minerals not studied yet by Mössbauer spectroscopy are being collected and investigated. In addition an identification system based on Artificial Neural Networks was implemented which is able to make a fast and precise mineral identification from the experimental Mössbauer parameters at a given temperature.

3. Field tests with the prototype Mars Rover FIDO

Field Integration Design Operations (FIDO) is a prototype Mars rover [5] devoted to "blind" science tests to evaluate the most efficient ways of traversing across diverse terrains while conducting reconnaissance-level field science allowing site characterization, and finding, approaching, and placing analytical instruments and a drilling system onto rock targets to characterize in detail sample materials that are representative of the sites traversed. This rover is designed to simulate the Athena Payload for 2003 Mars Exploration Rover mission. During August 2002, a FIDO test was conducted at JPL facilities in California. The rover itself was placed in the desert in Arizona, USA. Figure 2 shows a Mössbauer spectrum recorded from the desert soil. One Mössbauer measurement on Mars will take approximately 12 hours and the FIDO measurements were taken during the equivalent time per source activity.

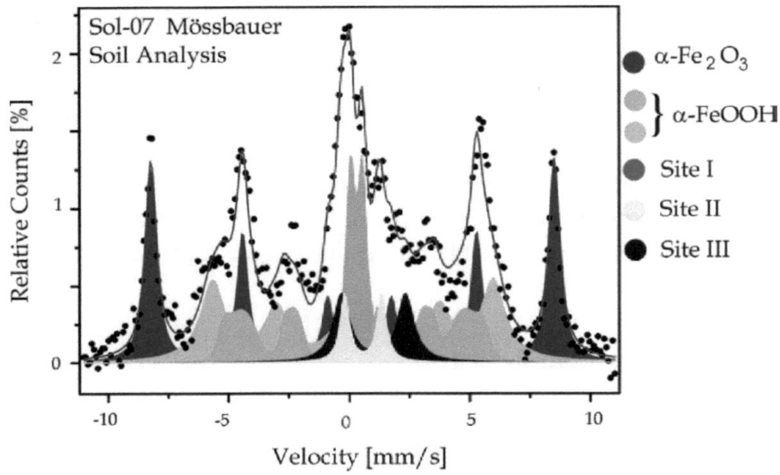

Figure 2. Mössbauer spectrum recorded from the desert soil. Hematite, goethite and Fe-bearing silicates were detected.

The first Mars Exploration Rover mission rocket is scheduled to launch in May 2003 from Cape Canaveral, Florida, and the second one in July 2003. The cruise to Mars takes approximately 7 months. The first rover should land on the Martian surface in January 2004. Each rover is expected to be working on Mars for at least three months, with the initial nominal ability to cross up to hundred meters per Sol (one Martian day).

The ESA Mars Express mission is scheduled to be launched in June 2003 from the Baikonur launch pad in Kazakhstan, arriving at Mars in December 2003 (around Christmas). The lander Beagle 2 is a stationary lander aimed to search for traces of extinct life. Its nominal mission duration on the surface is about 6 months.

4. Selected terrestrial applications

4.1. SURFACE ANALYSIS

MIMOS II can be used for the depth selective analysis of a certain category of samples where the surface layers of interest have a thickness of several tens of Angstroms up to a couple of hundred Angstroms with detection limits depending on the total iron content in those layers. The depth selectivity, which is certainly limited in respect to conversion electron Mössbauer spectroscopy (CEMS), is achieved by detecting in backscattering geometry both 6.4 keV X-rays and the 14.4 keV Mössbauer radiation simultaneously, as well as the transmission spectrum in cases of thin samples. An example is shown in Figure 3, demonstrating in a qualitative way that capability of "MIMOS II Terrestrial" which has an additional detector for transmission measurements.

Table I. Mössbauer parameters for the desert soil (Arizona State) measured for the 7th FIDO simulated Mars day (Sol-07). The preliminary Fe-phase identification is given

δ^* [mm/s]	Δ [mm/s]	B_{hf} [tesla]	Area [%]	Preliminary phase identification Confidence level: 1: Sure; 2: Tentative; 3: Uncertain
0.40	−0.25	51.9	30.9	Hematite (well crystallized, large grain sizes)[1]
0.37	−0.22	36.1	27.5	Goethite (well crystallized, large grain sizes)[1]
0.42	0.50		11.6	Phyllosillicate[2], lepidochrosite[3]
0.70	1.55		4.5	Fe-smectite[2], garnet (melanite)[3], ilmenite[3]
1.12	2.66		8.1	Phyllosillicate[2], clinopyroxene (augite)[2], ilmenite[3]
$\langle\delta\rangle^*$ [mm/s]	$\langle\Delta\rangle$ [mm/s]	$\langle B_{hf}\rangle$ [tesla]	Area [%]	
0.41	−0.24		17.4	Goethite (poorly crystallized, possible Al substitution)[1]

Ratios: $\frac{Fe^{2+}}{Fe^{2+} + Fe^{3+}} = 12.6\%$; $\quad Fe\frac{\text{Oxides} + \text{Oxyhydoxides}}{\text{Silicates}} = \frac{30.9 + 44.9}{24.2} = 3.13.$

Background counts = 1 702 090; *Related to α-Fe standard, Fe-content (from APXS) = 6.4%.

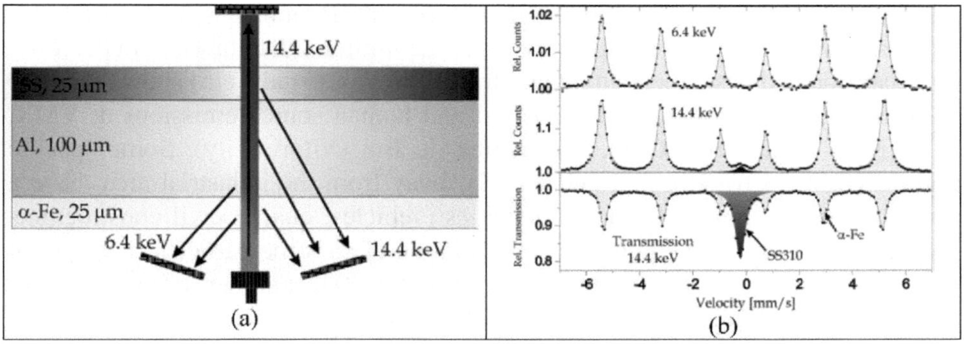

Figure 3. Demonstration of the depth selective surface analysis. (a) Experimental setup: 14.4 keV in standard transmission geometry and both 6.4 keV and 14.4 keV in backscattering arrangement. The room temperature Mössbauer spectra are recorded simultaneously using MIMOS II in backscattering (four detectors) and in transmission (b). Example spectra of a target with composition as given in (a).

4.2. AIR POLLUTION STUDIES

The miniaturized MB equipment was installed inside a particle sampler collecting dust and other fine particles present in the atmosphere. Genetic algorithms, fuzzy logic and artificial neural networks were used to perform the automatic analysis of Mössbauer spectroscopy [2]. The experimental setup is portable, easy to use and of low cost. The preliminary results obtained seem to be very promising and encourage further development of the system. The most interesting and promising

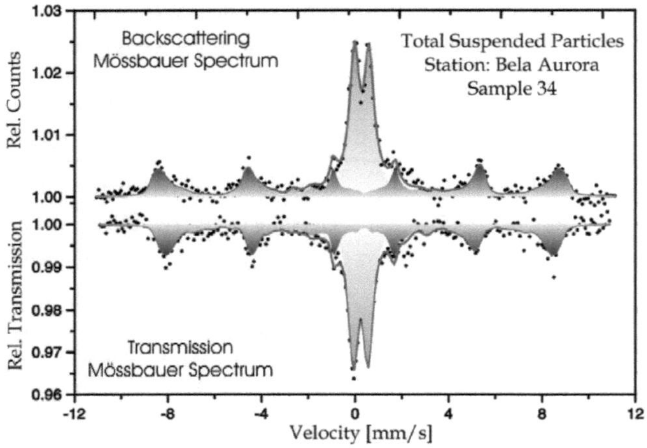

Figure 4. Room temperature Mössbauer spectra recorded simultaneously in transmission and in backscattering *in situ* at Vitória metropolitan region (Brazil) using MIMOS II.

results from the on-line and *in situ* airborne particles characterization are the fast and automatic answers obtained. The recent developments have already gained application and produced many relevant information for both Brazilian industries and environmental agencies. The MIMOS II instrument was installed in Vitória metropolitan region (VMR), southeastern Brazil. VMR embraces five cities, holds an area of 1461 km^2, has a population over 1.2 million inhabitants, is undergoing an expansion and has to bear an intense traffic. Previous studies [5] show that there is an equivalent distribution of industrial and human source emissions at VMR, at EAMES (Vila Velha City), and at Ilha do Boi (Vitória city). Some specific regions (Dório Silva Hospital and UFES), away from the industrial area, have a stronger contribution from human activities (vehicles, quarry, civil constructions and vegetation fire), but equivalent concentration of airborne particles. The system will be installed in these selected places during the 2002 summer (critical season). An assessment of industrial sources of TSP is given in Table II.

4.3. A SELECTED ARCHAEOLOGICAL STUDY

Wall paintings of banqueting, acting, traveling and horse riding scenes were discovered in China in the tombs group of the periods of the Han and Wei dynasty. These painting show great scenes and hectic atmosphere and reflect the fact that at that time the material richness and cultural life reached remarkably high standards. The samples originate from an unpublished excavation, that has been accomplished together with a Chinese Institute. The wall paintings come from a grave of a high military person, dating to the Eastern-Han Dynasty. The site is in the district of Xunyi, Province Shaanxi, China. A square sample of 1 inch-square in area of an ocher wall-painting piece was studied by 14.4 keV Fe Mössbauer spectroscopy. A backscattering geometry at both room temperature and 200 K, with a 100 mCi

Table II. Summary statistics for industrial sources of total suspended particles at Ilha do Boi, Vitória (samples collected from 1995 to 1999)

Fe-bearing phase	δ [mm/s]	Δ [mm/s]	B_{hf} [tesla]	Relative area [%]	Remarks
Hematite	0.37	−0.21	51.8	85 to 96	Always present
Ferric Stain	0.37	0.50		1.1 to 8	Industrial emission
Goethite	0.36	0.40	38.1	3 to 15	Soil
Pyrite	0.31	0.61		0.7 to 4.7	Mineral coal
Silicates				1 to 7.3	Soil
Magnetite	0.36	0.01	49.2	upto 6	Steelwork plant
	0.67	−0.01	46.0		

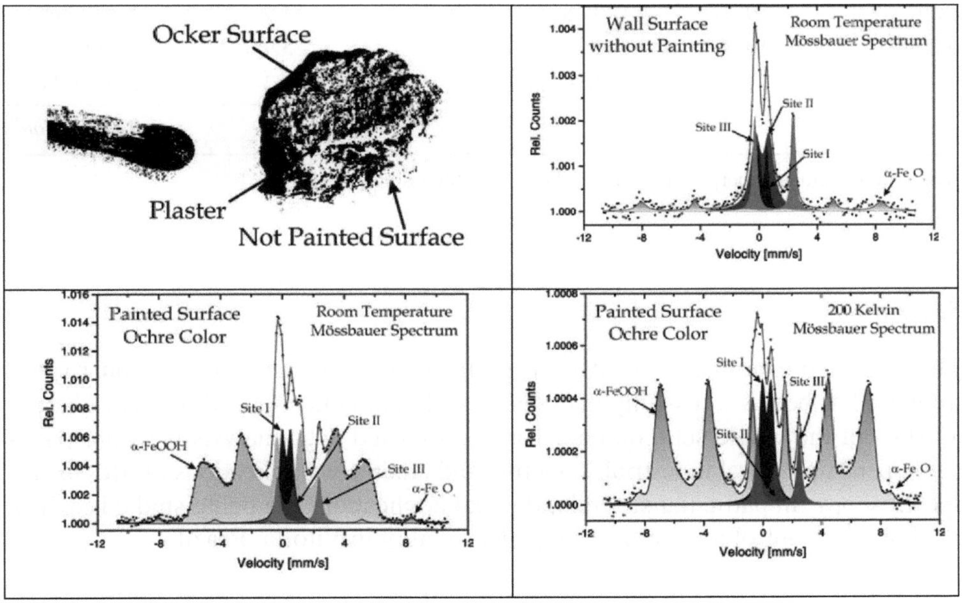

Figure 5. The analyzed piece of the wall painting. The upper part of the piece is the ochre painting, the lower is the wall itself (not painted) and the white layer between these surfaces is a carbonate plaster. The RT MS of the wall surface without any painting is shown in the upper right.

^{57}Co(Rh) source, was used for both painted and non-painted sides for Fe-bearing mineral investigations. The obtained Mössbauer parameters are given in Table III. The interpretation of the results in context of the archaeological findings is in progress.

Table III. Square fitted Mössbauer parameters of the Chinese wall painting at room temperature and 200 K

Region	δ [mm/s]	Δ [mm/s]	B_{hf} [tesla]	Relative area [%]	Phase
Not painted	0.36	0.53		15.7	Site I: Fe^{3+}
at room	0.33	0.91		45.1	Site II: Fe^{3+}
temperature	1.14	2.63		25.8	Site III: Fe^{2+}
	0.37	−0.21	50.9	13.4	α-Fe_2O_3
Ochre color	0.36	0.54		12.3	Site I: $Fe^{3+\dagger}$
at room	0.33	0.96		6.9	Site II: Fe^{3+}
temperature	1.11	2.68		5.8	Site III: Fe^{2+}
	0.37	−0.21	50.9	1.7	α-Fe_2O_3
				73.3	*Distribution*
Ochre color	0.36	0.58		14.9	Site I: $Fe^{3+\dagger}$
at 200 K	0.38	1.12		5.9	Site II: Fe^{3+}
	1.03	2.86		7.2	Site III: Fe^{2+}
				72.0	*Distribution*‡

†Site I may include superparamagnetic α-FeOOH; ‡Including α-Fe_2O_3.

5. Outlook

The presented examples show that the portable and miniaturized Mössbauer spectrometer has been successfully operated in different field applications. The characteristics of the whole set (equipment and automated data analysis software package) opens up several industrial, scientific and other new applications of Mössbauer spectroscopy. In particular several industrial applications are being studied such as steel quality control and Fe-bearing pellets quality in Vitória, Brazil.

Acknowledgements

This work was funded by the German Space Agency DLR under contract (50QM92081), by the Brazilian Education Ministry (CAPES; PASJ 142/1999), Companhia Siderúrgica de Tubarão and Companhia Vale do Rio Doce. The archaeological artifacts were kindly provided by the *Römisch-Germanisches Zentralmuseum* of Mainz, Germany. We also gratefully acknowledge financial support from the Fonds der Chemischen Industrie and the Materialwissenschaftliches Forschungszentrum of the University of Mainz.

References

1. Klingelhöfer, G., Fegley, Jr., B., Morris, R. V., Kankeleit, E., Held, P., Evlanov, E. and Priloutskii, O., *Planet. Space Sci.* **44** (1996), 1277.
2. de Souza, Jr., P. A., *Hyp. Interact.* **113** (1998), 47.
3. de Souza, Jr., P. A., *Laboratory Robotics and Automation* **11** (1999), 3.
4. de Souza, Jr., P. A. and Klingelhöfer, G., *Meteorit. Planet. Sci. A* **36** (2001), 139.
5. www://mars.jpl.nasa.gov/mer/fido.
6. de Souza, Jr., P. A., Morimoto, T., Guimarães, A. F., De Queiroz, R. S., Garg, V. K. and Klingelhöfer, G., *Precise Indication of Air Pollution Sources, Hyp. Interact.* (2002), in press.
7. Klingelhöfer, G., da Costa, G. M., Prous, A. and Bernhardt, B., *Rock Paintings from Minas Gerais, Brasil, Investigated by in situ Mössbauer Spectroscopy, Hyp. Interact.* (2002), in press.

Hyperfine Interactions **144/145:** 381–390, 2002.

Industrial Applications of Mössbauer Spectroscopy

FRANK J. BERRY
Department of Chemistry, The Open University, Walton Hall, Milton Keynes, MK7 6AA, United Kingdom

Abstract. The historical development of the use of Mössbauer spectroscopy in industrial applications is briefly outlined. The power of the technique for the study of commercially important materials and its capacity to make contributions as a research tool, in quality control, and for in-service evaluation are reviewed. The disadvantages of the technique in the industrial setting are considered. The power of Mössbauer spectroscopy when used to approach specific industrial problems is illustrated by its use in monitoring the nature of corrosion resistant coated steel for automobile manufacture and the *in situ* characterization of Fischer–Tropsch catalysts.

Key words: Mössbauer spectroscopy, industry, corrosion resistance, Fischer–Tropsch catalysts.

1. Introduction

Mössbauer spectroscopy has developed as an important technique for the study of materials of industrial interest. This development has occurred alongside the use of other methods such as X-ray powder diffraction, electron microscopy, e.s.r.-, n.m.r-, infrared-, Raman- and X-ray photoelectron spectroscopy for the study of these materials. However, many of these latter techniques require the samples to be in a specially prepared form such as single crystals whereas many materials of commercial interested are, at least, poorly crystalline and sometimes are multiphasic. Furthermore, other techniques often operate under conditions such as ultrahigh vacuum which are far removed from the gaseous atmospheres at elevated temperatures and pressures that characterize the working environment of industrial materials such as heterogeneous catalysts.

In contrast, Mössbauer spectroscopy has a special place amongst other techniques for the study of solids of industrial interest because the γ-radiation can readily penetrate a gaseous environment to enable *in situ* studies and the technique is suitable for the study of crystalline and non-crystalline solids, including high dispersions. It is also capable of both bulk and surface characterization, is non-destructive, and can be used as a fingerprint technique.

2. Historical perspective

It is interesting to note that all this was appreciated in the early 1970's when Mössbauer spectroscopy was only beginning to develop as a quantitative method of spectroscopic evaluation. At a meeting in Chicago in 1970, Mössbauer spectroscopy was described as "coming of age" and able to make contributions to industry as a research tool, in quality control, and for in-service evaluation [1]. It seems some thirty years on that this prediction has not been fully fulfilled. Most of the papers published on industrial applications of Mössbauer spectroscopy relate to its use as a research tool, there are few examples of a Mössbauer spectrometer being used for quality control or in-service evaluation. This is despite the development of Mössbauer instrumentation in the 1970's for industrial application and demonstrations of its use in monitoring iron-phase distributions in pellets during the reduction of ores [2], analyzing iron-bearing minerals in fossil fuels [3], and examining the structure and composition of wear layers [4].

The lack of evidence for the use of Mössbauer spectroscopy for quality control and in-service evaluation suggests that there are drawbacks to the technique. It is clear that a device to be used in industry must be able to function without the need for an expert to interpret its responses. Mössbauer spectroscopy is a relatively sophisticated technique and the interpretation of results is by no means always easy. There is also the problem of radioactive sources in an industrial setting, the limited number of Mössbauer isotopes, the need for cryogenic facilities and, by no means insignificant to an industrial environment, the technique does involve long sample turn-around times. Hence, within the context of quality control and in-service evaluation, which often require repetitive determinations of the same property, the full power of Mössbauer spectroscopy is not needed and it is this overqualification which has led to the technique not being used on the workshop floor.

3. Present situtation

A survey of current Mössbauer spectroscopy literature shows the technique being used extensively in chemical processing, studies of catalysts and batteries, in the study of electronic-related materials including thin films and ion implantation, in the steel and nuclear fuel industry for examining coatings, alloys, corrosion, in mining for the examination of ores and coals, in communication for the study of amorphous materials, nanophase materials, magnetic alloys, and in medical applications such as dentistry and implant surgery. Indeed, it has recently been estimated [5] that close to 80% of all Mössbauer-related publications involve materials that may have significant benefits to the industrial environment and society as a whole. In this sense the potential for Mössbauer spectroscopy as an important research probe to characterize industrially-important materials is unlimited.

4. Illustrations of use

Mössbauer spectroscopy is most powerfully used as a research technique for materials of industrial importance when a problem is clearly defined and the power of the technique, sometimes together with other methods, is used to solve the problem. This can be illustrated by two examples taken from the recent literature.

4.1. GALVANNEAL

Over ten years ago the United States automobile industry needed a new corrosion-resistant coated sheet steel to improve the life of automobiles. One product to be developed was galvanneal which is produced by post-dip annealing of zinc-coated sheet steel. In this process, the sheet steel is passed through a molten zinc bath and then a galvannealing furnace between 500 and 700 °C with an annealing time of about 30 seconds which requires the sheet steel to be processed at about three metres per second. The galvannealing stage is therefore expensive and minimization of the annealing temperature and time is important.

The work of Cook *et al.* using Mössbauer spectroscopy illustrates the power of the technique for addressing an industrial problem [5, 6]. A typical cross section of a galvanneal coating is shown in Figure 1. The conversion electron Mössbauer spectra, which examine the outermost rough surface layer to a depth of *ca.* 300 Å showed the presence of a zinc-rich zeta iron–zinc phase containing only *ca.* 5–6 wt.% iron whilst the thin layer on the steel substrate was composed of a gamma phase containing between *ca.*16 and 28 wt.% iron. Transmission- and

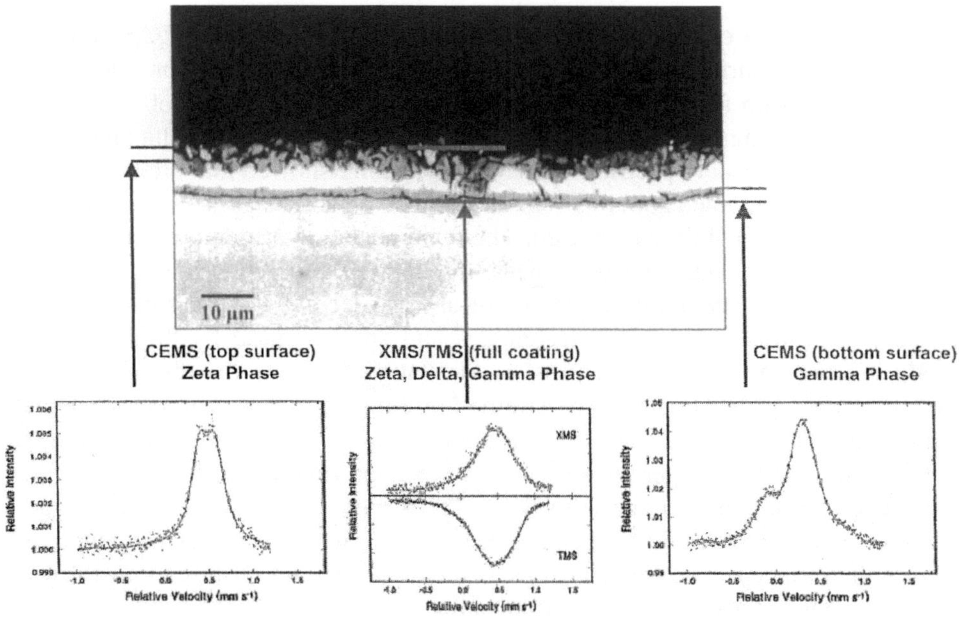

Figure 1. Cross-sectional micrograph of a typical galvanneal coating containing four iron–zinc alloy phases. Mössbauer analysis shows layering of the phases identified as zeta at the coating surface, delta- and gamma-1- phases in the bulk of the coating, and a gamma phase at the steel substrate. (Reproduced with permission from *Hyperfine Interactions*, in press.)

X-ray-Mössbauer spectroscopy were used to investigate the composition of the *ca.* 20 μm total layer and identified the presence of the delta iron–zinc phase containing between 7 and 12% iron and a gamma-1 phase containing between 17 and 21 wt.% iron.

The nature of these iron–zinc phases is of interest to the automobile industry because coatings with too much zeta phase with a high zinc content are difficult to weld whereas too much gamma phase, which has a low ductility and is brittle, gives rise to coating cracking and loss of adhesion when the steel sheet is moulded. In contrast, the delta iron–zinc phase is easily weldable and has a high ductility. Therefore, a desirable galvanneal coating would contain no zeta phase, very little gamma phase and as much delta phase as is possible.

To find out which annealing conditions would give the desired coating, galvanneal was produced at annealing temperatures of 482 and 538 °C using annealing times of 0, 4, 10, 15 and 25 seconds. The coatings were then examined by Mössbauer spectroscopy.

The galvanneal coating produced at the lower temperature of 482 °C (Figure 2) contained a large fraction of the zeta phase which decreased to zero after 25 seconds of annealing. The gamma ray- and conversion electron-Mössbauer spectra of the coating annealed for 4 seconds showed nearly equal fractions (*ca.* 40%) of the zeta and delta phases with the delta phase reaching about 80% after 25 seconds. The gamma phase remained reasonably constant (*ca.* 20%) after annealing for 4 seconds.

The production of galvanneal at the higher temperature of 538 °C resulted in a very different surface coating (Figure 3). The conversion electron Mössbauer spectra showed no formation of the zeta phase on the top surface of the coating following annealing for 4 seconds. The surface alloy was the pure delta phase. The gamma-ray Mössbauer spectrum of the entire coating showed that after 4 seconds the galvanneal contained about 70% of the delta phase and that this remained nearly constant as the anneal time increased. The remainder was the gamma phase.

Hence the Mössbauer spectra indicate that the optimum conditions correspond to an annealing temperature of 538 °C for about 4 seconds. These conditions had resisted determination by other techniques and are now used by the galvanneal producers and the coatings found to be very suitable for the needs of the automobile industry. The conditions determined by Mössbauer spectroscopy have also permitted a significant increase in the production rate of galvanneal from 3 metres per second to 25 metres per second and have enabled smaller producers to develop the process using smaller furnaces of about 15 metre length which represents a significant saving on capital expenditure. Furthermore, the increased productivity has resulted in the use of galvanneal as a corrosion resistant coating in a variety of different applications such as steel frames for residential and commercial building construction and for reinforced concrete.

The Mössbauer effect remains today the only technique which has revealed and can reliably monitor the iron–zinc phase composition in galvanneal steel.

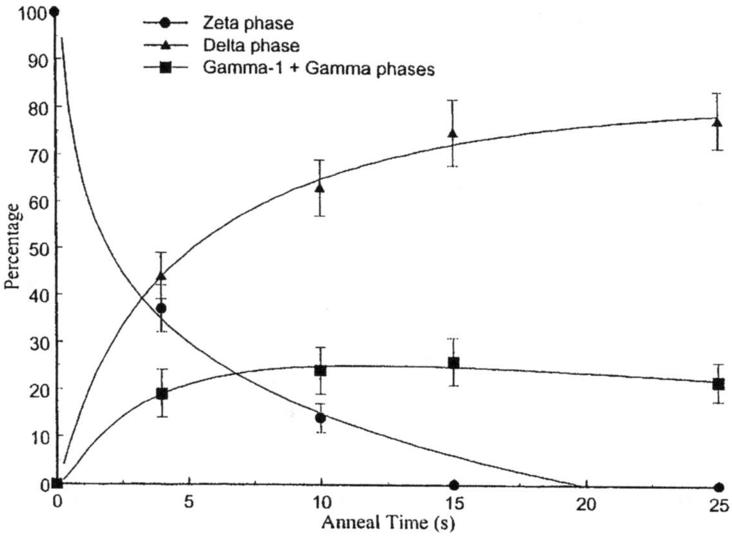

(a)

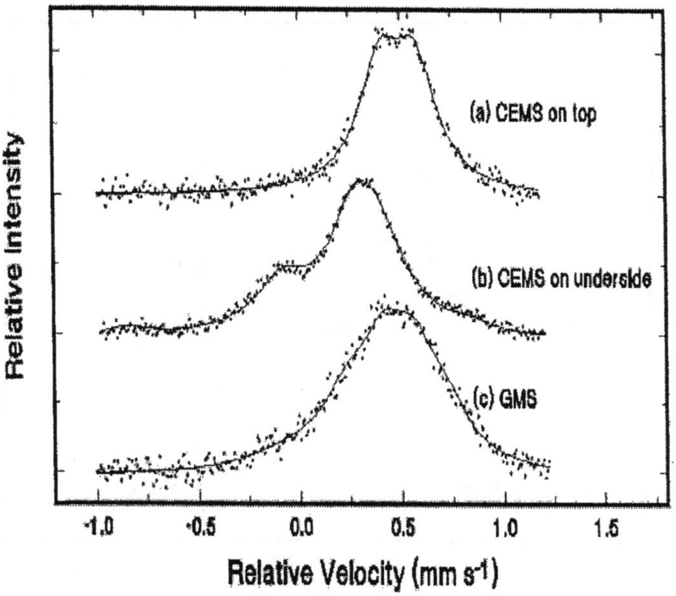

(b)

Figure 2. Mössbauer spectra recorded from galvanneal produced at an annealing temperature of 482 °C for 4 seconds along with the phase composition of the coatings as a function of the anneal time. The zeta phase forms initially and decreases with increasing anneal time. (Reproduced with permission from *Hyperfine Interactions*, in press.)

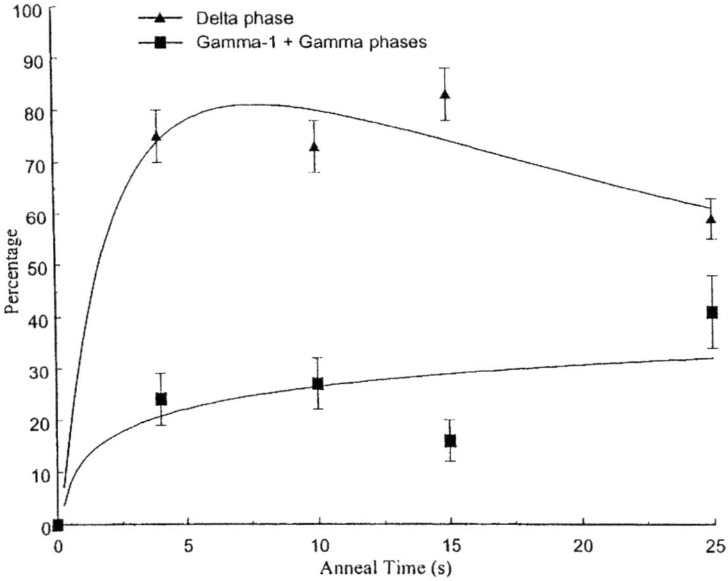

(a)

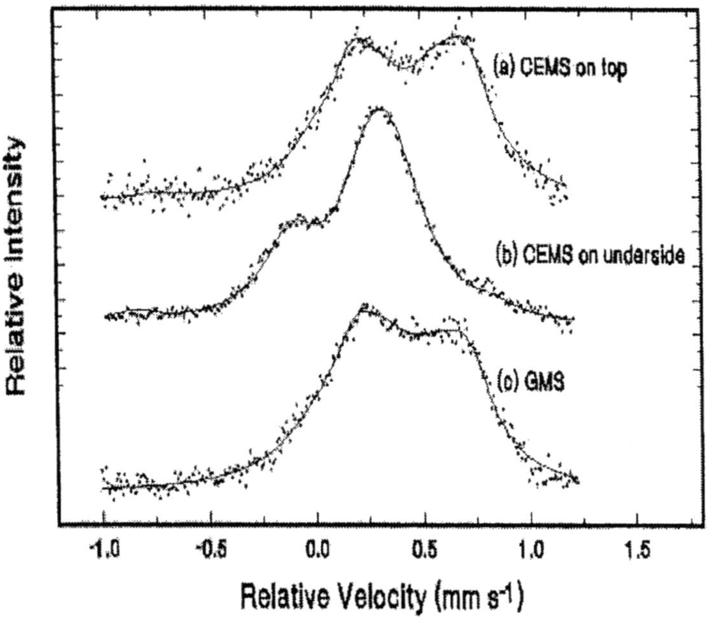

(b)

Figure 3. Mössbauer spectra recorded from the galvanneal produced at an annealing temperature of 538 °C for 4 seconds along with the phase composition of the coatings as a function of the annealing time. The zeta phase does not form and the delta phase reaches a maximum after an anneal time of 4 seconds. (Reproduced with permission from *Hyperfine Interactions*, in press.)

4.2. CATALYSTS

The power of Mössbauer spectroscopy for investigating a real industrial problem under authentic working conditions is illustrated well by Van der Kraan's use [7] of Mössbauer emission spectroscopy to examine industrial cobalt-based Fischer–Tropsch catalysts.

The Fischer–Tropsch process concerns the conversion of a mixture of carbon monoxide and hydrogen which can be obtained from coal or natural gas to fuels, chemical feedstocks, and speciality chemicals. The process is used in South Africa where there are large amounts of easily mineable coal which can be converted to oil to make the country independent of external oil suppliers. The oil crisis of the 1970's and the realization that oil supplies are not inexhaustible has focused attention on this process in the western world. Although this process has been used for several decades using iron-based catalysts, these have a limited lifetime and more recently supported cobalt catalysts have attracted widespread attention although these too may deactivate during extended periods of continuous operation by a number of processes including oxidation by means of water.

Van de Kraan and his co-workers [7] have built a special cell (Figures 4 and 5) in which industrial Fischer–Tropsch catalysts can be studied by Mössbauer emission spectroscopy *in situ* at pressure up to 20 bar and 10 bar steam injection at temperatures up to 723 K and able to handle the production of wax and continuous catalyst test runs of up to two months. The cell was used to test speculation that oxidation of the cobalt catalyst does not occur during Fischer–Tropsch synthesis when a pres-

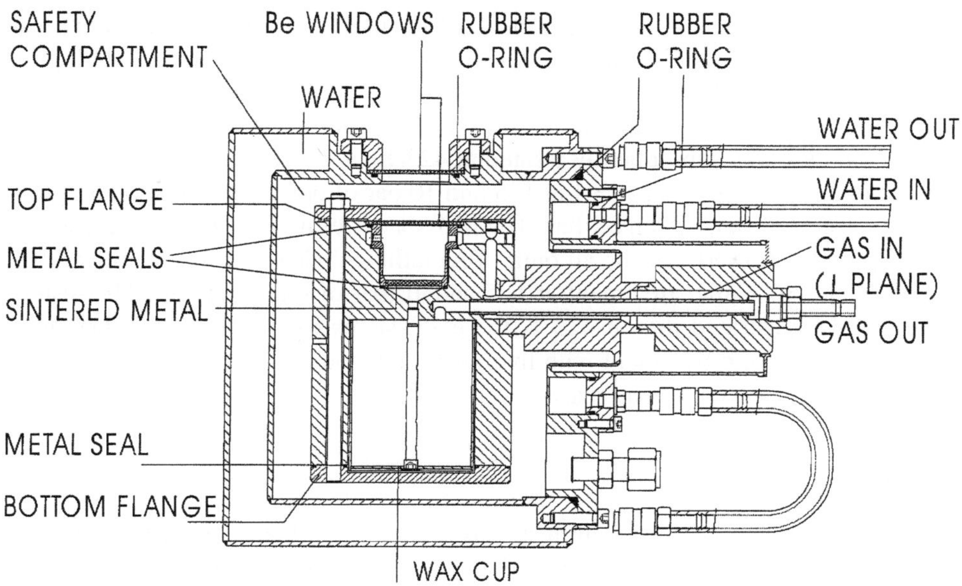

Figure 4. Schematic representation of the high pressure Mössbauer emission spectroscopy cell. (Reproduced with permission from *Catalysis Today* **71** (2002), 369.)

Figure 5. Photograph of the high-pressure Mössbauer emission cell. (Reproduced with permission from *Catalysis Today* **71** (2002), 369.)

sure of hydrogen to water ratio (P_{H_2}/P_{H_2O}) of 1 is used and the absolute water pressure is between 4.6 and 7.6 bar. A catalyst was prepared by adding *ca.* 1000 ppm of ^{57}Co to a 15Co/0.025Pt/100 Al_2O_3 catalyst which was reduced in hydrogen at 653 K for 16 h. The Mössbauer emission spectrum showed about 70% reduction.

The catalyst was subsequently treated at 423 K in the cell in a hydrogen-water gas mixture with (P_{H_2}/P_{H_2O}) = 1 at 1, 5 and 10 bar of pressure corresponding to a water pressure of 0.5, 2.5 and 5 bar. The Mössbauer emission spectra recorded *in situ* (Figure 6) showed that the amount of metallic cobalt increased from *ca.* 70% to 76, 78 and 86% after treatment at 1, 5 and 10 bar of total pressure. The results therefore demonstrated that at 423 K the catalyst resists oxidation and becomes more reduced during treatment with hydrogen and water as the total pressure, and therefore water pressure, is increased.

5. Conclusion

Mössbauer spectroscopy has demonstrated a good record as a research technique and is able to contribute powerfully to the resolution of industrial problems. Its use in quality control and for in-service evaluation is less extensive mainly because it is overqualified for such applications in the working environment.

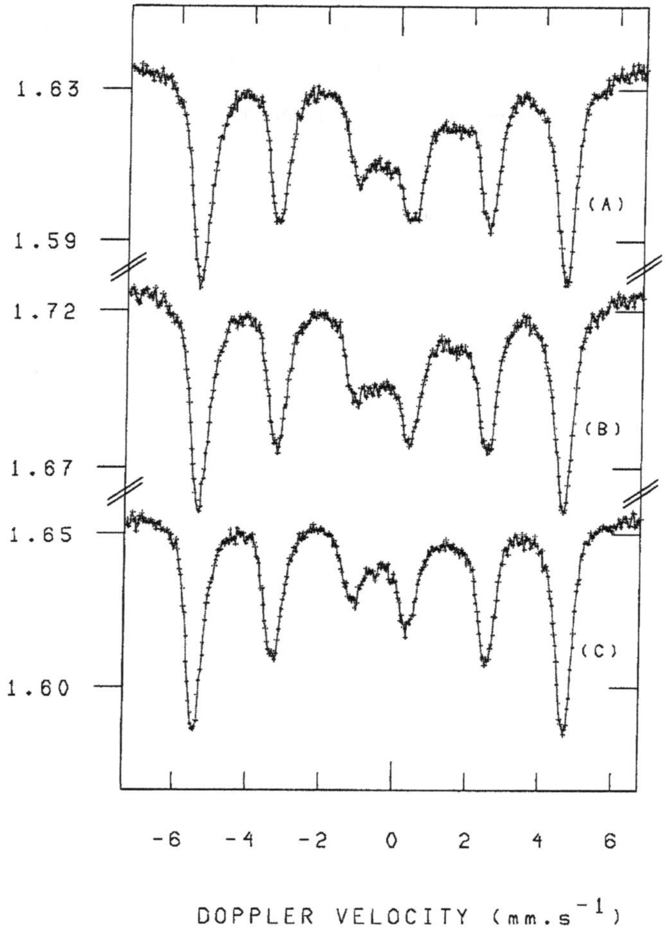

Figure 6. ^{57}Co Mössbauer emission spectra recorded at 423 K during treatment in a H_2/H_2O gas mixture with $(P_{H_2}/P_{H_2O}) = 1$ and at (a) 1 bar, (b) 5 bar, (c) 10 bar of total pressure. (Reproduced with permission from *Catalysis Today* **71** (2002), 369.)

It is important that good communication be maintained between industrial scientists and Mössbauer spectroscopists so that the unique properties of the technique and the contribution it can make to industrial research are identified and fully realized.

References

1. Schwartz, L. H., *Int. J. Nondestructive Testing* **1** (1970), 353.
2. Huffman, G. P. and Huggins, F. E., In: *Mössbauer Spectroscopy and Its Chemical Applications*, A.C.S., Washington, 1981, p. 265.
3. Karl, R. E. and Zuckerman, J. J., In: *Mössbauer Spectroscopy and Its Chemical Applications*, A.C.S., Washington, 1981, p. 221.

4. Company, R. G., Thornton Research Centre Paper 2198.R, Thornton Research Centre, England.
5. Cook, D. C., Grant, R. G. and Cook, P. S., U.S. Patent Number 5,628,044 (1997).
6. Cook, D. C., *Hyp. Interact.*, in press.
7. Crajé, M. W. J., van der Kraan, A. M., Van der Loosdrecht, J. and van Berge, P. J., *Catalysis Today* **71** (2002), 369.

Author Index to Volume 144/145/1–4 (2002)

MIX
Papier aus verantwortungsvollen Quellen
Paper from responsible sources
FSC® C105338

If you have any concerns about our products,
you can contact us on
ProductSafety@springernature.com

In case Publisher is established outside the EU,
the EU authorized representative is:
Springer Nature Customer Service Center GmbH
Europaplatz 3, 69115 Heidelberg, Germany

Printed by Libri Plureos GmbH
in Hamburg, Germany